Smart Nanomaterials Technology

Series Editors

Azamal Husen, Department of Biotechnology, Smt. S. S. Patel Nootan Science and Commerce College, Sankalchand Patel University, Gujarat, Gujarat, India

Mohammad Jawaid, Department of Chemical and Petroleum Engineering, United Arab Emirates University, Al Ain, United Arab Emirates

Nanotechnology is a rapidly growing scientific field and has attracted a great interest over the last few years because of its abundant applications in different fields like biology, physics and chemistry. This science deals with the production of minute particles called nanomaterials having dimensions between 1 and 100 nm which may serve as building blocks for various physical and biological systems. On the other hand, there is the class of smart materials where the material that can stimuli by external factors and results a new kind of functional properties. The combination of these two classes forms a new class of smart nanomaterials, which produces unique functional material properties and a great opportunity to larger span of application. Smart nanomaterials have been employed by researchers to use it effectively in agricultural production, soil improvement, disease management, energy and environment, medical science, pharmaceuticals, engineering, food, animal husbandry and forestry sectors.

This book series in Smart Nanomaterials Technology aims to comprehensively cover topics in the fabrication, synthesis and application of these materials for applications in the following fields:

- Energy Systems—Renewable energy, energy storage (supercapacitors and electrochemical cells), hydrogen storage, photocatalytic water splitting for hydrogen production
- Biomedical—controlled release of drugs, treatment of various diseases, biosensors,
- Agricultural—agricultural production, soil improvement, disease management, animal feed, egg, milk and meat production/processing,
- Forestry—wood preservation, protection, disease management
- Environment—wastewater treatment, separation of hazardous contaminants from wastewater, indoor air filters.

Md. Meraj Ansari · Anil K. Suresh ·
Nadeem Akhtar
Editors

Emergence of Sustainable Biomaterials in Tackling Inflammatory Diseases

Springer

Editors
Md. Meraj Ansari
Department of Animal Biotechnology
Faculty of Biotechnology
College of Applied Life Sciences
Jeju National University
Jeju Special Self-Governing Province
Korea (Republic of)

Anil K. Suresh
Department of Biological Sciences
School of Engineering and Applied Sciences
SRM University
Mangalagiri, Andhra Pradesh, India

Nadeem Akhtar
CBS Bio Platforms Inc.
Calgary, AB, Canada

ISSN 3004-8273 ISSN 3004-8281 (electronic)
Smart Nanomaterials Technology
ISBN 978-981-96-2111-8 ISBN 978-981-96-2112-5 (eBook)
https://doi.org/10.1007/978-981-96-2112-5

This Springer imprint is published by the registered company Springer Nature Singapore Pte Ltd.
The registered company address is: 152 Beach Road, #21-01/04 Gateway East, Singapore 189721, Singapore

About This Book

Long-term, low-grade inflammation is a silent killer that increases the risk of serious conditions such as type 2 diabetes, cancer, rheumatoid arthritis, osteoarthritis, neuroinflammatory diseases, and cardiovascular disease. It is concerning that three out of every five deaths worldwide can be attributed to inflammatory illnesses, which are characterized by dysregulated inflammatory responses.

Sustainable biomaterials can enhance the absorption of anti-inflammatory medications while reducing their negative effects, thereby improving their efficacy in treating dysregulated inflammation. This book focuses on the emergence of sustainable biomaterials for various anti-inflammatory treatment strategies. Biomaterials can be designed to scavenge danger signals, such as reactive oxygen and nitrogen species, at the onset of inflammation. Additionally, anti-inflammatory drugs can be delivered directly to sites of inflammation using nanoscale platforms.

While biomaterials show great potential in treating inflammatory diseases, they do have limitations. This book provides a comprehensive overview of the advantages and disadvantages of biomaterial applications in inflammation treatment. It also presents a current overview of recent developments in the fields of nanomedicine and nano-based drug delivery systems. Students studying pharmaceutical sciences, researchers, educators, and professionals in the pharmaceutical industry will benefit from this book's insights into the complexities of novel drug research and innovations aimed at combating inflammatory diseases.

Contents

About the Editors

Dr. Md. Meraj Ansari is an outstanding representation of inventiveness in the field of inflammatory research, fully committed to unlocking the mysteries of the illness. His journey from Saran to a leading position in research has been marked by a strong commitment to academia and an unyielding pursuit of scientific knowledge. Raised in a family that prioritized education and inquiry, Dr. Ansari was born in Arna, a village in the Saran district of Bihar. His mother is a homemaker, while his father worked as a teacher. He attended J.M.H.S. Raipura in Sarna for his early education and later completed his undergraduate studies in B.Sc. Biotechnology at B. N. College, part of Patna University.

Dr. Md. Meraj Ansari is a Research Professor in the Department of Animal Biotechnology at Jeju National University in South Korea. He earned his Ph.D. in Toxicology from Jamia Hamdard University, New Delhi, India. His primary research interests focus on developing nano-drug delivery systems to address inflammatory diseases.

Dr. Ansari has received a Young Scientist Award from the Department of Health Research, Ministry of Health and Family Welfare, Government of India. He previously worked as a Young Scientist at the Centre of Pharmaceutical Nanotechnology in the Department of Pharmaceutics at the National Institute of Pharmaceutical Education and Research, SAS Nagar, Mohali, Punjab, India. He has also contributed to research at the Institute of Nano Science and Technology, Mohali, Punjab.

In his research, he has focused on nanoscale systems for the controlled delivery of therapeutics, achieving sustained release of molecules over extended periods. His work has targeted conditions such as rheumatoid arthritis, ulcerative colitis, and the toxicity of anti-cancer drugs in animal models. Dr. Ansari has authored several book chapters and published numerous research papers in high-impact journals. e-mail: ansarimm@jejunu.ac.kr

Dr. Anil K. Suresh is an Associate Professor at the Department of Biotechnology, SRM University-AP, Amaravati, India. He received his M.Sc. in Biotechnology (2001) from Pondicherry Central University, India by qualifying JNU Merit Scholarship Award (1999). He obtained his Ph.D. in Biotechnology (2007) from the

National Chemical Laboratory affiliated with Pune University, India upon qualifying CSIR-UGC Fellowship Award (2001–2002). He was trained as a Visiting Scholar/ Postdoc (2007–2009) at McGill University, Canada, and as a Research Associate (2009–2011) at Oak Ridge National Laboratory, USA. He then worked as a Staff Scientist (2011–2013) at the City of Hope, California, USA.

He returned to India as a Ramalingawsami Fellow (2013), a prestigious competitive award given by the Department of Biotechnology. He is currently the Principal Investigator for his Biomedical Nanotechnology AND Sustainable Laboratory. His research focuses on biomedical applications of engineered nanoparticles; cell-imaging, targeted drug delivery, nano-toxicology, early diagnosis of diseases, cancer therapeutics and clinical medicine, supported catalysis, and environmental remediation.

Dr. Anil has authored 45 peer-reviewed high-impact international publications; portions of his research expertise are also published as 13 book chapters and 2 brief books by various international publishers like Elsevier, Springer, etc. He is a lifetime member of the Indian Science Congress Association, the American Association for Science and Technology, the Geochemical Society and the American Nano Society. e-mail: anil.s@srmap.edu.in

Dr. Nadeem Akhtar a scientist, is currently making significant strides in the animal health and longevity sector for a leading Canadian company, CBS Bio Platforms Inc. His impressive background demonstrates that he has held a variety of important positions in academia and research throughout his career.

Dr. Akhtar was a Research Associate at the University of Guelph, Canada, where the objective of his work was to improve companion animal's health in order to promote a superior and more health-conscious lifestyle. Prior to assuming this position, he excelled as a postdoctoral researcher at the Department of Chemical Engineering, University of Waterloo, Canada (2019–2021) and the Department of Animal Biosciences, University of Guelph, Canada (2015–2019).

His accomplishments in the National Eligibility Test (NET-2010) and Graduate Aptitude Test (GATE) in Life Sciences (2010) and Biotechnology (2011) further emphasize his academic excellence. He earned his Ph.D. from the esteemed Department of Biotechnology at the Thapar Institute of Engineering and Technology in India.

Dr. Akhtar's scientific contributions are nothing short of remarkable, with over 28 articles published in peer-reviewed international journals as well as 12 book chapters and 4 edited books. His abilities extend to patent accomplishments, as he has one granted patent to his name. In addition to his research, he has shown his dedication to knowledge advancement by serving as a reviewer for prestigious journals such as RSC Advances, the Journal of Applied Microbiology, Scientific Reports, and Poultry Science.

Dr. Akhtar's career exemplifies dedication and excellence, demonstrating his invaluable contributions to the fields of animal science and biotechnology. His journey continues to motivate and shape the global scientific landscape.

An Overview of Inflammatory Diseases

Md. Meraj Ansari, Nadeem Akhtar, Zamurrad Nayab, Shambhu Kumar, and Nemat Ali

Abstract Inflammation is the immunological response of the immune system. Our body's immune system responds to foreign invaders, injured cells, and poisonous substances. When our immune system works well, it cleans up debris, repairs wounds, and drives out alien invaders. However, excessive inflammation may cause a wide range of inflammatory diseases. Inflammation may be a symptom of chronic diseases. Experts believe that inflammation may play a significant role in a wide range of chronic diseases, such as ulcerative colitis, cancer, autoimmune diseases such as rheumatoid arthritis, and neurological disorders. People can be affected by these inflammatory diseases at an early or late stage. Therefore, basic knowledge about these inflammatory diseases may be helpful for management in an efficient manner. The aim of this chapter is to highlight basic knowledge about inflammatory disease, its epidemiology, risk factors associated with inflammatory diseases, and disease pathophysiology.

Md. M. Ansari (✉)
Department of Animal Biotechnology, Faculty of Biotechnology, College of Applied Life Sciences, Jeju National University, Jeju Special Self-Governing Province, Jeju-Si, South Korea
e-mail: rajmeraj385@gmail.com

Interdisciplinary Graduate Program in Advanced Convergence Technology and Science, Jeju National University, Jeju Special Self-Governing Province, Jeju-Si, South Korea

N. Akhtar
Department of Animal Biosciences, University of Guelph, Guelph, Canada

Z. Nayab
Department of Biotechnology, Science College Patna, Patna University, Patna, Bihar, India

S. Kumar
Department of Zoology, Ram Bahadur Singh College, Andaur, Samastipur, Bihar, India

Department of Zoology, L. N. M. University, Darbhanga, Bihar, India

N. Ali
Department of Pharmacology and Toxicology, College of Pharmacy, King Saud University, Riyadh, Saudi Arabia

Md. M. Ansari et al. (eds.), *Emergence of Sustainable Biomaterials in Tackling Inflammatory Diseases*, Smart Nanomaterials Technology,
https://doi.org/10.1007/978-981-96-2112-5_1

Keywords Inflammation · Cancer · Rheumatoid arthritis · Chronic inflammatory disease

1 Introduction

Inflammation is the immunological response of our immune system. The body's immune defense systems produce a protective immune response against invaders to maintain tissue homeostasis. Innate and adaptive immunity serves as protective immunity with the removal of dangerous stimuli such as injured cells, irritants, infections, and sterile lesions [1, 2]. The inflammatory immune response against cancer disease progressions is not resolving whereas the inflammatory immune response against infection and wound healing is resolving in nature. Moreover, inflammatory autoimmune disorders, microbiological infections, excess obesity, cigarette smoking, particulate matters of asbestos, and pro alcoholic are known to cause tumor-extrinsic inflammation. These factors have also been shown to raise the risk of cancer and hasten the course of malignancy. It is well recognised that inflammation serves as the host's defensive mechanism against infection and tissue damage. It can stop infections from spreading or aid in the healing of damaged tissue [3, 4]. Tissue macrophages or mast cells identify pathogen-associated molecular patterns (PAMPs) during the very early or acute condition of inflammation. This recognition triggers the release of inflammatory mediators in the form of cytokines, certain chemokines, vasoactive amines, and eicosanoids, consequently, they strengthen the body's immune response [5–7]. It is well recognized that these pro-inflammatory mediators raise vascular permeability, which causes a huge influx of plasma that contains soluble substances and antibodies [8]. Furthermore, it has been demonstrated that the injury site releases a range of signaling molecules that attract neutrophils and monocytes, such as chemokines, cytokines, eicosanoids, and adhesion molecules [9]. Monocytes and lymphocytes gather in the inflammatory areas when the inflammatory response intensifies in order to neutralize pathogens. Following this, inflammatory cells die and are removed by macrophages. The inflammatory process is replaced by the tissue repair process at the end of the inflammatory cascade, which reduces inflammation and restores tissue to normal physical and chemical reactions [10, 11]. As a result, several of the cell types and mediators which have the ability to control the cellular chemotaxis, migration of cells, and cellular proliferation in a highly programmed way are involved in the inflammatory consequence. The whole inflammatory process can be classified into mainly two categories, acute and chronic inflammation. Acute and chronic classification is on the basis of how long the condition of inflammation has been present. The first reaction to damaging stimuli is acute inflammation, which lasts for a few days or weeks. Granulocytes make up the bulk of the inflammatory cells that infiltrate during acute inflammation [12–14]. Tissues that are in chronic inflammation simultaneously undergo tissue death and repair. The primary immune cells that infiltrate areas of persistent inflammation are lymphocytes and macrophages [15, 16]. Acute inflammation will result in tissue fibrosis, necrosis, autoimmune, and chronic stage inflammation if the pro-inflammatory stimulus is not eliminated. It is found that the main causes of chronic inflammation are damage to tissue as well as inflammatory factors [17]. It has also been shown that persistent intracellular viral infections, including those caused by Mycobacterium tuberculosis, can cause chronic inflammation. Despite being less virulent, some viruses have been shown to trigger immunological reactions even in the absence of acute inflammation as a clinical symptom. Chronic inflammatory autoimmune diseases such as rheumatoid arthritis may result from persistent and long-term exposure to non-degradable but potentially very harmful chemicals such as silicosis or from a protracted body immune response against the self-tissues. Numerous chronic inflammatory diseases, including neurological disorders, osteoporosis, myocardial infarction, atherosclerosis, chronic heart failure, diabetes, psoriasis, and cancer, have been either directly or indirectly associated with chronic inflammation [18, 19].

2 Epidemiology of Inflammatory Diseases

The two types of inflammation that are typically studied are acute and chronic. From various studies, it is observed that the dysregulation of signalling pathways in chronic inflammation such as NF-KB, signal transducer and activator of transcription-3 and many others may initiate a variety of chronic diseases such as autoimmune disease, respiratory-related asthma, inflammatory bowel disease, neurological disorders, cardiovascular disease, diabetes, and cancer. These chronic inflammatory diseases may cause early-age morbidity and mortality. Globally, chronic inflammatory disorders account for the majority of fatalities. As per WHO, chronic inflammation is the major risk factor for human health. For the next 30 years, it is projected that the prevalence of illnesses linked to chronic inflammation will rise steadily in the United States. 61 million (21%) of the approximately 125 million people of America were living with chronic inflammatory diseases and had more than one in 2000. As per a 2014 estimate by the Rand Corporation, 12% of people had five or more chronic cases of the disease, 42% of people had more than one, and over 60% people of Americans had at least one severe chronic condition. Globally it is estimated that three deaths out of five from different types of chronic inflammatory diseases. In this section, we have discussed the prevalence of some common inflammatory diseases one by one.

- **Rheumatoid arthritis (RA)**: 18 million persons worldwide suffered from rheumatoid arthritis in 2019 [20]. As per [20], RA affects over 70% of the female population, and 55% of those people aged over 55. 13 million populations are impacted with RA from moderate to severe intensity that may be recovered or benefit from rehabilitation [21]. RA is a chronic autoimmune disease that mostly affects joints [22].

- **Osteoarthritis (OA)**: As per GBD estimate in 2019, 528 million people were affected around the globe with OA and this increase was 113% as compared to 1990. Approximately, 60% of women were affected by OA and 73% of people were those whose age was over 55. In the 365 million cases, knee parts are mostly affected by OA followed by the body hand, and hip [21]. 344 million people are affected by OA with severity from moderate to severe levels that make rehabilitation beneficial [22].
- **Psoriasis**: It is believed that psoriasis affects 2–3% of people worldwide [23]. Psoriasis is more common in polar areas of the globe but its impact is also on tropical or subtropical nations. Variation in environmental and genetic factors, the prevalence of psoriasis can differ from region to region in a nation as diverse as India. Only six studies from part of North India that estimated the occurrence of disease in the midst of adult dermatologic patients were available, majority of them were conducted in hospital settings [24–29]. There have been reports of a higher prevalence in men, with a peak age at onset in the 3rd and 4th decades of life [30, 31].
- **Diabetes**: As per WHO, 108 million people were affected with diabetes in 1980 but this value became 422 million in 2014. It has been observed that rapidly increasing prevalence of diabetes in low and middle-income countries as compared to high-income countries. The age-related mortality rate increased by 3% between 2000 and 2019.
- **Liver cancer**: The American Cancer Society estimated that approximately 41,630 cases of primary liver cancer will be detected in the United States in 2024 (28,000 in men and 13,630 in women). The fatal liver cancers will claim the lives of about 29,840 persons (19,120 males and 10,720 women). It is observed that the occurrence of liver cancer has more than tripled since 1980 whereas the fatality rate has more than doubled.
- **Cardiovascular diseases**: According to an updated 2017 data from the American Heart Association, roughly 800,000 deaths in the US are caused by cardiovascular diseases (CVDs), which account for one out of every three deaths. Cardiovascular disease (CVD) is responsible for 31% of deaths worldwide.
- **Chronic obstructive pulmonary disease (COPD)**: It was estimated that in 2014, COPD was the third most prevalent cause of death in the US, with over 15.7 million people (6.4%) reported to have received a diagnosis.
- **Allergies**: These impact over 50 million Americans annually and are among the sixth main causes of chronic human diseases in the US. In the US, asthma affects about 24 million individuals, including over 6 million children. In 2015, hay fever was detected in 8.2% of adults and 8.4% of children.
- **Cancer**: Global cancer statistics for the most prevalent cancer types in 2022, the most recent year for which GLOBOCAN has provided data. GLOBOCAN estimated 19,976,499 cancer cases in 2022 including non-melanoma skin cancer (NMSC). This number decreased to 18,741,966 cancer cases around the globe in 2022 after NMSC was taken out of the equation.

3 Risk Factors Associated to Inflammation

Diet: Overweight or diabetes people having a high trans fat diet, saturated fats, or refined sugar is strongly linked to enhanced production of pro-inflammatory chemicals. Taking too much processed food regularly has greater potential to inflame. They have the potential to change the microorganisms that are living in our stomach for better health, which has the potential to interact with the body's immune system and finally set them off in a way that results in persistent inflammation.

Obesity: Nowadays, obesity may be considered a hallmark for various inflammatory disorders. The overabundance of macronutrients in adipose tissues causes them to produce less adiponectin and initiates the release of pro-inflammatory and mediators of oxidative stress like IL-6 & TNF- α. According to certain research, a person's body mass index and the quantity of cytokines they release are directly correlated.

Stress: Studies reveal that stress, the body's reaction to being pushed to the limit or threatened causes or exacerbates a number of illnesses, including cancer, heart disease, depression, and neurodegenerative diseases. It is found that there is a strong correlation between the inflammatory cytokines released to both physical as well as emotional stress. It is also observed that disturbances in the sleep cycle are also considered an independent risk factor for chronic inflammation.

Smoking: It has been shown that smoking cigarettes reduces the synthesis of anti-inflammatory chemicals and increases inflammation. Smoking causes a rise in proinflammatory microorganisms, which in turn causes inflammation in the body. Compared to smokers, non-smokers had greater abundance of anti-inflammatory bacteria, which can create short-chain fatty acids that reduce inflammation.

Age: The innate immune system mediated inflammatory responses intensify and prolong with age, making older people more vulnerable to inflammatory diseases and tissue-damaging inflammation. With age, the body's immune system becomes weak and accelerates age-related diseases such as OA, it is found that enhanced levels of inflammatory mediators are absolutely linked with advancing age. The age-related rise in mediators of inflammation could be caused by several age-related variables such as age-related mitochondrial dysregulation and accumulation of free radicals throughout time.

4 Inflammatory Cells

The blood and tissues include a type of cell called an inflammatory cell, which is involved in both acute and chronic inflammatory reactions. Monocytes/macrophages, lymphocytes, and neutrophils are a few types of inflammatory cells. During inflammation, the site of tissue damage attracts inflammatory cells (Fig. 1). Cytokines are released by inflammatory cells in large quantities, which affect the behaviour of the

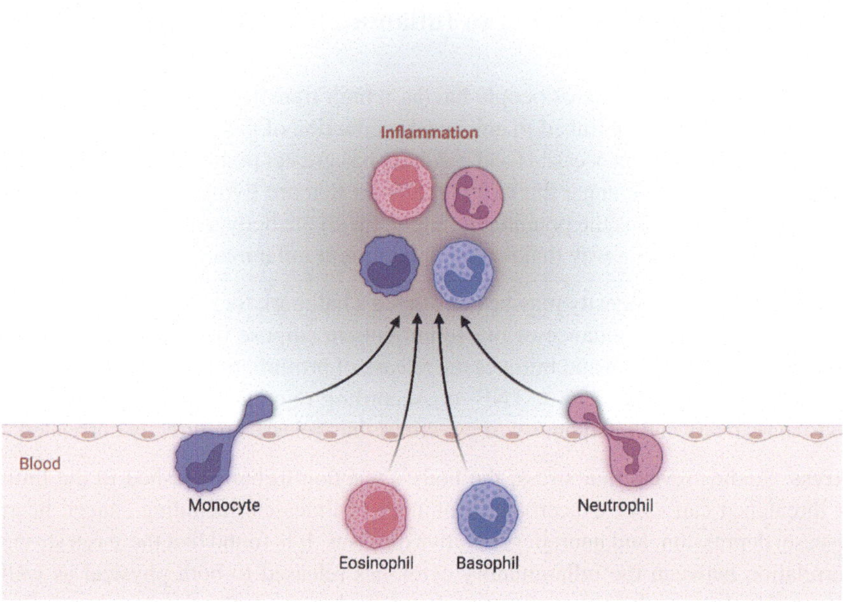

Fig. 1 Schematic representation of blood cell recruitment at the site of inflammation. Image created with Biorender.com

epithelium that covers them. The type of inflammatory cell is determined by the extent and type of injury, similar to those in other body parts. Polymorphonuclear leukocytes are the predominant cell type in acute situations, while lymphocytes, plasma cells, monocytes, and macrophages are linked to more chronic disorders. In this section, we have explained some inflammatory cells in detail.

4.1 *Neutrophils*

Neutrophils serve as soldiers of the body. It is the most abundant blood cell type and is also referred to as polymorphonuclear (PMN) leukocytes. In the bone marrow, around 10^{11} cells are created daily. In the course of a day, neutrophils under homeostatic conditions enter the circulation; they can infiltrate at the site of inflammation and are finally destroyed by macrophages. As effector cells in the innate arm of the immune system, neutrophils play a crucial role [32]. The neutrophils are the frontline soldiers of the innate immune system [33]. A significant portion of the global population suffers from various illness conditions that include chronic inflammation. Low-grade, persistent inflammation is the hallmark of chronic inflammation, which is more common among elderly people. When acute inflammation arises, neutrophils migrate at the site of inflammation for clearance of invaders or injury. Nevertheless, the function of neutrophils against inflammation is poorly known and has been characterised as either advantageous or disadvantageous, resulting in tissue destruction or boosting the immune response. According to new research, neutrophils may have a significant role in a number of chronic inflammatory diseases like autoimmune disorders, and others [34].

As we know neutrophils are the most prevalent type of blood cells, and they are widely known for being the first to respond to acute inflammation. Neutrophils have a role in resolution and tissue repair under normal circumstances. They phagocytose necrotic cells to prevent from drawing in additional immune cells, release inflammatory mediators the encourage angiogenesis and growth, and produce resolvins & protectins [35]. As we know neutrophils have confirmed their potentiality against chronic inflammation in recent decades.

Neutrophils migrate to the site of injury or inflammation, neutrophils take part in a key role in advancing the process by activating other immune cells. They also release serine proteases, and form neutrophil extracellular traps (NETs). NETs play a significant role in chronic inflammation. In Fig. 2, neutrophil extracellular trap formation pathway (lytic and non-lytic NETosis) has been given.

NETosis is a NET-dependent cell death, in which PMN constructs net-like structures of decondensed chromatin as well as proteases. By immobilising infections and containing antibiotic chemicals in them, these structures restrict them from further spreading. Furthermore, it is observed that antimicrobial role of NETs in the innate immune system in normal conditions, there is growing evidence that NETosis also

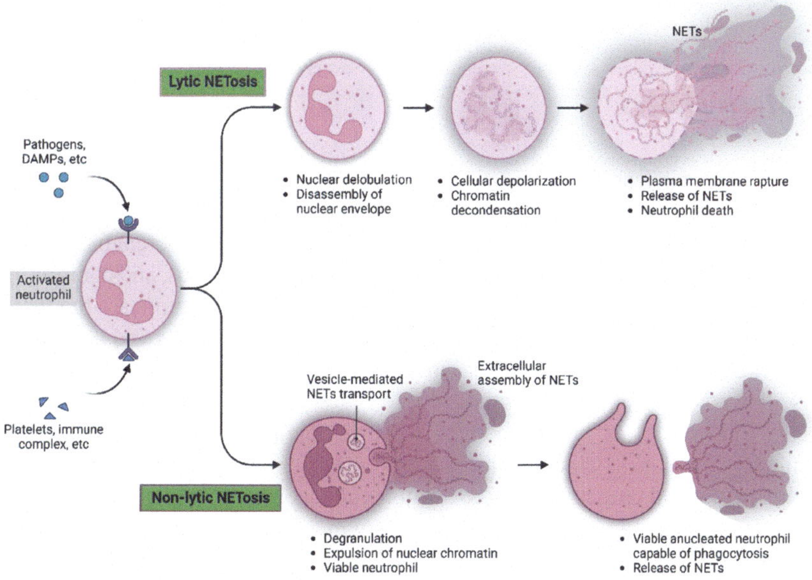

Fig. 2 Neutrophil extracellular traps (NETs) formation pathways. Image created with Biorender.com

takes place during the pathogenic process of other disease types, such as stroke, Alzheimer's, atherosclerosis, and airway inflammation [36]. It is found that NETs as immunomodulatory candidates and have a strong link to chronic inflammatory diseases. Furthermore, due to NET-dependent oxidised deoxyribonucleic acid and antimicrobial peptides, they can prepare some other immune cells to generate sterile inflammation by activating responses of the interferon [37]. NETs can also catch metastatic tumours, impede vital organ regions, induce thrombosis to occlude the vascular, and slow down the healing of diabetic wounds.

4.2 Monocytes

For the diagnosis of several hematologic, inflammatory, and immunological diseases, the presence or absence of monocytes might provide crucial diagnostic cues. Rapid monocyte mobilisation and the differentiation programmes that allow these cells to differentiate into macrophages or dendritic cells are mediated by environmental signals at the site of inflammation [38]. Monocytes take part in the processes of tissue repair, pathogen and dead cell removal, and the start of adaptive immunity. Fortunately, the development of chronic inflammatory illnesses including atherosclerosis and infections may also be influenced by recruited monocytes. The process is believed to follow a generic model of leukocyte adhesion and trafficking and to depend on the interactions of several adhesion molecules [39]. Monocytes in circulation enter tissues during both inflammatory and homeostatic processes.

4.3 Macrophages

Within the mononuclear phagocyte system, macrophages play a crucial role, which is comprised of closely similar bone marrow-derived cells such as tissue macrophages and blood monocytes. Monocytes from the circulation move into different tissues and change macrophages. Presenting antigens, phagocytosing infections, and controlling the immune system by releasing various cytokines and growth factors are the three primary functions of macrophages in inflammation. When mediators and inflammatory effectors cells are eliminated or deactivated, inflammation is suppressed and the host is allowed to repair injured tissues [40].

Active macrophages are deactivated by cytokine antagonists and anti-inflammatory cytokines, which are mostly produced by macrophages. Macrophages take part in the inflammatory process' autoregulatory loop [40]. Since a variety of physiologically active chemicals produced by macrophages have been linked to both positive and negative outcomes in inflammation, targeting these cells and their products with therapeutic interventions may be helpful in management of disease.

4.4 *Eosinophils and Basophils*

By using eosin staining, Paul Ehrlich discovered blood eosinophils in a number of species in 1879, including humans, dogs, and rabbits [41]. Following this initial discovery, other investigations have detailed the various biological roles played by eosinophils. The homeostasis harmony maintenance, immunological control, anti-inflammatory, and anti-tumorigenic actions are only a few of the more intricate intrinsic roles of eosinophils [42]. Through eosinophil-derived cytotoxic mediators that are implicated in eosinophilic inflammation, eosinophils also play a substantial role in tissue destruction, as demonstrated in Th2-high asthma and other eosinophilic-associated inflammatory disorders [43]. The identification of the primary route underlying eosinophilic inflammation in a single patient with an inflammatory illness linked with eosinophils using surrogate biomarkers may aid in the selection of a course of treatment. Due to their relative scarcity and phenotypic resemblance to mast cells, basophils have generally been disregarded in both clinical and experimental settings. However, recent findings of studies on human and animal models of mice have shown that basophils perform non-redundant effectors functions and are crucial for the start and progression of inflammation mediated by TH2 cytokines [44]. Type 2 inflammations and a host of skin conditions, including urticaria, atopic dermatitis, and helminth infection, have been linked to basophils. While basophils from the haematological compartment have been the focus of classical investigations on mast cells, they are comparable in structure and function to tissue-resident mast cells [45]. Nonetheless, a growing body of research in tissues such as the skin indicates that basophils can adopt specific traits in response to distinct developmental, chemotactic, and activation cues. In Fig. 3, we have represented basophils mediated skin inflammation. How basophils are pulled into the skin is unknown. It is widely acknowledged that basophils rapidly aggregate into peripheral tissues, including the skin, in a variety of circumstances, including helminth infection, tick bite, and allergic inflammation [46–48].

4.5 *Mast Cells*

Paul Ehrlich originally described mast cells (MCs) in 1878 [49]. They have been considered agents of IgE-mediated antiparasitic or allergic reactions. It is commonly known that mast cells have a role in allergy and anaphylactic reactions, as well as in innate and acquired immunity. An increasing amount of evidence connects mast cells to inflammatory diseases, where they are triggered by non-allergic substances such neuropeptides and cytokines, often in combination with IL-33 [50]. Mast cell functions as the first line of defense, releases diverse granule contents and mediates inflammation (Fig. 4).

Mast cells serve specialised roles in nearly every organ due to their strategic perivascular placement, sensitivity to changes in the microenvironment, and great

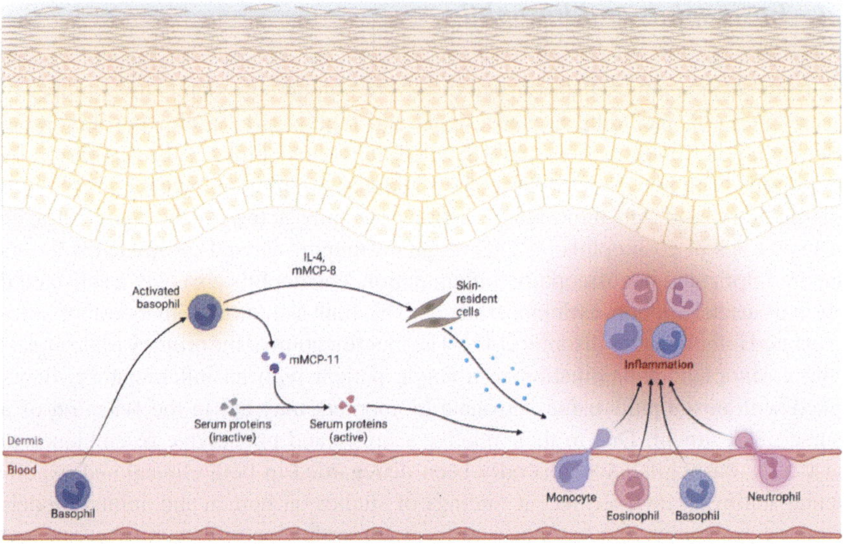

Fig. 3 Basophil-mediated skin inflammation. Image created with Biorender.com

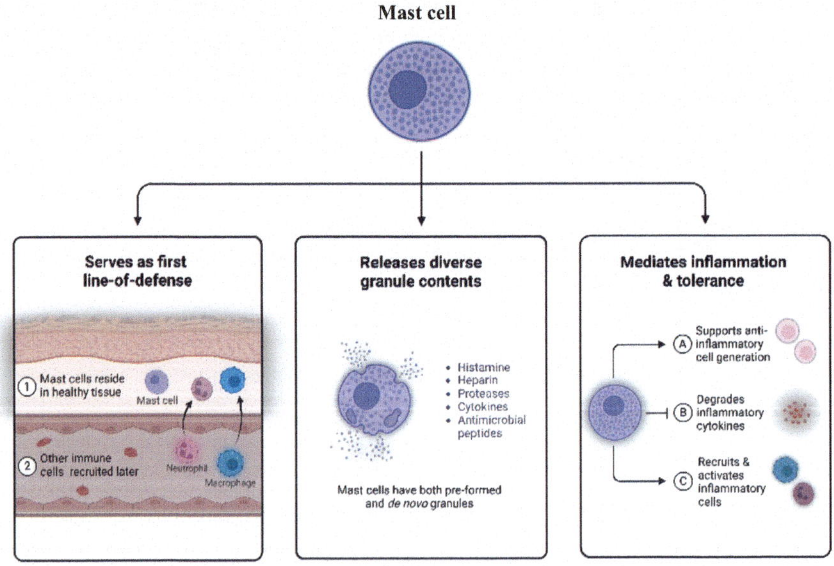

Fig. 4 Mast cell function as the first line of defense, release diverse granule content and mediate inflammation. Image created with Biorender.com

potentiality to release a wide range of effector chemicals that have been previously produced or are newly synthesised. Because they express a wide variety of surface and cytoplasmic receptors, mast cells can also respond to a broad range of cytokines, medications, and infections. Mast cells can also infiltrate at the site of inflammation and can be visualized by Toluidine blue staining. The mast cell's role as a cellular interface between the internal and external environments, as well as between the vasculature and tissues, is crucial for defence and healing. Mast cell interactions with other immune and nonimmune cells may favour chemicals that promote disease through the release of inflammatory mediators [51].

Important similarities exist between allergy and autoimmune illnesses. They are both the outcome of immune responses that are "hypersensitive" to naturally innocuous antigens [52]. Furthermore, a number of conditions that are currently recognised as autoimmune disorders include encephalomyelitis, rheumatoid arthritis, and bullous pemphigoid (Fig. 5). It is acknowledged that the adaptive immune system's cells control autoimmune reactions [53]. Furthermore, innate immune cells contribute a significant role in maintaining pathogenic response [54, 55].

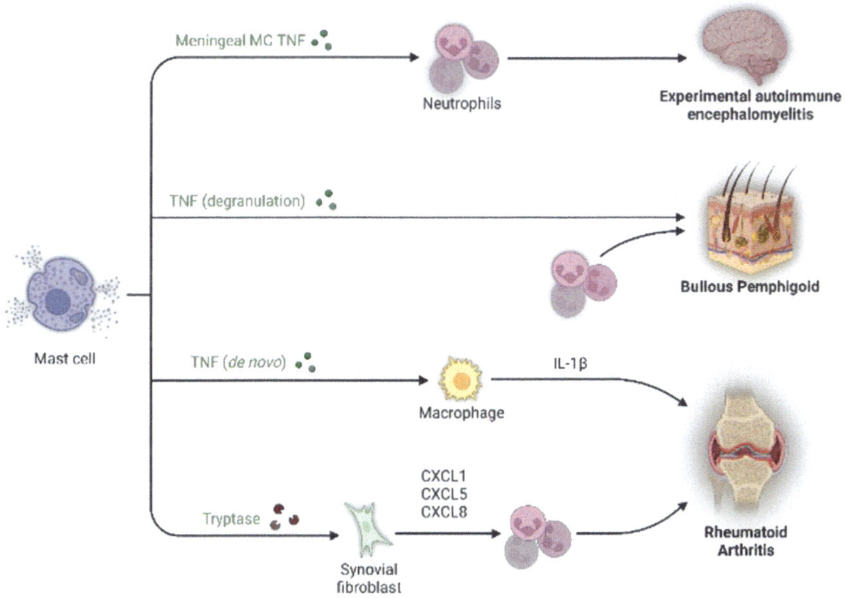

Fig. 5 Role of mast cells in autoimmune diseases. Image created with Biorender.com

4.6 Natural Killer and B Cells

Natural killer cells (NK cells), a subset of white blood cells, destroy contaminated and damaged cells, including malignant ones. They are members of the same lymphocyte class as B- and T-cells. The spread of viruses and malignant cells can be halted by NK cells, which have the power to eliminate harmful cells at an early stage. NK cells, for example, are crucial immunosurveillance cells that use their activation receptors, NKp46 and NKG2D, to identify contaminated, altered, or stressed cells [56]. When NK cells are activated, they become cytotoxic and kill microorganisms by releasing lytic granules (granzymes, perforin), generating death signals [57]. B cells utilise a process called antibody-dependent cell-mediated cytotoxicity, which involves turning into plasma cells, to kill germs by secreting antibodies. It has also been demonstrated that DCs, B cells, and macrophages can activate T cells by antigen cross-presentation [58]. Fortunately, the chemotactic mechanisms underlying the recruitment of monocytes to repair tissues when inflammation intensifies remain mostly unclear [59]. Nevertheless, it has been shown that inflammatory cytokines and the phenotype of monocytes at the site of inflammation are actively regulated by mediators. According to reports, these pro- and anti-inflammatory substances produced subpopulations of macrophages with distinct functional traits that influenced angiogenesis, matrix metabolism, fibroblast cell activity, and tissue repair processes [60, 61].

4.7 T Cells

T cells are an element of the immune system that are created from bone marrow stem cells. They assist in infection prevention in addition to aiding the body's battle against cancer. Also known as a thymocyte or T lymphocyte. By producing cytokines, T cells are essential for antiviral responses [62]. T cells activate in an inflammatory environment. Then, depending on the cytokines generated at the sites of inflammation, these cells differentiate into different T-cell subsets, including Th1, Th2, Th17, and regulatory T (Treg) cells. On the other hand, TGF-β stimulates the growth of Treg cells, which release cytokines that inhibit the immune system, such as IL-10 and TGF-β. TNF-α, IL-6, and IL-1β are among the inflammatory mediators that IL-17 is known to increase, but it has also been demonstrated that Treg cells are an efficient way to control how quickly inflammation resolves. Furthermore, it has been documented that CD4 + T cells stimulate B cells to aid in the development of virus-specific antibodies, On the other hand, CD8 + T lymphocytes can produce IFN-γ and TNF-α as well as kill virus-infected cells [63]. The inflammatory response to viral infections is subsequently heightened by the expression and secretion of additional chemokines and cytokines by each of these immune cells [64].

4.8 Dendritic Cells

A unique class of immune cells found in tissues, particularly skin, activates the immune system by presenting antigens to other immune system cells. Part of the phagocyte and antigen-presenting cell (APC) subpopulation are dendritic cells. Dendritic cells (DCs) hold a pivotal role in both immune tolerance and immune defence as they are the principal mediator between innate immunity and adaptive immunity. Different DC subsets must transit between lymphoid and non-lymphoid organs in order for DC-dependent activation to regulate immunity and inflammation. Interactions between chemokines and their receptors cause DC chemotaxis and migration, which are then tissue-specifically controlled by a variety of intracellular processes such as cytoskeletal reorganisation, metabolic remodelling, epigenetic reprogramming, and protein modification. Dysregulation of DC migration can result in aberrant DC activation or placement, which can then cause immune responses to become unbalanced and even immunological pathologies, such as tumours, autoimmune disorders, infectious diseases, and allergy reactions [65].

4.9 Vascular Endothelial Cells

Vascular endothelial cells are important participants in and controllers of inflammatory responses. Endothelial cells in a resting state limit inflammation, regulate blood flow, stop clotting, and stop blood proteins from entering tissues. Nitric oxide (NO) production is involved in these processes, and insufficient NO production is a primary factor in endothelial cell dysfunction. Vascular endothelial growth factor A (VEGFA) is one angiogenic stimulus that endothelial cells in chronic inflammation respond to by generating new blood vessels in order to sustain an inflammatory neo-tissue, such as pannus in rheumatoid arthritis. In addition, endothelial cells have the ability to grow into high endothelial venules in response to lymphotoxin-β, It facilitates the growth of tertiary lymphoid organs [66]. Acute and long-term changes occur concurrently in many inflammatory conditions. This could be the case because acute inflammation mediators, like TNF, influence the characteristics of endothelial cells linked to chronic inflammation, and acute inflammation mediators, like VEGFA, may also influence the behaviours of endothelial cells linked to acute inflammation [66].

4.10 Cytokines

Cytokines are the primary signalling molecules that are released by inflammatory cells and have a wide range of applications. They are divided into two groups: pro-inflammatory cytokines (IL-1, IL-6, TNF-α, IFN-γ, etc.) and anti-inflammatory cytokines (IL-4, IL-10, TGF-β, etc.) [67]. Among them, TNF-α is one of the oldest and most significant inflammatory mediators and is mostly produced by mast cells and macrophages. There are several roles that TNF-α plays in the inflammatory response; these include adhesion molecules, triggering the production of enzymes in the mechanism of prostaglandin synthesis (e.g., cyclooxygenase-2), nitric oxide synthase induction, which stimulates white blood cells and the endothelium [68, 69]. It has also been documented to stimulate lymphocytes and neutrophils, raise vascular endothelial cells' permeability, control other tissues' metabolic processes, and encourage the production and release of additional cytokines. It has also been documented to stimulate lymphocytes and neutrophils, raise vascular endothelial cells' permeability, control other tissues' metabolic processes, and encourage the production and release of additional cytokines [70, 71]. Consequently, it can be concluded that the well-known interleukins IL-1 and IL-6 play a role in the production of reactive oxygen species (ROS) and RNS as well as inflammatory molecules such integrins, chemokines, and matrix metalloproteinase (MMP). T cells and macrophages are these interleukins' two main cell origins. The JAKs-STAT and NF-κB pathways are triggered when IL-1 and IL-6 attach to their corresponding IL-1R and IL-6R receptors. Moreover, it has been demonstrated that IL-6 promotes B cell development for antibody production as well as T cell activation, proliferation, and differentiation [72, 73].

4.11 Type 2 Cytokines

The type 2 cytokine response has numerous essential host-protective functions, but it can also cause dysregulation, persistence, or hyperreactivity in type 2 immunity, which can result in disease (Fig. 6). It is possible for type 2 cytokines to promote cancer cell growth and carcinogenesis. Important to the pathogenesis of many allergic and fibrotic illnesses, they also inhibit the establishment of protective type 1 immunity against a range of bacterial, viral, and protozoan pathogens [74]. Together, a number of endogenous regulatory systems prevent or reduce the harmful effects of persistent type 2 immunity. Tissue macrophages and inflammatory monocytes have been recognised as important type 2 immune response modulators. Interfering with the recruitment, expansion, or maintenance of critical myeloid cell populations may be a novel therapeutic approach for a variety of type 2-driven illnesses. Therapeutic treatments targeting type 2 cytokine signalling pathways, prostaglandins, eosinophil production and recruitment, alarmins produced from epithelial cells, and regulatory T (TReg) cell activity in type 2-driven disease are detected at different phases of development.

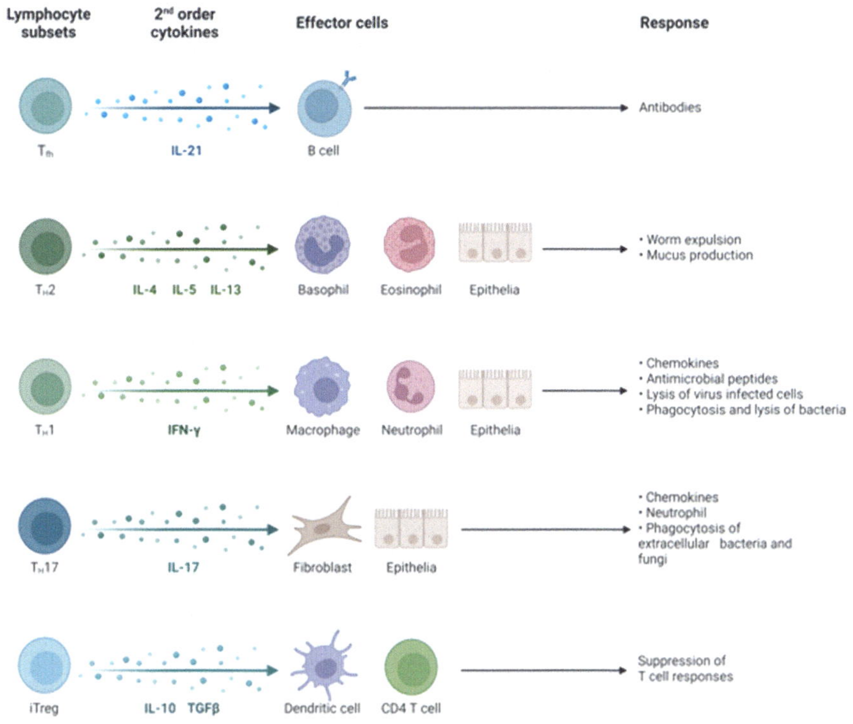

Fig. 6 Effectors response induced by second-order cytokines. Image created with Biorender.com

5 Pathophysiology of Inflammatory Diseases

The inflammatory disease may be due to acute and chronic inflammation. In the case of acute inflammation, neutrophils migrate from blood capillaries to infected tissues. Furthermore, monocytes infiltrated at the inflammation site and differentiated into macrophages or dendritic cells in which dendritic cells professionally present antigens. In chronic inflammation, infiltration of several inflammatory cells at the site of inflammation occurs, and they produce inflammatory cytokines and growth factors that contribute to the development of secondary repair and the progression of tissue damage. In this section, we have discussed the pathophysiology of some very common inflammatory diseases.

5.1 Rheumatoid Arthritis

RA is a chronic inflammatory disease with unknown etiology. On the basis of the presence or absence of anti-citrullinated protein antibodies, RA is classified into two major forms. About 67% of RA patients have ACPAs, which makes them a useful diagnostic tool for people with early, undifferentiated arthritis and a predictor of the risk that the condition will precede RA [75, 76]. Compared to the ACPA-negative category, the clinical profile of the ACPA-positive fraction of RA is more aggressive [77]. Compared to the ACPA-positive subgroup, ACPA-negative RA has been demonstrated to exhibit different genetic association patterns and immune cell responses to citrullinated antigens12 [78, 79]. In RA, the environment initiates the synthesis of ACPA, and genes and the environment are combined in epigenetic regulation. Gene-environment interactions impact autoantibodies' reactivity to citrullinated antigens in RA [80]. ACPAs can be diagnosed prior to the start of joint discomfort.

5.2 *Osteoarthritis*

The pathophysiology of osteoarthritis (OA) has been extensively studied in the last several decades [81]. The OA is characterised by the alteration of the synovial joint but the exact pathophysiological mechanism underlying the OA initiation and progression remain unclear. Recent studies revealed that age-related subchondral bone sclerosis, aberrant bone remodelling due to dysregulation of osteoblast and osteoclast play significant roles in OA [82, 83]. Increased subchondral bone porosity and remodelling, decreased bone density, and uneven matrix organisation in the bone mineralization were observed in the early stages of osteoarthritis (OA). These findings were attributed to subchondral pores and vascular invasion causing bone-cartilage crosstalk [84, 85]. These changes to the subchondral bone happened either before or concurrently with the early loss of articular cartilage [86].

5.3 *Diabetes*

Diabetes is a chronic disease that develops when the body either cannot use the insulin that the pancreas makes properly or does not create enough of it. The vast majority of diabetics are classified into one of two basic pathogenetic categories: type 1 or type 2. Diabetes mellitus is a multifaceted, chronic metabolic disease with an intricate pathophysiology. In a condition frequently referred to as hyperglycemia, elevated blood glucose levels are brought on by alterations in insulin secretion, insulin action, or both. The primary biomarker used to diagnose diabetes is hyperglycemia. 10% of instances of diabetes are type I, which is caused by immune-mediated selective β-cell death brought on by unidentified environmental stimuli in those with a polygenetic susceptibility. 90% of instances of diabetes are type II (T2D), which is characterised by decreased insulin sensitivity and decreased insulin production. In T2D, inflammation triggered by IL-1β is a major factor in β-cell loss. Inflammation in the liver and adipose tissues, especially via the IL-1β-inflammasome pathway, is the cause of insulin resistance in type 2 diabetes. Targeting inflammation to enhance disease prevention and management has gained attention again due to the growing role that inflammation plays in the pathophysiology of type 1 and type 2 diabetes as well as related metabolic disorders.

5.4 *Intestinal Inflammation*

Acute intestinal inflammation eliminates harmful chemicals and initiates healing, making it a protective mechanism. Due to recurrent acute damage or inadequate inflammation resolution, chronic inflammation in the gastrointestinal tract can result in chronic pancreatitis, peptic ulcer, and inflammatory bowel disease. Although the precise aetiology of Crohn's disease and ulcerative colitis is unknown, treatment aims to lessen inflammation and the symptoms it causes. Corticosteroids can reduce inflammation but do not typically lead to long-term remission.

5.5 *Liver Disease*

Liver is a vital metabolic organ that plays a significant role in the metabolism and detoxification of xenobiotics. Pro-inflammatory cytokines induce hepatocyte necrosis and stimulate regeneration in acute hepatocellular injury. Hepatocyte necrosis causes the release of IL-1α, which draws macrophages and monocytes from bone marrow. Macrophages are drawn from the peritoneum quickly and seem to have protective qualities. Both alcoholic and non-alcoholic fatty liver diseases are distinguished by substantial lipid accumulation and sterile inflammation.

5.6 *Atherosclerosis*

Atherosclerosis is known as the gradual hardening of the arterial wall with the deposition of plaque. Trans fat and cholesterol are responsible for plaque which causes inflammatory response and turbulent flow of blood inside the artery and may be the leading cause of cardiovascular diseases. Atherosclerosis involves the activation of endothelial cells in the artery lumen, which recruit monocytes through adhesion molecules. After maturing from monocytes into macrophages, these cells multiply and take up modified lipoprotein particles, producing inflammatory foamy macrophages [87]. Elevated blood cholesterol levels in the subintimal region can cause crystal precipitation, which triggers the NLRP3 inflammasome and releases active IL-1β [88]. Adaptive immune cells, like T and B lymphocytes, penetrate the arterial wall during atherogenesis and control the innate immune cells' harmful activities in plaques [89].

5.7 Neurological Disorders

Innate immune processes play a vital role in both normal brain aging and neuronal disease [90, 91]. Mutations in immune protein genes raise the chance of neuronal disease. The activation of microglia in response to abnormal nucleic acids, neuronal debris, misfolded proteins, or β-sheet proteins [92] can cause neuronal dysfunction, structural damage, and cell death. Alzheimer's disease is typified by β-amyloid fibrils that trigger the NLRP3 inflammasome and cause microglia to produce IL-1β through PRR activation [93]. Microglia express neurotransmitter and neurotrophin receptors that are distinct to each brain region. Norepinephrine and acetylcholine reduce excessive inflammation in the brain [94].

5.8 Skin Inflammation

In atopic dermatitis, a compromised skin barrier allows allergens to penetrate and trigger an excessive T_H2 response [95]. Skin barrier degeneration is exacerbated by local immunological imbalance because a vicious loop is created when important genes like filaggrins are downregulated by IL-4 and IL-13. Innate and adaptive immunity are involved in psoriasis, a disease driven by T cells. The signalling pathways of NF-kB and T_H17- and T_H1-type cytokines are what generate inflammation. Psoriasis causes systemic inflammation and often leads to rheumatic or cardiovascular complications [96].

5.9 Cancer

The tumor microenvironment contains the inflammatory response components, but they may have 'corrupted' activities. Cancer-related inflammation, including tumor-associated macrophages, has a significant impact on cancer growth, genomic instability, angiogenesis, and metastasis [97–99]. The macrophages associated with tumour inhibit cancer immunity by generating prostaglandins from the tryptophan metabolism of the tryptophan and expressing the checkpoint blockade triggers. Cancer-related inflammation involves various types of blood cells. Furthermore, chemokines as inflammatory mediators, TNF, IL-6, and IL-1 play a crucial role in chronic inflammation related to cancer [98]. The humoral arm of innate immunity contributes to cancer-related inflammation. An improved prognosis is associated with T cell-driven chronic inflammation that has an interferon signature. [100].

6 Conclusions and Future Perspectives

Inflammation has been classified into acute and chronic inflammation. The first type of inflammation is acute inflammation which is an instant and short-term reaction to injury or illness that may be marked by redness, heat, and swelling. It is nonspecific and affects the innate immunity. Chronic inflammation is associated with long-term inflammation, specialised reaction that requires developed immunity. There are several factors involved in acute and chronic inflammation. The inflammatory response varies depending on the condition and organ and can have both beneficial and negative effects on the patient. Inflammatory diseases can be treated based on their disease pathophysiology. Understanding inflammatory pathways is crucial for future therapeutic advancements that target inflammation. Nanomedicine is an emerging field of biological science. Targeting inflammatory disease with the next gene nanomaterials may be a ray of hope for tackling inflammatory disease.

Conflict of Interest The author declares no competing interest.

References

1. Coussens LM, Werb Z (2002) Inflammation and cancer. Nature 420(6917):860–867. https://doi.org/10.1038/nature01322
2. Nathan C, Ding A (2010) Nonresolving inflammation. Cell 140(6):871–882. https://doi.org/10.1016/j.cell.2010.02.029
3. Dorward DA, Lucas CD, Rossi AG, Haslett C, Dhaliwal K (2012) Imaging inflammation: molecular strategies to visualize key components of the inflammatory cascade, from initiation to resolution. Pharmacol Ther 135(2):182–199. https://doi.org/10.1016/j.pharmthera.2012.05.006
4. Sun SC (2017) The non-canonical NF-κB pathway in immunity and inflammation. Nat Rev Immunol 17(9):545–558. https://doi.org/10.1038/nri.2017.52
5. Medzhitov R (2008) Origin and physiological roles of inflammation. Nature 454(7203):428–435. https://doi.org/10.1038/nature07201
6. Kono H, Rock KL (2008) How dying cells alert the immune system to danger. Nat Rev Immunol 8(4):279–289. https://doi.org/10.1038/nri2215
7. Nathan C (2002) Points of control in inflammation. Nature 420(6917):846–852. https://doi.org/10.1038/nature01320
8. Headland SE, Norling LV (2015) The resolution of inflammation: principles and challenges. Semin Immunol 27(3):149–160. https://doi.org/10.1016/j.smim.2015.03.014
9. Wright HL, Moots RJ, Bucknall RC, Edwards SW (2010) Neutrophil function in inflammation and inflammatory diseases. Rheumatology (Oxford) 49(9):1618–1631. https://doi.org/10.1093/rheumatology/keq045
10. Rossi AG, Sawatzky DA, Walker A, Ward C, Sheldrake TA, Riley NA, Caldicott A, Martinez-Losa M, Walker TR, Duffin R, Gray M, Crescenzi E, Martin MC, Brady HJ, Savill JS, Dransfield I, Haslett C (2006) Cyclin-dependent kinase inhibitors enhance the resolution of inflammation by promoting inflammatory cell apoptosis. Nat Med 12(9):1056–1064. https://doi.org/10.1038/nm1468

11. Gilroy DW, Lawrence T, Perretti M, Rossi AG (2004) Inflammatory resolution: new opportunities for drug discovery. Nat Rev Drug Discov 3(5):401–416. https://doi.org/10.1038/nrd1383
12. Savill JS, Wyllie AH, Henson JE, Walport MJ, Henson PM, Haslett C (1989) Macrophage phagocytosis of aging neutrophils in inflammation. Programmed cell death in the neutrophil leads to its recognition by macrophages. J Clin Invest 83(3):865–875. https://doi.org/10.1172/JCI113970
13. Xu H, Barnes GT, Yang Q, Tan G, Yang D, Chou CJ, Sole J, Nichols A, Ross JS, Tartaglia LA, Chen H (2003) Chronic inflammation in fat plays a crucial role in the development of obesity-related insulin resistance. J Clin Investig 112(12):1821–1830. https://doi.org/10.1172/JCI19451
14. Kumar R, Clermont G, Vodovotz Y, Chow CC (2004) The dynamics of acute inflammation. J Theor Biol 230(2):145–155. https://doi.org/10.1016/j.jtbi.2004.04.044
15. Serhan CN (2008) Controlling the resolution of acute inflammation: a new genus of dual anti-inflammatory and proresolving mediators. J Periodontol 79(8 Suppl):1520–1526. https://doi.org/10.1902/jop.2008.080231
16. Ploeger HE, Takken T, de Greef MH, Timmons BW (2009) The effects of acute and chronic exercise on inflammatory markers in children and adults with a chronic inflammatory disease: a systematic review. Exerc Immunol Rev 15:6–41
17. Boyle M, Chun C, Strojny C, Narayanan R, Bartholomew A, Sundivakkam P, Alapati S (2014) Chronic inflammation and angiogenic signaling axis impairs differentiation of dental-pulp stem cells. PLoS ONE 9(11):e113419. https://doi.org/10.1371/journal.pone.0113419
18. Bozkurt B, Mann DL, Deswal A (2010) Biomarkers of inflammation in heart failure. Heart Fail Rev 15(4):331–341. https://doi.org/10.1007/s10741-009-9140-3
19. Glass CK, Saijo K, Winner B, Marchetto MC, Gage FH (2010) Mechanisms underlying inflammation in neurodegeneration. Cell 140(6):918–934. https://doi.org/10.1016/j.cell.2010.02.016
20. GBD (2019) Global burden of 369 diseases and injuries in 204 countries and territories, 1990–2019: a systematic analysis for the global burden of disease study 2019. https://vizhub.healthdata.org/gbd-results/.
21. Cieza A, Causey K, Kamenov K, Hanson SW, Chatterji S, Vos T (2021) Global estimates of the need for rehabilitation based on the global burden of disease study 2019: a systematic analysis for the global burden of disease study 2019. Lancet (London, England) 396(10267):2006–2017. https://doi.org/10.1016/S0140-6736(20)32340-0
22. Long H, Liu Q, Yin H, Wang K, Diao N, Zhang Y, Lin J, Guo A (2022) Prevalence trends of site-specific osteoarthritis from 1990 to 2019: findings from the global burden of disease study 2019. Arthritis Rheumatol (Hoboken, N.J.) 74(7):1172–1183. https://doi.org/10.1002/art.42089
23. Pariser DM, Bagel J, Gelfand JM, Korman NJ, Ritchlin CT, Strober BE, Van Voorhees AS, Young M, Rittenberg S, Lebwohl MG, Horn EJ, Foundation NP (2007) National psoriasis foundation clinical consensus on disease severity. Arch Dermatol 143(2):239–242. https://doi.org/10.1001/archderm.143.2.239
24. Okhandiar RP, Banerjee BN (1963) Psoriasis in the tropics: an epidemiological survey. J Indian Med Assoc 41:550–556
25. Bedi TR (1977) Psoriasis in north India: geographical variations. Dermatologica 155(5):310–314. https://doi.org/10.1159/000250983
26. Kaur I, Kumar B, Sharma KV, Kaur S (1986) Epidemiology of psoriasis in a clinic from North India. Indian J Dermatol Venereol Leprol 52(4):208–212
27. Bedi TR (1995) Clinical profile of psoriasis in North India. Indian J Dermatol Venereol Leprol 61(4):202–205
28. Kaur I, Handa S, Kumar B (1997) Natural history of psoriasis: a study from the Indian subcontinent. J Dermatol 24(4):230–234. https://doi.org/10.1111/j.1346-8138.1997.tb02779.x
29. Asokan N, Prathap P, Ajithkumar K, Betsy A, Binesh VG (2011) Pattern of psoriasis in a tertiary care teaching hospital in South India. Indian J Dermatol 56(1):118–119. https://doi.org/10.4103/0019-5154.77575

30. Lal S (1966) Clinical pattern of psoriasis in Punjab. Indian J Dermat Venereol 32(1):5–8
31. Mehta TK, Shah RN, Marquis L (1976) A study of 300 cases of psoriasis. Indian J Dermatol Venereol Leprol 42(2):67–70
32. Mayadas TN, Cullere X, Lowell CA (2014) The multifaceted functions of neutrophils. Annu Rev Pathol 9:181–218. https://doi.org/10.1146/annurev-pathol-020712-164023
33. Rosales C (2018) Neutrophil: a cell with many roles in inflammation or several cell types? Front Physiol 9:113. https://doi.org/10.3389/fphys.2018.00113
34. Herrero-Cervera A, Soehnlein O, Kenne E (2022) Neutrophils in chronic inflammatory diseases. Cell Mol Immunol 19(2):177–191. https://doi.org/10.1038/s41423-021-00832-3
35. Peiseler M, Kubes P (2019) More friend than foe: the emerging role of neutrophils in tissue repair. J Clin Investig 129(7):2629–2639. https://doi.org/10.1172/JCI124616
36. Cahilog Z, Zhao H, Wu L, Alam A, Eguchi S, Weng H, Ma D (2020) The role of neutrophil NETosis in organ injury: novel inflammatory cell death mechanisms. Inflammation 43(6):2021–2032. https://doi.org/10.1007/s10753-020-01294-x
37. Papayannopoulos V (2018) Neutrophil extracellular traps in immunity and disease. Nat Rev Immunol 18(2):134–147. https://doi.org/10.1038/nri.2017.105
38. Ingersoll MA, Platt AM, Potteaux S, Randolph GJ (2011) Monocyte trafficking in acute and chronic inflammation. Trends Immunol 32(10):470–477. https://doi.org/10.1016/j.it.2011.05.001
39. Shi C, Pamer EG (2011) Monocyte recruitment during infection and inflammation. Nat Rev Immunol 11(11):762–774. https://doi.org/10.1038/nri3070
40. Fujiwara N, Kobayashi K (2005) Macrophages in inflammation. Curr Drug Targets Inflamm Allergy 4(3):281–286. https://doi.org/10.2174/1568010054022024
41. Kay AB (2015) The early history of the eosinophil. Clin Exp Allergy J Br Soc Allergy Clin Immunol 45(3):575–582. https://doi.org/10.1111/cea.12480
42. Kanda A, Yun Y, Bui DV, Nguyen LM, Kobayashi Y, Suzuki K, Mitani A, Sawada S, Hamada S, Asako M, Iwai H (2021) The multiple functions and subpopulations of eosinophils in tissues under steady-state and pathological conditions. Allergol Int Off J Japan Soc Allergol 70(1):9–18. https://doi.org/10.1016/j.alit.2020.11.001
43. Lombardi C, Berti A, Cottini M (2022) The emerging roles of eosinophils: Implications for the targeted treatment of eosinophilic-associated inflammatory conditions. Curr Res Immunol 3:42–53. https://doi.org/10.1016/j.crimmu.2022.03.002
44. Siracusa MC, Kim BS, Spergel JM, Artis D (2013) Basophils and allergic inflammation. J Allergy Clin Immunol 132(4):789–788. https://doi.org/10.1016/j.jaci.2013.07.046
45. Shibuya R, Kim BS (2022) Skin-homing basophils and beyond. Front Immunol 13:1059098. https://doi.org/10.3389/fimmu.2022.1059098
46. Ohnmacht C, Schwartz C, Panzer M, Schiedewitz I, Naumann R, Voehringer D (2010) Basophils orchestrate chronic allergic dermatitis and protective immunity against helminths. Immunity 33(3):364–374. https://doi.org/10.1016/j.immuni.2010.08.011
47. Giacomin PR, Siracusa MC, Walsh KP, Grencis RK, Kubo M, Comeau MR, Artis D (2012) Thymic stromal lymphopoietin-dependent basophils promote Th2 cytokine responses following intestinal helminth infection. J Immunol (Baltimore, Md.: 1950) 189(9):4371–4378. https://doi.org/10.4049/jimmunol.1200691
48. Obata-Ninomiya K, Ishiwata K, Tsutsui H, Nei Y, Yoshikawa S, Kawano Y, Minegishi Y, Ohta N, Watanabe N, Kanuka H, Karasuyama H (2013) The skin is an important bulwark of acquired immunity against intestinal helminths. J Exp Med 210(12):2583–2595. https://doi.org/10.1084/jem.20130761
49. da Silva EZ, Jamur MC, Oliver C (2014) Mast cell function: a new vision of an old cell. J Histochem Cytochem Off J Histochem Soc 62(10):698–738. https://doi.org/10.1369/0022155414545334
50. Theoharides TC, Alysandratos KD, Angelidou A, Delivanis DA, Sismanopoulos N, Zhang B, Asadi S, Vasiadi M, Weng Z, Miniati A, Kalogeromitros D (2012) Mast cells and inflammation. Biochem Biophys Acta 1822(1):21–33. https://doi.org/10.1016/j.bbadis.2010.12.014

51. Dileepan KN, Raveendran VV, Sharma R, Abraham H, Barua R, Singh V, Sharma R, Sharma M (2023) Mast cell-mediated immune regulation in health and disease. Front Med 10:1213320. https://doi.org/10.3389/fmed.2023.1213320
52. Brown MA, Hatfield JK (2012) Mast cells are important modifiers of autoimmune disease: with so much evidence, why is there still controversy? Front Immunol 3:147. https://doi.org/10.3389/fimmu.2012.00147
53. Lohr J, Knoechel B, Nagabhushanam V, Abbas AK (2005) T-cell tolerance and autoimmunity to systemic and tissue-restricted self-antigens. Immunol Rev 204:116–127. https://doi.org/10.1111/j.0105-2896.2005.00241.x
54. Bach JF, Bendelac A, Brenner MB, Cantor H, De Libero G, Kronenberg M, Lanier LL, Raulet DH, Shlomchik MJ, von Herrath MG (2004) The role of innate immunity in autoimmunity. J Exp Med 200(12):1527–1531. https://doi.org/10.1084/jem.20042110
55. Tenner AJ (2004) Influence of innate immune responses on autoimmunity. Autoimmunity 37(2):83–84. https://doi.org/10.1080/08916930410001690067
56. Strandmark J, Steinfelder S, Berek C, Kühl AA, Rausch S, Hartmann S (2017) Eosinophils are required to suppress Th2 responses in Peyer's patches during intestinal infection by nematodes. Mucosal Immunol 10(3):661–672. https://doi.org/10.1038/mi.2016.93
57. Tosello-Trampont A, Surette FA, Ewald SE, Hahn YS (2017) Immunoregulatory role of NK cells in tissue inflammation and regeneration. Front Immunol 8:301. https://doi.org/10.3389/fimmu.2017.00301
58. Hofmann U, Beyersdorf N, Weirather J, Podolskaya A, Bauersachs J, Ertl G, Kerkau T, Frantz S (2012) Activation of CD4+ T lymphocytes improves wound healing and survival after experimental myocardial infarction in mice. Circulation 125(13):1652–1663. https://doi.org/10.1161/CIRCULATIONAHA.111.044164
59. Nahrendorf M, Swirski FK, Aikawa E, Stangenberg L, Wurdinger T, Figueiredo JL, Libby P, Weissleder R, Pittet MJ (2007) The healing myocardium sequentially mobilizes two monocyte subsets with divergent and complementary functions. J Exp Med 204(12):3037–3047. https://doi.org/10.1084/jem.20070885
60. Frantz S, Hofmann U, Fraccarollo D, Schäfer A, Kranepuhl S, Hagedorn I, Nieswandt B, Nahrendorf M, Wagner H, Bayer B, Pachel C, Schön MP, Kneitz S, Bobinger T, Weidemann F, Ertl G, Bauersachs J (2013) Monocytes/macrophages prevent healing defects and left ventricular thrombus formation after myocardial infarction. FASEB J Off Publ Fed Am Soc Exp Biol 27(3):871–881. https://doi.org/10.1096/fj.12-214049
61. Zhao Y, Zou W, Du J, Zhao Y (2018) The origins and homeostasis of monocytes and tissue-resident macrophages in physiological situation. J Cell Physiol 233(10):6425–6439. https://doi.org/10.1002/jcp.26461
62. Cecere TE, Todd SM, Leroith T (2012) Regulatory T cells in arterivirus and coronavirus infections: do they protect against disease or enhance it? Viruses 4(5):833–846. https://doi.org/10.3390/v4050833
63. Li G, Fan Y, Lai Y, Han T, Li Z, Zhou P, Pan P, Wang W, Hu D, Liu X, Zhang Q, Wu J (2020) Coronavirus infections and immune responses. J Med Virol 92(4):424–432. https://doi.org/10.1002/jmv.2568
64. Zabetakis I, Lordan R, Norton C, Tsoupras A (2020) COVID-19: the inflammation link and the role of nutrition in potential mitigation. Nutrients 12(5):1466. https://doi.org/10.3390/nu12051466
65. Liu J, Zhang X, Cheng Y, Cao X (2021) Dendritic cell migration in inflammation and immunity. Cell Mol Immunol 18(11):2461–2471. https://doi.org/10.1038/s41423-021-00726-4
66. Pober JS, Sessa WC (2007) Evolving functions of endothelial cells in inflammation. Nat Rev Immunol 7(10):803–815. https://doi.org/10.1038/nri2171
67. Nissinen L, Kähäri VM (2014) Matrix metalloproteinases in inflammation. Biochem Biophys Acta 1840(8):2571–2580. https://doi.org/10.1016/j.bbagen.2014.03.007
68. Courtois G, Gilmore TD (2006) Mutations in the NF-kappaB signaling pathway: implications for human disease. Oncogene 25(51):6831–6843. https://doi.org/10.1038/sj.onc.1209939

69. Koehne CH, Dubois RN (2004) COX-2 inhibition and colorectal cancer. Semin Oncol 31(2 Suppl 7):12–21. https://doi.org/10.1053/j.seminoncol.2004.03.041
70. Szlosarek PW, Balkwill FR (2003) Tumour necrosis factor alpha: a potential target for the therapy of solid tumours. Lancet Oncol 4(9):565–573. https://doi.org/10.1016/s1470-2045(03)01196-3
71. Üçeyler N, Urlaub D, Mayer C, Uehlein S, Held M, Sommer C (2019) Tumor necrosis factor-α links heat and inflammation with Fabry pain. Mol Genet Metab 127(3):200–206. https://doi.org/10.1016/j.ymgme.2019.05.009
72. Jordan SC, Choi J, Kim I, Wu G, Toyoda M, Shin B, Vo A (2017) Interleukin-6, A cytokine critical to mediation of inflammation, autoimmunity and allograft rejection: therapeutic implications of IL-6 receptor blockade. Transplantation 101(1):32–44. https://doi.org/10.1097/TP.0000000000001452
73. Nguyen DP, Li J, Tewari AK (2014) Inflammation and prostate cancer: the role of interleukin 6 (IL-6). BJU Int 113(6):986–992. https://doi.org/10.1111/bju.12452
74. Wynn TA (2015) Type 2 cytokines: mechanisms and therapeutic strategies. Nat Rev Immunol 15(5):271–282. https://doi.org/10.1038/nri3831
75. Nishimura K, Sugiyama D, Kogata Y, Tsuji G, Nakazawa T, Kawano S, Saigo K, Morinobu A, Koshiba M, Kuntz KM, Kamae I, Kumagai S (2007) Meta-analysis: diagnostic accuracy of anti-cyclic citrullinated peptide antibody and rheumatoid factor for rheumatoid arthritis. Ann Intern Med 146(11):797–808. https://doi.org/10.7326/0003-4819-146-11-200706050-00008
76. Bizzaro N, Bartoloni E, Morozzi G, Manganelli S, Riccieri V, Sabatini P, Filippini M, Tampoia M, Afeltra A, Sebastiani G, Alpini C, Bini V, Bistoni O, Alunno A, Gerli R, Forum Interdisciplinare per la Ricerca nelle Malattie Autoimmuni (FIRMA Group) (2013) Anti-cyclic citrullinated peptide antibody titer predicts time to rheumatoid arthritis onset in patients with undifferentiated arthritis: results from a 2-year prospective study. Arthritis Res Ther 15(1):R16. https://doi.org/10.1186/ar4148
77. Malmström V, Catrina AI, Klareskog L (2017) The immunopathogenesis of seropositive rheumatoid arthritis: from triggering to targeting. Nat Rev Immunol 17(1):60–75. https://doi.org/10.1038/nri.2016.124
78. Padyukov L, Seielstad M, Ong RT, Ding B, Rönnelid J, Seddighzadeh M, Alfredsson L, Klareskog L, Epidemiological Investigation of Rheumatoid Arthritis (EIRA) study group (2011) A genome-wide association study suggests contrasting associations in ACPA-positive versus ACPA-negative rheumatoid arthritis. Ann Rheum Dis 70(2):259–265. https://doi.org/10.1136/ard.2009.126821
79. Schuerwegh AJ, Ioan-Facsinay A, Dorjée AL, Roos J, Bajema IM, van der Voort EI, Huizinga TW, Toes RE (2010) Evidence for a functional role of IgE anticitrullinated protein antibodies in rheumatoid arthritis. Proc Natl Acad Sci USA 107(6):2586–2591. https://doi.org/10.1073/pnas.0913054107 (Retraction published Proc Natl Acad Sci USA 110(50):20345 (2013). https://doi.org/10.1073/pnas.1320459110)
80. van der Woude D, Alemayehu WG, Verduijn W, de Vries RR, Houwing-Duistermaat JJ, Huizinga TW, Toes RE (2010) Gene-environment interaction influences the reactivity of autoantibodies to citrullinated antigens in rheumatoid arthritis. Nat Genet 42(10):814–816. https://doi.org/10.1038/ng1010-814
81. Tong L, Yu H, Huang X, Shen J, Xiao G, Chen L, Wang H, Xing L, Chen D (2022) Current understanding of osteoarthritis pathogenesis and relevant new approaches. Bone Res 10(1):60. https://doi.org/10.1038/s41413-022-00226-9
82. Han X, Cui J, Chu L, Zhang W, Xie K, Jiang X, He Z, Du J, Ai S, Sun Q, Wang L, Wu H, Yan M, Yu Z (2022) Abnormal subchondral trabecular bone remodeling in knee osteoarthritis under the influence of knee alignment. Osteoarthritis Cartilage 30(1):100–109. https://doi.org/10.1016/j.joca.2021.10.005
83. Hu W, Chen Y, Dou C, Dong S (2021) Microenvironment in subchondral bone: predominant regulator for the treatment of osteoarthritis. Ann Rheum Dis 80(4):413–422. https://doi.org/10.1136/annrheumdis-2020-218089

84. Hu Y, Chen X, Wang S, Jing Y, Su J (2021) Subchondral bone microenvironment in osteoarthritis and pain. Bone Res 9(1):20. https://doi.org/10.1038/s41413-021-00147-z
85. Hügle T, Geurts J (2017) What drives osteoarthritis?-synovial versus subchondral bone pathology. Rheumatology (Oxford) 56(9):1461–1471. https://doi.org/10.1093/rheumatology/kew389
86. Ding R, Zhang N, Wang Q, Wang W (2022) Alterations of the subchondral bone in osteoarthritis: complying with Wolff's law. Curr Rheumatol Rev 18(3):178–185. https://doi.org/10.2174/1573397118666220401104428
87. Robbins CS, Hilgendorf I, Weber GF, Theurl I, Iwamoto Y, Figueiredo JL, Gorbatov R, Sukhova GK, Gerhardt LM, Smyth D, Zavitz CC, Shikatani EA, Parsons M, van Rooijen N, Lin HY, Husain M, Libby P, Nahrendorf M, Weissleder R, Swirski FK (2013) Local proliferation dominates lesional macrophage accumulation in atherosclerosis. Nat Med 19(9):1166–1172. https://doi.org/10.1038/nm.3258
88. Duewell P, Kono H, Rayner KJ, Sirois CM, Vladimer G, Bauernfeind FG, Abela GS, Franchi L, Nuñez G, Schnurr M, Espevik T, Lien E, Fitzgerald KA, Rock KL, Moore KJ, Wright SD, Hornung V, Latz E (2010) NLRP3 inflammasomes are required for atherogenesis and activated by cholesterol crystals. Nature 464(7293):1357–1361. https://doi.org/10.1038/nature08938
89. Ketelhuth DF, Hansson GK (2016) Adaptive response of T and B cells in atherosclerosis. Circ Res 118(4):668–678. https://doi.org/10.1161/CIRCRESAHA.115.306427
90. Heneka MT, Golenbock DT, Latz E (2015) Innate immunity in Alzheimer's disease. Nat Immunol 16(3):229–236. https://doi.org/10.1038/ni.3102
91. Heneka MT, Kummer MP, Latz E (2014) Innate immune activation in neurodegenerative disease. Nat Rev Immunol 14(7):463–477. https://doi.org/10.1038/nri3705
92. Kettenmann H, Hanisch UK, Noda M, Verkhratsky A (2011) Physiology of microglia. Physiol Rev 91(2):461–553. https://doi.org/10.1152/physrev.00011.2010
93. Halle A, Hornung V, Petzold GC, Stewart CR, Monks BG, Reinheckel T, Fitzgerald KA, Latz E, Moore KJ, Golenbock DT (2008) The NALP3 inflammasome is involved in the innate immune response to amyloid-beta. Nat Immunol 9(8):857–865. https://doi.org/10.1038/ni.1636
94. Pocock JM, Kettenmann H (2007) Neurotransmitter receptors on microglia. Trends Neurosci 30(10):527–535. https://doi.org/10.1016/j.tins.2007.07.007
95. Palmer CN, Irvine AD, Terron-Kwiatkowski A, Zhao Y, Liao H, Lee SP, Goudie DR, Sandilands A, Campbell LE, Smith FJ, O'Regan GM, Watson RM, Cecil JE, Bale SJ, Compton JG, DiGiovanna JJ, Fleckman P, Lewis-Jones S, Arseculeratne G, Sergeant A, McLean WH et al (2006) Common loss-of-function variants of the epidermal barrier protein filaggrin are a major predisposing factor for atopic dermatitis. Nat Genet 38(4):441–446. https://doi.org/10.1038/ng1767
96. Noda S, Krueger JG, Guttman-Yassky E (2015) The translational revolution and use of biologics in patients with inflammatory skin diseases. J Allergy Clin Immunol 135(2):324–336. https://doi.org/10.1016/j.jaci.2014.11.015
97. Balkwill F, Mantovani A (2001) Inflammation and cancer: back to virchow? Lancet (London, England) 357(9255):539–545. https://doi.org/10.1016/S0140-6736(00)04046-0
98. Mantovani A, Allavena P, Sica A, Balkwill F (2008) Cancer-related inflammation. Nature 454(7203):436–444. https://doi.org/10.1038/nature07205
99. Mantovani A, Marchesi F, Malesci A, Laghi L, Allavena P (2017) Tumour-associated macrophages as treatment targets in oncology. Nat Rev Clin Oncol 14(7):399–416. https://doi.org/10.1038/nrclinonc.2016.217
100. Gajewski TF, Schreiber H, Fu YX (2013) Innate and adaptive immune cells in the tumor microenvironment. Nat Immunol 14(10):1014–1022. https://doi.org/10.1038/ni.2703

Smart Nanomaterials and Nanocarriers: Basic Properties and Formulation Strategies

Anas Ahmad and Mohammad Imran

Abstract An extensive variety of functional nanoplatforms for clever applications in biotechnology and nanomedicine are made possible by the studies of nanostructured drug delivery systems, which enable the development of novel platforms for the effective transportation as well as controlled or regulated or sustained release of pharmacological compounds in the harsh micro-environmental conditions of diseased or abnormal tissue-systems of living organisms. This chapter describes the latest developments in the formulation and development of smart nanocarriers made of organic materials such as liposomes, dendrimers, hydrogels, and polymeric micelles and vesicles, and inorganic materials such as quantum dots, gold, and mesoporous silica-oxide nanocarriers and nanomaterials. Even with the amazing advancements in recent synthetic techniques, the majority of nanocarriers' actions are linked to a number of undesirable side effects that reduce their usefulness in applications related to biotechnology and nanomedicine. This chapter draws attention to certain important challenges in the formulation and designing and engineering of nanoparticulate drug delivery systems for biotechnological applications in general and anti-inflammatory applications in particular, because of the complex environment and multiform interactions produced inside the specific biological medium.

Keywords Nanomaterials · Drug delivery · Inflammation · Nanoparticles · Multifunctional nanocarriers

A. Ahmad (✉)
Julia McFarlane Diabetes Research Centre (JMDRC), Department of Microbiology, Immunology and Infectious Diseases, Snyder Institute for Chronic Diseases, Hotchkiss Brain Institute, Cumming School of Medicine, Foothills Medical Centre, University of Calgary, Calgary, AB, Canada
e-mail: anasahmad78689@gmail.com

M. Imran
Faculty of Health, Medicine and Behavioral Sciences, Frazer Institute, The University of Queensland, Brisbane, Australia

Md. M. Ansari et al. (eds.), *Emergence of Sustainable Biomaterials in Tackling Inflammatory Diseases*, Smart Nanomaterials Technology,
https://doi.org/10.1007/978-981-96-2112-5_2

1 Introduction

The goal of the fields of nanomedicine and nanotherapeutics is to address the inherent limitations of traditional medicine and diagnosis. However, despite the increased attention and appealing qualities of nanotechnologies, difficulties persist in their therapeutic use, leading some to contend that they have not yet realized their full potential [1, 2]. Though many distinct nanoparticles have been described in a significant body of research as possible therapeutic agents, only a small number have received approval from the United States Food & Drug Administration (FDA). Delivery of drug-payloads and imaging in cancer, infection, and anti-inflammatory therapy patients have been made possible by nanomedicines. Nanocarriers can be linked to a wide range of medications to facilitate the targeted delivery of drug-payloads, improving pharmacological efficacy as well as lowering the frequency and intensity of adverse drug reactions [3, 4]. Nanoscale systems can be used as a platform technology for intracellular delivery, which may facilitate the development of novel treatment approaches in the domains of gastroenterology, oncology, hepatology, and other relevant sciences. Additional inquiries into these technological platforms are necessary due to the inadequate understanding of the toxicological behaviours of novel nanoformulations [5, 6].

The application of tiny objects in medicine, fondly called nanoparticle-based medicines or nanomedicine, has great promise for enhancing patient care. Doxorubicin-loaded liposomal nanoformulations, which could be successfully utilized in anti-cancer therapies for about ten years, are arguably the most well-known example [7, 8]. These tiny vesicles hold the anticancer drug, which can enhance their concentrations in cancerous tissues while lowering the drug's accidental organ-to-organ transfer. As a result, cardiotoxicity—the primary adverse effect of doxorubicin—may manifest itself as much less severe [9, 10]. Since the 1980s, when nanotechnology was first applied to the treatment of cancer, new uses have been found, leading to significant breakthroughs in other medicinal arena, viz. enhanced bio-imaging or theranostics, as well as more sophisticated drug targeting techniques. Given the wide range of structural variations in the systems involved, defining the term "nanomedicine" is difficult, and most definitions that exist today concentrate mostly on the systems' sizes. Liposomal nanoformulations, polymeric nanocarriers or lipid-based nanoparticles, micellar formulations, dendrimer-based drug delivery platforms, quantum dots, gold NPs, carbon nano-tubes, and cyclodextrins are examples of nanoscale materials with potential use in medicine [11, 12]. A schematic summary of the most prevalent structures seen in the literature on nanomedicine to date is given in Fig. 1.

For the sake of clarity, supra-molecular assembly of some hundred nanometers for therapy or bio-imaging purposes is defined in this discussion using the words "nanocarriers", "nanoscale objects", and "nanosystems", as previously proposed. Therapeutic proteins are ignored here despite their similar sizes because of how they behave so differently in the physiological milieu [14, 15].

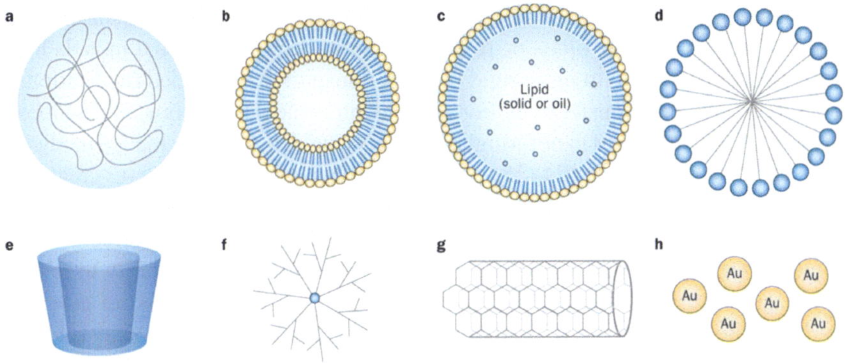

Fig. 1 Various kinds of nano-carriers and nanostructures designed for the formulation development of nanomedicines. **a** Polymeric nanocarriers **b** Liposomal neoformations are vesicular structures formulated from phospholipid bilayers mimicking the structures of the cellular membranes. **c** Lipid-based nanocarriers and nano-emulsions contain the inner oil phase surrounded by the surfactant. **d** Nano-micelles are the supra-molecular assemblies of surfactant molecules that form the inner lipophilic portion. **e** Cyclodextrins possess inner lipophilic portions that load the drugs and are composed of cyclic oligosaccharides. **f** Dendrimer-based nanostructures have branched molecular chains that usually associate smaller drug molecules at their extremities. **g** Carbon nanotubes. **h** Inorganic nanocarriers viz. gold-nanoparticles (Au-NPs). Adapted with permission from [13]

Natural and modified biomolecules, such as peptides, nucleic acids, and polysaccharides are being investigated to create the nanostructures in anti-inflammatory and anti-cancer nanomedicine. These biomolecules are frequently intrinsically bioactive as targeted and/or therapeutic agents, and they are typically biocompatible and biodegradable [16, 17]. While liposomes and polymeric particles continue to be the most successful nanoplatforms in terms of translational research, biomolecule-based building blocks have unique characteristics that these synthetic materials are unable to easily replicate. For instance, it is possible to accurately control the supra-molecular interaction between peptides and nucleic acids, and natural protein-based nanoformulations can replicate the molecular recognition and enzymatic functions of proteins [18, 19]. Crucially, biomolecule-derived components can be used to create multifunctional methods, each with unique properties and constraints. In the wider context of cancer nanomedicine, a comprehensive understanding of biomolecule-based nanostructures is still unclear. Some of the biomolecule-based nanomaterials have been addressed by researchers, emphasizing particular multifunctional properties in the context of cancer targeting and treatment [20, 21].

Delivering as much medication as possible to the specific site of inflammation is essential for the effective treatment of inflammatory illnesses. Liposomes and polymer micelles make up a significant percentage of the several nanoparticulate systems employed for medication delivery focused on inflammation. The development of drug delivery methods that target inflammation has made extensive use of liposomes [22, 23]. Because they live inside the liposomal lipid bilayer and are hydrophobic, many anti-inflammatory medications are prone to transfer to other hydrophobic regions in the blood, such as blood proteins. In comparison to liposomes with more flexible bilayers, research reports employing animal models have shown that liposomes with more rigid bilayers transported the loaded medication to lipoproteins at higher transfer rates. For all other intravenously administered nanocarriers as well as lipid-based nanoformulations, it is crucial to keep the loaded medicine inside the delivery vehicle [24, 25]. Therefore, it becomes imperative to create so-called "smart drug delivery systems", which are drug delivery vehicles that release drugs slowly into the bloodstream but quickly release drugs when triggered by environmental stimuli. When stimulated, smart nanoparticles have the ability to release additional drug molecules into their surroundings. The stimuli consist of elements that are biological (enzymes, antibodies, and tiny molecules), chemical (pH and ions), and physical (temperature, light, magnetic field, and electricity) [26, 27].

Systems for delivering smart drugs have been designed to react to outside stimuli like light, heat, radiofrequency, ultrasound, and light. High spatial and temporal resolution can be achieved while delivering ultrasound, a pressure wave. When ultrasound interacts with tissues, energy is lost, which can result in both heating and non-thermal mechanical impacts. Microbubbles and ultrasound have been utilized in medication delivery [28, 29]. In this chapter, a detailed discussion on the synthesis, properties and strategies for the formulation and characterization as well as applications of nanoparticles in various diseases and disorders in general and inflammatory disorders in particular is provided.

2 Types of Nanomaterials and Formulation Strategies

2.1 Methods of Preparation and Characterization of Lipid-Based Nanocarriers

2.1.1 High-Pressure Homogenization Method

Nanoemulsions, solid lipid nanoparticles, and nanostructured lipid carriers all benefit from the application of high-pressure homogenization [30]. The approach is based on the reduction of droplet and particle sizes under conditions of increased pressure. This technique provides a trustworthy and effective means of manufacturing many SLNs and NLCs. One research focused on the synthesis of NLCs containing coenzyme Q10 at a pace of 25 kg/h, while another employed this technology to create stavudine-laden SLNs in batches as large as 60 kg [31, 32]. First, coarse lipid-based nanoparticles are formulated followed by putting them through homogenization under high pressure with a few microns of gaps between them [33]. When liquid mixtures are subjected to pressures in the range of 100 to 2000 bar, they are propelled to speeds more than 1000 km/h. Droplet size is drastically reduced by cavitation forces and shear stresses caused by the fast motion [34]. Droplet size and homogeneity targets are reached using recurrent homogenization processes. Reduced production times streamlined manufacturing procedures, removal of organic solvents, and possible scalability are among the advantages of high-pressure homogenization (HPH) [31, 34]. When making SLNs and NLCs using high-pressure HPH, manufacturers may choose between two separate approaches: hot homogenization and cold homogenization. Combining drugs with molten lipids at the temperature of 5–10 degrees over the melting temperature of the solid lipid is a standard procedure in the context of hot melt extrusion. Therapeutic agents are either made to dissolve or distributed evenly throughout the molten lipid system during this procedure [35]. The resultant solid lipid nanoparticles and nano lipid carrier dispersions typically have a lipid content of between 5 and 10%. Surfactants in the aqueous phases are heated separately to reach the same temperature as the lipid melt. After that, the solution is added to molten lipids all the while stirring constantly to create a hot pre-emulsion. Under constant temperature control, the resulting pre-emulsions are homogenized in the piston-gap homogenizers [36]. Generally, a series of 3–5 homogenization cycles at pressures ranging from 500 to 1500 bars may be used to get appropriate SLNs and NLCs. However, it has been shown that particle coalescence under highly kinetic settings may create larger PS when the number of cycles and homogenization pressure are increased [37]. Homogenization is followed by a cooling phase in which lipid crystallization occurs, yielding SLNs and NLCs, respectively. For example, celecoxib-loaded SLNs were formulated by combining the lipid component glyceryl palmitostearate (Precirol® ATO 5) with the emulsifier polyoxyl-35 castor oil (Cremophor® EL). Further, using this method, submicron lipid nanoparticles were reported as a potential treatment alternative for breast cancer and leukaemia [38]. These lipid nanoparticles had a PS of less than 200 nm and could provide sustained release of drugs. In another study, atorvastatin was incorporated into the SLNs by means of high-pressure homogenization (HPH). The glyceryl behenate solid lipid Compritol-888 was used as the lipid-based matrix, while the glyceryl behenate phospholipid poloxamer-188 was used as the emulsifying agent [39]. However, hot HPH does have certain restrictions that should be taken into consideration. Some pharmaceutical agents may be destroyed by being exposed to high temperatures, while others may be lost to a water-phase during the process of homogenization [40].

To overcome the above-mentioned limits of hot high-pressure homogenization, an investigation on cold high-pressure homogenization was initiated. In this method, by employing dry ice or liquid nitrogen, therapeutic compounds are quickly chilled after being solubilized or dispersed within molten lipids. Through this method, drug particles were formed evenly dispersed within the lipid matrix due to the quick cooling rate [34]. To achieve a particle size PS range of 50–100 nm, the lipid-drug combinations are further comminated in a ball mill or mortar. Cold temperatures, generally between 0 and 4 °C, are used for homogenization, during which the lipid-based microparticles can be submerged in a cooled water solution which contains the surfactant. It is common practice to do 5–10 cycles of homogenization, each at a pressure of 500 bars [41]. Encapsulation efficiencies of 77.2% and 98.5% for calf-thymus DNA and TRP siRNA, respectively, in SLNs demonstrate the technique's effectiveness. Compritol-888 and stearic acid were utilized to create the lipid phase used for the above research. In addition, a combination of 1% w/v Phospholipon-90H and 2%

w/v poloxamer 188 as emulsifiers were also used [42]. In another investigation, it was observed that employing Compritol 888 as the solid lipid component and poloxamer 188 as an emulsifying agent was the key to successfully incorporating toad venom extract into SLNs. The enhanced formulation showed substantial effectiveness in suppressing tumor development, with an EE of 93% [43, 44]. In a separate study, the synthesis of SLNs containing the hydrophilic chemical didanosine was compared using hot and cold high-pressure homogenization (HPH) methods. Results showed that encapsulation efficiency (EE) was 51.6% for cold HPH and 32.5% for hot HPH, while drug loading (DL) was 3.39% for cold HPH and 0.08% for hot HPH [44]. Similar results were observed by other researchers [45]. The values of EE and DL may be increased using several possible ways. For example, drugs whose solubilization is pH-dependent could be lost to the polar (aqueous or hydrophilic) phase less often if the pH of aqueous phases gets corrected. Energy savings of up to 90.9% have been attributed to this method [46, 47]. However, cold high-pressure homogenization (HPH) produces bigger particles with a higher degree of diameter variations compared to hot HPH, thus it's necessary to be aware of this constraint while choosing this method to prepare SLNs [34].

2.1.2 High-Speed Stirring and Ultra-Sonication Method

Ultra-sonication and high-speed stirring (sometimes called the high-shear homogenizing process) are common techniques for the dispersion of the drugs to nanoparticulate sizes. To produce solid lipid nanoparticles and nano-lipid carriers, high-speed stirring is a simple and cost-effective technique [40, 48]. First, the lipids are heated to temperatures 5–10°C above their melting point, and then the drugs or therapeutics agents are added and mixed well throughout the liquid. The lipid-drug melt is then mixed with an aqueous surfactant solution kept at the same temperature. Next, a high-shear mixer is used to distribute the resultant mixture evenly. Shearing action induced by particularly turbulent eddies leads to the production of an oil-in-water (o/w) emulsion. Next, this dispersion is cooled to produce SLNs and NLCs [49]. After the high-speed stirring, ultrasonication is often employed to further break up droplets by processes including bubble formation, growth, and implosive collapse [50, 51]. In a study it was reported that [52, 53], high-speed stirring at 20,000 r.p.m. for 1 min followed by ultra-sonication for approximately 4 min successfully produced SLNs and NLCs incorporating lycopene. These formulations had PDIs between 0.22 and 0.32, and their PSs were under 200 nm. Monostearin and medium-chain triglycerides were the lipids tested for the study [52]. Employing the stearic acid as the solid-lipid source, another report used a similar procedure to synthesize SLNs with a PDI of 0.215 [53]. Linagliptin-loaded SLNs were prepared by first stirring at 15,000 rpm for 10 min, followed by ultra-sonicating the mixture for around 3 min. Results showed that by utilizing this method, SLNs may be tuned to have a PS of ~225 nms, polydispersity index of 0.18, and encapsulation efficiency of 73.8%. When compared to the administration of linagliptin solution in the rat model, these improved solid-lipid nanoparticles showed a three-fold improvement in drug permeability and oral bioavailability, respectively [54–56]. However, both approaches have the downside of exposing drugs to high temperatures for extended periods of time [57, 58]. Further, contamination of the final product with metals from the sonicator probes is another problem that has been linked to the ultrasonication method [34]. Bath sonication could be one approach to addressing this issue; however, it needs to be in combination with the other methods for the successful lowering of the PS and polydispersity index levels [48]. Further, even if the total lipid concentrations are modest, a great deal of surfactant is required when high-speed stirring or the ultrasonication technique is used [48].

2.1.3 Microemulsion Method

This method's inception dates to the early 1990s [57, 59], and it has since attracted the attention of numerous researchers [57]. To create a nanoemulsion, a microemulsion is first diluted in the cold aqueous-based media. The next step is the formulation of solid lipid nanocarriers and nano-lipid carriers by precipitation of lipids. In brief, the medication is dissolved in lipids that have been heated over their melting point. Then, an aqueous phase is injected, consisting of water and surfactant (pre-heated to the same temperature), while gently agitating the mixture [60]. This creates a microemulsion that is both transparent and thermodynamically stable. The microemulsion is then mixed into a cold aqueous solution between 2 and 10 °C and then gently agitated by mechanical mixing [40]. Volumes of cold aqueous phases are 25–50 times more than the volumes of the hot emulsion [34]. The production of SLNs or NLCs follows the fast crystallization of lipids after the generation of a nanoemulsion following dilution [61]. Curcumin-containing SLNs were successfully manufactured using this method; they exhibit very small PSs (200 nm) and lower polydispersity indices (0.2). P-glycoprotein-induced resistance to doxorubicin in the triple-negative breast tumours was overcome by using SLNs, which were shown to be 5–10 times more efficient than free curcumin [62]. The advantages of the microemulsion method are its ease of use, consistency, and potential for expansion. In the context of industrial-scale production, a large, temperature-controlled tank may be used to prepare it. The microemulsion is then moved to a tank of cold water, where the lipids may be precipitated [11]. Notably, the use of solvents is also not compulsory for this method. However, this process have limitations due to their high dilution requirements and high surfactant consumption [63].

2.1.4 Solvent Emulsification Diffusion Method

Solvent emulsification and diffusion are often employed to make polymer-based nanocarriers. In 2003, researchers made SLNs and NLCs using this approach [64]. The operation typically consists of the use of organic solvent which becomes moderately miscible with an aqueous phase. Since the organic solvents and aqueous phase become saturated, they are in thermodynamic equilibrium. A water-based solvent solubilizes both lipids and drugs. The same solvent emulsifies in a saturated stabilized water phase. Afterward, the oil-in-water (o/w) emulsion is generated by spinning this combination. The solvent incorporating drugs enters the continuous phase more readily by diluting the emulsion with water at a 1:5 to 1:10 ratio. SLNs and NLCs spontaneously develop after solvent removal, usually by lyophilization or vacuum distillation [65]. A study reported the generation of tamoxifen-loaded SLNs through utilizing the technique. Tamoxifen and beta-lipoic acid-stearyl amine conjugation were dissolved in water. Thereafter, 2% w/v stabilizer was added, then the mixture was homogenized at higher speeds in a solvent-saturated water-based solution. In the last step, SLNs were produced by mixing the emulsion with water and removing organic solvents by employing a low-pressure rotary evaporator. This formulation exhibited a PDI of 0.114 and a PS of 233 nm [66]. This approach may decrease the degradation of the drug due to heat and stress during quick stirring as explained in the HPH method. Further, previous research has shown that this technique might promote the smooth development of solid lipid nanoparticles and nano lipid carriers [67]. Moreover, this method works with both hydrophilic and hydrophobic therapeutic compounds with equal efficacy. Insulin, tretinoin, and cyclosporine have been integrated into SLNs and NLCs using the approach.

2.1.5 Double Emulsion Method

Hydrophilic pharmaceutical ingredients and biomolecules like peptides and proteins may be encapsulated effectively using the double emulsion technique. The current study focuses on adding insulin to SLNs and NLCs matrices for oral administration by utilizing this method [68–70]. An active pharmaceutical ingredient and a stability-enhancing substance (stabilizers) are dissolved in water to form an aqueous solution, which is then mixed with an aqueous phase insoluble organic medium which contains the lipid to form an emulsion [71] or lipids alone to form a solvent-free system [72]. The primary emulsion gets dispersed in the water phase using a hydrophilic emulsifying agent for the generation of the w/o/w emulsion. Next, lipid precipitation creates SLNs and NLCs dispersions after solvent evaporation [73]. Further, stabilizers prevent medications from becoming "partitioned" into the aqueous phase as the solvent evaporates. The raloxifene-loaded SLNs were prepared using a two-fold emulsification method. The product showed 180 nm PS and 4.5% EE. SLNs were more cytotoxic than unencapsulated raloxifene against MCF-7 cells [71, 74]. In another study, the investigators produced diethyldithiocarbamate SLNs using a similar method. Prepared SLNs had EEs of 80% and PSs under 200 nm [72]. Notably, the double emulsion approach eliminates molten lipids, a major benefit in respect to the thermal degradation of drugs. However, hydrophilic drugs leaking into the water phase lower SLNs and NLCs encapsulation efficiency and drug loading [50].

2.1.6 Phase Inversion Temperature Method

Phase inversion temperature approach has developed nanoemulsions, SLNs, and NLCs [75, 76]. Inversion of the water-in-oil as well as the oil-in-water emulsions owing to temperature changes is key. This method requires temperature-dependent non-ionic polyoxyethylated surfactants. High hydrophilic-lipophilic balance (HLB) surfactants have substantially hydrated hydrophilic moieties at low temperatures. Due to ethoxy group dehydration, surfactants become more lipophilic and have lower HLB values as temperatures increase [75]. The "PIT" is the temperature at which a surfactant's affinity for the aqueous and lipid phases is balanced [77, 78]. At temperatures above the phase inversion temperature, surfactants promote water-in-oil (w/o) emulsions. On the other hand, oil-in-water (o/w) emulsions are produced below the PIT [79]. Water-in-oil (w/o) emulsions may be made from an oil–water-surfactant combination heated and stirred over the phase PIT to make SLNs and NLCs. The specimens are then rapidly cooled and shaken to break down water-in-oil microemulsions and generate oil-in-water nano-emulsions. The advantage of this process is that it uses minimal energy, and no organic solvents are involved in the manufacturing procedure. However, nanoemulsion produced by this method may have stability issues and require numerous temperature cycles, such as three between 60 and 90 °C [80, 81].

2.1.7 Membrane Contactor Method

This approach uses a membrane contactor to make SLNs and NLCs. The process involves driving a lipid phase through the holes of the membranes at the temperatures above the melting points of the solid lipid while applying pressure. This process creates tiny droplets. At the same time point, through the opposite side of a module's internal membrane flows a surfactant-based liquid phase. The fluid's tangential motion across the membrane effectively washes away pore-opening droplets. Solid-lipid nanocarriers and nano-lipid carrier formulations are made by cooling the heated emulsions to room temperature [82]. However, this strategy has drawbacks. This strategy requires a complicated system to utilize efficiently. Clogging in membranes utilized to produce SLNs and NLCs in this approach is another major concern.

2.1.8 Supercritical Fluid-Based Methods

Several solid lipid nanoparticles and nano lipid carriers manufacturing processes use supercritical fluids like CO2. Supercritical fluid extraction of an emulsion begins with the production of oil-in-water emulsions and then extraction of organic solvent. The usual method injects the emulsion into the extraction column's top half and supplies supercritical CO_2 counter-currently from the bottom. SFEE extracts solvents more efficiently than the other method which uses rotary evaporation, dilutions, and diffusion. Lipids precipitate when solvents get readily and entirely removed. Previous studies showed that SLNs and NLCs exhibit a steady and uniform PS distribution with this method [83, 84]. O/W emulsion preparation through this method employs completely miscible and partly miscible solvents. In a study, using this method, SLNs with indomethacin and ketoprofen were synthesized. A model drug was mixed with chloroform and lipids, viz. tripalmitin, tristearin, or Gelucire-50/13 was dissolved to make emulsions. After adding sodium glycocholate-containing water to the organic solution, was applied. The oil-in-water emulsion was injected at 2 mL/min onto a 1 m long, 4 L extraction column. Meanwhile, 40 g/min supercritical CO2 was introduced. In manufactured SLNs, drug loadings (DLs) were 10 to 20% and PSs were < 30 nm [83]. Uniform distributions of the particle size and greater and more efficient extraction of the solvents are hallmarks of supercritical fluid-based techniques.

2.1.9 Coacervation Method

Coacervation produces polymeric nanoparticles [85], and researchers initially used it to make SLNs in 2010. This technique uses alkaline fatty acid salts from acidification-induced precipitation. This experiment used sodium stearate, an alkaline fatty salt. After adding polymer-based stabilizers like poly-vinyl alcohol or hydroxy propyl methyl cellulose to the aqueous solution, the salt gets dispersed in the aqueous solution of the polymer-based stabilizers such as polyvinyl siloxane or hydroxy propyl methyl cellulose [86]. API may get dissolved in the organic phase [87] or could be incorporated in the blank solid lipid nanoparticles later [88]. This process is started by the dissolution of drugs in ethanol preceded by their absorption into the lipid phase. When heated above the Krafft point, the combination becomes a colourless, odourless, and tasteless micellar solution of fatty alkaline salts. Adding a tiny amount of acidifying coacervating solution precipitates the lipids. After making a suspension, it was cooed in a water bath to reach the temperature of 15 °C while stirring at 300 rpm. The suspension is then chilled and agitated in a water bath to precipitate SLNs or NLCs. In one study, SLNs and NLCs containing insulin or glargine insulin were made utilizing this method. The investigator used sodium stearate, polyvinyl alcohol (PVA) 9000, and lactic acid solution (1M) as coacervators. After oral treatment in rats, glargine-insulin-loaded nano lipid carriers demonstrated excellent *ex-vivo* absorption and 6% absolute oral bioavailability. This formulation also lowered the blood glucose in healthy rats [89]. In another report, temozolomide-loaded solid lipid nanoparticles were made from sodium-behenate and polyvinyl alcohol. Optimizing solid lipid nanoparticles demonstrates higher anti-proliferative efficacies on cells of skin cancer. No adverse toxic effects were seen in mice administered with these SLNs [88]. Therefore, this method produces solid lipid nanoparticles and nano lipid carriers in absence of the solvents or other advanced equipment. However, is sometimes only employed for those lipids that form fatty acids and are not appropriate for pH-sensitive pharmaceutical drugs [90].

2.1.10 Solvent Injection Method

Researchers first documented solvent injection for preparing the solid lipid nanoparticles and nano lipid carriers in 2003 [91]. This method dissolves the lipid or medication in a water-miscible medium viz. acetone, isopropanol, methanol, ethanol or their combinations. The water phase is generally made by mixing water or a buffer solution with an emulsifier or combination. A needle injects the organic phase into the aqueous phase quickly while mechanical stirring is maintained [92]. Notably, the solvent emulsification-diffusion process and this approach share similar fundamental concepts. Two key processes enhance SLNs and NLC formation concurrently and synergistically after injection. Initially, the solvent undergoes diffusion from droplets in the water phase, leading to a decrease in the size of the droplets. Consequently, the concentration of lipids inside the droplets is elevated, resulting in the creation of localized areas that are supersaturated and stabilized by emulsifiers present in the aqueous phase. Furthermore, the emulsifiers serve to decrease the interfacial tension existing between water and solvent, so facilitating the generation of minute droplets consisting of solvent and lipid at the point of injection. The phenomenon of droplet fragmentation into smaller droplets with similar lipid contents is attributed to interfacial pulsation and turbulence that occurs during solvent diffusion. The energy that is released reallocating the solvent to their equilibriums, is responsible for providing the necessary energy for the division of droplets. Hence, within the context of the solvent injection technique, the process of solvent diffusion gives rise to the creation of minuscule droplets and the precipitation of lipids. Emulsifiers are important in the characterization of particle sizes and size distributions.

2.2 Methods of Preparation and Characterization of Polymeric Nanocarriers

Dispersing completed polymers using a variety of techniques, such as solvent evaporation, nanoprecipitation, salting-out, dialysis, and supercritical fluid technologies, can yield polymeric nanoparticles [93, 94]. These methods are comparable in that they employ a water phase with stabilizers and an organic phase containing the nanoparticle ingredients. The other similarity is the inefficient encapsulation of medications that leak from the organic phase into the aqueous phase and are partially and readily water soluble (i.e., proteins and peptides) [95, 96].

2.2.1 Solvent Evaporation

Solvent evaporation is the most used technique for creating polymeric nanoparticles for the delivery of active ingredients. Using biodegradable polymers, this technique creates polymeric nanoparticles that are utilized in medication delivery systems.

Using an aqueous solution of surfactants, emulsions are created when polymers and medications are made to dissolve in the organic phase. The hydrophobic medication was previously dissolved in a solvent such as dichloromethane or chloroform. However, due to ethyl acetate's toxicological characteristics, dichloromethane and chloroform are no longer utilized [97, 98].

Using an aqueous solution of surfactants, emulsions are created when polymers and medications are made to dissolve in the organic phase. The hydrophobic medication was previously dissolved in a solvent such as dichloromethane or chloroform. However, due to ethyl acetate's toxicological characteristics, dichloromethane and chloroform are no longer utilized. Following that, the organic solvent is evaporated, and ultracentrifugation is used to gather the nanoparticles. The nanoparticles are rinsed with distilled water to get rid of surfactants. Following all stages, nanoparticles can be lyophilized. The solvent evaporation approach is a straightforward technique for producing polymeric nanoparticles, although it requires outside energy. The strength of external energy, the length of time required, and the potential for nanodroplets to clump together during the evaporation process can all have an effect on nanoparticle shape and size [99, 100].

2.2.2 Salting Out

Salting out is an additional method for producing polymeric nanoparticles. Acetone is a solvent that is miscible in water, and this procedure uses the salting out action to separate it. Put another way, the salting out strategy alters the process of emulsification and/or solvent diffusion by starting with the emulsions. To form the organic phase, the polymer is dissolved in the solvents that are entirely miscible with an aqueous phase, such as ethanol, tetrahydrofuran, or acetone. These organic phases are then mixed with water, which also contains the salting-out agents and stabilizers. Choices of salting-out agents, which can either be electrolytes like calcium chloride or magnesium chloride or some other non-electrolytes like sucrose, have a significant impact on how well the medicine is encapsulated. Following this procedure, the mixture is combined with a sufficient amount of water and cross-flow filtering is used to collect the resulting nanoparticles [101, 102].

The salting out method's main advantage is that it makes protein encapsulants less stressed. An additional advantage is that there is no rise in temperature. Consequently, heat-sensitive substances can be treated using this method. Despite its advantages, this method has certain disadvantages, namely the requirement for extensive cleaning processes and its limited application to lipophilic medications [103, 104].

2.2.3 Dialysis

One preparation technique that offers a simple and efficient way to prepare tiny PNPs is dialysis. Prior to being placed into the dialysis tubes to produce nanoparticles through dialysis, the polymer is first dissolved in organic solvents. The dialysis tube's compatibility with the nanoparticle's molecular weight is a crucial consideration. The solvent becomes less soluble during dialysis due to displacement. This causes the polymer to gradually aggregate, resulting in a uniform suspension of nanoparticles [105, 106].

2.2.4 Super Critical Fluid Technology

Supercritical fluids are widely understood to be fluids that, in spite of pressure changes, do not change phase. The most commonly utilized supercritical fluid is carbon dioxide, which is compatible with the critical state, nontoxic, non-flammable, and reasonably priced. It is anticipated that supercritical fluid technology will provide an intriguing and practical means of producing particles while circumventing many of the drawbacks of conventional techniques. The fact that there is no solvent in the precipitated product when using supercritical fluid technology to manufacture polymeric nanoparticles is by far its greatest benefit [107, 108]. The supercritical anti-solvent method and the rapid growth of supercritical solution are the most utilized techniques. The medications are then made to dissolve in organic solvents and then get released into supercritical fluids in fast expansion of the supercritical solution method. Filterable nanoparticles are left behind when the organic phase is rapidly dissolved by the supercritical solvent. Another method involves dissolving the polymer and active component in a supercritical solvent at high pressure [109, 110].

2.2.5 Characterization Techniques of Polymeric Nanocarriers

Evaluation of the nanoparticle sizes, zeta potential, PDI values, shape, surface areas, and compositions forms the basis for characterizing nanoparticles. Particle size and size distribution are the most crucial aspects of nanoparticles since they affect drug loading, drug release, and stability. Another significant characteristic of nanoformulations is their surface zeta potential, which is utilized to evaluate the surface charge properties. A variety of advanced microscopic methods, including transmission electron microscopy, scanning electron microscopy, and atomic force microscopy, are used in the characterization of nanoparticles [111–113]. For surface chemistry analysis of nanoparticle suspension, X-ray photoelectron spectroscopy is employed; for morphological studies, atomic force microscopic analysis, as well as scanning electron microscopic evaluation, can be utilized; and for size and size distribution measurement, laser or dynamic light scattering techniques can also be employed [114, 115].

2.3 Methods of Preparation and Characterization of Lipoprotein-Based Nanocarriers

Lipoproteins have been reported to as a class of natural bio-nanocarriers that are primarily involved in the transportation of triglycerides and some other hydrophobic compounds in a targeted manner with the aid of the bloodstream. These are composed of some phospholipids, apo-lipoproteins, and some of the highly hydrophobic cores made up of the cholesteryl ester. They are also able to identify particular receptors on both healthy and aberrant cells. Lipoproteins' inherent biocompatibility, biodegradability, nanoscale size, and receptor-mediated absorption make them an important tool for the targeted administration of cancer diagnostics and treatments [116, 117]. There are five distinct kinds of circulating lipoproteins, each with a unique lipid and protein makeup. Two important subclasses of lipoproteins, low-density lipoproteins and high-density lipoproteins, have been extensively manipulated to provide appealing and adaptable delivery systems for anticancer medications. The potential uses of high- and low-density lipoprotein bio-nanocarriers for targeted drug delivery to cancer cells have been emphasized in a number of research studies [118, 119].

In clinics, blood-based lipoprotein assays are commonly utilized for diagnosis. Preclinical studies are presently being conducted on other lipoprotein applications, like drug delivery using nanocarriers. On the other hand, synthetic liposomes have been used in clinical settings to treat fungal and cancerous disorders since the 1990s. Liposomes and lipoproteins differ primarily in that liposomes have an aqueous core while lipoproteins do not. When compared to liposomes, which have short circulation durations unless coated with polyethylene glycol and might trigger the immune system upon repeated administration, lipoproteins are a valuable substitute that may have unfavourable effects [120, 121]. A wide variety of compounds can be loaded onto lipoproteins due to their heterogeneous nature. Hydrophilic compounds, for instance, can be chemically attached to lipid or protein components, whereas tiny lipophilic molecules can be passively loaded in lipoproteins using straightforward mixing procedures. Payloads can attach to lipid or protein components on the lipoprotein surface or be integrated into the lipoprotein's core [122, 123]. Lipid core reconstitution, for instance, is a widely used technique that involves chemically replacing the medicinal drug with the lipoprotein core lipids. Since their tiny size (less than 25 nm) allows for tissue penetration and receptor-mediated uptake, low-density lipoproteins and high-density lipoproteins have been the most often employed lipoproteins for drug delivery. This allows for selective transport to target tissues. Chylomicrons, both synthetic and natural, have also been suggested as lymphatic transport delivery systems [124, 125].

In general, substances that are more lipophilic have a greater propensity for binding to lipoprotein and are thought to be excellent choices for the mediation of the metabolism of drugs, which is the key route of excretion from bodies. Three chief techniques for encapsulating medicinal and diagnostic substances in lipoprotein nanoparticles have been reported. The first method involves inserting pharmaceutical agents into the lipoprotein's lipid core, also known as "core loading". The second method involves intercalating into the lipoprotein's phospholipid monolayer, also known as "surface loading". The third method involves covalently attaching pharmaceutical compounds to the amino acid residue of the apolipoprotein, a process known as "protein loading" [117, 126]. After lyophilization and organic extraction, hydrophobic chemicals can be added to the lipoprotein core by substituting the lipids with these agents [127]. Using lipoprotein component pieces in combination with sonication or cholate dialysis to create nanoparticles is another method that has been documented [128, 129].

The noncovalent intercalation of the medicinal chemicals onto the lipoprotein surface is the foundation of the surface loading strategy. Weaker interactions like van der Waals force control how much the agent intercalates into the particle surface. Because of the amphiphilic nature of the intercalated agents, their structures can partly be buried in lipoprotein surfaces, shielding hydrophilic regions from hydrogen bonds or the electrostatic interaction with water [117, 130, 131]. Balances between these factors affect lipoprotein stability and loading efficiency. Conjugations of pharmacological or bio-imaging substances to surfaces of apolipoproteins are referred to as protein loading. The typical amino acids used for conjugation are the residues of arginine, cysteine, tyrosine, and lysine. Drug binding ability, however, can be modified under a variety of physiological or pathological conditions [117, 132, 133].

Despite the fact that lipoproteins have been the subject of several research investigating their potential as a delivery system, their application in clinical settings has been restricted because ex vivo loading of medicinal drugs requires the extraction of these particles from fresh blood plasma. It is hard to rely on desired donor plasma to produce lipoproteins since the processes for isolating lipoproteins are time-consuming, the quality of the lipoprotein samples varies from batch to batch, and it is challenging to produce significant amounts of lipoproteins [117, 134]. Naturally, by loading native and circulating lipoproteins in vivo without removing these from bodies by sustained or targeted administration of drugs, nearly all of these issues can be resolved. Using particular monoclonal antibodies that have been previously linked to the drugs or that could be conjugated to drugs through appropriate linker systems, like avidin–biotin linkers, and have been targeted actively, against the special kinds of apolipoproteins that are featured on surfaces of lipoproteins, is one innovative, safe, and adaptable method for loading the natural as well as the circulating lipoproteins without removing these from body. Therefore, by choosing the proper anti-apo-antibodies like single-chained variable segments, it is possible to steer the appropriate circulating lipoprotein subclasses [135]. To enhance the membrane association of pharmaceuticals, for instance, biotinylated monoclonal antibodies and biotinylated medications connected via streptavidin molecule have been suggested. Note that a description of the more précised homing device could also be specified for more precise targeting of the lipoproteins to the chosen substitute receptor rather than the own receptors [117, 136].

2.4 *Preparation and Characterization of Viral Nanocarriers*

Viral nanoparticles have been used in a variety of fields, including energy, electronics, and next-generation bio-medical devices. Viral nanotechnology is among the young, highly multidisciplinary sciences. Viral nanoparticles have sometimes been called programmable nanocarriers due to their ability to be altered and functionalized through a variety of methods. They have been created as potential new materials. The ideas and procedures—such as film and hydrogel formation, encapsulation strategies, bioconjugation chemistries, and mineralization tactics—that enable the engineering of viral nanoparticles have been explored by a number of researchers. With all of these methods at our disposal, the only restriction on the possible uses of viral nanoparticles is your creativity.

2.4.1 Encapsulation of Therapeutic Compounds and Drug-Payload in Viral Nanoparticles

One of the several roles played by viral capsids is the formation of the shells or tubes that protect viral genomes, which is essentially its natural payload. Another role is the delivery of this drug to cellular targets, which completes the replication cycles of the viruses. To preserve the virus coat proteins' inherent ability to self-assemble into shells and tubes while also enabling the encapsulation of synthetic polymers, medications, imaging agents, other proteins, and inorganic nanoparticles, researchers working on the development of viral nanoparticles aim to modify these properties [137–139]. Viral nanoparticles can be isolated as complete particles with nucleic acids or as viral nanoparticles without genetic material from a native host or from the heterogenous expressing system. To reconstruct hybrid viral nanoparticles encasing the cargo, a number of techniques have been devised that enable viral nanoparticles to their deconstruction into coating protein monomer in-vitro and combined with the artificial therapeutic payloads [137, 140, 141].

Additionally, viral nanoparticles can be used as confined reaction vessels for material production inside the capsid that is spatially limited. Lastly, tiny molecules can diffuse between the inside of the capsid and the exterior medium thanks to holes in the capsid structure. One of three methods—covalent attachment, interaction with the encapsulated nucleic acids, or trapped inside the capsid by taking advantage of swelling processes reliant on pH and metal ions—can be used to keep the molecules inside the capsid [142, 143].

Viral nanoparticles have been used for trapping synthetic nucleic acid as well as polyanions because they spontaneously have the capability of self-assembly from their coating monomeric proteins and nucleic acid having negative charges. As demonstrated using coat proteins from Hibiscus-chlorotic ring-spot viruses and Cowpea-chlorotic mottle virus, hybrid viral nanoparticles form with a polymer payload when purified coating protein are combined with polymers possessing the negative charges. By using this method, cytotoxic medications like doxorubicin and complexes of negatively charged polymers have been encapsulated, enabling the use of viral nanoparticles for targeted drug delivery [144, 145].

Viral nanoparticles with varying sizes and unique symmetries are produced by varying core particle sizes. Comparatively speaking, viral nanoparticles produced from bigger gold cores are larger than those formed from smaller nanoparticle cores. Transmission electron microscopic analyses and three-dimensional reconstruction of these images were used to analyze BMV viral nanoparticles encasing various-sized gold cores in detail. The results showed that the viral nanoparticles with cores of 6 nm, exhibited T-1 symmetrical proportions, the ones of 9 nm cores resembled the pseudo T-2 symmetrical proportions, and the ones of 12 nm cores displayed T-3 symmetrical proportions. These studies demonstrate the adaptability of viral nanoparticles as well as coating protein: despite the seeming stiffness, templating enables the assembly of a wide variety of hybrid structures of various sizes [137, 146, 147].

Since viral nanotechnology is considered a multi-disciplinary area that is at the intersection of chemical and material sciences, along with medicine, and virology, it has made it easier for ideas and methods from various fields to be widely shared. Numerous techniques and approaches have been created and put into practice to alter viral nanoparticles for use in biomedical applications. Site-specific ligand-conjugation for the targeting of specific tissues, medical bio-imaging, or administration of pharmacological compounds is made possible by bioconjugation chemistry [148, 149]. Viral nanoparticles may then be tuned and designed to fit any desired purpose using other techniques like encapsulation and mineralization. Viral nanoparticle hydrogels, which can be applied in tissue engineering, are created when the particles can show self-assembly into distinct systems and possess self-organization capacities into thin films or arrays. Viral nanotechnology has entered a new phase where viruses may be altered to create custom nanoscale construction materials [150, 151]. As more platforms for viral nanoparticles and chemistries become accessible, the only thing stopping this field's advancement is creativity. The creation of viral nanoparticles as diagnostic tools, vaccinations, imaging modalities, and targeted therapeutics are some of the recent advances in medicine. Because viral engineering techniques are widely established, this discovery holds significant potential for the field's quick transition to in-vivo stages through pre-clinical model systems and clinical studies. The topic of viral nano-biotechnology is exciting and rapidly evolving, with significant potential for the creation of next-generation medical devices and treatments [137, 141].

The recent surge in publications detailing tissue engineering using viral nanoparticles highlights the enormous potential of these biogenic materials. Although tissue regeneration, tailoring and customizations are still in their early stages, viral nanoparticles may be crucial to their development. For the accurate mimicking of the extracellular matrix and trigger differentiation of the stem cells, multiple growth factors will need to be presented [152–154]. However, since viral nanoparticles can be chemically and genetically modified in a variety of ways, these nanoscale platforms may eventually provide a special template to regulate tissue differentiation. Biocompatible viral nanoparticles will probably be essential to the creation of new tissue engineering materials as the area develops [137, 144, 155].

2.5 Preparation and Characterization of Protein-Based Nanocarriers

Targeted biodistribution of sensitive anti-inflammatory pharmacological compounds and inflammation-suppressing drug-payloads is impacted by a number of variables, including as harsh processing, storage and environmental conditions, and ambient temperature. These elements pose serious obstacles to the stability and target delivery of anti-inflammatory substances [156, 157]. Protein-based nanocarriers, however, may be able to overcome these difficulties. Because protein nanocomposites are small, nontoxic, easily synthesized at a low cost, and rapidly eliminated from the human body, they are particularly interesting for the delivery of bioactive compounds. Many techniques, such as protein encapsulation, template-induced method, nano-spray drying, emulsion method, nanoprecipitation nano-spraying, crosslinking, adsorption, and disintegration, can be used to manufacture them. Bioactive components are delivered using a variety of protein-based nanoparticles. Scholars have deliberated over diverse varieties of protein nanocarriers, their synthesis, characterization, and their uses [157–159].

The many physicochemical properties of proteins are combined with nanoscale technologies in a rapidly emerging field of research known as protein nanotechnology. A novel class of nanoparticles known as protein nanoparticles, often referred to as protein-based nanoparticles, was created because of this field's integration with medicinal formulations [160]. Table 1 presents a few strategies for creating protein-based nanoparticles. Protein nanoparticles may be created in a variety of sizes and forms as they are solid colloidal particles. They are between 10 and 100 nm in size. It is possible to regulate particle size, shape, and interfacial characteristics of nanocarriers. Drug delivery that is site-specific is possible using protein nanoparticles. The choice of the protein nanoparticles' substance is determined by several aspects. These ask for required particle size, the drug's inherent qualities, biodegradation capabilities and compatibilities, the intended drug-release profiles, surface charge, and hydrophobicity, as well as the product's toxicity and antigenicity [123, 161]. Many applications, including bio-sensing devices, medicines, energy-harvestation, diagnostics, thin film coatings, purification platforms, and packaging materials, have made extensive use of nanotechnology. It holds a unique role in the biomedical industry since it has contributed to the development of several imaging agents and medication delivery systems. Enhancing factors like bioavailability is the goal of applying nanotechnology in the biopharmaceutical business to improve how therapeutic medications operate in vivo [148, 162].

Because protein-based nanoparticles are so tiny, they can enter cells through the process of endocytosis. In biological products, proteins have specific roles and have special qualities. As a foundation, they can be utilized to create nanoparticles. Because of their durability, biodegradability, controllability over particle size, and surface changes, protein nanoparticles provide several benefits over other forms of nanoparticles. Moreover, they are less immunogenic and hazardous [170, 171]. Furthermore, most of the protein-based nanoparticles' stability, half-life, and activity

Table 1 Approaches and techniques for the preparation of protein-based nanoformulations (reprinted with permission from [157]

Process	Principle	Benefits	Limits	References
Encapsulation	Organizes positions of various functional proteins	Increase the capability of specific targeting, biostability, and packing proficiency of nanoparticles	Natural structures of proteins can be damaged	[163]
Template-induced method	Works on soft and hard templates	Effective in designing the nanoparticles and controlling the position of synthetic polymers	Protein strain potential occurs	[164]
Nano spray drying	Nanoparticles processed through carbon dioxide gas	Step-by-step process and cost-effective	Production of protein nanoparticles on a small scale	[165]
Emulsion/solvent extraction method	Single emulsion preparation and double emulsion preparations	Highly stable process	An unstable process requiring stabilizers and surfactants	[158]
Nanoprecipitation	Miscible solvents used for protein and drug	Rapid technique. Helpful for small-sized particles	The concentration and nature of the surfactant affect the size of the particles	[166]
Electro-spraying	Protein and drugs dispersed in a solvent to form protein nanoparticles	High drug loading efficiency and self-dispersion	Not suitable for complex proteins	[167]
Crosslinking	Chemical reagents can be used for the crosslinking process which incorporates reactivity	Sustained drug delivery. Increased stability of proteins	Time-consuming	[168]

(continued)

Table 1 (continued)

Process	Principle	Benefits	Limits	References
Adsorption/ Grafting	Non-covalent interactions to attach proteins to surfaces	Low cost. Efficient. Mildest immobilization method	The protein can be desorbed	[169]
Desolvation	A desolvating agent is added and proteins are dehydrated and change their conformation from stretched to coil conformation	Extensively used in the preparation of nanoparticles	Toxic crosslinkers are used	[167]

are shielded from renal clearance and enzyme breakdown. Since protein-based nanoparticles are naturally non-antigenic, they may be utilized in a variety of targeted drug delivery applications, including cancer therapy, tumour therapy, vaccinations, and lung administration [172].

Numerous critical bodily processes, including homeostasis, defence, tissue repair, and the removal of dead cells, depend on the human immune system. The cells that always make up our immune system travel all over the body, screening each and every cell. These cells look for defective or malignant cells as well as invasive infections. Malfunctioning cells are removed as soon as they are identified. Innate or adaptive immune cells belong to one of the two branches of the immune system [173]. It is certain that innate immune cells will react to pathogen invasion right away. They have receptors that use molecular patterns to identify the pathogen. After phagocytosing the pathogens, these cells release substances that enable a prompt defence against infections. Within innate immunity, neutrophils and macrophages are prominent cells. B cells as well as T cells, T-helper CD4 + cells, and CD8 + killer cells make up the adaptive immune system. Natural Killer cells, B cells, and innate cells are all regulated by cytokines that are generated by CD4 + helper T cells. On the other hand, contaminated cells are eliminated by CD8 + killer cells. When T and B cells encounter an antigen, they become clonal. These cells support the innate defence system, which eradicates invasive infections, and trigger an immune response [174, 175].

One of the best therapies for treating cancer is immunotherapy. Using the patient's immune system to fight an illness is known as immunotherapy. Immunotherapy is categorized as either activation or suppression immunotherapy depending on whether it stimulates or inhibits the immune response. Immunotherapy is emerging as a cutting-edge cancer treatment that makes use of adoptive or monoclonal cell therapy. Adoptive cell transfer treatment, cancer vaccinations, and cytokine injection are the three major pillars of cancer immunotherapy. There are three primary kinds of nanoparticle-based immunotherapies, each with specific goals [176, 177]. Using delivery mechanisms for nanoparticles is the first tactic. Nanoparticles can function as vaccinations or as nano-carriers. By focusing on the T cells, this approach seeks to specifically trigger a cytotoxic response against the tumour. Reaching the lymph nodes is one way to do this. Furthermore, adoptive cell transfer treatment (ACT) is another tactic. The patient's immune cells are isolated during this treatment. Following an ex-vivo treatment, these cells are reinfused. Therapeutic medications are infused into the tumour location as part of the third tactic [178, 179].

Systematic medication delivery by nanoparticles is used to treat a number of disorders, but the main use of nanoparticle-mediated delivery of drugs. It is very desirable to deliver medications targeted via nano-based devices as therapeutic platforms for the treatment of inflammatory diseases. Nanoparticles have been among the highly efficacious therapeutic drug carriers for pharmacological compounds, and they are designed for use in disease diagnostics and therapies. With the least possible harm to healthy tissues, cancerous cells can be eradicated in such focused settings. Usually, it is better to use nanocarriers to target tumours in order to boost the lethal dose while lowering systemic toxicity. Nevertheless, immunotherapy was not the initial application of nanoparticles in medicine [180, 181]. A few years ago, scientists dubbed immunotherapy the "Breakthrough of the Year". After that, nanoparticles were first employed as chemotherapeutic drugs and then as immunotherapeutic agents. In addition, several anticancer drugs based on nanoparticle formulations have been approved by the FDA and are now undergoing clinical studies. Due to significant advancements in nanotechnology and material science, a variety of nanoparticle carriers, including nano-micelles, dendrimer-based nanocarriers, liposomal nanoparticles, inorganic nanocarriers, polymer-based nanocarriers, and metal-oxide formulations, are employed in the field of inflammation-targeting nanomedicines [182, 183].

Serum proteins have drawn attention because they are under evaluation as natural plagiarized drug delivery systems possessing optimum biocompatibility and biodegradability with a wide range of bio-medical uses, including anti-inflammatory nanomedicines. Serum proteins, especially those that are endogenously produced, are frequently used as nano-carriers for bio-imaging and small molecular medicinal formulations because of their low immunogenicity and noxiousness. They are also outstanding candidates for vaccination excipients and medicines. Cancer immunotherapy makes use of a variety of protein-based nanoparticles, each with a unique purpose. Prominent proteinaceous nanoformulations that are used in immune therapy include albumin-based nanoparticles, bovine serum albumin (BSA), casein-based nanoformulations, and lactoferrin nanoformulations which encapsulate various drugs and medicines [184, 185].

Because albumin is a biological transporter that can move molecules across endothelial membranes, albumin-based nanocarrier systems have been developed. Caveolae-mediated transcytosis is used. Numerous investigations have shown that albumin builds up in solid tumours, which makes it a crucial transporter of anticancer drugs. Through the caveolae route, malignant cells can absorb albumin nanoparticles. A paclitaxel formulation is called Abraxane. It has been reported that albumin protein can intrinsically target inflammatory tissues. When using protein nanoparticles, premedication with corticosteroids and antihistamines is not necessary. Anti-inflammatory therapy using cationic bovine serum albumin is also being investigated [186, 187]. It is a brand-new method of delivering siRNA that uses siRNA to create stable nanoparticles. In the lungs, siRNA accumulates more readily when exposed to CBSA. When silencing RNAs against Bcl-2 were injected into the bloodstream via protein nanoparticles, genetic silencing effects which cancer cells to undergo apoptotic cell death were observed. Orally given gliadin nanoparticles are utilized in pharmacological treatment for colon cancer. Within 48 h, this nano-system delivers the medication gradually and induces cancer cells to undergo apoptosis [188, 189]. These cancer nano-immunotherapies have a great deal of promise for cancer therapy. However, there are a few disadvantages to this nano-immunotherapy method. Nanoparticles exhibit detrimental interactions with endothelial systems, such as macrophages, and colloidal instability under physiological settings, as well as a reduction in blood circulation time. However, a major drawback of nanoparticle-based therapeutics is the formation of protein-corona on nanocarriers. Therapy outcomes are impacted by nano-sized materials with protein-corona, which could change physico-chemical properties. This issue is crucial for immunotherapies based on the surface of the nanoparticle molecule recognition [190, 191]. A schematic representation of the formulation of protein nanocarriers has been depicted in Fig.**2**.

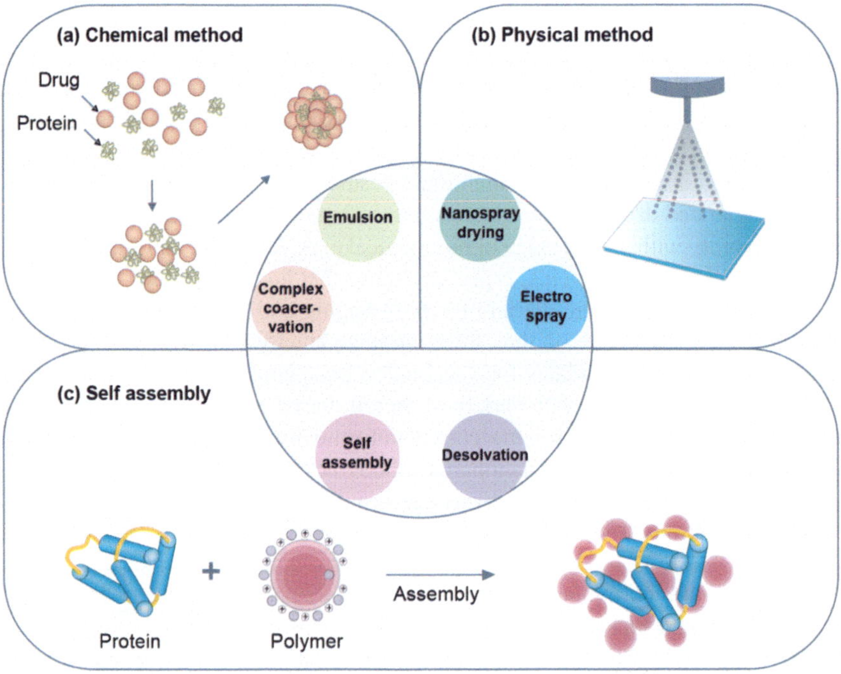

Fig. 2 Categories for the methods of preparation of the protein nanocarriers. (a) Chemical method implicates the formulation of protein nanocarriers by employing chemical reactions (b) Physical method involves the physical aggregation after the separation of proteins into the nanoparticles and implicates the electro-spraying techniques and nanoscale spray-drying technique. (c) Self-assembly technique is the technique for formulating the nanocarriers by agglutination of the proteins that undertakes the desolvation approach. Reprinted with permission from [158]

2.6 Preparation and Characterization of Nucleic Acid-Based Nanocarriers

A safe, effective, targeted, and non-pathogenic method of delivering gene therapy materials is becoming more and more necessary. Due to their capacity to be tuned with a wide range of physico-chemical characteristics, nanocarriers for the delivery of nucleic acid provide an unparalleled chance to get beyond these limitations. They can be easily functionalized with any kind of biomolecule or moisture for targeted distribution. The use of nucleic acid treatments to modify the expression of DNA or RNA, viz. anti-sense DNAs, mRNAs, si RNAs, or miRNAs, has been extensively studied [192, 193]. Remarkably, the combination of gene treatments with nanoscale delivery technologies has expanded the therapeutical and biological uses of these biomolecules, including vaccinations, genetic silencing, bioanalysis, and protein replacements. The formulation and development of intelligent nucleic-acids-based delivery techniques that offer efficiency and usefulness in the architecture of biomolecular diagnostic and therapeutical platforms has been the subject of discussions among several researchers. Generalized designing concerns of nucleic-acid-based delivery nanocarriers, their remarkable capabilities, and structural and functional interactions of nanomaterial with the biological tissues and diseased cells have always been important to highlight [194, 195].

Polymeric nanocarriers are mostly composed of polycations, which frequently have amine groups that can be ionizable. These compounds interact through attractive coulombic interactions with the negatively charged nucleic acids which contain the phosphate group, so facilitating self-assembling to produce contracted nanoscale platforms known as polyplex. Depending on the intended use, selecting a suitable polymer (such as polyethyleneimine or poly-L-lysine) should be considered while developing the delivery system. As an example, branching polyethyleneimine has many accessible amine groups that support the higher buffering capacities and help endosomal escape inside the cell, which releases the RNA into the cytoplasm. Nevertheless, it has been demonstrated to cause some in vivo and in vitro toxicity [196, 197].

Another delivery method that has emerged recently is peptide-based formulations, which contain five to thirty amino acids that are often referred to as homing or cell-penetrating peptides. Both could engage covalently or non-covalently with nucleic acids, which promotes cell internalization. Nano-vectors based on inorganic compounds offer distinctive, multifunctional delivery systems. This adaptable category includes lanthanides as nucleic acid carriers, carbon-nanotubes, cholesterol-based conjugate systems, iron-oxide, silica-oxide, and noble metals (like gold or silver) [198, 199]. Therapeutic payloads can be efficiently targeted and transported using custom-made inorganic nano-vectors. As part of a prospective therapy, mesoporous silica nanocarriers coated with polymer-based coatings, for instance, have been demonstrated to effectively transport siRNA and downregulate genes relevant to osteoporosis [200, 201].

Numerous studies have demonstrated that the combination of multiple delivery categories enhances treatment efficiency, considering the advantages and disadvantages of each class. Alongside the advancement of new technologies, gene-based delivery nano-carriers have achieved great strides toward clinical deployment; yet, numerous issues remain about whether these carriers can target the given nucleic acids towards a particular type of cell. Therefore, it is necessary to reevaluate the pursuit of a formulation platform that fits all users [202, 203]. When similar methods yield unexpectedly different results, the same methodology becomes applicable to the genetic payload. Using a lipid nanoparticle composition akin to the successfully performing vaccines, CureVac's microRNA lipid-nanoparticles-based vaccinations against viruses of COVID-19, or CVnCoV, was a viable option. Regretfully, the efficacy of the two-dose CVnCoV vaccine in avoiding the illness was just 47%. The distinction between CureVac's unaltered messenger RNA payloads and Pfizer and Moderna's changed payload, as well as the requirement to modify the particle for the particular RNA sequence, are likely conjectures related to these results [204–206].

This primer's main goal is to highlight various facets of employing nanoparticles for the delivery of nucleic acids. Primers describe the testing of several delivery paradigms, including polymeric and inorganic nanoparticles, liposomes, and lipid nanoparticles, highlighting the main factors taken into account while creating delivery platforms and potential production techniques. The primer covers several methods for describing the characteristics of nanomaterials, talks about data analysis, and aids in determining the biological ramifications down the line [207, 208]. Relevant examples have been used by several researchers to explain and clarify the main uses of nucleic-acid-based nanocarriers, which include bioanalysis, nano-barcoding, silencing of particular genes and their editing, vaccinations, and immune therapies. The field's optimization and constraints are taken into consideration while reviewing data deposition and repeatability [14, 209].

For the transport of nucleic acids and imaging, inorganic nanoparticles such as gold, silica, and iron oxide have all been investigated. They can be designed to have certain shapes, sizes, and/or structures. Gold and iron oxide nanoparticles are among the inorganic nanoparticles that are often thought to be non-toxic. It's interesting to note that studies reveal inorganic nanoparticles have the ability to either increase or decrease immune responses, and that surface chemistry plays a major role in determining how well-suited they are for the immune system [210, 211]. It is commonly recognized that surface modification, size, and solubility all affect how biocompatible nanoparticles are and how useful they are in biological applications. Consequently, it is crucial to conduct experimental testing of the interactions between nanoparticles and biology and to alter the nanoparticles to optimize their biocompatibility with cells [212, 213].

2.7 Preparation and Characterization of Nanohydrogels

A class of polymeric materials known as hydrogel products is able to retain a lot of water in its three-dimensional networks due to its hydrophilic structure. It is crucial that these goods are used extensively in a variety of industrial and environmental application areas. It was to be expected that synthetic hydrogels, with their greater capacity for absorbing water, longer shelf life, and greater range of basic chemical resources, would eventually replace natural hydrogels. There is a growing body of literature on this topic, particularly in the scientific fields of study [214, 215]. To provide an overview of the technological features of this expanding multidisciplinary field of research, a number of articles and technical reports pertaining to hydrogel products from an engineering perspective have been reported. Scholars have examined the many classification schemes for hydrogels, the physical and chemical properties of these materials, and the technical viability of their application. Numerous other publications have also covered the methods used to produce hydrogels, as well as the implications for process design, block diagrams, and ideal conditions for the preparatory phase [216, 217].

The synthesis of hydrogels can be done via several classical chemical methods. These comprise both one-step processes, such as polymerization and the parallel cross-linking of multifunctional monomers, and multi-step processes, such as the synthesis of reactive group-containing polymer molecules and the subsequent cross-linking of those molecules, potentially through the reaction of polymers with appropriate cross-linking agents [218]. A polymer engineer may design and create polymer networks with specific attributes like mechanical strength, biodegradation, and chemical and biological reaction to stimuli, as well as molecular-scale control over structures like cross-linking density. Because of their biocompatibility, tunability, and reactivity to different stimuli, hydrogels—three-dimensional networks of hydrophilic polymers with the capacity to expand and hold water—have emerged as potential materials for drug administration in the treatment of inflammatory bowel disease [219, 220].

Natural biomaterials are often derived from living things such as microbes, plants, or animals. Since many of them naturally occur in human bodies or other organisms, they have strong biocompatibility and biodegradability. The production of natural biomaterials may be done in two main ways: (i) collected straight from biological tissues or organisms and then subjected to further processes such as enzymatic hydrolysis and purification. Alginate and chitosan are frequently prepared using this technique; ii) by utilizing the fermentation process, which is a particular microorganism's life activity. This process is typically utilized to make hyaluronic acid and dextran. Furthermore, natural biomaterials that were utilized to create stimulus-responsive hydrogels could be sharply separated into two categories: protein biomaterials and polysaccharide biomaterials, based on the kind of component molecule [221, 222].

2.7.1 Formulations Based on Chitosan and Chitosan-Derivatives

Chitin has been known as the only naturally occurring cationic polysaccharide, which is widely distributed in insects and crustaceans. On the other hand, chitosan, an N-deacetylated derivative of chitin, has superior mechanism properties, reduced toxicity, and a positive charge. Furthermore, chitosan has a limited ability to inhibit the development and multiplication of bacteria. Chitosan has three key advantages when it comes to medication distribution. (i) A broad variety of pharmacological compounds, from tiny molecules to supramolecular proteins, may be loaded into chitosan hydrogels. (ii) Chitosan is readily digested by the human body upon medication release [223, 224]. (iii) As previously said, because chitin is widely distributed, it is inexpensive and simple to collect the raw ingredients needed to make chitosan. Chitosan has been frequently used in the formulation of pH-sensitive drug-release formulations because of its variable solubility at different pH values. While chitosan-based hydrogel has numerous potential uses in medicine, its single administration is insufficient to effectively treat inflammation or promote wound healing. Therefore, the most common methods for obtaining functionalized chitosan hydrogel are physical and chemical cross-linkings [225, 226].

Theoretically, substances that have the ability to react or connect with chitosan's reactive amino group might be used to modify the substance on demand; however, substances with a high potential for toxicity should be avoided. The chemical processes of acylation, alkylation, and carboxylation are frequently used to generate stimuli-responsive hydrogels based on chitosan. For example, carboxymethyl chitosan has been shown to aid in wound healing by encouraging the restoration of healthy skin and preventing keloid fibroblast development [156]. Researchers reported on an injectable multifunctional hydrogel based on carboxymethyl chitosan. After adding zinc oxide nanorods to carboxymethyl chitosan to improve the hydrogel system's antibacterial capabilities, they injected it into the wound in an uneven form. Ultimately, a notable decrease in inflammation and enhancement in wound healing were noted. The low toxicity and slow degradation rate of genipin have made it an ideal natural cross-linking agent in recent times [227].

2.7.2 Formulations Based on Hyaluronic Acid and Its Hyaluronic Acid Derivatives

Hyaluronic acid is a naturally occurring polysaccharide that is widely available as a raw material and has strong biocompatibility, much like alginate. However, in contrast to the aforementioned types of stimuli-responsive hydrogel materials, hyaluronic acid is an essential component of the human body and is required for tissue healing and organ composition (such as skin, synovial fluid, and the vitreous body). The widespread usage of hyaluronic acid is dependent on its production by the fermentation of microbes rather than extracting directly from animal tissue, and that has changed with the advancement of biotechnology and gene engineering. Hyaluronic acid has emerged as a popular hydrogel biomaterial in the medical field in recent years. It is included in many medical products, including eye drops, dermal fillers, and viscoelastic agents for ocular surgery [228, 229].

Hyaluronic acid has drawn a lot of attention in the fields of drug transporters, tissue engineering, and wound dressing since it is an essential component of tissue or organs, viz. the joint tissue, eyes, etc. To create multifunctional, stimuli-sensitive hydrogels with improved pharmacological efficacy and minimal adverse drug reactions, most recent work has focused on mixing hyaluronic acid-based hydrogel with a variety of substances, such as anti-inflammatory medications, nucleic acid, or cells. For instance, scientists suggested creating a plasma hydrogel rich in platelets and hyaluronic acid [230, 231]. The stimuli-responsive hydrogels, by encasing manganese oxide (a nanozyme that may catalyze the dissociation of reactive oxygen species), might remain in joints longer, encourage cartilage repair, and reduce inflammation, so inhibiting the progression of osteoarthritis. To accomplish local and sustained genetic drug delivery, researchers looked into a hydrogel film based on hyaluronic acid and implanted with poly(lactic-co-glycolic acid) nanoparticles [232]. The inflammatory response-related genes cyclooxygenase-1 and cyclooxygenase-2 can be efficiently downregulated by the tailored microRNA plasmid. Consequently, tendinitis in the wounded area might be managed, which would ultimately lead to a decrease in postoperative tendon adhesion. To treat bacterial keratitis, researchers created a hyaluronic acid hydrogel that contains ferrous sulphate. Following the hydrogel injection, the iron was liberated, which in turn caused S. aureus to undergo ferroptosis and cure the keratitis [233].

2.7.3 Formulations Based on Alginate and Alginate Derivatives

Alginates are often found as salts of cations including sodium, barium, and magnesium and are derived from brown algae. Sodium alginate is the alginate agent that is most frequently utilized in a variety of sectors. Aside from its advantageous qualities such as biocompatibility, mucoadhesiveness, and degradability, sodium alginate-based hydrogel's sensitivity to pH makes it an important component in drug release system research. Researchers have undertaken a description of the process underlying its pH-responsive nature. In an aqueous solution, a mass of carboxylate radicals behaves as a polyanion. The carboxylate radicals react with hydrogen ions in an acidic environment to form carboxyl groups. These cause the molecular chains to constrict and the hydrophobicity to weaken, which causes the hydrogel to shrink [234, 235].

Because of the repulsive interaction between carboxyl groups, they gradually dissociate as pH rises, which ultimately causes the hydrogel to inflate. Substance molecules are then discharged. A pH-sensitive sodium alginate–cellulose hydrogel was created by researchers in order to administer aspirin and methotrexate simultaneously. According to their research, this hydrogel exhibits disparate drug release behaviour based on the different internal environment pHs of the stomach, small intestine, colon, rectum, etc. [236]. At pH 1.2 in the stomach, the hydrogel formulation could prevent unwanted leaking of methotrexate because of the shrinking of molecular chain; approximately two hours post the oral ingestion, releasing of the aspirin could be significantly aided in small intestine having a pH of around 7.4, while methotrexate showed its primary release in colorectal regions having a pH of 6.5. Hence, these pH-sensitive hydrogels aided in chemical therapeutics as well as the relief from pain simultaneously [237].

2.7.4 Formulations Based on Dextran and Dextran Derivatives

Glucans are the higher molecular weight polymers that contain glucose. The glycosidic linkages can either be beta- or alpha-linked, which means that they can be loosely categorized as alpha-glucan (or dextran) and beta-glucan. The primary sources of beta-glucans include bacteria, fungi, algae, and oats. Additionally, it has been shown to be beneficial in the control of blood sugar, cholesterol metabolism, tumour suppression, and immune system strengthening. While stimuli-responsive hydrogels are often made with dextran and its derivatives [238].

In the natural world, several lactic acid bacteria mostly create dextran extracellularly from sucrose, however, leuconostoc mesenteroides fermentation is often the basis for its industrial production. For almost eight decades, dextran has been utilized in medicine as a blood plasma replacement, a testament to its exceptional biocompatibility in clinical settings. Moreover, dextran has many hydroxyl groups, which are advantageous for cross-linking and other chemical or physical modifications. Dextran as well as dextran-derivatives are therefore the promising biomaterials to formulate the hydrogels that respond to stimuli [239]. Several dextran-based hydrogels-drug-releasing systems were formulated recently for various applications, including antibacterial, insulin release, and cancer treatment. For example, researchers described the use of dextran phosphate to make a nanohydrogel for cancer therapy. Because phosphate groups are present in hydrogels, they developed, it is pH-responsive, lowering their swelling ratios in acidic environments while releasing their payloads in neutral ones [240].

Since its discovery in the nineteenth century, dextran has been used in a wide range of sectors, including food and medicine. However, more thorough studies and advancements in the field of stimuli-responsive hydrogels are needed.

2.7.5 Formulations Based on Cellulose and Cellulose Derivatives

Made mainly of glucose, cellulose is a linear polymer and the most prevalent polysaccharide on the planet. Cellulose may be divided into two categories based on its source: cellulose produced by bacteria and cellulose derived from plants. Since cellulose's molecular chains include a large number of hydroxyls and carboxyl, this biomaterial is easily cross-linked or reacts with modifying agents. Additionally, the hydrogel made of cellulose has improved mechanical properties and stability due to intermolecular hydrogen bonding. Researchers have focused a lot of attention on cellulose-based stimuli-responsive hydrogels because of these qualities as well as their high biodegradability and renewability. Most recent research has concentrated on using it as scaffolds for tissue engineering, medication transporters, and wound dressings [241, 242].

Cellulose is typically processed into cellulose nanofibrils or cellulose nanocrystals to increase its specific surface area and achieve a more uniform particle size, making it better suited to the formation of hydrogels. Whereas cellulose nanocrystals are highly crystalline and have a reduced aspect ratio, they are stiffer, cellulose nanofibrils are semicrystalline and have a high aspect ratio. In order to achieve the intended effect, cellulose nanocrystals and nanofibrils can be added to different hydrogel matrices using a particular cross-linked technique or utilized to create self-assembling hydrogels based on distinct requirements [243].

2.7.6 Formulations Based on Gelatine and Gelatine Derivatives

Gelatine is a kind of protein that is found in both humans and animals. It is made by hydrolysing collagen. Gelatine is a great material for studying drug transporters because of its excellent hydrophilicity, low cost, and simplicity of engineering. In the meanwhile, gelatine is frequently mixed with other groups or ions to create stimulus-responsive hydrogels that may be used for a variety of purposes, including prolonged drug release and stimuli response [244].

Hydrogels are frequently prepared using gelatine methacryloyl, which is the derivative of gelatine altered with the light-based cross-linkable group of methacrylamide. Methacrylic anhydride and gelatine are often used in the manufacture of gelatine methacryloyl. Gelatine methacryloyl preserves all of the benefits of natural gelatine, including its high water-holding capacity, biodegradability, and exceptional film-forming ability. It is also less costly, more biocompatible, and less immunogenic. Gelatine methacryloyl will change from a liquid to a solid because of the chemical reactions of these groups of the methacrylamide, when exposed to ultraviolet light; this process is known as photocuring [245]. This characteristic allows the material to be used as bioink for the burgeoning 3-dimensional bioprinting industry. Furthermore, it has been demonstrated in regenerative medicine that nanomaterials-combined gelatine methacryloyl, which is regarded as next-generation treatment paradigm for the regeneration of tissues, is both practical and efficacious. In order to obtain physiological signals for medical wearables, researchers developed gelatine methacryloyl-based tactile sensors. These sensors can be modified by different groups or ions, and gelatine methacryloyl can be designed to respond to stimuli such as pH, reactive oxygen species, etc., or to release drugs to treat tumours or inflammations [246].

2.7.7 Peptide-Based Stimulus-Responsive Hydrogels

Another substance that has a lot of promise for hydrogel synthesis is peptides, typically oligopeptides, which are created by dehydrating and condensing many amino groups. Peptides are common biomolecules found in many different types of animals.

Because they are essential to physiological function, they have the following properties: biocompatibility, biodegradability, bioactivity, and biosafety. Furthermore, peptides may self-assemble into a variety of structures and form order via noncovalent interactions such as hydrogen bonds, hydropathy properties, stacking of pi-pi, and electrostatic effects. These nanostructures can be created using peptides. In addition, enough opportunities for further functionalization are provided by the presence of carboxyl, amino, and quantity side chains. Peptide-based hydrogels are exceptionally flexible in terms of both design and modification thanks to these characteristics [247, 248].

Owing to these advantages, the peptide-based hydrogel has been created and is used in cell culture, tissue engineering, 3-D printing, and as a dressing and medication carrier. Jelleine-1, a naturally occurring amphiphilic peptide, was employed by researchers to create the peptide-based hydrogel-formulation which can be injected for the purposes of wound healing. Jelleine-1 was initially found in royal jelly and has a broad antibacterial range. Its nanofiber gels by self-assembly in phosphate-buffered saline. They went on to say that using this hydrogel with a Jelleine-1 basis helped mice with wounds from the burnings, which become infected by the methicillin-resistant S. aureus recover more quickly [249]. Smac-TLR7/8 hydrogel was suggested by researchers for immunoregulation in the tumour microenvironment. In their study, they conjugated TLR7/8a, a toll-like receptor agonist, with Smac-N7 peptides using the succinic-acid linkers, which may increase the radiosensitivity of tumours. Through an amidation process, this oligopeptide was further linked to the self-assembly of the peptide KEF-9 to create Smac TLR-7/8 nano-sized fibres. Subsequently, these nanofibers may self-assemble in a sodium chloride solution to generate hydrogel. Smac TLR-7/8 hydrogel possesses the capability to cause tumor-associated macrophages of the M2-type to become activated M1-type after injection. Consequently, enhances the results of immunotherapy and radiation therapy concurrently [250]. Physico-chemical and mechanical features of hydrogels have been depicted in Fig. 3.

2.8 *Biomimetic and Cell-Membrane-Based Nanocarriers*

If administered intravenously, the majority of nanomedicines that are not PEGylated are quickly absorbed and destroyed by the immune system. Biomimetic nanoparticles are hybrid nanostructures with a top layer resembling the membrane of a cell, such as platelets, immune cells, cancer cells, or erythrocytes. This group includes liposomes made with cell membrane proteins and cell membrane-coated nanoparticles [251, 252]. These biomimetic nanoparticles have a membrane covering that possesses the inherent properties of the mother cells, which lessens the possibility that the immune system will recognize them. This approach also demonstrates an active targeting capability and extended nanoparticle circulation. Excellent papers have been written on this topic; this list just includes some of the most recent lectures [253].

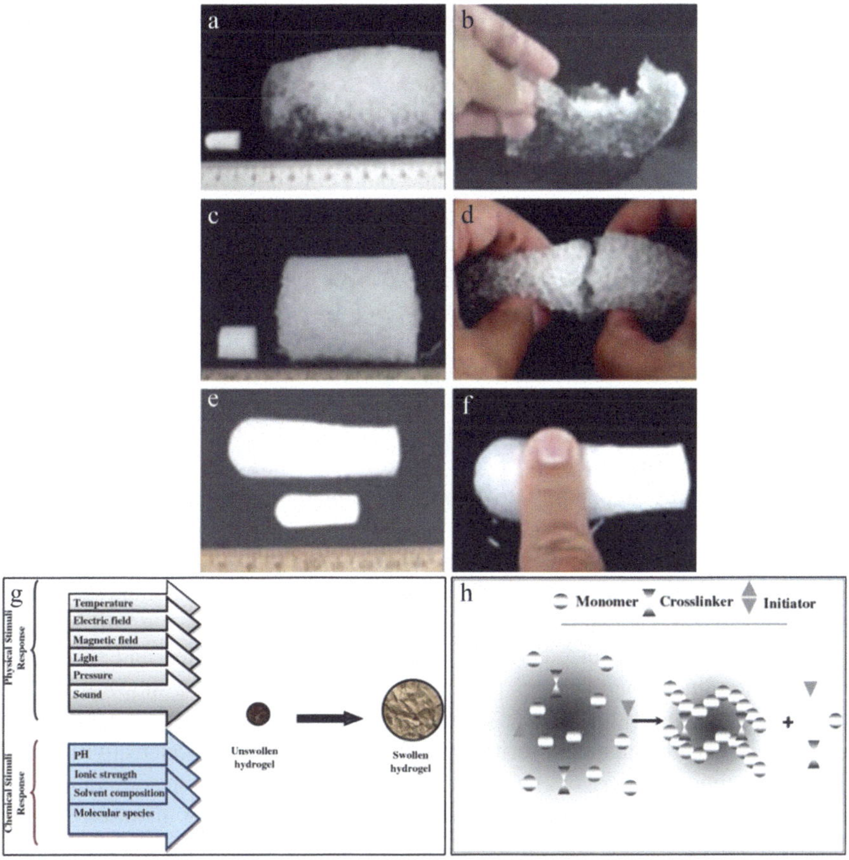

Fig. 3 Distinct swelling as well as mechanical features of hydrogels have been presented here of: first (**a** and **b**), second (**c** and **d**) and third (**e** and **f**) SPH generations. **g** Stimuli-responsive swelling behaviours of the hydrogel formulations. **h** Schematic representation of the hydrogel formulation. Adapted with permission from [214]

Among the anti-cytokine biologics employed in the current therapeutic therapy of RA, which majorly targets autoimmune inflammation responses, are those that inhibit TNF-alpha and interleukins. The rates of these responses are still below average. Researchers developed an interesting method to fuse neutrophil membranes onto a polymer core and then inject these hybridized bio-mimetic nanocarriers into knee joints of transgenic mice models of auto-immune rheumatoid arthritis and the mice model of collagen-induced arthritis. Neutrophils are essential for lowering inflammation and mending tissue damage, which is why this was done [254]. Through the suppression of overall arthritis severity and the reduction of joint destruction, the results demonstrated a strong therapeutic effectiveness. Notably, the authors prepared the core of the nanoparticles using poly-lactic-co-glycolic acid, an FDA-approved biodegradable polymer. Alternatively, endotoxin-binding and neutralizing macrophage membranes coated polymer-based nanocarriers have been demonstrated to suppress systemic inflammation in the mice model of lipopolysaccharide-induced inflammation following the i.v. administration [255].

In a separate study, bio-mimetic nanocarriers containing rapamycin were formulated by employing the membranous protein extracted from the J774 macrophage cells in their activated states. Parenteral injections of hybridized nanoparticles were demonstrated to reduce the proliferating aorta cells in an ApoE-/- animal model of atherosclerotic plaques. Other techniques were developed, which include liposomal nanoformulations that may pass through layers of inflamed endothelial cells and imitate leukocytes. Upon intravenous injection, macrophage-derived macrovesicles coated poly lactic co-glycolic acid nanocarriers demonstrated a notable improvement in the therapeutic effectiveness in a collagen-induced arthritis animal model, specifically targeting rheumatoid arthritis. In a similar vein, tacrolimus-encapsulated macrophage-coated nanoparticles greatly slowed the development of rheumatoid arthritis in mice [256].

All of these research points to hybrid biomimetic nanomedicines as a novel and promising strategy because of their variable designs resulting from adjustable nanoparticle cores and distinctive biodistribution features displayed by the various cell types employed. Some significant issues, such as potential immunological responses and complement activation following their administration, are still present, though. Also, some advancements are needed for the sterilization and scaling-up of these biomimetic nanodevices [257, 258]. A few studies have shown the pharmacological capability of cellular-membrane-coated bio-mimetic nanoparticles in illnesses such as rheumatoid arthritis, type-2 immunological hypersensitivities, and IBD, despite insufficiency of explorations in the field of autoimmune inflammation. Various bio-mimetic nanocarriers have exhibited the functioning in various illness states as binder decoys for pathways that trigger chronic inflammation and as tissue-damage-repairing and inflammation-resolving mimics of native cells. Employing the processes of T-cell attraction in the development of inflammatory bowel illness, synthetic leukocyte membrane-emulating nanoparticles were created to increase the expression of α-4β-7, crucial integrin proteins on the T lymphocytes, enabling them to adhere to irritated tissues of the mucosa [22, 253]. These specified leukosomes exhibited much higher binding to inflammatory endothelia as a result of this overexpression. Moreover, administering these biomimetic nanoparticles to DSS-induced IBD mice led to decreased CD45 + immune cells, enhanced crypt formation, and suppression of edema. It was hypothesized that the specialized ability of these nanoparticles to connect to receptors that the disease-causing immune cells would normally bind to was the reason for the therapeutic benefits shown after therapy [259, 260]. Their findings of decreased immune cell infiltration and lower levels of inflammation in the colons of mice given nanoparticle treatments provided evidence in favour of this theory. As a result, these nanoparticles acted as competitive binders for receptors that may otherwise negatively overstimulate the immune response. Red blood cells mimicking nanoparticles are also employed in showing this strategy for the removal of pathogenic antibodies. More precisely, these biomimetic nanoparticles based on red blood cells served as binding decoys for antibodies that would normally attach to native red blood cells and identify them for hemolysis outside of the blood vessel [261, 262]. Mice treated with these red blood cell nanocarriers in an induced anaemia paradigm developed normal haemoglobin levels and red blood cell counts. On the other hand, mice that were not given the nanoparticles had a two-fold drop in haemoglobin levels and a 60% fall in the number of red blood cells. Finally, it was shown that neutrophils mimicking nanoparticles significantly improve rheumatoid arthritis therapy outcomes. The purpose of this investigation was to determine if nanoparticles might block the negative immunological response that the disease's development incited by fusing their membranes with neutrophils [263, 264]. Like the examples that were previously mentioned, these nanoparticles also functioned as biological compounds that target neutrophils as semblances. This is particularly crucial for the treatment of rheumatoid arthritis since a decrease in the recruitment of neutrophils into synovial fluid could be associated with the condition's reversal. In fact, these nanoparticles were used to reduce joint damage and inhibit proinflammatory cytokines in two mouse models of arthritis. The aforementioned research showcases the adaptability of biomimetic nanoparticles in identifying and modifying the fundamental pathways that underpin and propel several autoimmune disorders [265, 266].

3 Physicochemical Characters of Nanomaterials

3.1 Chemical Composition of Nanocarriers

Although many of the novel chemical compounds being created for the treatment of local illnesses are very effective, it is challenging to precisely regulate the size, shape, and composition of nanostructures because of their limited solubility in conventional local delivery carriers. The toxicological profile of nanoparticles can be influenced by their physicochemical properties, which include size, shape, chemical content, surface charge, and aggregation state. This might have implications for the usage of nanoparticles in medicine [267, 268].

According to their chemical makeup, nanoparticles may often be divided into two categories: organic and inorganic. Inorganic nanoparticles include silica, metal, carbon nanotubes, quantum dots, and other materials, whereas organic nanoparticles are mostly polymers, exosomes, liposomes, and protein-based nanoparticles. However, in certain instances, a more intricate structure and a blend of many chemical components will be required to accomplish multiple activities [269]. For instance, to preserve the stability of the structure within the host body, metal cores will be used for the physical attributes together with a bio-organic coating on the surface. The physical attributes of nanoparticles, such as their size, shape, elasticity, surface charge, and surface functionalization, in addition to their inherent chemical compositions, can greatly influence how they function within the body. This offers an almost limitless range of applications for nanotechnology in biomedicine. Many malignancies, autoimmune disorders, bone regeneration, and other conditions have been linked to nanoparticle use recently, and hundreds of formulae are undergoing clinical and experimental studies [270].

According to some research, the chemical makeup of the particles affects how well endothelium function is affected by nanoparticles. Compared to Y2O3 and zinc oxide nanoparticles, cerium oxide nanoparticles, also known as nanoceria, appear to be more innocuous and cause just a very modest inflammatory response in human aortic endothelial cells. Considering everything said above, more research is required to determine how various nanoparticles affect vascular reactivity, endothelial dysfunction, and toxicity [271, 272].

Usually, a core plus a shell makes up a drug nanocarrier. While the shell regulates the drug's interactions with living medium, inhibits immune system recognition, allows targeting of sick tissues and organs, and provides bio-adhesion qualities, the cores retain the medications and release them in a regulated manner. Organic compounds continue to be the most commonly utilized building blocks for drug inclusion and multifunctional shell creation [273]. Moreover, the incorporation of metals into drug nanocarrier compositions confers unique features such as radioactive enhancement, antimicrobial or antifungal activities, or imaging capacities for tailored therapies. Nanoscale hybrid metal–organic frameworks have recently attracted interest as flexible drug delivery and therapeutic materials due to their intrinsic properties [274].

It is important to evaluate the homogeneity of the nanoparticle formulations regarding both size and content. Adverse changes in composition i.e., drug loading, may cause just a portion of the nanoparticles to have therapeutic activity. Indeed, the content and/or placement of the medication within each drug-loaded nanoparticle can vary significantly. These variables are crucial for quality control, but they also have a big impact on the release mechanism and how biological media interact with it. Although the size distribution and polydispersity of nanoparticles are often assessed using well-established techniques, the methods for evaluating the chemical composition of individual nanoparticles are still in their infancy [275]. Nonetheless, the task of chemically mapping the positions of the targeting ligands and integrated medicines is crucial but extremely difficult. For instance, the drug release mechanism (associated with diffusion, desorption, and degradation) is determined by the drug's position, which can be in the top layers of nanoparticles or incorporated in the cores. The destiny of nanoparticles in the biological medium is determined by several important factors, such as the uniformity of the coatings and the effective grafting of targeting ligands onto the surface of each nanoparticle. Given their small size (usually less than 200 nm) and complex core–shell composition, mapping the chemical composition and structure (morphology, crystallinity) of individual nanoparticles is a difficult task, but one that is essential to ensuring the successful transfer of scientific knowledge in nanomedicine to industrial real-world applications [161, 276].

3.2 *Hydrodynamic Diameter, Particle Size, and Dispersity Indices of Nanocarriers*

Using a dynamic light scattering, particle size analyzer, the hydrodynamic diameter of surface-coated nanoparticles is determined. The hydrodynamic diameter is more significant for biological applications than the particle diameter determined by high-resolution transmission electron microscopy. The presence of ligand and solvated water molecules causes the hydrodynamic diameter to be typically bigger than the particle diameter [277, 278]. It is well known that the body can eliminate nanoparticles through the renal system if their hydrodynamic diameter is less than 5 nm. Clinical applications need full renal elimination since most nanoparticles are harmful. Like the particle diameter distribution, the hydrodynamic diameter distribution has a log-normal distribution. Fitting the measured hydrodynamic diameters to a log–normal function yields an average hydrodynamic diameter. The hydrodynamic diameter distribution of monodispersed nanoparticles will be extremely narrow [279, 280]. These hydrodynamic dimensions are ligand-dependent; the hydrodynamic diameters are modest for small ligands and high for big ligands, such as polymers. Nonetheless, when ligand size rises, the steric hindrance also grows, leading to an increase in colloidal stability. This suggests that to minimize a hydrodynamic diameter for renal excretion, ligands of the appropriate size should be utilized for the surface coating. Strong surface charges on ligand-coated nanoparticles can cause them to aggressively resist one another. For biomedical applications, surface coating using ligands that are tiny in size yet have a high surface charge would be very helpful [281, 282].

The mean primary diameter of an individual nanocarrier, which is called an ultrafine nanoparticle, is 1 μm, but that of a microparticle, also known as a tiny particle, is 0. Larger particles often form from the agglomeration of primary particles. The lung deposition pattern of primary and agglomerated particles is determined by their size as these move through air from points of their creation to points of their exposure. Assessment of the size of nanoparticles in airborne agglomerations will provide particle-size of agglomerate rather than original nanoparticles. Depending on particle sizes of agglomerate, various methods could have been used to measure their size [283, 284]. Aerodynamic particle sizers and optical particle sizers can detect particles with aerodynamic sizes between 300 and 10 μm and 500 and 10 μm, respectively. The mass median aerodynamic diameter is the most common way to express the aerodynamic particle size. Depending on the type of SMPS being utilized, particle sizes which range from 2.5 nm to 1 μm can have their electrical mobility diameter measured using a derivative mobilities analyser and scanning mobilities particle-sizers. The count median diameter is the most common way to express the electrical mobility diameter. While electrical mobilities sizes are only dependent on the physical sizes of the particles, the aerodynamic size is also dependent on the density of the particles [285, 286]. Particles with agglomeration particle sizes varying from 0.1 to 1 μm will not deposit in the alveolar area as effectively as particles with primary or agglomerative nanoparticle sizes between 10 and 100 nm. Since there is little airflow in the alveoli, the method of deposition in the lungs for nanoparticles between 10 and 100 nm is diffusion. Particles of lower agglomeration sizes show higher deposition in alveoli than particles of higher agglomerative sizes, according to several in vivo inhalation studies. Similar deposition fractions are displayed by particles with different primary particle sizes but the same agglomeration diameter [287, 288].

The size of nanoparticles in solution may be determined using a very small number of methods, even fewer of which are generally accessible or reasonably priced for routine on-site examination. Dynamic light scattering, also known as quasi-elastic light scattering or photon correlation spectroscopy in the literature, is one method that may accurately estimate the hydrodynamic size of nanoparticles. The heterogeneity of molecular or particle sizes in a mixture is measured by the dispersity or polydispersity index [289, 290]. If every item in the collection is the same in terms of mass, shape, or size, it is referred to as uniform. Non-uniform refers to a sample of objects with irregular mass distribution, size, and shape. Any type of chemical dispersion can include the items, including crystals in a rock, particles in a colloid, droplets in a cloud, and polymer macromolecules in a solid polymer mass or solution. The molecular mass distribution of polymers, the size, surface area, and/or mass distribution of a population of particles, and the film thickness distribution of thin films can all be used to characterize them [291, 292].

Drug delivery methods based on nanoparticles, also known as nanocarriers, are widely used to increase the bioavailability of poorly soluble medications. Both lipophilic and hydrophilic molecules may be incorporated by them, and they can shield them from destruction both in vivo and in vitro. The safety, stability, effectiveness, and behaviour of these nanocarriers both in vitro and in vivo are determined by a variety of their physical characteristics [123, 293]. These consist of the polydispersity index, which indicates how well-quality they are in relation to the size distribution, and the average particle size/diameter. Nanocarrier formulations' appropriateness for a certain drug delivery pathway is contingent upon several factors, including their average diameter, polydispersity index, and size stability. The polydispersity index and particle size distribution of nanocarriers is a crucial physical property that needs to be taken into account when producing products intended for food or medicinal use. These characteristics of the lipidic nanocarriers can impact the final product's appearance, processability, stability, and bulk qualities [294]. Nanocarriers need to be characterized after preparation to ensure that they are suitable for both in vitro and in vivo uses. The term "polydispersity index" refers to a parameter that is utilized in particle size distribution characterisation to specify the size range of lipidic nanocarrier systems. The degree of non-uniformity of a particle size distribution is referred to as polydispersity or dispersity as suggested by the International Union of Pure and Applied Chemistry. Polydispersity has two distinct features in molecular science by utilizing chromatography techniques, nanotechnology, and nanoparticle research by employing light scattering, depending on the attribute of interest [295]. For a comparative analysis, various types of nanocarriers and nano-formulations and their relief sizes have been depicted in Fig. 4.

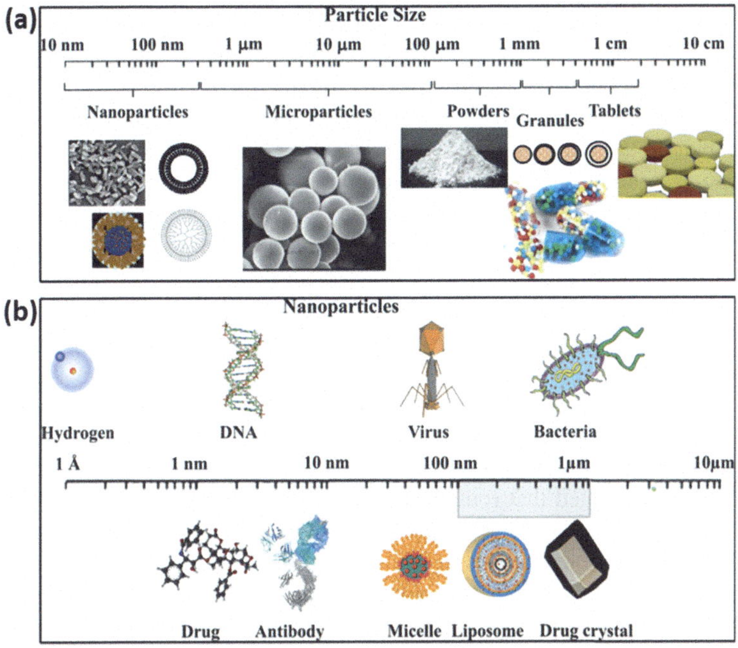

Fig. 4 **a** Relative particle sizes of pharmaceutical formulations arraying from the nanoparticles to tablets. **b** Illustrations of various types of nanocarrier formulations and their relative particle sizes (Adapted with permission form [296]

3.3 *Surface Morphological Features Including Surface Coatings and Surface Roughness*

The biological characteristics of nanostructured surfaces on biomaterials are highly promising and are generating significant attention in fundamental explorations, implant industry developments, and bio-engineering employments. However, there is much more to learn about the biological and therapeutic consequences of nano-scale interaction at both cellular as well as molecular levels. In order to assess how well-nanostructured surfaces of nanocarriers can lead to induction of particular cellular interaction in all the pertinent pharmacological, reparative, anti-inflammatory, anti-microbial, and clinical applications, researchers have examined surface development, nano-structuration, and other nanoengineered drug delivery systems. The primary cellular traits involved in surface/cell interaction, the techniques and physio-chemical characteristics involved in nano-topography performance, and an overview of findings and future directions shown in the research utilizing nanostructured surface and nanoparticulate drug delivery platforms have also been thoroughly examined. Many researchers have also looked into the commercial translation and future possibilities of this emerging discipline, with a focus on multifunctional nanostructured surfaces and their implications for medicine. Nanostructured biomedical implant surfaces can act as the immunomodulating surface that targets macrophage cells and reduces the expression of pro-inflammatory cytokines. They can also promote fibroblasts and/or epithelial cellular proliferation and adherence, inhibit the proliferation of microbes and bio-film accumulations, and improve the process of osteogenesis thereby targeting osteoblast cells, osteocytes, and mesenchymal stem cells. Furthermore, a number of methodological choices have been made accessible for the creation of drug-delivering platforms on the surface of the metallic implants; nevertheless, clinical applications are still lacking. Pre-clinical and clinical investigations, as well as the bio-engineering solution for commercialized translations and approvals of governing bodies, can further illuminate the efficacy of which nano-structured nanoparticles-based drug-delivering surface may lead to targeting certain cellular interactions and favour clinical results.

Although surface roughness features could be expressed in nanoscale, microscale or macroscale, there are nano-morphologies that exhibit high roughness values and nano-morphologies that have roughness values that are near to a smooth pattern. It has been documented that roughness has a direct impact on the pace of healing as well as the adherence and proliferation of bacterial cells. Uneven surfaces allow cells to invade and condense inside the surface roughness, increasing intercellular contact—a crucial process for nutritious and cellular evolutions. Moreover, a rougher surface affects the surface's free energy. Because osteoprogenitor cells are isolated between huge peaks and valleys characterized by noticeable rough morphologies on surfaces, excessive roughness levels are actually reported to be detrimental to these cells. As a result, cells are unable to interact with one another, establish a rapid union, and build the ideal repaired tissue. However, high roughness surfaces are beneficial to produce multispecies bio-films because they allow bacteria to build bio-films in the layer which generally transcend peaks and harsh valleys while also impeding the growth of repairing cells.

In reference to the anti-inflammatory surface coating of nanoparticles, techniques for the nanostructuring surface in the material have been created using a variety of procedures, including anodizing, acid etching, laser usage, and coating deposition. The functionalization and surface bioactivity of nanostructured surfaces have become infinitely customizable by adding bioactive coating, the inclusion of nanocarriers, or the combination of several techniques to create drug delivery systems. In animal studies, using mouse models, researchers created titanium specimens that were modified with titanium oxide nano-tubes and also performed functionalization with the coatings of the bone-morphogenetic protein-2 from humans. They then demonstrated that the combined effects of the two proteins greatly enhanced osseointegration [297]. As suggested, several investigations have demonstrated encouraging antibacterial properties at nanostructured implant surfaces with characteristics inspired by nature, such as nano coatings, nano doping, and nano-tubular surfaces. The hybrid balanced microbes and osteoblast cellular competitions on the surface of the implants enhanced the anti-bacterial effects to significant extent, thereby reducing the adhesions of the bacteria, down-regulating the cellular responses against the inflammation, and simultaneously speeding up the formation of the bones, according to other *in-vivo* report assessing the zinc oxide-polydopamine-arginine-glycine-aspartic acid-cysteine nano-rods arrays' coating on surface of titanium [298].

Furthermore, bio-active coating, which is composed of both synthetic as well as non-synthetic elements, can be used to accomplish the tasks of surfaces with several applications. Studies have demonstrated the antibacterial and osteoinductive effects of polysaccharide coatings containing nano-encapsulated bone morphogenetic protein-2 and vancomycin [299]. In a similar vein, other writers have detailed a technique that aims for a double positive without sacrificing biocompatibility by releasing bone morphogenetic protein-2 and silver nanoparticles from a hydroxyapatite covering. Crucially, multifunctional surface applications utilizing combination medications or bioactive compounds for cell development show promise for acting in concert to provide antibacterial, anti-inflammatory, and quick osseointegration effects [300, 301]. In the meantime, more research is required to translate this versatile technology into implantable devices. However, further research is required to fully understand aspects including the right medication dosage, the ideal chemical release duration, and durability procedures.

3.4 Surface Functionalization

Since these processes are intimately tied to the surface makeup of the nanoparticles, surface modification is a potent technique to address or mitigate concerns related to nanoparticle toxicity and absorption. The method of surface functionalization of nanoparticles tries to enhance and/or add characteristics beneficial for the application of nanoparticles in medicine. Various kinds of nanomaterials can be employed in the initial stages of functionalization because of their exposed functional groups and distinctive chemical characteristics [302]. Usually, earlier stages of interfacial modifications involve the addition of the organic functional groups—such as amino or carboxylic groups—that are effective for binding biological molecules. This is accomplished by employing homogenous or heterogenous bi-functional crosslinking agents. Among the various linkers for silica-based nanoparticles are aminosilanes, which add the amino functional group to the surfaces of the particles to prepare these for the next bio-conjugation [303].

Using crosslinkers with thiol or amino functional groups that can react with the metal and form a covalent bond, noble metals like gold can be functionalized. These bifunctional linkers, which include thiocarboxylic acids, include functional groups at other ends that are employed in the binding of the ligand. By employing a ligand exchange technique that involves the replacement of the original surface of metal-oxide with an amine-based functional group or, thiol, diol, and carboxylic acids that are helpful in the next stages, metal-oxide may be readily transformed. The carbon-based nanomaterials include a significant amount of hybridized carbon atoms that can be utilized to form functional groups. Oxidation can create carboxylic and hydroxyl groups on the nanoparticle surface; halogenation can provide halogenated carbon, which can subsequently be modified further, for example by interacting with an amine group; and cycloaddition can introduce other functional groups [304].

The two methods of non-covalent and covalent conjugation have been employed for modifying the surfaces of the nanoparticles. The non-covalent method works well with metal-based and silica-based nanoparticles and undertakes multitudes of weaker interactions, which also include electrostatic interactions, ionic interactions, van der Walls forces of interactions and hydrophobic interactions, absorptions, and hydrogen bonding. The usage of non-covalent bonds has the benefits of comparative straightforwardness, as they have no effect on the structures of the molecule involved or the way they have interactions with other biological target tissues or cells. On the other hand, non-covalent changes are readily affected by a variety of factors, including ionic strength and pH [305].

Depending on the nanoparticle composition, several other techniques may be employed for the creation of a covalent bond approach. Furthermore, by employing successive functionalization, this technique permits alterations at several levels. The theranostic approach may be used by utilizing this paradigm for the creation of a structure that serves several intentions, e.g., disease diagnosis and therapeutics. Generally, several cross-linking molecules have been employed for the creation of the covalent interaction between ligands and the surface of the nanoparticles. As an illustration, consider polyethylene glycol, which may be produced with specific functional moieties at the end and utilized as a homologous or heterologous bi-functional linker for carrying out various functionalization procedures [270].

According to research, the synthesis of certain heterobifunctional polyethylene glycol molecules is a current area of interest and is necessary for the construction of effective nano-platforms for use in medicine. Polyethylene glycol has the ability to function as a spacer for molecules with very large molecular weights due to its polymeric properties. Distributing the molecules evenly throughout the surface of the nanoparticle and lowering steric hindrance caused by bound ligands, this promotes high-density bioconjugation. Covalent connections are frequently used to bind ligands necessary to conduct targeting and/or to manage the toxicity of nanoparticles. Non-covalent connections are used to load nanoparticles with compounds that must be released in target cells, such as pharmaceuticals or RNA interference [306]. Lately, there has been a lot of research done on the utilization of sensitive bonds, including heat or pH sensitivity, to create nanoplatforms for regulated medication release. Specifically, in contrast to the normal milieu, the tumour microenvironment is very acidic. It is possible to create pH-sensitive nanoplatforms that are particularly activated by the acidity of the tumour environment to allow for regulated medication release. In a recent study, doxorubicin-loaded liposome nanoparticles were coated with a pH-sensitive cationic polymer, Poly(beta-amino ester). Moreover, hyaluronic acid was added to the nanoparticle surface to enable active targeting via CD44. The outcomes of the in vitro studies were validated by in vivo tests, which showed that the doxorubicin-loaded nanoparticles lessened adverse effects and more effectively suppressed tumour development than the medication alone [307].

4 Conclusion and Future Perspective

The materialistic and pharmacological benefits of natural and nature-derived derived building blocks are integrated by smarter anti-inflammatory nanostructures, opening up new possibilities for the multi-functional designs of anti-inflammatory medication nano-carriers. At the same time, the clinical translation becomes more difficult, the more complicated the materials are. In addition, to fully realize the promise of anti-inflammatory nanostructures, a comprehensive and in-depth understanding of nanomaterial-tissue interactions needs to be elucidated. In the in-vivo settings, smart anti-inflammatory nanomaterials can interact and obstruct biological pathways due to their wide molecular and supramolecular structural variability. Consequently, system optimization can be difficult because the molecular makeup of these anti-inflammatory nanostructures' building components might have a significant impact on how effective they are. The rational design of these smart anti-inflammatory nanostructures is made possible by computational simulation and statistical models. For instance, hydrogel-forming tripeptides have been screened and identified using molecular dynamics simulations; additional in-silico techniques have been employed for the simulation of the adsorption of proteins, extravasation, cellular uptakes, and release of the therapeutic payload, as well as the development of the structure–activity relationship for evaluations and assessments of the safety of these nanocarriers. In contrast to inorganic or polymeric anti-inflammatory nanostructures, these anti-inflammatory nanoparticulate systems are more challenging to anticipate regarding their in vivo behaviour because of the intricacies of these biomolecules-based anti-inflammatory nanocarriers and the paucity of systematic evidences of their experimentations in-vivo.

Systems with extremely complex spatial structures and responsive mechanisms have been built in order to investigate the multifunctional benefit of these anti-inflammatory nanoparticles. However, there are still few reliable and repeatable production techniques available for these kinds of intricate structures. Compared to proteins, nucleic acids, and peptides may have an advantage because of their reproducible supramolecular behaviour and ability to create structures. The production and characterization of anti-inflammatory nanomaterials can also be automated and monitored with the use of high-throughput and microfluidic techniques.

Many anti-inflammatory nanoparticle designs that are tailored for therapeutic delivery and designed to get past the various biological barriers that are present in different patient groups and diseases have been covered in this chapter. Variations in the stages of disease progression, distinct physiologies, and patient comorbidities all contribute to these delivery hurdles. Nanoparticles created for various patient demographics, inflammatory diseases, or combinations of the two can be used to address this wide range of demands. Nanoparticle platforms provide a range of tunable characteristics that may be selected for optimization of delivering drugs for particular applications. These characteristics include particle size, nanoparticle shapes, surface or interfacial charges, qualities of the surfaces, and their responsiveness. By enabling newer patients who can exhibit qualifications for the existing therapeutic paradigms with newer and improved drug-delivery processes, this customization may be employed in concert with the precision-medicines for expanding the accessibility of precision-based nanomedicines, improve patient stratification techniques when screening nanoparticle platforms, and finally, enhance the overall pharmacological efficiencies of both precision-based nanomedicines and anti-inflammatory nanoparticulate drug delivery platforms.

Size and form are two of these nanoparticle properties that have been thoroughly researched in a variety of biological states. In certain situations, tendencies that can be applied to intelligent nanoparticle design have been found. For instance, while targeting surface markers is more important in applications where specific cell types must uptake anti-inflammatory nanoparticles, such as in many inflammatory disorders and immune-therapeutic applications, nanoparticle charge is particularly important in muco-penetrating applications and intracellular applications that require endosomal escape. But as design issues get more intricate, it is accompanied by the attempts for the generalization of trends across larger population, compromising precisions of results in the smaller groups for the hope of producing a delivery principle that applies to all situations. In order to increase the specificity of these claims, more in-depth analyses of anti-inflammatory nanoparticle design investigations and the ensuing interactions within the human body are required. This is especially true when we begin to stratify patient populations in order to identify the best anti-inflammatory nanoparticle platforms for various subgroups. Researchers can gather information and analyse results to contribute to the expanding body of knowledge about design-function relationship trends in anti-inflammatory nanomedicine by continuing to explore anti-inflammatory nanoparticle technologies in lab settings. Before making any attempt to generalize the results, it is necessary to put the trends seen in research settings into context. This is because even seemingly insignificant variations in anti-inflammatory nanoparticle composition, animal models, and pathology can have a significant impact on nanoparticle performance, and these factors must be taken into account when advancing nanoparticle technology toward clinical translation.

Most of the present clinical achievements with anti-inflammatory nanocarriers in precision medicine are seen in the therapeutic domain, including the use of biomarker molecules for determining which therapy is more suitable for the patients or the capacity to identify earlier phases of the diseases by particular ligand-receptor interaction. The degree of improved permeability and retention effect, for instance, can be used to predict how well an anti-inflammatory nanoparticle therapy would accumulate at the site of inflammation in a patient. However, because these approaches possess the capabilities to greatly increase the efficacies of anti-inflammatory precision medicines but could not yet see the clinical translations. This chapter has concentrated on the probable therapeutical applicability of anti-inflammatory nanoparticles, particularly their applications in precision-based nanomedicines. The reason for this lack of clinical advancement is probably that platforms for anti-inflammatory nanoparticles are tested for effectiveness in large populations, where the wide range of biological barriers across patients may obscure the possibility of treating marginally smaller subgroups with success. As was previously indicated, since stratified populations are expected to react to anti-inflammatory nanoparticle treatment more consistently, the presentation of the graded patient population may help hasten therapeutic advancements. More graded and ranked trials are required since we are presently unable to forecast which anti-inflammatory nanoparticle approaches would be the most helpful in precision-based application because ongoing clinical studies have been typically not stratified.

Naturally, adopting a precision lens for screening anti-inflammatory nanoparticles will result in fewer patients being approved for medicine, which will lower the potential market size of each therapy based on anti-inflammatory nanoparticles. This decrease may cause concern given the high development costs connected with sophisticated anti-inflammatory nanoparticle designs and, consequently, the elevated financial risks related to a nanoparticle formulation's possible failure in clinical translation. Nonetheless, anti-inflammatory nanoparticulate drug delivery systems that demonstrate efficacy in particular patient populations may find use in the administration of many medicines, including both customized and universal ones. Therefore, the creation of a single, highly effective anti-inflammatory nanoparticle platform for a stratified population may result in several fruitful clinical uses. Additionally, compared to anti-inflammatory nanoparticles produced for broad populations, precision nanoparticle designs may allow for improved therapeutic efficacy, and notable improvements in quality of life, and even dosing may be able to justify these precision delivery systems' higher price points.

These findings may have an impact on the rational design of anti-inflammatory drug carriers in the future for a variety of treatments, including personalized and generic ones. This could help a wide range of cargo, such as proteins, nucleotides, and small molecules, as more sophisticated anti-inflammatory nanocarriers designs are investigated. Additionally, the authorization of additional nanocarrier platforms may facilitate the route to the clinic for innovative uses of these anti-inflammatory nanocarriers for the delivery of therapeutic cargos that have already received approval. We are moving toward designing more modular patient therapies and opening the door to the possibility of futuristic prescription that not only admit the best therapies but also complement these with the most effective drug-delivery platforms by striving to develop more anti-inflammatory nanocarriers-based platforms and precision medicines for food and drug administration approval and clinical use. In the clinical arena, the idea of providing a single therapy across several delivery platforms is not new. Medications that are often prescribed, like birth control, are already available in a variety of forms to accommodate patients' lifestyles, including oral pills, injections, and implants. The growth of seemingly narrow markets will be made possible by expanding both the precision-nanomedicines and boosted anti-inflammatory nanocarrier approaches, which shall support ongoing clinic-based advancement of tailored medications.

The use of anti-inflammatory nanocarriers in precision medicine-related applications enables more individualized therapy, such as multidrug treatments, better cellular targeting, and fewer off-target effects. All of this may be accomplished by customizing anti-inflammatory nanocarriers for the intended use, enhancing accumulation at the target location, and adding responsiveness for on-demand drug release. This will reduce undesired toxicities and allow for a wider range of doses or combination therapies. The impacts of precision medicine therapies can also be enhanced by increasing the specificity and local activity of anti-inflammatory nanoparticle delivery systems, which will benefit a wider range of people and enhance overall patient outcomes. The studies provided in this chapter demonstrate how intelligent anti-inflammatory nanoparticle design might improve precision medicine overall. Furthermore, the knowledge obtained from precision medicine such as patient classification and genetic profiling can direct the rational choice of a nanoparticle platform, resulting in the development of the best possible anti-inflammatory nanoparticle-based precision therapy.

References

1. Ventola CL (2012) The nanomedicine revolution. PT 37:512–525
2. Xu M, Han X, Xiong H, Gao Y, Xu B, Zhu G, Li J (2023) Cancer nanomedicine: emerging strategies and therapeutic potentials. Molecules 28:5145. https://doi.org/10.3390/molecules28135145
3. Yusuf A, Almotairy ARZ, Henidi H, Alshehri OY, Aldughaim MS (2023) Nanoparticles as drug delivery systems: a review of the implication of nanoparticles' physicochemical properties on responses in biological systems. Polymers (Basel) 15:1596. https://doi.org/10.3390/polym15071596
4. Wolfram J, Zhu M, Yang Y, Shen J, Gentile E, Paolino D, Fresta M, Nie G, Chen C, Shen H et al (2015) Safety of nanoparticles in medicine. Curr Drug Targets 16:1671–1681
5. Chehelgerdi M, Chehelgerdi M, Allela OQB, Pecho RDC, Jayasankar N, Rao DP, Thamaraikani T, Vasanthan M, Viktor P, Lakshmaiya N et al (2023) Progressing nanotechnology to improve targeted cancer treatment: overcoming hurdles in its clinical implementation. Mol Cancer 22:169. https://doi.org/10.1186/s12943-023-01865-0
6. Patra JK, Das G, Fraceto LF, Campos EVR, Rodriguez-Torres MDP, Acosta-Torres LS, Diaz-Torres LA, Grillo R, Swamy MK, Sharma S et al (2018) Nano based drug delivery systems: recent developments and future prospects. J Nanobiotechnol 16:71. https://doi.org/10.1186/s12951-018-0392-8
7. Wu D, Si M, Xue H-Y, Wong H-L (2017) Nanomedicine applications in the treatment of breast cancer: current state of the art. Int J Nanomedicine 12:5879–5892. https://doi.org/10.2147/IJN.S123437
8. Rommasi F, Esfandiari N (2021) Liposomal nanomedicine: applications for drug delivery in cancer therapy. Nanoscale Res Lett 16:95. https://doi.org/10.1186/s11671-021-03553-8
9. Thorn CF, Oshiro C, Marsh S, Hernandez-Boussard T, McLeod H, Klein TE, Altman RB (2011) Doxorubicin pathways: pharmacodynamics and adverse effects. Pharmacogenet Genomics 21:440–446. https://doi.org/10.1097/FPC.0b013e32833ffb56
10. Kuerban K, Gao X, Zhang H, Liu J, Dong M, Wu L, Ye R, Feng M, Ye L (2020) Doxorubicin-loaded bacterial outer-membrane vesicles exert enhanced anti-tumor efficacy in non-small-cell lung cancer. Acta Pharm Sin B 10:1534–1548. https://doi.org/10.1016/j.apsb.2020.02.002
11. Jin C, Wang K, Oppong-Gyebi A, Hu J (2020) Application of nanotechnology in cancer diagnosis and therapy—a mini-review. Int J Med Sci 17:2964–2973. https://doi.org/10.7150/ijms.49801
12. Gavas S, Quazi S, Karpiński TM (2021) Nanoparticles for cancer therapy: current progress and challenges. Nanoscale Res Lett 16:173. https://doi.org/10.1186/s11671-021-03628-6

13. Lamprecht A (2015) Nanomedicines in gastroenterology and hepatology. Nat Rev Gastroenterol Hepatol 12:195–204. https://doi.org/10.1038/nrgastro.2015.37
14. Pelaz B, Alexiou C, Alvarez-Puebla RA, Alves F, Andrews AM, Ashraf S, Balogh LP, Ballerini L, Bestetti A, Brendel C et al (2017) Diverse Applications of nanomedicine. ACS Nano 11:2313–2381. https://doi.org/10.1021/acsnano.6b06040
15. A foray into the multifaceted world of nanotechnologies (2016). https://doi.org/10.1039/9781782628873-00212
16. Wang J, Li Y, Nie G (2021) Multifunctional biomolecule nanostructures for cancer therapy. Nat Rev Mater 6:766–783. https://doi.org/10.1038/s41578-021-00315-x
17. Zeng Y, Xiang Y, Sheng R, Tomás H, Rodrigues J, Gu Z, Zhang H, Gong Q, Luo K (2021) Polysaccharide-based nanomedicines for cancer immunotherapy: a review. Bioact Mater 6:3358–3382. https://doi.org/10.1016/j.bioactmat.2021.03.008
18. Joseph TM, Kar Mahapatra D, Esmaeili A, Piszczyk Ł, Hasanin MS, Kattali M, Haponiuk J, Thomas S (2023) Nanoparticles: taking a unique position in medicine. Nanomaterials (Basel) 13:574. https://doi.org/10.3390/nano13030574
19. Sun L, Liu H, Ye Y, Lei Y, Islam R, Tan S, Tong R, Miao Y-B, Cai L (2023) Smart nanoparticles for cancer therapy. Sig Transduct Target Ther 8:1–28. https://doi.org/10.1038/s41392-023-01642-x
20. Kashyap BK, Singh VV, Solanki MK, Kumar A, Ruokolainen J, Kesari KK (2023) Smart nanomaterials in cancer theranostics: challenges and opportunities. ACS Omega 8:14290–14320. https://doi.org/10.1021/acsomega.2c07840
21. Saleh HM, Hassan AI (2023) Synthesis and characterization of nanomaterials for application in cost-effective electrochemical devices. Sustainability 15:10891. https://doi.org/10.3390/su151410891
22. Brusini R, Varna M, Couvreur P (2020) Advanced nanomedicines for the treatment of inflammatory diseases. Adv Drug Deliv Rev 157:161–178. https://doi.org/10.1016/j.addr.2020.07.010
23. Yasmin F, Najeeb H, Shaikh S, Hasanain M, Naeem U, Moeed A, Koritala T, Hasan S, Surani S (2022) Novel drug delivery systems for inflammatory bowel disease. World J Gastroenterol 28:1922–1933. https://doi.org/10.3748/wjg.v28.i18.1922
24. van Alem CMA, Metselaar JM, van Kooten C, Rotmans JI (2021) Recent advances in liposomal-based anti-inflammatory therapy. Pharmaceutics 13:1004. https://doi.org/10.3390/pharmaceutics13071004
25. Al-Jipouri A, Almurisi SH, Al-Japairai K, Bakar LM, Doolaanea AA (2023) Liposomes or extracellular vesicles: a comprehensive comparison of both lipid bilayer vesicles for pulmonary drug delivery. Polymers (Basel) 15:318. https://doi.org/10.3390/polym15020318
26. Sousa A, Phung AN, Škalko-Basnet N, Obuobi S (2023) Smart delivery systems for microbial biofilm therapy: dissecting design, drug release and toxicological features. J Control Release 354:394–416. https://doi.org/10.1016/j.jconrel.2023.01.003
27. Bhadran A, Shah T, Babanyinah GK, Polara H, Taslimy S, Biewer MC, Stefan MC (1977) Recent advances in polycaprolactones for anticancer drug delivery. Pharmaceutics 2023:15. https://doi.org/10.3390/pharmaceutics15071977
28. Pitt WG, Husseini GA, Staples BJ (2004) Ultrasonic drug delivery—a general review. Expert Opin Drug Deliv 1:37–56. https://doi.org/10.1517/17425247.1.1.37
29. Wang X, Li C, Wang Y, Chen H, Zhang X, Luo C, Zhou W, Li L, Teng L, Yu H et al (2022) Smart drug delivery systems for precise cancer therapy. Acta Pharm Sin B 12:4098–4121. https://doi.org/10.1016/j.apsb.2022.08.013
30. Amasya G, Aksu B, Badilli U, Onay-Besikci A, Tarimci N (2019) QbD guided early pharmaceutical development study: production of lipid nanoparticles by high pressure homogenization for skin cancer treatment. Int J Pharm 563:110–121. https://doi.org/10.1016/j.ijpharm.2019.03.056
31. Hu C, Qian A, Wang Q, Xu F, He Y, Xu J, Xia Y, Xia Q (2016) Industrialization of lipid nanoparticles: from laboratory-scale to large-scale production line. Eur J Pharm Biopharm 109:206–213. https://doi.org/10.1016/j.ejpb.2016.10.018

32. Shegokar R, Singh KK, Müller RH (2011) Production & stability of stavudine solid lipid nanoparticles—from lab to industrial scale. Int J Pharm 416:461–470. https://doi.org/10.1016/j.ijpharm.2010.08.014
33. Esposito E, Sguizzato M, Drechsler M, Mariani P, Carducci F, Nastruzzi C, Cortesi R (2017) Progesterone lipid nanoparticles: scaling up and in vivo human study. Eur J Pharm Biopharm 119:437–446. https://doi.org/10.1016/j.ejpb.2017.07.015
34. Mehnert W, Mäder K (2012) Solid lipid nanoparticles: production, characterization and applications. Adv Drug Deliv Rev 64:83–101. https://doi.org/10.1016/j.addr.2012.09.021
35. Suter F, Schmid D, Wandrey F, Zülli F (2016) Heptapeptide-loaded solid lipid nanoparticles for cosmetic anti-aging applications. Eur J Pharm Biopharm 108:304–309. https://doi.org/10.1016/j.ejpb.2016.06.014
36. Štecová J, Mehnert W, Blaschke T, Kleuser B, Sivaramakrishnan R, Zouboulis CC, Seltmann H, Korting HC, Kramer KD, Schäfer-Korting M (2007) Cyproterone acetate loading to lipid nanoparticles for topical acne treatment: particle characterisation and skin uptake. Pharm Res 24:991–1000. https://doi.org/10.1007/s11095-006-9225-9
37. Gupta S, Kesarla R, Chotai N, Misra A, Omri A (2017) Systematic approach for the formulation and optimization of solid lipid nanoparticles of efavirenz by high pressure homogenization using design of experiments for brain targeting and enhanced bioavailability. Biomed Res Int 2017:e5984014. https://doi.org/10.1155/2017/5984014
38. Üner M, Yener G, Ergüven M (2019) Design of colloidal drug carriers of celecoxib for use in treatment of breast cancer and leukemia. Mater Sci Eng, C 103:109874. https://doi.org/10.1016/j.msec.2019.109874
39. Yadav M, Schiavone N, Guzman-Aranguez A, Giansanti F, Papucci L, Perez de Lara MJ, Singh M, Kaur IP (2020) Atorvastatin-loaded solid lipid nanoparticles as eye drops: proposed treatment option for age-related macular degeneration (AMD). Drug Deliv Transl Res 10:919–944. https://doi.org/10.1007/s13346-020-00733-4
40. Das S, Chaudhury A (2011) Recent advances in lipid nanoparticle formulations with solid matrix for oral drug delivery. AAPS PharmSciTech 12:62–76. https://doi.org/10.1208/s12249-010-9563-0
41. Duan Y, Dhar A, Patel C, Khimani M, Neogi S, Sharma P, Kumar NS, Vekariya RL (2020) A brief review on solid lipid nanoparticles: part and parcel of contemporary drug delivery systems. RSC Adv 10:26777–26791. https://doi.org/10.1039/D0RA03491F
42. Sharma G, Chopra K, Puri S, Bishnoi M, Rishi P, Kaur IP (2018) Topical Delivery of TRPsiRNA-loaded solid lipid nanoparticles confer reduced pain sensation via TRPV1 silencing, in rats. J Drug Target 26:135–149. https://doi.org/10.1080/1061186X.2017.1350857
43. Zhang S-J, Zhang Y-T, Zhao J-H, Shen L-N, Shi F, Feng N-P (2013) Preparation and in vitro anti-tumor properties of toad venom extract-loaded solid lipid nanoparticles. Die Pharmazie Int J Pharm Sci 68:653–660. https://doi.org/10.1691/ph.2013.2216
44. Kasongo KW, Müller RH, Walker RB (2012) The use of hot and cold high pressure homogenization to enhance the loading capacity and encapsulation efficiency of nanostructured lipid carriers for the hydrophilic antiretroviral drug, didanosine for potential administration to paediatric patients. Pharm Dev Technol 17:353–362. https://doi.org/10.3109/10837450.2010.542163
45. Wang D, Zhao P, Cui F, Li X (2007) Preparation and characterization of solid lipid nanoparticles loaded with total flavones of hippophae rhamnoides (TFH). PDA J Pharm Sci Technol 61:110–120
46. Duong V-A, Nguyen T-T-L, Maeng H-J, Chi S-C (2019) Data on optimization and drug release kinetics of nanostructured lipid carriers containing ondansetron hydrochloride prepared by cold high-pressure homogenization method. Data Brief 26:104475. https://doi.org/10.1016/j.dib.2019.104475
47. Neupane YR, Sabir MD, Ahmad N, Ali M, Kohli K (2013) Lipid drug conjugate nanoparticle as a novel lipid nanocarrier for the oral delivery of decitabine: ex vivo gut permeation studies. Nanotechnology 24:415102. https://doi.org/10.1088/0957-4484/24/41/415102

48. Duong V-A, Nguyen T-T-L, Maeng H-J, Chi S-C (2019) Preparation of ondansetron hydrochloride-loaded nanostructured lipid carriers using solvent injection method for enhancement of pharmacokinetic properties. Pharm Res 36:138. https://doi.org/10.1007/s11095-019-2672-x
49. Souto EB, Doktorovova S, Zielinska A, Silva AM (2019) Key production parameters for the development of solid lipid nanoparticles by high shear homogenization. Pharm Dev Technol 24:1181–1185. https://doi.org/10.1080/10837450.2019.1647235
50. Severino P, Santana MHA, Souto EB (2012) Optimizing SLN and NLC by 22 full factorial design: effect of homogenization technique. Mater Sci Eng, C 32:1375–1379. https://doi.org/10.1016/j.msec.2012.04.017
51. Xie S, Zhu L, Dong Z, Wang X, Wang Y, Li X, Zhou W (2011) Preparation, characterization and pharmacokinetics of enrofloxacin-loaded solid lipid nanoparticles: influences of fatty acids. Colloids Surf, B 83:382–387. https://doi.org/10.1016/j.colsurfb.2010.12.014
52. Akhoond Zardini A, Mohebbi M, Farhoosh R, Bolurian S (2018) Production and characterization of nanostructured lipid carriers and solid lipid nanoparticles containing lycopene for food fortification. J Food Sci Technol 55:287–298. https://doi.org/10.1007/s13197-017-2937-5
53. Fazly Bazzaz BS, Khameneh B, Namazi N, Iranshahi M, Davoodi D, Golmohammadzadeh S (2018) Solid Lipid nanoparticles carrying eugenia caryophyllata essential oil: the novel nanoparticulate systems with broad-spectrum antimicrobial activity. Lett Appl Microbiol 66:506–513. https://doi.org/10.1111/lam.12886
54. Zhu C, Li W, Wang X, Xue J, Zhao L, Song Y, Zhou T, Zhang M (2020) Thiopental sodium loaded solid lipid nano-particles attenuates obesity-induced cardiac dysfunction and cardiac hypertrophy via inactivation of inflammatory pathway. Drug Deliv 27:1188–1200. https://doi.org/10.1080/10717544.2020.1803449
55. Shah P, Chavda K, Vyas B, Patel S (2021) Formulation development of linagliptin solid lipid nanoparticles for oral bioavailability enhancement: role of P-Gp inhibition. Drug Deliv Transl Res 11:1166–1185. https://doi.org/10.1007/s13346-020-00839-9
56. Jannin V, Blas L, Chevrier S, Miolane C, Demarne F, Spitzer D (2018) Evaluation of the digestibility of solid lipid nanoparticles of glyceryl dibehenate produced by two techniques: ultrasonication and spray-flash evaporation. Eur J Pharm Sci 111:91–95. https://doi.org/10.1016/j.ejps.2017.09.049
57. Cortesi R, Esposito E, Luca G, Nastruzzi C (2002) Production of lipospheres as carriers for bioactive compounds. Biomaterials 23:2283–2294. https://doi.org/10.1016/S0142-9612(01)00362-3
58. Wissing SA, Kayser O, Müller RH (2004) Solid lipid nanoparticles for parenteral drug delivery. Adv Drug Deliv Rev 56:1257–1272. https://doi.org/10.1016/j.addr.2003.12.002
59. Shah RM, Eldridge DS, Palombo EA, Harding IH (2017) Microwave-assisted microemulsion technique for production of miconazole nitrate- and econazole nitrate-loaded solid lipid nanoparticles. Eur J Pharm Biopharm 117:141–150. https://doi.org/10.1016/j.ejpb.2017.04.007
60. Masiiwa WL, Gadaga LL (2018) Intestinal permeability of artesunate-loaded solid lipid nanoparticles using the everted gut method. J Drug Deliv 2018:e3021738. https://doi.org/10.1155/2018/3021738
61. Souza LG, Silva EJ, Martins ALL, Mota MF, Braga RC, Lima EM, Valadares MC, Taveira SF, Marreto RN (2011) Development of topotecan loaded lipid nanoparticles for chemical stabilization and prolonged release. Eur J Pharm Biopharm 79:189–196. https://doi.org/10.1016/j.ejpb.2011.02.012
62. Fathy Abd-Ellatef GE, Gazzano E, Chirio D, Ragab Hamed A, Belisario DC, Zuddas C, Peira E, Rolando B, Kopecka J, Assem Said Marie M et al (2020) Curcumin-loaded solid lipid nanoparticles bypass P-glycoprotein mediated doxorubicin resistance in triple negative breast cancer cells. Pharmaceutics 12:96. https://doi.org/10.3390/pharmaceutics12020096
63. Igartua M, Saulnier P, Heurtault B, Pech B, Proust JE, Pedraz JL, Benoit JP (2002) Development and characterization of solid lipid nanoparticles loaded with magnetite. Int J Pharm 233:149–157. https://doi.org/10.1016/S0378-5173(01)00936-X

64. Trotta M, Debernardi F, Caputo O (2003) Preparation of solid lipid nanoparticles by a solvent emulsification-diffusion technique. Int J Pharm 257:153–160. https://doi.org/10.1016/S0378-5173(03)00135-2
65. Hu FQ, Yuan H, Zhang HH, Fang M (2002) Preparation of solid lipid nanoparticles with clobetasol propionate by a novel solvent diffusion method in aqueous system and physicochemical characterization. Int J Pharm 239:121–128. https://doi.org/10.1016/S0378-5173(02)00081-9
66. Dhaundiyal A, Jena SK, Samal SK, Sonvane B, Chand M, Sangamwar AT (2016) Alpha-lipoic acid-stearylamine conjugate-based solid lipid nanoparticles for tamoxifen delivery: formulation, optimization, in-vivo pharmacokinetic and hepatotoxicity study. J Pharm Pharmacol 68:1535–1550. https://doi.org/10.1111/jphp.12644
67. Patravale VB, Mirani AG (2019) Preparation and characterization of solid lipid nanoparticles-based gel for topical delivery. In: Weissig V, Elbayoumi T (eds) Pharmaceutical nanotechnology: basic protocols. Springer, New York, NY, pp 293–302. ISBN 978-1-4939-9516-5
68. Fonte P, Nogueira T, Gehm C, Ferreira D, Sarmento B (2011) Chitosan-coated solid lipid nanoparticles enhance the oral absorption of insulin. Drug Deliv Transl Res 1:299–308. https://doi.org/10.1007/s13346-011-0023-5
69. Shudo J, Pongpeerapat A, Wanawongthai C, Moribe K, Yamamoto K (2008) In vivo assessment of oral administration of probucol nanoparticles in rats. Biol Pharm Bull 31:321–325. https://doi.org/10.1248/bpb.31.321
70. Sarmento B, Martins S, Ferreira D, Souto EB (2007) Oral insulin delivery by means of solid lipid nanoparticles. Int J Nanomed 2:743–749. https://doi.org/10.2147/IJN.S2.4.743
71. Nabi-Meibodi M, Vatanara A, Najafabadi AR, Rouini MR, Ramezani V, Gilani K, Etemadzadeh SMH, Azadmanesh K (2013) The effective encapsulation of a hydrophobic lipid-insoluble drug in solid lipid nanoparticles using a modified double emulsion solvent evaporation method. Colloids Surf, B 112:408–414. https://doi.org/10.1016/j.colsurfb.2013.06.013
72. Mazur KL, Feuser PE, Valério A, Poester Cordeiro A, de Oliveira CI, Assolini JP, Pavanelli WR, Sayer C, Araújo PHH (2019) Diethyldithiocarbamate loaded in beeswax-copaiba oil nanoparticles obtained by solventless double emulsion technique promote promastigote death in vitro. Colloids Surf B Biointerfaces 176:507–512. https://doi.org/10.1016/j.colsurfb.2018.12.048
73. Gallarate M, Trotta M, Battaglia L, Chirio D (2009) Preparation of solid lipid nanoparticles from W/O/W emulsions: preliminary studies on insulin encapsulation. J Microencapsul 26:394–402. https://doi.org/10.1080/02652040802390156
74. Becker Peres L, Becker Peres L, de Araújo PHH, Sayer C (2016) Solid lipid nanoparticles for encapsulation of hydrophilic drugs by an organic solvent free double emulsion technique. Colloids Surf B Biointerfaces 140:317–323. https://doi.org/10.1016/j.colsurfb.2015.12.033
75. Heurtault B, Saulnier P, Pech B, Proust J-E, Benoit J-P (2002) A novel phase inversion-based process for the preparation of lipid nanocarriers. Pharm Res 19:875–880. https://doi.org/10.1023/A:1016121319668
76. Roger K, Cabane B, Olsson U (2010) Formation of 10–100 nm size-controlled emulsions through a sub-PIT cycle. Langmuir 26:3860–3867. https://doi.org/10.1021/la903401g
77. Montenegro L, Sarpietro MG, Ottimo S, Puglisi G, Castelli F (2011) Differential scanning calorimetry studies on sunscreen loaded solid lipid nanoparticles prepared by the phase inversion temperature method. Int J Pharm 415:301–306. https://doi.org/10.1016/j.ijpharm.2011.05.076
78. Montenegro L, Sinico C, Castangia I, Carbone C, Puglisi G (2012) Idebenone-loaded solid lipid nanoparticles for drug delivery to the skin: in vitro evaluation. Int J Pharm 434:169–174. https://doi.org/10.1016/j.ijpharm.2012.05.046
79. Parhi R, Suresh P (2012) Preparation and characterization of solid lipid nanoparticles-a review. Curr Drug Discov Technol 9:2–16. https://doi.org/10.2174/157016312799304552
80. Carbone C, Tomasello B, Ruozi B, Renis M, Puglisi G (2012) Preparation and optimization of PIT solid lipid nanoparticles via statistical factorial design. Eur J Med Chem 49:110–117. https://doi.org/10.1016/j.ejmech.2012.01.001

81. Izquierdo P, Feng J, Esquena J, Tadros TF, Dederen JC, Garcia MJ, Azemar N, Solans C (2005) The influence of surfactant mixing ratio on nano-emulsion formation by the pit method. J Colloid Interface Sci 285:388–394. https://doi.org/10.1016/j.jcis.2004.10.047
82. Charcosset C, Fessi H (2005) Preparation of nanoparticles with a membrane contactor. J Membr Sci 266:115–120. https://doi.org/10.1016/j.memsci.2005.05.016
83. Chattopadhyay P, Shekunov BY, Yim D, Cipolla D, Boyd B, Farr S (2007) Production of solid lipid nanoparticle suspensions using supercritical fluid extraction of emulsions (SFEE) for pulmonary delivery using the AERx system. Adv Drug Deliv Rev 59:444–453
84. Andrade LN, Oliveira DML, Chaud MV, Alves TFR, Nery M, da Silva CF, Gonsalves JKC, Nunes RS, Corrêa CB, Amaral RG et al (2019) Praziquantel-solid lipid nanoparticles produced by supercritical carbon dioxide extraction: physicochemical characterization, release profile, and cytotoxicity. Molecules 24:3881. https://doi.org/10.3390/molecules24213881
85. Wong CY, Al-Salami H, Dass CR (2020) Lyophilisation improves bioactivity and stability of insulin-loaded polymeric-oligonucleotide nanoparticles for diabetes treatment. AAPS PharmSciTech 21:108. https://doi.org/10.1208/s12249-020-01648-6
86. Chirio D, Gallarate M, Peira E, Battaglia L, Muntoni E, Riganti C, Biasibetti E, Capucchio MT, Valazza A, Panciani P et al (2014) Positive-charged solid lipid nanoparticles as paclitaxel drug delivery system in glioblastoma treatment. Eur J Pharm Biopharm 88:746–758. https://doi.org/10.1016/j.ejpb.2014.10.017
87. Battaglia L, Muntoni E, Chirio D, Peira E, Annovazzi L, Schiffer D, Mellai M, Riganti C, Salaroglio IC, Lanotte M et al (2017) Solid lipid nanoparticles by coacervation loaded with a methotrexate prodrug: preliminary study for glioma treatment. Nanomedicine 12:639–656. https://doi.org/10.2217/nnm-2016-0380
88. Clemente N, Ferrara B, Gigliotti CL, Boggio E, Capucchio MT, Biasibetti E, Schiffer D, Mellai M, Annovazzi L, Cangemi L et al (2018) Solid lipid nanoparticles carrying temozolomide for melanoma treatment, preliminary in vitro and in vivo studies. Int J Mol Sci 19:255. https://doi.org/10.3390/ijms19020255
89. Muntoni E, Marini E, Ahmadi N, Milla P, Ghè C, Bargoni A, Capucchio MT, Biasibetti E, Battaglia L (2019) Lipid nanoparticles as vehicles for oral delivery of insulin and insulin analogs: preliminary ex vivo and in vivo studies. Acta Diabetol 56:1283–1292. https://doi.org/10.1007/s00592-019-01403-9
90. Battaglia L, Gallarate M, Peira E, Chirio D, Solazzi I, Giordano SMA, Gigliotti CL, Riganti C, Dianzani C (2015) Bevacizumab loaded solid lipid nanoparticles prepared by the coacervation technique: preliminary in vitro studies. Nanotechnology 26:255102. https://doi.org/10.1088/0957-4484/26/25/255102
91. Schubert MA, Müller-Goymann CC (2003) Solvent injection as a new approach for manufacturing lipid nanoparticles—evaluation of the method and process parameters. Eur J Pharm Biopharm 55:125–131. https://doi.org/10.1016/S0939-6411(02)00130-3
92. Das MK, Bhattacharya A, Ghosal SK (2006) Effect of penetration enhancers on skin permeation of trazodone hydrochloride from matrix type transdermal formulation through mouse and human cadaver epidermis. Acta Pol Pharm 63:535–541
93. Rao JP, Geckeler KE (2011) Polymer nanoparticles: preparation techniques and size-control parameters. Prog Polym Sci 36:887–913. https://doi.org/10.1016/j.progpolymsci.2011.01.001
94. Krishnamoorthy K, Mahalingam M (2015) Selection of a suitable method for the preparation of polymeric nanoparticles: multi-criteria decision making approach. Adv Pharm Bull 5:57–67. https://doi.org/10.5681/apb.2015.008
95. Lee BK, Yun Y, Park K (2016) PLA micro- and nano-particles. Adv Drug Deliv Rev 107:176–191. https://doi.org/10.1016/j.addr.2016.05.020
96. Procopio A, Lagreca E, Jamaledin R, La Manna S, Corrado B, Di Natale C, Onesto V (2022) Recent fabrication methods to produce polymer-based drug delivery matrices (experimental and in silico approaches). Pharmaceutics 14:872. https://doi.org/10.3390/pharmaceutics14040872

97. Zielińska A, Carreiró F, Oliveira AM, Neves A, Pires B, Venkatesh DN, Durazzo A, Lucarini M, Eder P, Silva AM et al (2020) Polymeric nanoparticles: production, characterization, toxicology and ecotoxicology. Molecules 25:3731. https://doi.org/10.3390/molecules25163731
98. Urbaniak T, Musiał W (2019) Influence of solvent evaporation technique parameters on diameter of submicron lamivudine-Poly-ε-caprolactone conjugate particles. Nanomaterials (Basel) 9:1240. https://doi.org/10.3390/nano9091240
99. Kammari R, Das NG, Das SK (2017) Chapter 6—nanoparticulate systems for therapeutic and diagnostic applications. In: Mitra AK, Cholkar K, Mandal A (eds) Emerging nanotechnologies for diagnostics, drug delivery and medical devices. Micro and Nano Technologies, Elsevier, Boston, pp 105–144. ISBN 978-0-323-42978-8
100. Lombardo D, Kiselev MA (2022) Methods of liposomes preparation: formation and control factors of versatile nanocarriers for biomedical and nanomedicine application. Pharmaceutics 14:543. https://doi.org/10.3390/pharmaceutics14030543
101. Hyde AM, Zultanski SL, Waldman JH, Zhong Y-L, Shevlin M, Peng F (2017) General principles and strategies for salting-out informed by the hofmeister series. Org Process Res Dev 21:1355–1370. https://doi.org/10.1021/acs.oprd.7b00197
102. Pulingam T, Foroozandeh P, Chuah J-A, Sudesh K (2022) Exploring various techniques for the chemical and biological synthesis of polymeric nanoparticles. Nanomaterials (Basel) 12:576. https://doi.org/10.3390/nano12030576
103. Singh S, Gupta R, Chawla S, Gauba P, Singh M, Tiwari RK, Upadhyay S, Sharma S, Chanda S, Gaur S (2022) Natural sources and encapsulating materials for probiotics delivery systems: recent applications and challenges in functional food development. Front Nutr 9:971784. https://doi.org/10.3389/fnut.2022.971784
104. Sun Q, Yin S, He Y, Cao Y, Jiang C (2023) Biomaterials and encapsulation techniques for probiotics: current status and future prospects in biomedical applications. Nanomaterials (Basel) 13:2185. https://doi.org/10.3390/nano13152185
105. Lu X, Liu T (2006) 16—polymer/calcium carbonate nanocomposites. In: Mai YW, Yu ZZ (eds) Polymer nanocomposites. Woodhead Publishing Series in Composites Science and Engineering; Woodhead Publishing, pp 412–439. ISBN 978-1-85573-969-7
106. Bhardwaj H, Jangde RK (2023) Current updated review on preparation of polymeric nanoparticles for drug delivery and biomedical applications. Next Nanotechnol 2:100013. https://doi.org/10.1016/j.nxnano.2023.100013
107. Manjare SD, Dhingra K (2019) Supercritical fluids in separation and purification: a review. Mater Sci Energy Technol 2:463–484. https://doi.org/10.1016/j.mset.2019.04.005
108. Clifford AA, Williams JR (2000) Introduction to supercritical fluids and their applications. In: Williams JR, Clifford AA (eds) Supercritical fluid methods and protocols. Humana Press, Totowa, NJ, pp 1–16. ISBN 978-1-59259-030-8
109. Liu G, Li J, Deng S (2021) Applications of supercritical anti-solvent process in preparation of solid multicomponent systems. Pharmaceutics 13:475. https://doi.org/10.3390/pharmaceutics13040475
110. Franco P, De Marco I (2020) Supercritical antisolvent process for pharmaceutical applications: a review. Processes 8:938. https://doi.org/10.3390/pr8080938
111. Ahmad A, Vaghasiya K, Kumar A, Alam P, Raza SS, Verma RK, Khan R (2021) Enema based therapy using liposomal formulation of low molecular weight heparin for treatment of active ulcerative colitis: new adjunct therapeutic opportunity. Mater Sci Eng, C 121:111851
112. Prakash R, Mishra RK, Ahmad A, Khan MA, Khan R, Raza SS (2021) Sivelestat-loaded nanostructured lipid carriers modulate oxidative and inflammatory stress in human dental pulp and mesenchymal stem cells subjected to oxygen-glucose deprivation. Mater Sci Eng, C 120:111700
113. Mishra RK, Selim A, Gowri V, Ahmad A, Nadeem A, Siddiqui N, Raza SS, Jayamurugan G, Khan R (2022) Thiol-functionalized cellulose-grafted copper oxide nanoparticles for the therapy of experimental colitis in swiss albino mice. ACS Biomater Sci Eng 8:2088–2095. https://doi.org/10.1021/acsbiomaterials.2c00124

114. Dar AH, Ahmad A, Kumar A, Gowri V, Jori C, Sartaliya S, Neethu KM, Khan R, Jayamurugan G (2023) Enhancing photophysical and photosensitizing properties of nanoaggregates of weakly emissive non-planar push-pull chromophore
115. Ahmad A (2021) Customized Polymeric Nanoparticles for Targeting Inflammatory Disorders. PhD Thesis, IISER Mohali
116. Di L, Maiseyeu A (2021) Low-density lipoprotein nanomedicines: mechanisms of targeting, biology, and theranostic potential. Drug Deliv 28:408–421. https://doi.org/10.1080/10717544.2021.1886199
117. Mahmoudian M, Salatin S, Khosroushahi AY (2018) Natural Low- and high-density lipoproteins as mighty bio-nanocarriers for anticancer drug delivery. Cancer Chemother Pharmacol 82:371–382. https://doi.org/10.1007/s00280-018-3626-4
118. Dai L, Li S, Hao Q, Zhou R, Zhou H, Lei W, Kang H, Wu H, Li Y, Ma X (2023) Low-density lipoprotein: a versatile nanoscale platform for targeted delivery. Nanoscale Adv 5:1011–1022. https://doi.org/10.1039/d2na00883a
119. Jain A, Jain K, Kesharwani P, Jain NK (1888) Low density lipoproteins mediated nanoplatforms for cancer targeting. J Nanopart Res 2013:15. https://doi.org/10.1007/s11051-013-1888-7
120. Busatto S, Walker SA, Grayson W, Pham A, Tian M, Nesto N, Barklund J, Wolfram J (2020) Lipoprotein-based drug delivery. Adv Drug Deliv Rev 159:377–390. https://doi.org/10.1016/j.addr.2020.08.003
121. Huang H, Cruz W, Chen J, Zheng G (2015) Learning from biology: synthetic lipoproteins for drug delivery. Wiley Interdiscip Rev Nanomed Nanobiotechnol 7:298–314. https://doi.org/10.1002/wnan.1308
122. Feingold KR (2000) Introduction to lipids and lipoproteins. In: Feingold KR, Anawalt B, Blackman MR, Boyce A, Chrousos G, Corpas E, de Herder WW, Dhatariya K, Dungan K, Hofland J, Kalra S, Kaltsas G, Kapoor N, Koch C, Kopp P, Korbonits M, Kovacs CS, Kuohung W, Laferrère B, Levy M, McGee EA, McLachlan R, New M, Purnell J, Sahay R, Shah AS, Singer F, Sperling MA, Stratakis CA, Trence DL, Wilson DP (eds) Endotext. MDText.com, Inc., South Dartmouth (MA)
123. Mitchell MJ, Billingsley MM, Haley RM, Wechsler ME, Peppas NA, Langer R (2021) Engineering precision nanoparticles for drug delivery. Nat Rev Drug Discov 20:101–124. https://doi.org/10.1038/s41573-020-0090-8
124. Ma X, Song Q, Gao X (2018) Reconstituted high-density lipoproteins: novel biomimetic nanocarriers for drug delivery. Acta Pharm Sin B 8:51–63. https://doi.org/10.1016/j.apsb.2017.11.006
125. Kontush A, Lindahl M, Lhomme M, Calabresi L, Chapman MJ, Davidson WS (2015) Structure of HDL: particle subclasses and molecular components. In: von Eckardstein A, Kardassis D (eds) High density lipoproteins: from biological understanding to clinical exploitation. Springer International Publishing, Cham, pp 3–51. ISBN 978-3-319-09665-0
126. Markovic M, Ben-Shabat S, Aponick A, Zimmermann EM, Dahan A (2020) Lipids and lipid-processing pathways in drug delivery and therapeutics. Int J Mol Sci 21:3248. https://doi.org/10.3390/ijms21093248
127. Reynolds L, Mulik RS, Wen X, Dilip A, Corbin IR (2014) Low-density lipoprotein-mediated delivery of docosahexaenoic acid selectively kills murine liver cancer cells. Nanomedicine 9:2123–2141. https://doi.org/10.2217/nnm.13.187
128. Lundberg B, Suominen L (1984) Preparation of biologically active analogs of serum low density lipoprotein. J Lipid Res 25:550–558. https://doi.org/10.1016/S0022-2275(20)37767-1
129. Matz CE, Jonas A (1982) Micellar complexes of human apolipoprotein A-I with phosphatidylcholines and cholesterol prepared from cholate-lipid dispersions. J Biol Chem 257:4535–4540. https://doi.org/10.1016/S0021-9258(18)34756-2
130. Ng KK, Lovell JF, Zheng G (2011) Lipoprotein-inspired nanoparticles for cancer theranostics. Acc Chem Res 44:1105–1113. https://doi.org/10.1021/ar200017e
131. Ollila OHS, Lamberg A, Lehtivaara M, Koivuniemi A, Vattulainen I (2012) Interfacial tension and surface pressure of high density lipoprotein, low density lipoprotein, and related lipid droplets. Biophys J 103:1236–1244. https://doi.org/10.1016/j.bpj.2012.08.023

132. Bergt C, Fu X, Huq NP, Kao J, Heinecke JW (2004) Lysine residues direct the chlorination of tyrosines in YXXK motifs of apolipoprotein A-I when hypochlorous acid oxidizes high density lipoprotein*. J Biol Chem 279:7856–7866. https://doi.org/10.1074/jbc.M309046200
133. Bricarello DA, Smilowitz JT, Zivkovic AM, German JB, Parikh AN (2011) Reconstituted lipoprotein: a versatile class of biologically-inspired nanostructures. ACS Nano 5:42–57. https://doi.org/10.1021/nn103098m
134. Chaudhary J, Bower J, Corbin IR (2019) Lipoprotein drug delivery vehicles for cancer: rationale and reason. Int J Mol Sci 20:6327. https://doi.org/10.3390/ijms20246327
135. Almer G, Mangge H, Zimmer A, Prassl R (2015) Lipoprotein-related and apolipoprotein-mediated delivery systems for drug targeting and imaging. Curr Med Chem 22:3631–3651. https://doi.org/10.2174/0929867322666150716114625
136. Salatin S, Yari Khosroushahi A (2017) Overviews on the cellular uptake mechanism of polysaccharide colloidal nanoparticles. J Cell Mol Med 21:1668–1686. https://doi.org/10.1111/jcmm.13110
137. Pokorski JK, Steinmetz NF (2011) The art of engineering viral nanoparticles. Mol Pharm 8:29–43. https://doi.org/10.1021/mp100225y
138. Gelderblom HR (1996) Structure and classification of viruses. In: Baron S (ed) Medical microbiology. University of Texas Medical Branch at Galveston, Galveston (TX). ISBN 978-0-9631172-1-2
139. Roos WH, Ivanovska IL, Evilevitch A, Wuite GJL (2007) Viral capsids: mechanical characteristics, genome packaging and delivery mechanisms. Cell Mol Life Sci 64:1484–1497. https://doi.org/10.1007/s00018-007-6451-1
140. Aljabali AA, Hassan SS, Pabari RM, Shahcheraghi SH, Mishra V, Charbe NB, Chellappan DK, Dureja H, Gupta G, Almutary AG et al (2021) The viral capsid as novel nanomaterials for drug delivery. Future Sci OA 7:FSO744. https://doi.org/10.2144/fsoa-2021-0031
141. Steinmetz NF (2010) Viral nanoparticles as platforms for next-generation therapeutics and imaging devices. Nanomedicine 6:634–641. https://doi.org/10.1016/j.nano.2010.04.005
142. Zhang Y, Dong Y, Zhou J, Li X, Wang F (2018) Application of plant viruses as a biotemplate for nanomaterial fabrication. Molecules 23:2311. https://doi.org/10.3390/molecules23092311
143. Xue Q, Swevers L, Taning CNT (2023) Plant and insect virus-like particles: emerging nanoparticles for agricultural pest management. Pest Manag Sci 79:2975–2991. https://doi.org/10.1002/ps.7514
144. Chung YH, Cai H, Steinmetz NF (2020) Viral nanoparticles for drug delivery, imaging, immunotherapy, and theranostic applications. Adv Drug Deliv Rev 156:214–235. https://doi.org/10.1016/j.addr.2020.06.024
145. Kwak M, Minten IJ, Anaya D-M, Musser AJ, Brasch M, Nolte RJM, Müllen K, Cornelissen JJLM, Herrmann A (2010) Virus-like particles templated by DNA micelles: a general method for loading virus nanocarriers. J Am Chem Soc 132:7834–7835. https://doi.org/10.1021/ja101444j
146. Mourdikoudis S, Pallares RM, Thanh NTK (2018) Characterization techniques for nanoparticles: comparison and complementarity upon studying nanoparticle properties. Nanoscale 10:12871–12934. https://doi.org/10.1039/C8NR02278J
147. Vonnemann J, Sieben C, Wolff C, Ludwig K, Böttcher C, Herrmann A, Haag R (2014) Virus inhibition induced by polyvalent nanoparticles of different sizes. Nanoscale 6:2353–2360. https://doi.org/10.1039/C3NR04449A
148. Malik S, Muhammad K, Waheed Y (2023) Nanotechnology: a revolution in modern industry. Molecules 28:661. https://doi.org/10.3390/molecules28020661
149. Rosi NL, Mirkin CA (2005) Nanostructures in biodiagnostics. Chem Rev 105:1547–1562. https://doi.org/10.1021/cr030067f
150. Zielińska A, Karczewski J, Eder P, Kolanowski T, Szalata M, Wielgus K, Szalata M, Kim D, Shin SR, Słomski R et al (2023) Scaffolds for drug delivery and tissue engineering: the role of genetics. J Control Release 359:207–223. https://doi.org/10.1016/j.jconrel.2023.05.042
151. Koudelka KJ, Pitek AS, Manchester M, Steinmetz NF (2015) Virus-based nanoparticles as versatile nanomachines. Ann Rev Virol 2:379–401. https://doi.org/10.1146/annurev-virology-100114-055141

152. Kim KR, Lee AS, Kim SM, Heo HR, Kim CS (2023) Virus-like nanoparticles as a theranostic platform for cancer. Front Bioeng Biotechnol 10:1106767. https://doi.org/10.3389/fbioe.2022.1106767
153. Raja IS, Kim C, Song S-J, Shin YC, Kang MS, Hyon S-H, Oh J-W, Han D-W (2019) Virus-incorporated biomimetic nanocomposites for tissue regeneration. Nanomaterials (Basel) 9:1014. https://doi.org/10.3390/nano9071014
154. Shan W, Wang C, Chen H, Ren L (2023) Rational design of virus-like particles for nanomedicine. Acc Mater Res 4:814–826. https://doi.org/10.1021/accountsmr.3c00050
155. Nooraei S, Bahrulolum H, Hoseini ZS, Katalani C, Hajizade A, Easton AJ, Ahmadian G (2021) Virus-like particles: preparation, immunogenicity and their roles as nanovaccines and drug nanocarriers. J Nanobiotechnol 19:59. https://doi.org/10.1186/s12951-021-00806-7
156. Mikušová V, Mikuš P (2021) Advances in chitosan-based nanoparticles for drug delivery. Int J Mol Sci 22:9652. https://doi.org/10.3390/ijms22179652
157. Saif A, Anjum L, Faisal Z, Akram N, Shah YA, Irfan R, Saeed F, Afzaal M, Asif Shah M (2023) Recent advances in protein-based nanoparticles and their applications in the delivery of bioactive compounds. Int J Food Prop 26:2866–2880. https://doi.org/10.1080/10942912.2023.2261662
158. Hong S, Choi DW, Kim HN, Park CG, Lee W, Park HH (2020) Protein-based nanoparticles as drug delivery systems. Pharmaceutics 12:604. https://doi.org/10.3390/pharmaceutics12070604
159. Zhang X, Li X, Zhao Y, Zheng Q, Wu Q, Yu Y (2022) Nanocarrier system: an emerging strategy for bioactive peptide delivery. Front Nutr 9. https://doi.org/10.3389/fnut.2022.1050647
160. Khan I, Saeed K, Khan I (2019) Nanoparticles: properties, applications and toxicities. Arab J Chem 12:908–931. https://doi.org/10.1016/j.arabjc.2017.05.011
161. Rizvi SAA, Saleh AM (2018) Applications of nanoparticle systems in drug delivery technology. Saudi Pharm J 26:64–70. https://doi.org/10.1016/j.jsps.2017.10.012
162. Malik S, Muhammad K, Waheed Y (2023) Emerging applications of nanotechnology in healthcare and medicine. Molecules 28:6624. https://doi.org/10.3390/molecules28186624
163. Vithani K, Goyanes A, Jannin V, Basit AW, Gaisford S, Boyd BJ (2018) An overview of 3D printing technologies for soft materials and potential opportunities for lipid-based drug delivery systems. Pharm Res 36:4. https://doi.org/10.1007/s11095-018-2531-1
164. Yun YH, Lee BK, Park K (2015) Controlled drug delivery: historical perspective for the next generation. J Control Release 219:2–7. https://doi.org/10.1016/j.jconrel.2015.10.005
165. Boyd BJ, Bergström CAS, Vinarov Z, Kuentz M, Brouwers J, Augustijns P, Brandl M, Bernkop-Schnürch A, Shrestha N, Préat V et al (2019) Successful oral delivery of poorly water-soluble drugs both depends on the intraluminal behavior of drugs and of appropriate advanced drug delivery systems. Eur J Pharm Sci 137:104967. https://doi.org/10.1016/j.ejps.2019.104967
166. Elzoghby AO, Vranic BZ, Samy WM, Elgindy NA (2015) Swellable floating tablet based on spray-dried casein nanoparticles: near-infrared spectral characterization and floating matrix evaluation. Int J Pharm 491:113–122. https://doi.org/10.1016/j.ijpharm.2015.06.015
167. Tiwari G, Tiwari R, Sriwastawa B, Bhati L, Pandey S, Pandey P, Bannerjee SK (2012) Drug delivery systems: an updated review. Int J Pharm Investig 2:2–11. https://doi.org/10.4103/2230-973X.96920
168. Calixto GMF, de Annunzio SR, Victorelli FD, Frade ML, Ferreira PS, Chorilli M, Fontana CR (2019) Chitosan-based drug delivery systems for optimization of photodynamic therapy: a review. AAPS PharmSciTech 20:253. https://doi.org/10.1208/s12249-019-1407-y
169. Castro E, Kumar A (2013) Nanoparticles in drug delivery systems. In: Nanomedicine in drug delivery. CRC Press. ISBN 978-0-429-18532-8
170. Miao Y, Yang T, Yang S, Yang M, Mao C (2022) Protein nanoparticles directed cancer imaging and therapy. Nano Converg 9:2. https://doi.org/10.1186/s40580-021-00293-4
171. Griffiths G, Gruenberg J, Marsh M, Wohlmann J, Jones AT, Parton RG (2022) Nanoparticle entry into cells; the cell biology weak link. Adv Drug Deliv Rev 188:114403. https://doi.org/10.1016/j.addr.2022.114403

172. Kaltbeitzel J, Wich PR (2023) Protein-based nanoparticles: from drug delivery to imaging, nanocatalysis and protein therapy. Angew Chem Int Ed 62:e202216097. https://doi.org/10.1002/anie.202216097
173. The innate and adaptive immune systems. In: InformedHealth.org [Internet]; Institute for Quality and Efficiency in Health Care (IQWiG) (2020)
174. Li D, Wu M (2021) Pattern recognition receptors in health and diseases. Sig Transduct Target Ther 6:1–24. https://doi.org/10.1038/s41392-021-00687-0
175. Alberts B, Johnson A, Lewis J, Raff M, Roberts K, Walter P (2002) Innate immunity. In: *M*olecular biology of the cell, 4th edn. Garland Science
176. Naran K, Nundalall T, Chetty S, Barth S (2018) Principles of immunotherapy: implications for treatment strategies in cancer and infectious diseases. Front Microbiol 9:3158. https://doi.org/10.3389/fmicb.2018.03158
177. Koury J, Lucero M, Cato C, Chang L, Geiger J, Henry D, Hernandez J, Hung F, Kaur P, Teskey G et al (2018) Immunotherapies: exploiting the immune system for cancer treatment. J Immunol Res 2018:9585614. https://doi.org/10.1155/2018/9585614
178. Bezbaruah R, Chavda VP, Nongrang L, Alom S, Deka K, Kalita T, Ali F, Bhattacharjee B, Vora L (1946) Nanoparticle-based delivery systems for vaccines. Vaccines (Basel) 2022:10. https://doi.org/10.3390/vaccines10111946
179. Zhou L, Zou M, Xu Y, Lin P, Lei C, Xia X (2022) Nano drug delivery system for tumor immunotherapy: next-generation therapeutics. Front Oncol 12. https://doi.org/10.3389/fonc.2022.864301
180. Yao Y, Zhou Y, Liu L, Xu Y, Chen Q, Wang Y, Wu S, Deng Y, Zhang J, Shao A (2020) Nanoparticle-based drug delivery in cancer therapy and its role in overcoming drug resistance. Front Mol Biosci 7:193. https://doi.org/10.3389/fmolb.2020.00193
181. Baranwal J, Barse B, Di Petrillo A, Gatto G, Pilia L, Kumar A (2023) Nanoparticles in cancer diagnosis and treatment. Materials (Basel) 16:5354. https://doi.org/10.3390/ma16155354
182. Wagner J, Gößl D, Ustyanovska N, Xiong M, Hauser D, Zhuzhgova O, Hočevar S, Taskoparan B, Poller L, Datz S et al (2021) Mesoporous silica nanoparticles as pH-responsive carrier for the immune-activating drug resiquimod enhance the local immune response in mice. ACS Nano 15:4450–4466. https://doi.org/10.1021/acsnano.0c08384
183. Byrne EH, Fisher DE (2017) Immune and molecular correlates in melanoma treated with immune checkpoint blockade. Cancer 123:2143–2153. https://doi.org/10.1002/cncr.30444
184. Lohcharoenkal W, Wang L, Chen YC, Rojanasakul Y (2014) Protein nanoparticles as drug delivery carriers for cancer therapy. BioMed Res Int 2014. https://doi.org/10.1155/2014/180549
185. Jain A, Singh SK, Arya SK, Kundu SC, Kapoor S (2018) Protein nanoparticles: promising platforms for drug delivery applications. ACS Biomater Sci Eng 4:3939–3961. https://doi.org/10.1021/acsbiomaterials.8b01098
186. Cho H, Jeon SI, Ahn CH, Shim MK, Kim K (2022) Emerging albumin-binding anti-cancer drugs for tumor-targeted drug delivery: current understandings and clinical translation. Pharmaceutics 14. https://doi.org/10.3390/pharmaceutics14040728
187. Hoogenboezem EN, Duvall CL (2018) Harnessing albumin as a carrier for cancer therapies. Adv Drug Deliv Rev 130:73–89. https://doi.org/10.1016/j.addr.2018.07.011
188. Goyal R, Chopra H, Singh I, Dua K, Gautam RK (2022) Insights on prospects of nano-siRNA based approaches in treatment of cancer. Front Pharmacol 13. https://doi.org/10.3389/fphar.2022.985670
189. Huang J, Xiao K (2022) Nanoparticles-based strategies to improve the delivery of therapeutic small interfering RNA in precision oncology. Pharmaceutics 14:1586. https://doi.org/10.3390/pharmaceutics14081586
190. Mikelez-Alonso I, Aires A, Cortajarena AL (2020) Cancer nano-immunotherapy from the injection to the target: the role of protein corona. Int J Mol Sci 21:519. https://doi.org/10.3390/ijms21020519
191. Kumari N, Choi SH (2022) Tumor-associated macrophages in cancer: recent advancements in cancer nanoimmunotherapies. J Exp Clin Cancer Res 41:68. https://doi.org/10.1186/s13046-022-02272-x

192. Mendes BB, Conniot J, Avital A, Yao D, Jiang X, Zhou X, Sharf-Pauker N, Xiao Y, Adir O, Liang H et al (2022) Nanodelivery of nucleic acids. Nat Rev Methods Primers 2:1–21. https://doi.org/10.1038/s43586-022-00104-y
193. Sahu KK, Pradhan M, Singh D, Singh MR, Yadav K (2023) Non-viral nucleic acid delivery approach: a boon for state-of-the-art gene delivery. J Drug Deliv Sci Technol 80:104152. https://doi.org/10.1016/j.jddst.2023.104152
194. Ma W, Zhan Y, Zhang Y, Mao C, Xie X, Lin Y (2021) The biological applications of DNA nanomaterials: current challenges and future directions. Sig Transduct Target Ther 6:1–28. https://doi.org/10.1038/s41392-021-00727-9
195. Mehta M, Bui TA, Yang X, Aksoy Y, Goldys EM, Deng W (2023) Lipid-Based nanoparticles for drug/gene delivery: an overview of the production techniques and difficulties encountered in their industrial development. ACS Mater Au 3:600–619. https://doi.org/10.1021/acsmaterialsau.3c00032
196. De R, Mahata MK, Kim K (2022) Structure-based varieties of polymeric nanocarriers and influences of their physicochemical properties on drug delivery profiles. Adv Sci (Weinh) 9:2105373. https://doi.org/10.1002/advs.202105373
197. Wickline SA, Hou KK, Pan H (2023) Peptide-based nanoparticles for systemic extrahepatic delivery of therapeutic nucleotides. Int J Mol Sci 24:9455. https://doi.org/10.3390/ijms24119455
198. Khairkhah N, Namvar A, Bolhassani A (2023) Application of cell penetrating peptides as a promising drug carrier to combat viral infections. Mol Biotechnol:1–16. https://doi.org/10.1007/s12033-023-00679-1
199. Zhou M, Zou X, Cheng K, Zhong S, Su Y, Wu T, Tao Y, Cong L, Yan B, Jiang Y (2022) The role of cell-penetrating peptides in potential anti-cancer therapy. Clin Transl Med 12:e822. https://doi.org/10.1002/ctm2.822
200. Xie X, Yue T, Gu W, Cheng W, He L, Ren W, Li F, Piao J-G (2023) Recent advances in mesoporous silica nanoparticles delivering siRNA for cancer treatment. Pharmaceutics 15:2483. https://doi.org/10.3390/pharmaceutics15102483
201. Jain D, Prajapati SK, Jain A, Singhal R (2023) Nano-formulated siRNA-based therapeutic approaches for cancer therapy. Nano Trends 1:100006. https://doi.org/10.1016/j.nwnano.2023.100006
202. Kamimura K, Suda T, Zhang G, Liu D (2011) Advances in gene delivery systems. Pharm Med 25:293–306. https://doi.org/10.2165/11594020-000000000-00000
203. Lee CS, Bishop ES, Zhang R, Yu X, Farina EM, Yan S, Zhao C, Zeng Z, Shu Y, Wu X et al (2017) Adenovirus-mediated gene delivery: potential applications for gene and cell-based therapies in the new era of personalized medicine. Genes Dis 4:43–63. https://doi.org/10.1016/j.gendis.2017.04.001
204. Buschmann MD, Carrasco MJ, Alishetty S, Paige M, Alameh MG, Weissman D (2021) Nanomaterial delivery systems for mRNA vaccines. Vaccines 9:65. https://doi.org/10.3390/vaccines9010065
205. Kim J, Eygeris Y, Gupta M, Sahay G (2021) Self-assembled mRNA vaccines. Adv Drug Deliv Rev 170:83–112. https://doi.org/10.1016/j.addr.2020.12.014
206. Wilson B, Geetha KM (2022) Lipid nanoparticles in the development of mRNA vaccines for covid-19. J Drug Deliv Sci Technol 74:103553. https://doi.org/10.1016/j.jddst.2022.103553
207. Chatterjee S, Kon E, Sharma P, Peer D (2024) Endosomal escape: a bottleneck for LNP-mediated therapeutics. Proc Natl Acad Sci USA 121:e2307800120. https://doi.org/10.1073/pnas.2307800120
208. Golubovic A, Tsai S, Li B (2023) Bioinspired lipid nanocarriers for RNA delivery. ACS Bio Med Chem Au 3:114–136. https://doi.org/10.1021/acsbiomedchemau.2c00073
209. Mukalel AJ, Riley RS, Zhang R, Mitchell MJ (2019) Nanoparticles for nucleic acid delivery: applications in cancer immunotherapy. Cancer Lett 458:102–112. https://doi.org/10.1016/j.canlet.2019.04.040
210. Luther DC, Huang R, Jeon T, Zhang X, Lee Y-W, Nagaraj H, Rotello VM (2020) Delivery of drugs, proteins, and nucleic acids using inorganic nanoparticles. Adv Drug Deliv Rev 156:188–213. https://doi.org/10.1016/j.addr.2020.06.020

211. Unnikrishnan G, Joy A, Megha M, Kolanthai E, Senthilkumar M (2023) Exploration of inorganic nanoparticles for revolutionary drug delivery applications: a critical review. Discover Nano 18:157. https://doi.org/10.1186/s11671-023-03943-0
212. Pandey RP, Vidic J, Mukherjee R, Chang C-M (2023) Experimental methods for the biological evaluation of nanoparticle-based drug delivery risks. Pharmaceutics 15:612. https://doi.org/10.3390/pharmaceutics15020612
213. Witika BA, Makoni PA, Matafwali SK, Chabalenge B, Mwila C, Kalungia AC, Nkanga CI, Bapolisi AM, Walker RB (2020) Biocompatibility of biomaterials for nanoencapsulation: current approaches. Nanomaterials (Basel) 10:1649. https://doi.org/10.3390/nano10091649
214. Ahmed EM (2015) Hydrogel: preparation, characterization, and applications: a review. J Adv Res 6:105–121. https://doi.org/10.1016/j.jare.2013.07.006
215. Caló E, Khutoryanskiy VV (2015) Biomedical applications of hydrogels: a review of patents and commercial products. Eur Polymer J 65:252–267. https://doi.org/10.1016/j.eurpolymj.2014.11.024
216. Cao H, Duan L, Zhang Y, Cao J, Zhang K (2021) Current hydrogel advances in physicochemical and biological response-driven biomedical application diversity. Sig Transduct Target Ther 6:1–31. https://doi.org/10.1038/s41392-021-00830-x
217. Sánchez-Cid P, Jiménez-Rosado M, Romero A, Pérez-Puyana V (2022) Novel trends in hydrogel development for biomedical applications: a review. Polymers (Basel) 14:3023. https://doi.org/10.3390/polym14153023
218. Ali F, Khan I, Chen J, Akhtar K, Bakhsh EM, Khan SB (2022) Emerging fabrication strategies of hydrogels and its applications. Gels 8:205. https://doi.org/10.3390/gels8040205
219. Mantha S, Pillai S, Khayambashi P, Upadhyay A, Zhang Y, Tao O, Pham HM, Tran SD (2019) Smart hydrogels in tissue engineering and regenerative medicine. Materials (Basel) 12:3323. https://doi.org/10.3390/ma12203323
220. Maiti S, Maji B, Yadav H (2024) Progress on green crosslinking of polysaccharide hydrogels for drug delivery and tissue engineering applications. Carbohyd Polym 326:121584. https://doi.org/10.1016/j.carbpol.2023.121584
221. Brovold M, Almeida JI, Pla-Palacín I, Sainz-Arnal P, Sánchez-Romero N, Rivas JJ, Almeida H, Royo Dachary P, Serrano-Aulló T, Soker S et al (2018) Naturally-derived biomaterials for tissue engineering applications. Adv Exp Med Biol 1077:421–449. https://doi.org/10.1007/978-981-13-0947-2_23
222. Liu S, Yu J-M, Gan Y-C, Qiu X-Z, Gao Z-C, Wang H, Chen S-X, Xiong Y, Liu G-H, Lin S-E et al (2023) Biomimetic natural biomaterials for tissue engineering and regenerative medicine: new biosynthesis methods, recent advances, and emerging applications. Mil Med Res 10:16. https://doi.org/10.1186/s40779-023-00448-w
223. Elieh-Ali-Komi D, Hamblin MR (2016) Chitin and chitosan: production and application of versatile biomedical nanomaterials. Int J Adv Res (Indore) 4:411–427
224. Casadidio C, Peregrina DV, Gigliobianco MR, Deng S, Censi R, Di Martino P (2019) Chitin and chitosans: characteristics, eco-friendly processes, and applications in cosmetic science. Mar Drugs 17:369. https://doi.org/10.3390/md17060369
225. Desai N, Rana D, Salave S, Gupta R, Patel P, Karunakaran B, Sharma A, Giri J, Benival D, Kommineni N (2023) Chitosan: a potential biopolymer in drug delivery and biomedical applications. Pharmaceutics 15:1313
226. Kołodziejska M, Jankowska K, Klak M, Wszoła M (2021) Chitosan as an underrated polymer in modern tissue engineering. Nanomaterials (Basel) 11:3019. https://doi.org/10.3390/nano11113019
227. Lu H, Li X, Zhang M, Xu C, Li W, Wan L (2022) Antibacterial cellulose nanocrystal-incorporated hydrogels with satisfactory vascularization for enhancing skin regeneration. Front Bioeng Biotechnol 10:876936. https://doi.org/10.3389/fbioe.2022.876936
228. Larrañeta E, Henry M, Irwin NJ, Trotter J, Perminova AA, Donnelly RF (2018) Synthesis and characterization of hyaluronic acid hydrogels crosslinked using a solvent-free process for potential biomedical applications. Carbohydr Polym 181:1194–1205. https://doi.org/10.1016/j.carbpol.2017.12.015

229. Xu Q, Torres JE, Hakim M, Babiak PM, Pal P, Battistoni CM, Nguyen M, Panitch A, Solorio L, Liu JC (2021) Collagen- and hyaluronic acid-based hydrogels and their biomedical applications. Mater Sci Eng R Rep 146:100641. https://doi.org/10.1016/j.mser.2021.100641
230. Jacob S, Nair AB, Shah J, Sreeharsha N, Gupta S, Shinu P (2021) Emerging role of hydrogels in drug delivery systems, tissue engineering and wound management. Pharmaceutics 13:357. https://doi.org/10.3390/pharmaceutics13030357
231. Kang H, Zuo Z, Lin R, Yao M, Han Y, Han J (2022) The most promising microneedle device: present and future of hyaluronic acid microneedle patch. Drug Deliv 29:3087–3110. https://doi.org/10.1080/10717544.2022.2125600
232. Yu H, Huang C, Kong X, Ma J, Ren P, Chen J, Zhang X, Luo H, Chen G (2022) Nanoarchitectonics of cartilage-targeting hydrogel microspheres with reactive oxygen species responsiveness for the repair of osteoarthritis. ACS Appl Mater Interfaces 14:40711–40723. https://doi.org/10.1021/acsami.2c12703
233. Aldrich JL, Panicker A, Ovalle R, Sharma B (2023) Drug delivery strategies and nanozyme technologies to overcome limitations for targeting oxidative stress in osteoarthritis. Pharmaceuticals (Basel) 16:1044. https://doi.org/10.3390/ph16071044
234. Lee KY, Mooney DJ (2012) Alginate: properties and biomedical applications. Prog Polym Sci 37:106–126. https://doi.org/10.1016/j.progpolymsci.2011.06.003
235. Abka-khajouei R, Tounsi L, Shahabi N, Patel AK, Abdelkafi S, Michaud P (2022) Structures, properties and applications of alginates. Mar Drugs 20:364. https://doi.org/10.3390/md20060364
236. Hu Y, Hu S, Zhang S, Dong S, Hu J, Kang L, Yang X (2021) A double-layer hydrogel based on alginate-carboxymethyl cellulose and synthetic polymer as sustained drug delivery system. Sci Rep 11:9142. https://doi.org/10.1038/s41598-021-88503-1
237. Bansal D, Gulbake A, Tiwari J, Jain SK (2016) Development of liposomes entrapped in alginate beads for the treatment of colorectal cancer. Int J Biol Macromol 82:687–695. https://doi.org/10.1016/j.ijbiomac.2015.09.052
238. Synytsya A, Novak M (2014) Structural analysis of glucans. Ann Transl Med 2:17. https://doi.org/10.3978/j.issn.2305-5839.2014.02.07
239. Dahiya D, Nigam PS (2023) Dextran of diverse molecular-configurations used as a blood-plasma substitute, drug-delivery vehicle and food additive biosynthesized by leuconostoc, lactobacillus and weissella. Appl Sci 13:12526. https://doi.org/10.3390/app132212526
240. Hu Q, Lu Y, Luo Y (2021) Recent advances in dextran-based drug delivery systems: from fabrication strategies to applications. Carbohydr Polym 264:117999. https://doi.org/10.1016/j.carbpol.2021.117999
241. Glasser WG (2008) Cellulose cellulose and associated heteropolysaccharides heteropolysaccharides. In: Fraser-Reid BO, Tatsuta K, Thiem J (eds) Glycoscience: chemistry and chemical biology. Springer, Berlin, Heidelberg, pp 1473–1512. ISBN 978-3-540-30429-6
242. Klemm D, Heublein B, Fink H-P, Bohn A (2005) Cellulose: fascinating biopolymer and sustainable raw material. Angew Chem Int Ed 44:3358–3393. https://doi.org/10.1002/anie.200460587
243. George J, Sabapathi S (2015) Cellulose nanocrystals: synthesis, functional properties, and applications. Nanotechnol Sci Appl 8:45–54. https://doi.org/10.2147/NSA.S64386
244. Santoro M, Tatara AM, Mikos AG (2014) Gelatin carriers for drug and cell delivery in tissue engineering. J Control Release:210–218. https://doi.org/10.1016/j.jconrel.2014.04.014
245. Yue K, Santiago GT, Alvarez MM, Tamayol A, Annabi N, Khademhosseini A (2015) Synthesis, properties, and biomedical applications of gelatin methacryloyl (GelMA) hydrogels. Biomaterials 73:254–271. https://doi.org/10.1016/j.biomaterials.2015.08.045
246. Fang W, Yang M, Wang L, Li W, Liu M, Jin Y, Wang Y, Yang R, Wang Y, Zhang K et al (2023) Hydrogels for 3D bioprinting in tissue engineering and regenerative medicine: current progress and challenges. Int J Bioprint 9:759. https://doi.org/10.18063/ijb.759
247. Chow D, Nunalee ML, Lim DW, Simnick AJ, Chilkoti A (2008) Peptide-based biopolymers in biomedicine and biotechnology. Mater Sci Eng R Rep 62:125–155. https://doi.org/10.1016/j.mser.2008.04.004

248. Yang S, Wang M, Wang T, Sun M, Huang H, Shi X, Duan S, Wu Y, Zhu J, Liu F (2023) Self-assembled short peptides: recent advances and strategies for potential pharmaceutical applications. Mater Today Bio 20:100644. https://doi.org/10.1016/j.mtbio.2023.100644
249. Binaymotlagh R, Chronopoulou L, Palocci C (2023) Peptide-based hydrogels: template materials for tissue engineering. J Funct Biomater 14:233. https://doi.org/10.3390/jfb14040233
250. Zhang Y, Feng Z, Liu J, Li H, Su Q, Zhang J, Huang P, Wang W, Liu J (2022) Polarization of tumor-associated macrophages by TLR7/8 conjugated radiosensitive peptide hydrogel for overcoming tumor radioresistance. Bioact Mater 16:359–371. https://doi.org/10.1016/j.bioactmat.2021.12.033
251. Zhao Z, Wang D, Li Y (2021) Versatile biomimetic nanomedicine for treating cancer and inflammation disease. Med Rev 3:123–151. https://doi.org/10.1515/mr-2022-0046
252. Gallo J, Villasante A (2023) Recent advances in biomimetic nanocarrier-based photothermal therapy for cancer treatment. Int J Mol Sci 24:15484. https://doi.org/10.3390/ijms242015484
253. Sushnitha M, Evangelopoulos M, Tasciotti E, Taraballi F (2020) Cell membrane-based biomimetic nanoparticles and the immune system: immunomodulatory interactions to therapeutic applications. Front Bioeng Biotechnol 8:627. https://doi.org/10.3389/fbioe.2020.00627
254. Magyari L, Varszegi D, Kovesdi E, Sarlos P, Farago B, Javorhazy A, Sumegi K, Banfai Z, Melegh B (2014) Interleukins and interleukin receptors in rheumatoid arthritis: research, diagnostics and clinical implications. World J Orthop 5:516–536. https://doi.org/10.5312/wjo.v5.i4.516
255. Wang Q, Qin X, Fang J, Sun X (2021) Nanomedicines for the treatment of rheumatoid arthritis: state of art and potential therapeutic strategies. Acta Pharm Sin B 11:1158–1174. https://doi.org/10.1016/j.apsb.2021.03.013
256. Boada C, Zinger A, Tsao C, Zhao P, Martinez JO, Hartman K, Naoi T, Sukhoveshin R, Sushnitha M, Molinaro R et al (2020) Rapamycin-loaded biomimetic nanoparticles reverse vascular inflammation. Circ Res 126:25–37. https://doi.org/10.1161/CIRCRESAHA.119.315185
257. Liu Q, Zou J, Chen Z, He W, Wu W (2023) Current research trends of nanomedicines. Acta Pharm Sinica B 13:4391–4416. https://doi.org/10.1016/j.apsb.2023.05.018
258. Gao C, Liu Y, Zhang T-L, Luo Y, Gao J, Chu J-J, Gong B-F, Chen X-H, Yin T, Zhang J et al (2023) Biomembrane-derived nanoparticles in alzheimer's disease therapy: a comprehensive review of synthetic lipid nanoparticles and natural cell-derived vesicles. Int J Nanomed 18:7441–7468. https://doi.org/10.2147/IJN.S436774
259. Jin K, Luo Z, Zhang B, Pang Z (2018) Biomimetic nanoparticles for inflammation targeting. Acta Pharm Sin B 8:23–33. https://doi.org/10.1016/j.apsb.2017.12.002
260. Martinez JO, Molinaro R, Hartman KA, Boada C, Sukhovershin R, De Rosa E, Kirui D, Zhang S, Evangelopoulos M, Carter AM et al (2018) Biomimetic nanoparticles with enhanced affinity towards activated endothelium as versatile tools for theranostic drug delivery. Theranostics 8:1131–1145. https://doi.org/10.7150/thno.22078
261. Nedelcu A, Mosteanu O, Pop T, Mocan T, Mocan L (2021) Recent advances in nanoparticle-mediated treatment of inflammatory bowel diseases. Appl Sci 11:438. https://doi.org/10.3390/app11010438
262. Shi C, Dawulieti J, Shi F, Yang C, Qin Q, Shi T, Wang L, Hu H, Sun M, Ren L et al (2022) A nanoparticulate dual scavenger for targeted therapy of inflammatory bowel disease. Sci Adv 8:eabj2372. https://doi.org/10.1126/sciadv.abj2372
263. Das M, Mondal S, Ghosh R, Biswas P, Moussa Z, Darbar S, Ahmed SA, Das AK, Bhattacharya SS, Pal D et al (2022) A nano erythropoiesis stimulating agent for the treatment of anemia and associated disorders. iScience 25:105021. https://doi.org/10.1016/j.isci.2022.105021
264. Pan DC, Myerson JW, Brenner JS, Patel PN, Anselmo AC, Mitragotri S, Muzykantov V (2018) Nanoparticle properties modulate their attachment and effect on carrier red blood cells. Sci Rep 8:1615. https://doi.org/10.1038/s41598-018-19897-8
265. Zhang Q, Dehaini D, Zhang Y, Zhou J, Chen X, Zhang L, Fang RH, Gao W, Zhang L (2018) Neutrophil membrane-coated nanoparticles inhibit synovial inflammation and alleviate joint

damage in inflammatory arthritis. Nat Nanotechnol 13:1182–1190. https://doi.org/10.1038/s41565-018-0254-4
266. Zhang L, Qin Z, Sun H, Chen X, Dong J, Shen S, Zheng L, Gu N, Jiang Q (2022) Nanoenzyme engineered neutrophil-derived exosomes attenuate joint injury in advanced rheumatoid arthritis via regulating inflammatory environment. Bioact Mater 18:1–14. https://doi.org/10.1016/j.bioactmat.2022.02.017
267. Afzal O, Altamimi ASA, Nadeem MS, Alzarea SI, Almalki WH, Tariq A, Mubeen B, Murtaza BN, Iftikhar S, Riaz N et al (2022) Nanoparticles in drug delivery: from history to therapeutic applications. Nanomaterials (Basel) 12:4494. https://doi.org/10.3390/nano12244494
268. Date AA, Hanes J, Ensign LM (2016) Nanoparticles for oral delivery: design, evaluation and state-of-the-art. J Control Release 240:504–526. https://doi.org/10.1016/j.jconrel.2016.06.016
269. Liu J, Liu Z, Pang Y, Zhou H (2022) The interaction between nanoparticles and immune system: application in the treatment of inflammatory diseases. J Nanobiotechnol 20:127. https://doi.org/10.1186/s12951-022-01343-7
270. Ahmad F, Salem-Bekhit MM, Khan F, Alshehri S, Khan A, Ghoneim MM, Wu H-F, Taha EI, Elbagory I (2022) Unique properties of surface-functionalized nanoparticles for bio-application: functionalization mechanisms and importance in application. Nanomaterials (Basel) 12:1333. https://doi.org/10.3390/nano12081333
271. Gojova A, Lee J-T, Jung HS, Guo B, Barakat AI, Kennedy IM (2009) Effect of cerium oxide nanoparticles on inflammation in vascular endothelial cells. Inhalation Toxicol 21:123–130. https://doi.org/10.1080/08958370902942582
272. Kennedy IM, Wilson D, Barakat AI, HEI Health Review Committee (2009) Uptake and inflammatory effects of nanoparticles in a human vascular endothelial cell line. Res Rep Health Eff Inst:3–32
273. Zhang D, Liu S, Guan J, Mou F (2022) "Motile-targeting" drug delivery platforms based on micro/nanorobots for tumor therapy. Front Bioeng Biotechnol 10:1002171. https://doi.org/10.3389/fbioe.2022.1002171
274. Zhang Q, Yan S, Yan X, Lv Y (2023) Recent advances in metal-organic frameworks: synthesis, application and toxicity. Sci Total Environ 902:165944. https://doi.org/10.1016/j.scitotenv.2023.165944
275. Andresen TL, Larsen JB (2020) Compositional inhomogeneity of drug delivery liposomes quantified at the single liposome level. Acta Biomater 118:207–214. https://doi.org/10.1016/j.actbio.2020.10.003
276. Chenthamara D, Subramaniam S, Ramakrishnan SG, Krishnaswamy S, Essa MM, Lin F-H, Qoronfleh MW (2019) Therapeutic efficacy of nanoparticles and routes of administration. Biomater Res 23:20. https://doi.org/10.1186/s40824-019-0166-x
277. Jia Z, Li J, Gao L, Yang D, Kanaev A (2023) Dynamic light scattering: a powerful tool for in situ nanoparticle sizing. Colloids Interfaces 7:15. https://doi.org/10.3390/colloids7010015
278. Maguire CM, Rösslein M, Wick P, Prina-Mello A (2018) Characterisation of particles in solution—a perspective on light scattering and comparative technologies. Sci Technol Adv Mater 19:732–745. https://doi.org/10.1080/14686996.2018.1517587
279. Chapter 5—hydrodynamics. In: Dhont JKG (ed) Studies in interface science: an introduction to dynamics of colloids, vol 2. Elsevier, pp 227–313
280. Jiang D, Rosenkrans ZT, Ni D, Lin J, Huang P, Cai W (2020) Nanomedicines for renal management: from imaging to treatment. Acc Chem Res 53:1869–1880. https://doi.org/10.1021/acs.accounts.0c00323
281. Sousa AA, Schuck P, Hassan SA (2021) Biomolecular interactions of ultrasmall metallic nanoparticles and nanoclusters. Nanoscale Adv 3:2995–3027. https://doi.org/10.1039/d1na00086a
282. Villaraza AJL, Bumb A, Brechbiel MW (2010) Macromolecules, dendrimers and nanomaterials in magnetic resonance imaging: the interplay between size, function and pharmacokinetics. Chem Rev 110:2921–2959. https://doi.org/10.1021/cr900232t

283. De Jong WH, Borm PJ (2008) Drug delivery and nanoparticles: applications and hazards. Int J Nanomed 3:133–149
284. Braakhuis HM, Park MV, Gosens I, De Jong WH, Cassee FR (2014) Physicochemical characteristics of nanomaterials that affect pulmonary inflammation. Part Fibre Toxicol 11:18. https://doi.org/10.1186/1743-8977-11-18
285. Vasilatou K, Wälchli C, Iida K, Horender S, Tritscher T, Hammer T, Rissler J, Gaie-Levrel F, Auderset K (2022) Extending traceability in airborne particle size distribution measurements beyond 10 μm: counting efficiency and unit-to-unit variability of four aerodynamic particle size spectrometers. Aerosol Sci Technol 57:24–34. https://doi.org/10.1080/02786826.2022.2139659
286. Reid JS, Jonsson HH, Maring HB, Smirnov A, Savoie DL, Cliff SS, Reid EA, Livingston JM, Meier MM, Dubovik O et al (2003) Comparison of size and morphological measurements of coarse mode dust particles from Africa. J Geophys Res Atmos 108. https://doi.org/10.1029/2002JD002485
287. Braakhuis HM, Gosens I, Krystek P, Boere JA, Cassee FR, Fokkens PH, Post JA, van Loveren H, Park MV (2014) Particle size dependent deposition and pulmonary inflammation after short-term inhalation of silver nanoparticles. Part Fibre Toxicol 11:49. https://doi.org/10.1186/s12989-014-0049-1
288. Löndahl J, Möller W, Pagels JH, Kreyling WG, Swietlicki E, Schmid O (2014) Measurement techniques for respiratory tract deposition of airborne nanoparticles: a critical review. J Aerosol Med Pulm Drug Deliv 27:229–254. https://doi.org/10.1089/jamp.2013.1044
289. Kato H, Suzuki M, Fujita K, Horie M, Endoh S, Yoshida Y, Iwahashi H, Takahashi K, Nakamura A, Kinugasa S (2009) Reliable size determination of nanoparticles using dynamic light scattering method for in vitro toxicology assessment. Toxicol In Vitro 23:927–934. https://doi.org/10.1016/j.tiv.2009.04.006
290. Takahashi K, Kato H, Saito T, Matsuyama S, Kinugasa S (2008) Precise measurement of the size of nanoparticles by dynamic light scattering with uncertainty analysis. Part Part Syst Charact 25:31–38. https://doi.org/10.1002/ppsc.200700015
291. Ivleva NP (2021) Chemical analysis of microplastics and nanoplastics: challenges, advanced methods, and perspectives. Chem Rev 121:11886–11936. https://doi.org/10.1021/acs.chemrev.1c00178
292. Ulusoy U (2023) A review of particle shape effects on material properties for various engineering applications: from macro to nanoscale. Minerals 13:91. https://doi.org/10.3390/min13010091
293. Dhiman N, Awasthi R, Sharma B, Kharkwal H, Kulkarni GT (2021) Lipid nanoparticles as carriers for bioactive delivery. Front Chem 9:580118. https://doi.org/10.3389/fchem.2021.580118
294. Danaei M, Dehghankhold M, Ataei S, Hasanzadeh Davarani F, Javanmard R, Dokhani A, Khorasani S, Mozafari MR (2018) Impact of particle size and polydispersity index on the clinical applications of lipidic nanocarrier systems. Pharmaceutics 10:57. https://doi.org/10.3390/pharmaceutics10020057
295. Gaikwad VL, Choudhari PB, Bhatia NM, Bhatia MS (2019) Chapter 2—characterization of pharmaceutical nanocarriers: in vitro and in vivo studies. In: Grumezescu AM (ed) Nanomaterials for drug delivery and therapy. William Andrew Publishing, pp 33–58. ISBN 978-0-12-816505-8
296. Lee BK, Yun YH, Park K (2015) Smart nanoparticles for drug delivery: boundaries and opportunities. Chem Eng Sci 125:158–164. https://doi.org/10.1016/j.ces.2014.06.042
297. Li Y, Song Y, Ma A, Li C (2019) Surface immobilization of TiO_2 nanotubes with bone morphogenetic protein-2 synergistically enhances initial preosteoblast adhesion and osseointegration. Biomed Res Int 2019:e5697250. https://doi.org/10.1155/2019/5697250
298. Li J, Tan L, Liu X, Cui Z, Yang X, Yeung KWK, Chu PK, Wu S (2017) Balancing bacteria-osteoblast competition through selective physical puncture and biofunctionalization of ZnO/polydopamine/arginine-glycine-aspartic acid-cysteine nanorods. ACS Nano 11:11250–11263. https://doi.org/10.1021/acsnano.7b05620

299. Wang Z, Wang K, Lu X, Li C, Han L, Xie C, Liu Y, Qu S, Zhen G (2015) Nanostructured architectures by assembling polysaccharide-coated BSA nanoparticles for biomedical application. Adv Healthc Mater 4:927–937. https://doi.org/10.1002/adhm.201400684
300. Kumeria T, Mon H, Aw MS, Gulati K, Santos A, Griesser HJ, Losic D (2015) Advanced biopolymer-coated drug-releasing titania nanotubes (TNTs) implants with simultaneously enhanced osteoblast adhesion and antibacterial properties. Colloids Surf, B 130:255–263. https://doi.org/10.1016/j.colsurfb.2015.04.021
301. Xie C-M, Lu X, Wang K-F, Meng F-Z, Jiang O, Zhang H-P, Zhi W, Fang L-M (2014) Silver nanoparticles and growth factors incorporated hydroxyapatite coatings on metallic implant surfaces for enhancement of osteoinductivity and antibacterial properties. ACS Appl Mater Interfaces 6:8580–8589. https://doi.org/10.1021/am501428e
302. Sanità G, Carrese B, Lamberti A (2020) Nanoparticle surface functionalization: how to improve biocompatibility and cellular internalization. Front Mol Biosci 7:587012. https://doi.org/10.3389/fmolb.2020.587012
303. Verma J, Warsame C, Seenivasagam RK, Katiyar NK, Aleem E, Goel S (2023) Nanoparticle-mediated cancer cell therapy: basic science to clinical applications. Cancer Metastasis Rev 42:601–627. https://doi.org/10.1007/s10555-023-10086-2
304. Singh R, Srinivas SP, Kumawat M, Daima HK (2024) Ligand-based surface engineering of nanomaterials: trends, challenges, and biomedical perspectives. OpenNano 15:100194. https://doi.org/10.1016/j.onano.2023.100194
305. Rana S, Yeh Y-C, Rotello VM (2010) Engineering the nanoparticle-protein interface: applications and possibilities. Curr Opin Chem Biol 14:828–834. https://doi.org/10.1016/j.cbpa.2010.10.001
306. Padín-González E, Lancaster P, Bottini M, Gasco P, Tran L, Fadeel B, Wilkins T, Monopoli MP (2022) Understanding the role and impact of poly (ethylene glycol) (PEG) on nanoparticle formulation: implications for covid-19 vaccines. Front Bioeng Biotechnol 10:882363. https://doi.org/10.3389/fbioe.2022.882363
307. AlSawaftah NM, Awad NS, Pitt WG, Husseini GA (2022) PH-responsive nanocarriers in cancer therapy. Polymers (Basel) 14:936. https://doi.org/10.3390/polym14050936

Application of Nanomaterials in Nanomedicine

Md. Nazrul Islam and Arijit Bag

Abstract Nanotechnology is an emerging field in the biological sciences. The development of nanomedicine is highly dependent on nanotechnology advancements. This interdisciplinary field combines knowledge of medicine and nanotechnology to innovate advanced approaches for disease diagnosis and treatment pathways. Nanomaterials exhibit novel physical, chemical, and biological characteristics that provide various application techniques in medicine and healthcare. Nowadays, different types of nanomaterials are being used against inflammatory diseases. Nanoparticles are in general made of metals, metal oxides, polymers, lipids, etc. Significant advancements in nanomedicine have raised the drug to a new level, with important implications for health care. Research on the enormous potential of nanotechnology in healthcare is necessary. In the medical field, a great deal of research is being conducted on optimal techniques and approaches. The conventional treatment has seen substantial advancement, and both nanotechnology and nanoparticle quality have improved and produced positive outcomes. The properties of nanomaterials that are explored for their medicinal applications are elaborated on in this chapter. For enhanced visibility, therapeutic nanoparticles are classified based on their application, chemical composition, and properties.

Keywords Nanomaterials · Nanoparticles · Inflammatory disease · Nanomedicine

Md. N. Islam
Chemistry Department, Jawahar Lal Nehru College, Chakradharpur, Jharkhand, India

A. Bag (✉)
Department of Applied Chemistry, Maulana Abul Kalam Azad University of Technology West Bengal, Simhat, Haringhata, Nadia, West Bengal, India
e-mail: bagarijit@gmail.com

Md. M. Ansari et al. (eds.), *Emergence of Sustainable Biomaterials in Tackling Inflammatory Diseases*, Smart Nanomaterials Technology,
https://doi.org/10.1007/978-981-96-2112-5_3

1 Introduction

The rapidly growing nanomedicine research combines nanotechnology [1–7] with medicine to improve drug-delivery techniques, activity, and efficiencies. It involves the synthesis of drugs and drug carriers at the nanoscale level. The nano drugs and drug-loaded nanocarriers are employed in various devices and methodologies to diagnose diseases [8–11], and treat or prevent them. Nanoforms of chemicals that are usable as drugs or nano-drug carriers [12–15] have the prosperous potential to revolutionize healthcare by providing alternative treatment pathways [16–22], drug-delivery systems [13, 17, 18], imaging techniques [9, 10, 14], etc. It also may be helpful to repair damaged tissues.

Nanoparticles possess unique physicochemical properties. Nanoparticles are exceptionally appreciated for their high surface-volume ratio and tunable properties. Their properties are tuned according to their applications in various arms of medicinal and clinical fields by controlling their size, shape, and particle density depending on their synthetic methods and conditions. To functionalize magnetic nanoparticles for cancer treatment specific ligands are attached to the nanostructures. Gold nanoparticles are functionalized appropriately for imaging purposes and photothermal therapy [7, 14, 17, 18].

2 Different Fields of Therapeutics Where Nanotechnology Could Be Implemented

Critical areas where nanotechnology is being implemented or has the potential to be implemented are listed below.

2.1 Cancer Therapy

Safe and painless treatment of cancer is a high-demand research area at present. Diversified techniques and clinical methodology have been implemented around the world to detect and treat cancer. However, intense importance is given to further research to enhance the success rate and minimize the cost and time duration. Nanoparticles draw the colossal attention of cancer researchers for several reasons. Clinical engineering specifically emphasizes on development of new and easier roots for cancer treatment [23–30]. It also focused on effective and specific drug delivery mediums [13, 31–33]. Advanced anticancer drugs, augmented with precisely designed carriers, increase target specificity by directly delivering drugs at the tumor site [34–41]. These methods are associated with minimum damage to healthy cells. As a result, adverse effects of chemotherapy could be under control.

2.2 Drug Vehicle

Target-specific drug delivery [42–50] is highly challenging and needs sophisticated scientific advancement in this field. The difficulties are enhanced if an efficient drug is water-insoluble or very unstable. The use of nano-drug carriers may resolve the problem. Nanoparticles can encapsulate drugs, proteins, or genes, protecting them from degradation and improving their biocompatibility. This allows for sustained and controlled release of therapeutics, improving treatment outcomes.

2.3 Gene Therapy

In spite of legal restrictions and ethical issues regarding gene therapy in the field of sports and agriculture, active research is continuing to date. Nanotechnology has a high potential to deliver therapeutic genes to specific cells, correct genetic abnormalities or promote the production of therapeutic proteins. This approach holds promise for the treatment of diseases related to genetic disorders.

2.4 Antibacterial Therapy

The known antibiotics are not enough for treating all kinds of bacterial diseases at present. In addition, the development of drug resistance by most of the common bacteria demands continuous up-gradation of anti-bacterial drugs. In this regard, nanoparticles with antibacterial properties can be designed to combat drug-resistant bacteria and improve the efficacy of antibiotics. They can also be used in wound dressings to prevent infections. Researchers are trying to develop anti-bacterial nanoparticles for clinical and medical use.

2.5 Cardiovascular Disease

Nanotechnology can assist in targeted drug delivery to the cardiovascular system to treat conditions like atherosclerosis and thrombosis. Additionally, nanoscale imaging techniques can help detect early-stage cardiovascular problems.

2.6 Neurological Disorders

Nano-drug vehicles easily cross the blood–brain interface and effectively deliver the drugs to the target specific areas of the brain. Thus to achieve great success against Alzheimer's and Parkinson's disease which are due to neurological disorders, the implementation of nano-drug carriers is the only way so far.

2.7 Ophthalmology

Glaucoma and macular degeneration are very common diseases for both young and aged persons. Nanotechnology can be effectively implemented for the treatment of these diseases.

2.8 Immunotherapy

Nanoparticles can enhance the effectiveness of immunotherapies by delivering antigens, adjuvants, or immune modulators directly to immune cells or tumor sites.

2.9 Vaccines

Nanotechnology can be used to design and deliver more effective vaccines, improving their stability, enhancing immune response, and enabling targeted delivery.

2.10 Regenerative Medicine

Nanomaterials can support tissue engineering and regenerative medicine by providing scaffolds for cell growth, aiding tissue repair and regeneration.

2.11 Inflammation and Autoimmune Disorders

Nanoparticles can deliver anti-inflammatory agents to specific sites, offering potential treatments for various autoimmune diseases.

2.12 *Pain Management*

Nanodrug carriers are employed for delivering analgesic drugs to nerves or pain receptors in any specific area that provides localized pain relief.

2.13 *Respiratory Diseases*

The treatment of asthma and other respiratory diseases like chronic obstructive pulmonary disease (COPD) may be improved by nanomedicines.

2.14 *Tissue Engineering*

Nanoparticles can be used to create scaffolds for growing new tissue. They can also be used to deliver drugs to damaged tissue, promoting healing.

3 Classification of Nanoparticles

Nanoparticles may be classified with respect to different properties shown in Table 1.

3.1 *Size and Shape*

3.1.1 Nanoparticles

Typically have dimensions between 1 and 100 nm (nm) in at least one dimension.

3.1.2 Nanospheres

Nanoparticles with a spherical shape.

3.1.3 Nanorods

Nanoparticles with an elongated rod-like shape.

Table 1 Classification of different types of nanoparticles based on different properties

Size and shape	Composition	Origin	Surface Modified	Surface Coatings	Charge
1. Nanoparticles 2. Nanospheres 3. Nanorods 4. Nanowires 5. Nanosheets	1. Metal 2. Semiconductor 3. Polymer 4. Lipid-based 5. Carbon-based	1. Natural 2. Engineered	1. Hydrophobic 2. Hydrophilic	1. Core–Shell 2. Surface-Functionalized	1. Positively Charged 2. Negatively Charged

3.1.4 Nanowires

Extremely elongated nanoparticles with a high aspect ratio.

3.1.5 Nanosheets

Nanoparticles with a two-dimensional, sheet-like structure.

3.2 Composition

Based on the chemical composition, nanoparticles may be classified as follows:

3.2.1 Metal Nanoparticles

Nanoparticles are made of various metals like gold, silver, platinum, etc. These metals often exhibit unique properties at the nanoscale. Metal nanoparticles have high demand in clinical and medical use. They are easy to synthesize. Most of the metal nanoparticles are water soluble.

3.2.2 Semiconductor Nanoparticles

Semiconductor Nanoparticles are composed of semiconductor materials like quantum dots. They are generally used in different imaging techniques. They emit light of different colors based on their size, making them valuable for bio-imaging and diagnostics.

3.2.3 Polymeric Nanoparticles

These consist of organic polymers. Polymeric Nanoparticles are excellent drug carriers. For the controlled and tunable release of drugs, nano-polymers are in high demand. Dendrimers, one kind of nano-polymers, are attractive drug vehicles. They are also applicable to imaging techniques. The main advantages of using these drug vehicles are hidden under their tunable physical properties (3sf -. size, shape, and surface functionality). Most drugs are encapsulated within the carriers. However, in some cases, drugs may be attached to the carrier surface. Thus, their use as a carrier is highly recommended.

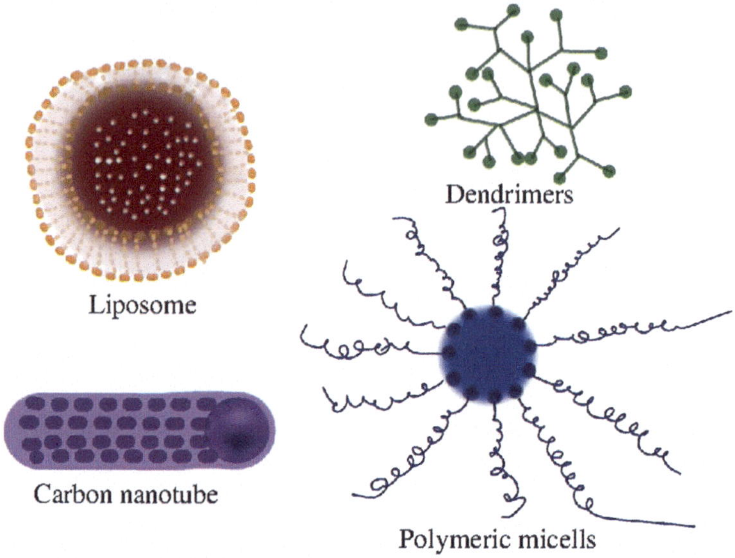

Fig. 1 Different types of nanodrug vehicles

3.2.4 Lipid-Based Nanoparticles

Lipid-based nanodrug carriers are mainly different types of liposomes (shown in Fig. 1). They are highly bio-compatible. Hence, they are in the center of interest of many researchers. Liposomes are hollow spheres. Due to their double lipid bilayer structure and the presence of significant hydrophobic and hydrophilic parts, liposomes can carry both kinds of drugs(hydrophobic drug and hydrophilic drug). Control drug release is also possible with this type of carrier. For gene delivery and vaccines, liposomes are promising carriers.

3.2.5 Carbon-Based Nanoparticles

Carbon nanomaterials have gained significant attention in nanomedicine. Among several carbon nanomaterials, carbon nanotubes (CNTs shown in Fig. 1) possess exceptional mechanical strength and electrical conductivity. Thus, CNTs are effectively applied for tissue engineering and biosensors. Other carbon nanomaterials include graphene, fullerene, carbon quantum dots, etc. Graphene, a single layer of carbon atoms, exhibits remarkable properties like high surface area and electrical conductivity, making it ideal for drug delivery, biosensing, and bioimaging.

3.3 *Origin*

3.3.1 Natural Nanoparticles

This kind of nanoparticles occur naturally. Thus, they may be collected from different natural sources including living organisms. A few examples of such nanoparticles include some minerals, proteins, or lipids with nanoscale dimensions.

3.3.2 Engineered Nanoparticles

Most of the useful nanoparticles are synthetically created in laboratories. They are designed for specific applications using various techniques like chemical synthesis, physical methods, or biological processes.

3.4 *Based on Surface Modification*

3.4.1 Hydrophobic Nanoparticles

They have a water-repellent surface, which can be helpful for specific applications.

3.4.2 Hydrophilic Nanoparticles

This kind of nanoparticles have a water-attracting surface, improving their solubility in aqueous solutions.

3.5 *Surface Coatings*

3.5.1 Core–Shell Nanoparticles

Consists of core material that is surrounded by a distinct coating, offering additional functionalities or protection.

3.5.2 Surface-Functionalized Nanoparticles

Surface-modified with various functional groups to enhance stability, improve targeting, or enable specific interactions.

3.6 *Charge*

3.6.1 Positively Charged Nanoparticles

Due to positive surface charge, these materials easily interact with negatively charged biomolecules and cell membranes.

3.6.2 Negatively Charged Nanoparticles

For positively charged molecules, negatively charged nanoparticles are also required as carriers.

4 Relevant Properties Required for Clinical and Medical Applications

For nanoparticles to be suitable for clinical and medical applications, they must possess specific properties that proclaim their safe and effective use. These properties ensure that nanoparticles can fulfill their intended therapeutic or diagnostic functions while minimizing potential risks and side effects. Some of the relevant properties of nanoparticles for clinical and medical applications are discussed below. Achieving the desired combination of these properties requires careful design, engineering, and testing of nanoparticles. Research and development in nanotechnology continue to focus on creating nanoparticles with tailored properties to address specific clinical needs and improve patient outcomes.

4.1 *Biocompatibility*

Nanoparticles must be biocompatible, meaning they do not cause harmful reactions or toxic effects when introduced into the body. Biocompatibility ensures that the nanoparticles can coexist with biological tissues and cells without causing significant adverse reactions.

4.2 *Stability*

Nanoparticles should have sufficient stability to maintain their structure and function under physiological conditions, such as changes in pH and temperature, to ensure consistent performance.

4.3 Targeting Capability

For targeted therapies, nanoparticles should be capable of specifically targeting the desired cells or tissues. To achieve target specificity, the surfaces of nanocarriers are modified with appropriate functional groups containing compounds. The selected functional groups bind specifically only to the target receptors of the target cells.

4.4 Controlled Drug Release

Controlled drug release, also known as sustained or extended drug release, refers to the deliberate and gradual release of a therapeutic agent over an extended period at a predetermined rate. This approach minimizes the side effects.

4.5 Size and Shape

The cellular uptake of a drug may be increased by influencing biodistribution through proper tuning of the size and shape of the nano drugs or drug carriers. It is also essential for the accumulation of drugs at the targeted site. The size and shape of the drugs or drug-loaded carriers are equally important for the circulation of drugs through blood.

4.6 Surface Charge

The charge on the surface of nano drugs or drug-loaded vehicles is a key deciding factor that regulates interactions of drugs with the host biological system. Neutral or slightly negatively charged nanoparticles are generally preferred to minimize non-specific interactions and potential immune responses.

4.7 Solubility

Nanoparticles intended for systemic administration should have good solubility in physiological fluids to avoid aggregation or unwanted precipitation.

4.8 Low Immunogenicity

To mitigate the likelihood of triggering immune responses and potential elimination by the immune system, nanoparticles should demonstrate minimal immunogenicity or possess characteristics reminiscent of stealth.

4.9 Imaging and Diagnostic Properties

Appropriate imaging characteristics are needed for imaging or diagnostic applications of nanomaterials. The common use of nanomaterials in medical and clinical imaging is due to their contrast activity. These properties are crucial for achieving precise detection and visualization of the intended target through MRI of X-ray tomography.

4.10 Biodegradability

Biodegradable nanoparticles can be advantageous for specific applications as they can break down into non-toxic byproducts, reducing long-term accumulation in the body.

4.11 Scale-Up and Manufacturing

The production of nanoparticles should be scalable, reproducible, and cost-effective to enable large-scale manufacturing for clinical use.

4.12 Regulatory Approval

Nanoparticles intended for medical applications must meet regulatory requirements for clinical use. The safety factor is the most important criterion that should be achieved before regulatory approval. Different kinds of toxicity test results are required for safety clearance.

5 General Technique of Synthesis

Nanoparticle synthesis involves controlling the size, composition, and shape of nanoparticles. Several techniques are available for nanoparticle synthesis. All of them are suitable for different materials and applications. Below is a general overview of some standard nanoparticle synthesis techniques:

5.1 Reduction Through Chemical Reaction

One of the easiest ways of synthesis of nanoparticles involves the reduction of metal salts. In this method, a selectively taken reducing agent is added to the metal slat solution. A stabilizing agent is required for the long-term stability of the prepared nanoparticles. Depending upon the reaction conditions like temperature, salt concentration, and addition time of the reducing agents the shape and size of the prepared nanoparticle changes dramatically.

5.2 Sol–Gel Method

For the preparation of nanoparticles of metal oxides, sol–gel methods are commonly employed. In this method, a colloidal solution (sol) is allowed to undergo gelation. As a result, a matrix of solid nanoparticles is obtained.

5.3 Precipitation Method

Nanoparticles may also be obtained by precipitation from a mixture of two specific solutions. This method is also very handy to control the particle size and shape by controlling the reaction conditions.

5.4 High-Energy Ball Milling

This mechanical method involves grinding bulk materials in a ball mill to produce nanoparticles by fracturing and welding powder particles. It is suitable for synthesizing various materials, including metals, alloys, and composite nanoparticles.

5.5 *Electrochemical Deposition*

Electrochemical methods involve the electrodeposition of metal ions onto an electrode surface, leading to the formation of nanoparticles. The electrodeposition process can be controlled to obtain nanoparticles with desired sizes and shapes.

5.6 *Microemulsion Method*

This method uses a microemulsion system of water, oil, surfactants, and co-surfactants to generate nanoparticles in the confined microdomains. It enables the synthesis of small and uniform nanoparticles.

5.7 *Laser Ablation*

Laser ablation involves irradiating a target material with a laser in a liquid medium, which leads to the formation of nanoparticles. This method includes vaporization of the liquid followed by condensation. This technique produces nanoparticles with high purity and narrow size distributions.

5.8 *Green Synthesis*

Green synthesis uses plant extracts, microorganisms, or natural compounds to produce nanoparticles as reducing or stabilizing agents. This eco-friendly approach is gaining popularity due to its sustainable nature.

5.9 *Gas-Phase Synthesis*

Gas-phase synthesis of nanoparticles involves the chemical vapor deposition (CVD) method. In this method, a specific compound is vaporized first and then condensed over the target substrate or thin film. Another gas-phase synthesis method involves aerosol formation and deposition. These methods are highly sensitive and purpose-oriented. These methods are mainly applied to sensor design and clinical technology.

It is important to note that each synthesis technique is associated with advantages and limitations. However, the selection of the method relies on the nanoparticles' specific material, size, shape, and application requirements. Additionally, nanoparticle synthesis requires careful consideration of safety, reproducibility, and scalability to ensure successful translation from the laboratory to practical applications (Table 2).

6 Commonly Used Nanomedicines

Presently, several nanomedicines are approved for commercial use. Researchers continue to explore emerging materials and methods for advancing nanomedicine. Below are some commonly used nanomedicines.

6.1 Doxil (Doxorubicin Liposomal)

Doxil is a doxorubicin-loaded liposome. It is a chemotherapy drug. It is clinically used for various cancer treatments.

6.2 Abraxane (Nab-Paclitaxel)

The key ingredient of Abraxane is paclitaxel. It is an albumin-bound nanomedicine. It is also a chemotherapy drug. Abraxane is a regular drug for breast cancer and pancreatic cancer. It is also applied to lung cancer treatment.

6.3 Ferumoxytol (Feraheme)

Ferumoxytol is a drug for iron deficiency anemia. It contains iron oxide nanoparticles. Due to severe kidney failure remarkable iron deficiency occurs in the patient's body which requires iron replacement. Ferumoxytol is used to fulfill this requirement. This drug is employed through intravenous injection.

6.4 Onivyde (Irinotecan Liposomal)

Onivyde is a chemotherapy drug. It contains irinotecan as the key ingredient. The main drug is loaded in liposomes. This drug is prescribed for metastatic pancreatic cancer as a combination drug.

Table 2 Synthesis technique of nanoparticles

Chemical reduction	Sol–gel method	Precipitation	Ball milling	Electrochemical deposition	Microemulsion method	Laser ablation	Green path way	Gas-phase synthesis
Metal slat solution + reducing agent + stabilizing agent	Colloidal solution (sol) => gelation	Mixture of two specific solutions => precipitation of nanoparticles	Grinding bulk materials => nanoparticles	Electrodeposition of metal ions on electrode surface	Microemulsion is used	Vaporization of the liquid => condensation	From plant extracts	Chemical vapor deposition/ aerosol formation and deposition

6.5 Visudyne (Verteporfin)

Visudyne is a nanoparticle-based photosensitizer. It is an ocular drug. It is prescribed for photodynamic therapy to treat macular degeneration.

6.6 Gene Therapy Drugs

Nanocarriers are being developed to deliver genes to cells. These are implemented for diseases including cancer, HIV, and cystic fibrosis.

These are just a few examples of the many nanomedicines currently being used or developed. As research in this area continues, we can expect to see even more innovative and effective nanomedicines in the future.

7 Emerging Materials and Methods

As the field of nanomedicine continues to advance, more innovative materials and methods are likely to emerge, enhancing therapeutic interventions' effectiveness, precision, and safety. It's important to note that developing and approving new nanomedicines often require rigorous testing and evaluation of safety-related issues before being introduced into clinical practice.

7.1 RNA Nanoparticles

Researchers are exploring RNA nanoparticles as a targeted drug delivery medium. RNA nanoparticles are equally important for gene therapy applications. RNA nanoparticles are engineered in such a way that self-assemble property is observed. As a result, they carry therapeutic payloads like siRNA for gene silencing or mRNA for protein expression.

7.2 Gold Nanoparticles

Medicinal applications of gold nanoparticles (GNPs) have been increasing day by day[18, 37, 51–69]. This is due to the high functionalization possibilities of GNPs. Employing proper modification of the ligand that functionalized the GNPs, the target specificity of the medicine may be enhanced [21, 70–89]. GNPs are also successfully used in photothermal therapy and clinical imaging for diagnosis.

7.3 Carbon Nanotubes

Carbon nanotubes (CNTs) have wide applications in the clinical and medicinal fields. CNTs are effective for gene and drug delivery. They are also applied for photothermal therapy. CNTs are attractive candidates for nanomedicine due to their unique properties. The thermal conductivity and surface area of CNTs are so high that facilities their application in photothermal therapy.

7.4 Metal–Organic Framework

A new addition to the nano-drug carriers is the metal–organic framework (MOF). MOFs are regular porous materials. The pore sizes are tunable. MOFs are very flexible too. Hence, the drug-loading capacity of MOFs could be controlled by applying the proper synthesis procedure. Except for therapeutic use as nano drug carriers, MOFs are also employed for live cell imaging. Compared to other drug carriers, the drug loading capacity of MOFs is higher.

7.5 DNA Origami

DNA origami involves designing and folding DNA molecules into specific shapes or structures. It holds promise for precise drug delivery and nano-scale devices for biomedical applications.

7.6 CRISPR-Cas Nanoparticles

Nanoparticle-based delivery systems are being developed to deliver CRISPR-Cas gene-editing tools to specific cells or tissues, facilitating gene correction for genetic disorders.

7.7 Extracellular Vesicles

Extracellular vesicles, such as exosomes, are natural nanoscale particles secreted by cells. They are being investigated as potential drug carriers and as mediators of intercellular communication.

7.8 Immune Cell-Derived Nanoparticles

Nanoparticles may also be obtained from immune cells. Examples contain macrophages. Researchers are exploring their use in immunotherapy.

7.9 3D Printing of Nanomedicines

3D printing of nanomedicines is being utilized to build personalized patient-specific nanomedicines. This allows to study of precise drug dosing and release profiles.

8 Application of Gold Nanoparticles in Medicine

Gold is a precious metal with significant economic value. In the field of biomedicine, gold nanoparticles (GNPs) at the nanoscale have gained importance due to their various biomedical use. The application of GNPs was mentioned in the Middle Era when it was known as "potable gold" or "aurum potable" in Latin. These gold nanoparticles played a crucial role in diagnostics for arthritis. It used to be employed for syphilis diagnoses. For the treatment of heart problems, hysteria, and tumors, GNPs were used in the Middle Era. Though, it was not known as nanoparticles at that time.

Modern biomedical applications of GNPs include radiotherapy, photothermal therapy, etc. [89–129]. Researchers have found fascinating applications of GNPs in CT imaging too. Designing drug vehicles using GNPs is a highly-rated research topic. It is also associated with nano-enzyme delivery and targeting.

Radiotherapy involves the use of high-intensity radiation beams, either externally or internally implanted. It targets tumor death by destroying stem cells which are intracellular components. For this reason, radiotherapy is applied to most cancer patients. In Photothermal therapy, light is absorbed by plasmonic nanoparticles. As a result, heat is generated. This thermal ablation destroys the affected cell. To obtain highly accurate 3D pictures of affected tissues, computed tomography (CT) employing X-ray is used. For this non-invasive technique of live cell imaging, GNPs are precisely recommended due to their ability to absorb X-rays. The integration of GNPs in various clinical and medical uses demonstrates their versatility and potential impact in enhancing diagnostics and therapeutic interventions. Continued research and development in this field hold promise for even more sophisticated and targeted approaches to medical treatments and imaging.

Specialized carriers are employed to transport drugs to targeted areas. Gold nanoparticles (GNPs) can serve as effective drug carriers, with their surface modification using functional groups or cationic polymers such as thiol, amine, or carboxyl groups to increase their drug-carrying capacity. These engineered surfaces of GNPs

[21, 37, 69–82] facilitate stable and efficient drug immobilization, protecting the drug from enzyme degradation. Encapsulation of drugs with GNP complexes, aided by serum proteins in the bloodstream, further enhances drug protection [100–113]. Specific molecules are linked to the GNP-drug complexes to achieve targeted specificity. GNPs possess exceptional physical, chemical, electronic, and optical properties, making them superior to other metals. Additionally, they exhibit better biocompatibility [114–125]. When functionalizing GNPs with small molecules, DNA, or proteins, surface coating techniques employ phosphate-containing agents, thiols, or amines. Surface modification enhances the efficiency, stability, and toxicity profile of GNPs while reducing their aggregation tendency. Such tailored modifications further improve the potential of GNPs as versatile drug carriers in various therapeutic applications.

8.1 Application in Radiotherapy

The critical factors which help to accept gold nanoparticles as a radiotherapy agent are the following:

1. Gold has a higher Z number. Thus, its photoelectric absorption coefficient is high.
2. GNPs have a larger surface/volume ratio. They are accepted as therapeutic agents and drug careers with their surface and delivered to the targeted zone for targeted drug delivery with diagnosis and combination therapy of tumors.
3. Compared to others, GNPs have less cytotoxicity which enhances their biocompatibility.
4. High contrast activity for clinical imaging.
5. Higher permeability and preferential deposit at the place of tumor sites.
6. Well-behaved size distribution.
7. Unique optical, chemical, and electrical properties.

8.2 Classification of GNPs

GNPs are classified into two types: Unorganized and Modified.

8.2.1 Unorganized GNPs

Unorganized GNPs in nanomedicine refer to gold nanoparticles used in their natural or unmodified form without specific functionalization or surface modifications [130]. While unorganized GNPs may not possess the same level of control and specificity as their organized and functionalized counterparts, they still hold certain advantages and find applications in various areas of nanomedicine. Some of the potential applications of unorganized GNPs in nanomedicine include.

Imaging Contrast Agents

Unmodified GNPs are used as photoacoustic contrast agents (CA). They are also used as CA for other imaging techniques like MRI and CT scans.

Drug Delivery

Unorganized GNPs are not target-specific drug carriers. However, as a low-cost medicine, these kinds of GNPs are used. Of course, their efficiency is less compared to organized GNPs.

Photothermal Therapy

In photothermal therapy, cell death occurs upon exposure to the laser pulse. Thus, for this purpose, unorganized GNPs are effectively applicable.

Diagnostic Biomarkers

Unorganized GNPs can be conjugated with specific antibodies or ligands to create diagnostic biomarkers [131–135]. These biomarkers can detect particular disease markers or pathogens, aiding in the early diagnosis of diseases.

Antimicrobial Agents

Unmodified GNPs have been shown to possess antimicrobial properties. They can combat bacterial and fungal infections, offering potential wound healing and infection control applications.

Anti-inflammatory Agents

Unorganized GNPs have demonstrated anti-inflammatory properties, making them promising candidates for inflammation conditions.

Biosensors

Unmodified GNPs can be incorporated into biosensing platforms to detect various analytes, including biomolecules, toxins, and environmental pollutants.

While unorganized GNPs lack the tailored functionalities of surface-modified nanoparticles, their simplicity and ease of synthesis make them attractive for specific applications. Their use in nanomedicine is an important research area. The engineering of nanoparticle surface modification techniques is continuing to expand the possibilities for their incorporation into targeted and precise nanomedicine applications.

8.2.2 Organized or Surface-Modified GNPs

Organized or surface-modified GNPs in nanomedicine refer to gold nanoparticles engineered with specific functionalities by functionalizing their surface with various molecules, ligands, or polymers. These modifications allow precise control over their properties, enabling targeted and controlled drug release, contrast activity, and other therapeutic uses. Some critical applications of organized GNPs in nanomedicine are discussed here.

Targeted Drug Delivery

Organized GNPs are used as target-specified drug vehicles. The functionalization of GNPs with antibodies or specific ligands increases their targeted receptor binding accuracy. This surface modification acts like a bookmark. This targeted functionalization allows selective drug delivery to diseased cells while sparing healthy tissues, reducing side effects, and improving treatment efficacy.

Image-Guided Therapy

With advanced therapeutic techniques, image-guided therapy could be performed. This approach enables real-time monitoring of the therapeutic process and helps optimize treatment outcomes. Attaching radioactive isotopes or fluorescent dyes to GNPs, such kind of monitoring treatment may be conducted.

Photothermal Therapy

Organized GNPs are equally applicable for photothermal therapy as unorganized GNPs. The additional advantage of the use of organized GNPs is its tunable property. The absorption wavelength of GNPs may be tuned to make it an absorber of near-infrared (NIR) light.

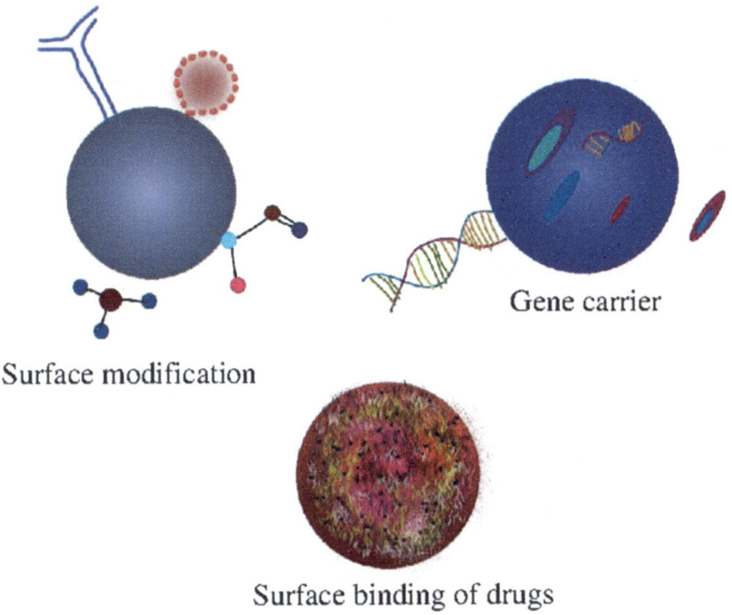

Fig. 2 Surface modification and drug loading

Biosensors and Diagnostics

Organized GNPs can detect specific biomolecules or pathogens. Functionalizing the surface with molecules that recognize the target analyte enables sensitive and rapid diagnostic testing.

Gene Delivery

Surface modifications of GNPs allow them to carry genetic material, such as siRNA or plasmid DNA, to target cells for gene therapy applications [136–138]. This holds promise for treating genetic disorders and modulating gene expression in various diseases.

Theranostics Organized

GNPs with combined therapeutic and diagnostic functionalities are known as theranostic agents [139–141]. These nanoparticles enable simultaneous treatment and monitoring of disease progression, leading to personalized and precise medicine (Fig. 2).

MRI Contrast Agents

Suitably Functionalized GNPs are excellent CA for MRI. It provides enhanced contrast and better visualization of tissues and organs.

Immunoassays

Organized GNPs can be used in immunoassays to detect antibodies or antigens in biological samples, contributing to medical diagnostics and research.

Nanovaccines

Surface-functionalized GNPs can serve as carriers for antigens and adjuvants in vaccine development, enhancing the immune response and improving vaccine efficacy.

The surface modification of GNPs is crucial in enhancing their biocompatibility, stability, and functionality. By engineering the surface properties of GNPs, researchers can create tailored nanoparticles with specific characteristics suited for various nanomedicine applications, contributing to the advancement of precision medicine and improved patient care.

8.3 Different Generations of GNPs

There were a few drawbacks of the first-generation nanoparticles. In vivo delivery was difficult for the first-generation GNP. Thus, the focus has shifted to surface modifications in the second and third generations of nanoparticles. These modifications enhance targeting and stealth effects, improve nanoparticle absorption, and enable more significant disposition in the targeted area. This also helps to achieve preferred and manageable biodistribution. Functionalized gold nanoparticles, in particular, offer a wide range of advantages.

1. They enhance blood circulation and cellular uptake capacity in tumor tissues, improving clearance and targeted inhibition of tumor metastasis and invasions.
2. Their coupling sites can aid in biomarker-based efficacy predictions and disease diagnosis.
3. Specific targeting enables reduced toxicity, improved clearance, and immune system-mediated surveillance, inhibiting multidrug resistance (MDR).
4. They facilitate controlled drug delivery to targeted areas, helping to minimize side effects.

By incorporating these changes, the text conveys a more explicit message about the benefits of functionalized gold nanoparticles in nanomedicine applications.

8.4 *Active and Passive Targeting by Gold Nanoparticles*

GNPs are usable for both active and passive targeting in medical applications, depending on the specific functionalization and properties of the nanoparticles. These targeting strategies are crucial in enhancing the effectiveness and specificity of therapeutic interventions. An overview is given below.

8.4.1 Passive Targeting

Passive targeting relies on the unique properties of nanoparticles to accumulate preferentially in certain areas, typically in tumor tissues, through their enhanced permeability as well as retention (EPR) effect. This effect occurs due to the leaky vasculature and poor lymphatic drainage in tumors, fascinating nanoparticles' accumulation at the tumor's microenvironment, passively. The EPR effect is more prominent in solid tumors, making passive targeting as an attractive strategy for delivering therapeutic agents specifically to tumor tissues.

8.4.2 Active Targeting

For active targeting, the surface functionalization of GNPs by specific ligands, antibodies, peptides, or other molecules that can recognize and bind to receptors or biomarkers overexpressed on the surface of target cells. By actively targeting these specific receptors, the functionalized gold nanoparticles can selectively deliver therapeutic agents to the targeted cells or tissues, increasing treatment precision and reducing off-target effects.

8.4.3 Advantages of Active and Passive Targeting

Passive targeting offers a relatively straightforward approach, relying on the tumor microenvironment's characteristics to achieve nanoparticle accumulation. It can be effective in targeting various solid tumors, regardless of specific biomarkers. However, active targeting provides a more tailored approach, allowing for selective delivery to cells expressing specific receptors or biomarkers. This enhances treatment specificity and efficacy, especially for tumors with known overexpressed receptors.

Combination Strategy.

Active and passive targeting strategies are combined to improve drug delivery and treatment outcomes in a better way. By functionalizing gold nanoparticles with ligands or antibodies for active targeting, nanoparticles can achieve enhanced accumulation in the tumor cells. While taking advantage of the EPR effect for passive targeting, the same may be achieved. GNPs hold great promise in medical applications thanks to their versatility. The choice of targeting strategy depends on the specific application.

8.5 *Different Kinds of Surface-Modified GNPs*

8.5.1 Polyethylene Glycol (PEG) Developed Gold Nanoparticles

GNPs coated or functionalized with polyethylene glycol molecules are also very important for specific applications[21, 37, 69–87]. PEG is a hydrophilic polymer widely used in nanomedicine due to its unique properties that enhance the biocompatibility and stability of nanoparticles, making it an attractive choice for surface modification.

PEGylation Process

The process of developing PEG-coated gold nanoparticles typically involves the following steps:

1. Synthesis of Gold Nanoparticles—Gold nanoparticles are first synthesized using various chemical or physical methods, resulting in nanoparticles with a desired size and shape.
2. PEG Functionalization—The surface of the gold nanoparticles is then modified by attaching PEG molecules to the nanoparticle surface. This can be achieved by various methods, such as thiol chemistry, amine coupling, or electrostatic adsorption, depending on the nature of the PEG molecules and the functional groups present on the nanoparticle surface.

PEGylation Benefits

Improved Biocompatibility: PEGylation imparts hydrophilicity to the gold nanoparticles, reducing their tendency to aggregate and enhancing their stability in biological fluids.

1. Reduced Immune Response—PEG-coated nanoparticles exhibit a stealth effect, as PEG molecules create a protective barrier around the nanoparticles, reducing their recognition and clearance by the immune system.
2. Extended Blood Circulation—The PEG coating can increase the circulation time of the nanoparticles in the bloodstream, enhancing their potential for drug delivery or imaging applications.
3. Reduced Non-Specific Binding—PEGylation helps minimize non-specific interactions with proteins and cells, reducing off-target effects and enhancing targeted delivery to specific tissues or cells.

Applications of PEG-Coated Gold Nanoparticles

1. Drug Delivery—PEG-coated gold nanoparticles can serve as drug carriers, allowing for controlled and targeted delivery of therapeutic agents to specific cells or tissues.
2. Imaging—PEGylated gold nanoparticles can be used as contrast agents in various imaging modalities, such as CT and photoacoustic.
3. Theranostics—PEG-coated gold nanoparticles with combined diagnostic and therapeutic functionalities are used in theranostic applications for personalized medicine.

The development of PEG-coated gold nanoparticles is an essential aspect of nanomedicine research, as it enables the creation of biocompatible and functionalized nanoparticles with significant potential for various medical applications. However, it's important to note that the choice of PEGylation method and nanoparticle design should be carefully optimized to achieve the desired therapeutic outcomes and minimize potential cytotoxicity or immunogenicity concerns.

8.5.2 Chitosan-Developed Gold Nanoparticles

Chitosan-developed gold nanoparticles[89–91] are gold nanoparticles that have been coated or functionalized with chitosan, a biocompatible and biodegradable polysaccharide derived from chitin, found in the shells of crustaceans like shrimp and crabs. Chitosan is widely used in biomedical and nanomedicine applications due to its unique properties, including its biocompatibility, low toxicity, and ability to form stable nanoparticles through self-assembly or complexation with metal ions.

Synthetic Process

The process of developing chitosan-coated gold nanoparticles typically involves the following steps:

1. Synthesis of Gold Nanoparticles—Gold nanoparticles are first synthesized using chemical or physical methods to obtain nanoparticles with the desired size and shape.
2. Chitosan Functionalization—Chitosan molecules are then introduced to the surface of the gold nanoparticles through adsorption or chemical conjugation. Chitosan has functional groups like amino and hydroxyl, facilitating its interaction with the gold surface.

Chitosan Benefits

1. Biocompatibility—Chitosan is biocompatible and can be safely used in various biological systems without causing significant adverse effects.
2. Stability—The chitosan coating enhances the strength of gold nanoparticles, preventing their aggregation and maintaining their integrity in physiological conditions.
3. Drug Delivery—Chitosan-coated gold nanoparticles can be utilized as drug carriers, providing controlled and targeted delivery of therapeutic agents.
4. Biomedical Imaging—Chitosan-coated gold nanoparticles can also serve as contrast agents in imaging techniques like CT and photoacoustic imaging.
5. Immune Response Modulation—Chitosan possesses immunomodulatory properties, which can benefit specific therapeutic applications.

Applications of Chitosan-Coated Gold Nanoparticles

1. Drug Delivery—Chitosan-coated gold nanoparticles are used as carriers for delivering drugs, genes, or other therapeutic agents to specific cells or tissues in a controlled manner.
2. Cancer Therapy—Chitosan-coated gold nanoparticles can be functionalized with targeting ligands to deliver therapeutic agents to tumor cells selectively, improving cancer treatment efficacy.
3. Wound Healing—Chitosan-coated gold nanoparticles have shown promise in wound healing applications due to their antimicrobial properties and ability to promote tissue regeneration.
4. Bioimaging—Chitosan-coated gold nanoparticles can be utilized as contrast agents in biomedical imaging techniques for improved visualization of tissues or organs.
5. Diagnostics – Chitosan-coated gold nanoparticles have been explored for their potential use in diagnostic assays and biosensors for detecting specific biomolecules or pathogens.

Chitosan-developed gold nanoparticles are a versatile and promising platform in nanomedicine, offering a wide range of potential applications in drug delivery, imaging, diagnostics, and therapeutic interventions. However, as with any nanomaterial, carefully considering the synthesis method, surface functionalization, and potential biocompatibility and toxicity issues is essential to ensure safe and effective use in biomedical applications.

8.5.3 Bovine Serum Albumin (BSA)-Developed Gold Nanoparticles

Bovine serum albumin (BSA)-developed gold nanoparticles [142–145] coated or stabilized with BSA, a common protein derived from bovine (cow) serum. BSA is widely used in various biological and nanomedicine applications due to its biocompatibility and ability to form stable complexes with metal nanoparticles, including gold nanoparticles.

Synthesis Process

The process of developing BSA-coated gold nanoparticles typically involves the following steps:

1. Synthesis of gold nanoparticles—Gold nanoparticles are first synthesized using various chemical or physical methods to obtain nanoparticles with the desired size and shape.
2. BSA coating—BSA molecules are then adsorbed onto the surface of the gold nanoparticles, forming a stabilizing layer around the nanoparticles. BSA has multiple functional groups, including amine and carboxyl groups, that can interact with the gold surface, providing stabilization and preventing nanoparticle aggregation.

Benefits

1. Biocompatibility—BSA is a naturally occurring protein and is generally considered safe for use in biomedical applications.
2. Stabilization—The BSA coating enhances the stability of gold nanoparticles, preventing their aggregation and maintaining their colloidal stability.
3. Biocompatible Surface—The presence of BSA on the nanoparticle surface can improve the biocompatibility of gold nanoparticles and reduce potential adverse effects in biological systems.
4. Drug Delivery—BSA-coated gold nanoparticles can be utilized as drug carriers, providing a biocompatible and stable platform for controlled drug delivery.
5. Bioimaging—BSA-coated gold nanoparticles can be used as contrast agents in various imaging techniques, such as CT and photoacoustic imaging.

Applications of BSA-Coated Gold Nanoparticles

1. Drug Delivery—BSA-coated gold nanoparticles can be used as carriers for delivering drugs, genes, or other therapeutic agents to specific cells or tissues in a controlled and targeted manner.
2. Biomedical Imaging—BSA-coated gold nanoparticles can serve as contrast agents in various imaging modalities, improving the visualization of tissues or organs in biomedical imaging.
3. Diagnostics—BSA-coated gold nanoparticles can be utilized in diagnostic assays and biosensors to detect specific biomolecules or pathogens.

BSA-developed gold nanoparticles offer a biocompatible and stable platform for various nanomedicine applications. The BSA coating provides enhanced stability and can be functionalized with specific ligands or targeting moieties for targeted drug delivery or imaging applications. However, as with any nanoparticle-based material, thorough characterization and assessment of potential biocompatibility and toxicity should be conducted to ensure their safety and efficacy in biomedical applications.

9 Polymeric Nanoparticles for Oral and Ocular Drug Delivery

Polymeric nanoparticles are also promising drug vehicles [146–156] for oral and ocular diseases. They have very unique properties that facilitate their drug loading and delivery with ease. These nanoparticles can encapsulate a wide range of drugs and protect them from degradation, enhancing their bioavailability and targeted delivery.

9.1 Oral Drug-Delivery

9.1.1 Enhanced Bioavailability

Polymeric nanoparticles can improve the solubility and stability of water-insoluble drugs, leading to enhanced absorption and safe transportation after oral administration.

9.1.2 Controlled Release

Polymeric nanoparticles could be modeled to release drugs in a controlled and sustained manner, ensuring prolonged therapeutic effects and reducing dosing frequency.

9.1.3 Gastrointestinal Protection

Polymeric nanoparticles can protect sensitive drugs from degradation in the harsh gastrointestinal environment, preserving their therapeutic efficacy.

9.1.4 Targeted Delivery

Surface functionalization of polymeric nanoparticles allows for targeted delivery to specific sites in the gastrointestinal tract, improving drug localization and reducing off-target effects.

9.1.5 Protection of Labile Drugs

Polymeric nanoparticles can protect labile drugs from enzymatic degradation, enabling the oral delivery of biologics and sensitive therapeutic agents.

9.2 Ocular Drug-Delivery

9.2.1 Prolonged Retention Time

Polymeric nanoparticles can prolong drug retention in the eye, allowing for sustained drug release and reduced dosing frequency.

9.2.2 Enhanced Penetration

Nanoparticles can improve the permeability of drugs across ocular barriers, such as the cornea and conjunctiva, enhancing drug absorption.

9.2.3 Targeted Delivery

Surface modification of polymeric nanoparticles enables targeted drug delivery to specific ocular tissues, such as the retina.

9.2.4 Treatment of Posterior Segment Diseases

Polymeric nanoparticles are particularly useful for delivering drugs to the posterior eye segment, where conventional drug delivery is challenging.

9.2.5 Reduced Frequency of Injections

Polymeric nanoparticles can offer an alternative to frequent intravitreal injections for treating conditions like diabetic retinopathy and AMD (age-related macular degeneration).

9.3 Challenges in Oral and Ocular Drug Delivery

Polymeric nanoparticles, however, do face some challenges in both oral and ocular drug delivery.

1. Biocompatibility and Safety—Polymeric nanoparticles' biocompatibility and long-term safety need to be thoroughly evaluated to avoid potential toxic effects.
2. Manufacturing Consistency—Ensuring reproducibility and scalability of polymeric nanoparticle production is critical for widespread use.
3. Regulatory Approval—Developing and obtaining regulatory approval for polymeric nanoparticle-based drug delivery systems can be complex and time-consuming.

Despite these challenges, present investigations and achievements in nanotechnology continuously improve the performance and safety of polymeric nanoparticles to make them usable for ocular and oral dose preparation. They hold great promise for overcoming the challenges of conventional methods of preparation of deliverable doses of drugs. This will also help to enable more effective and targeted treatments for various diseases affecting the oral and ocular regions.

10 Prospectives of Nanomedicine

Nanomedicine holds great promise for revolutionizing medical treatments and diagnostics, but it also faces several challenges that must be addressed for successful implementation. An optimistic approach includes advantages and difficulties of nanomedicines as follows.

10.1 Targeted Drug Delivery

Nanoparticles can be engineered to deliver drugs selectively to specific cells or tissues, reducing off-target effects and enhancing treatment efficacy.

10.2 Personalized Medicine

Nanomedicine allows customized therapies based on an individual's unique biological profile, leading to more precise and effective treatments.

10.3 Enhanced Imaging

Nanoparticles can serve as contrast agents for various imaging modalities, improving the sensitivity and accuracy of medical diagnostics.

10.4 Combination Therapies

Nanoparticles can carry multiple therapeutic agents, enabling combination therapies to tackle complex diseases with different mechanisms of action.

10.5 Minimized Side Effects

By improving drug delivery and reducing systemic exposure, nanomedicine can potentially minimize side effects associated with conventional treatments.

10.6 Theranostics

Nanomedicine facilitates theranostic applications, combining diagnostics and therapy in a single platform for real-time monitoring and treatment.

10.7 Regenerative Medicine

Nanotechnology can aid tissue engineering and regenerative medicine by providing scaffolds and nanomaterials that promote tissue repair and regeneration.

10.8 Challenges of Nanomedicine

10.8.1 Safety Concerns

Nanomaterials' potential toxicity and long-term effects on human health and the environment need a thorough evaluation.

10.8.2 Manufacturing and Scalability

Large-scale production of nanomedicines with consistent quality remains challenging and costly.

10.8.3 Regulatory Approval

Nanomedicines require rigorous testing for safety and efficacy, and navigating regulatory pathways can be complex and time-consuming.

10.8.4 Biodistribution and Clearance

Understanding how nanomedicines interact with the body, their biodistribution, and clearance is essential for optimizing their performance.

10.8.5 Immune Response

Nanoparticles may trigger immune responses, affecting their effectiveness and safety.

10.8.6 Nanoparticle Stability

Nanoparticles can undergo aggregation or degradation, affecting their stability and performance.

10.8.7 Cost

The development and production of nanomedicines can be expensive, potentially limiting their accessibility and affordability.

10.8.8 Ethics and Social Implications

Ethical considerations surrounding nanomedicine, such as privacy, equity, and societal acceptance, need careful examination.

Addressing these challenges requires collaboration among scientists, engineers, clinicians, regulators, and policymakers to ensure nanomedicine's safe and successful integration into mainstream healthcare. With continued research, advancements, and responsible implementation, nanomedicine has the potential to revolutionize medical practices and improve patient outcomes.

11 Conclusion

The diverse range of nanomaterials holds immense potential for revolutionizing the field of nanomedicine. From nanoparticles and carbon-based nanomaterials to liposomes, dendrimers, and quantum dots, each type of nanomaterial offers unique advantages for various applications in diagnostics, imaging, drug delivery, and therapy. Further research and development in this field will likely lead to the emergence of novel nanomedicine approaches, providing improved treatments and personalized healthcare options. However, it is crucial to continue exploring the safety aspects and potential toxicity of nanomaterials to ensure their safe and effective use in clinical settings.

Conflict of Interest The author declares no competing interest.

References

1. Song G, Cheng L, Chao Y, Yang K, Liu Z (2017) Emerging nanotechnology and advanced materials for cancer radiation therapy. Adv Mater 29(32):1700996
2. Liu Y, Zhang P, Li F et al (2018) Metal-based *nanoenhancers*for future radiotherapy: radiosensitizing and synergistic effects on tumor cells. Theranostics 8(7):1824–1849. https://doi.org/10.7150/thno.22172
3. Laprise-Pelletier M, Simão T, Fortin MA (2018) Gold nanoparticles in radiotherapy and recent progress in nanobrachytherapy. Adv Healthc Mater 7(16):1701460
4. Brun E, Sanche L, Sicard-Roselli C (2009) Parameters governing gold nanoparticle X-ray radiosensitization of DNA in solution. Colloids Surf B Biointerfaces 72(1):128–134. https://doi.org/10.1016/j.colsurfb.2009.03.025
5. Baumann M, Krause M, Overgaard J et al (2016) Radiation oncology in the era of precision medicine. Nat Rev Cancer 16(4):234–249
6. Shanmugam V, Selvakumar S, Yeh CS (2014) Near-infrared lightresponsive nanomaterials in cancer therapeutics. Chem Soc Rev 43(17):6254–6287
7. Alkilany AM et al (2012) Gold nanorods: their potential for photothermal therapeutics and drug delivery, tempered by the complexity of their biological interactions. Adv Drug Deliv Rev 64(2):190–199
8. Jakhmola A, Anton N, Vandamme TF (2012) Inorganic nanoparticles based contrast agents for x-ray computed tomography. Adv Healthc Mater 1(4):413–431
9. Liu Y, Ai K, Lu L (2012) Nanoparticulate x-ray computed tomography contrast agents: from design validation to in vivo applications. Acc Chem Res 45(10):1817–1827
10. Lee N, Choi SH, Hyeon T (2013) Nano-sized CT contrast agents. Adv Mater 25(19):2641–2660
11. Lusic H, Grinstaff MW (2013) X-ray-computed tomography contrast agents. Chem Rev 113(3):1641–1666
12. Her S, Jaffray DA, Allen C (2017) Gold nanoparticles for applications in cancer radiotherapy: Mechanisms and recent advancements. Adv Drug Deliv Rev 109:84–101
13. Jeong EH, Jung G, Hong CA, Lee HJ (2013) Gold nanoparticle (AuNP)-based drug delivery and molecular imaging for biomedical applications. Arch Pharmacal Res 37:53–59
14. Huang X, El-Sayed MA (2010) Gold nanoparticles: optical properties and implementations in cancer diagnosis and photothermal therapy. J Adv Res 1(1):13–28

15. Huang HCR, Grandhi T, Potta T, Rege K (2010) Gold nanoparticles in cancer imaging and therapeutics. Nano Life 01(03n04):289–307
16. Qin Z, Bischof JC (2012) Thermophysical and biological responses of gold nanoparticle laser heating. Chem Soc Rev 41(3):1191–1217
17. Dreaden EC et al (2012) The golden age: gold nanoparticles for biomedicine. Chem Soc Rev 41(7):2740–2779
18. Haume K, Rosa S, Grellet S et al (2016) Gold nanoparticles for cancer radiotherapy: a review. Cancer Nanotechnol 7(1):8
19. Her S, Jaffray DA, Allen C (2017) Gold nanoparticles for applications in cancer radiotherapy: mechanisms and recent advancements. Adv Drug Deliv Rev 109:84–101. https://doi.org/10.1016/j.addr.2015.12.012
20. Nagi NMS, Khair YAM, Abdalla AME (2017) Capacity of gold nanoparticles in cancer radiotherapy. Jpn J Radiol 35(10):555–561
21. Ngwa W, Kumar R, Sridhar S et al (2014) Targeted radiotherapy with gold nanoparticles: current status and future perspectives. Nanomedicine (Lond) 9(7):1063–1082. https://doi.org/10.2217/nnm.14.55
22. Ishikura S (2012) Optimal radiotherapy for non-small-cell lung cancer: current progress and future challenges. Gen Thorac Cardiovasc Surg 60(3):127–131. https://doi.org/10.1007/s11748-011-0832-y
23. Shrestha S, Cooper LN, Andreev OA, Reshetnyak YK, Antosh MP (2016) Gold nanoparticles for radiation enhancement in vivo. Jacobs J Radiat Oncol 3(1):026
24. Linam J, Yang LX (2015) Recent developments in radiosensitization. Anticancer Res 35(5):2479–2485
25. Mesbahi A (2010) A review on gold nanoparticles radiosensitization effect in radiation therapy of cancer. Rep Pract Oncol Radiother 15(6):176–180. https://doi.org/10.1016/j.rpor.2010.09.001
26. Guo Z, Zhu S, Yong Y et al (2017) Synthesis of BSA-coated BiOI@BiS semiconductor heterojunction nanoparticles and their applications for radio/photodynamic/photothermal synergistic therapy of tumor. Adv Mater 29(44):1704136. https://doi.org/10.1002/adma.201704136
27. Ma N, Jiang YW, Zhang X et al (2016) Enhanced radiosensitization of gold nanospikes via hyperthermia in combined cancer radiation and photothermal therapy. ACS Appl Mater Interfaces 8(42):28480–28494. https://doi.org/10.1021/acsami.6b10132
28. Zhang XD, Wu D, Shen X et al (2012) Size-dependent radiosensitization of PEG-coated gold nanoparticles for cancer radiation therapy. Biomaterials 33(27):6408–6419. https://doi.org/10.1016/j.biomaterials.2012.05.047
29. Kobayashi K, Usami N, Porcel E, Lacombe S, Le Sech C (2010) Enhancement of radiation effect by heavy elements. Mutat Res 704(1–3):123–131. https://doi.org/10.1016/j.mrrev.2010.01.002
30. Yong Y, Cheng X, Bao T et al (2015) Tungsten sulfide quantum dots as multifunctional nanotheranostics for in vivo dual-modal image-guided photothermal/radiotherapy synergistic therapy. ACS Nano 9(12):12451–12463. https://doi.org/10.1021/acsnano.5b05825
31. Wen L, Chen L, Zheng S et al (2016) Ultrasmall biocompatible WO3-x nanodots for multimodality imaging and combined therapy of cancers. Adv Mater 28(25):5072–5079. https://doi.org/10.1002/adma.201506428
32. Song G, Ji C, Liang C et al (2017) TaOx decorated perfluorocarbon nanodroplets as oxygen reservoirs to overcome tumor hypoxia and enhance cancer radiotherapy. Biomaterials 112:257–263. https://doi.org/10.1016/j.biomaterials.2016.10.020
33. Liu J, Yang Y, Zhu W et al (2016) Nanoscale metal-organic frameworks for combined photodynamic & radiation therapy in cancer treatment. Biomaterials 97:1–9. https://doi.org/10.1016/j.biomaterials.2016.04.034
34. Wu H, Lin J, Liu P et al (2015) Is the autophagy a friend or foe in the silver nanoparticles associated radiotherapy for glioma? Biomaterials 62:47–57. https://doi.org/10.1016/j.biomaterials.2015.05.033

35. Kim SR, Kim EH (2018) Feasibility study on the use of gold nanoparticles in fractionated kilovoltage X-ray treatment of melanoma. Int J Radiat Biol 94(1):8–16. https://doi.org/10.1080/09553002.2018.1393579
36. Dou Y, Guo Y, Li X et al (2016) Size-tuning ionization to optimize gold nanoparticles for simultaneous enhanced CT imaging and radiotherapy. ACS Nano 10(2):2536–2548. https://doi.org/10.1021/acsnano.5b07473
37. Chang MY, Shiau AL, Chen YH, Chang CJ, Chen HH, Wu CL (2008) Increased apoptotic potential and dose-enhancing effect of gold nanoparticles in combination with single-dose clinical electron beams on tumor-bearing mice. Cancer Sci 99(7):1479–1484. https://doi.org/10.1111/j.1349-7006.2008.00827.x
38. Singh P, Pandit S, Mokkapati VRSS, Garg A, Ravikumar V, Mijakovic I (2018) Gold nanoparticles in diagnostics and therapeutics for human cancer. Int J Mol Sci 19(7):1979. https://doi.org/10.3390/ijms19071979
39. Yang YS, Carney RP, Stellacci F, Irvine DJ (2014) Enhancing radio therapy by lipid nanocapsule-mediated delivery of amphiphilic gold nanoparticles to intracellular membranes. ACS Nano 8(9):8992–9002. https://doi.org/10.1021/nn502146r
40. Kuncic Z, Lacombe S (2018) Nanoparticle radio-enhancement: principles, progress and application to cancer treatment. Phys Med Biol 63(2):02TR01. https://doi.org/10.1088/1361-6560/aa99ce
41. Silva F, Paulo A, Pallier A et al (2020) Dual imaging gold nanoplat forms for targeted radiotheranostics. Materials (Basel) 13(3):513. https://doi.org/10.3390/ma13030513
42. Siddique S, Chow JCL (2020) Application of nanomaterials in biome dical imaging and cancer therapy. Nanomaterials (Basel) 10(9):E1700
43. Yeh YC, Creran B, Rotello VM (2012) Gold nanoparticles: preparation, properties, and applications in bionanotechnology. Nanoscale 4(6):1871–1880. https://doi.org/10.1039/C1NR11188D
44. Mieszawska AJ, Mulder WJ, Fayad ZA, Cormode DP (2013) Multifunctional gold nanoparticles for diagnosis and therapy of disease. Mol Pharm 10(3):831–847. https://doi.org/10.1021/mp3005885
45. Rana S, Bajaj A, Mout R, Rotello VM (2012) Monolayer coated gold nanoparticles for delivery applications. Adv Drug Deliv Rev 64(2):200–216. https://doi.org/10.1016/j.addr.2011.08.006
46. Arvizo R, Bhattacharya R, Mukherjee P (2010) Gold nanoparticles: opportunities and challenges in nanomedicine. Expert Opin Drug Deliv 7(6):753–763. https://doi.org/10.1517/17425241003777010
47. Zheng T, Bott S, Huo Q (2016) Techniques for accurate sizing of gold nanoparticles using dynamic light scattering with particular application to chemical and biological sensing based on aggregate formation. ACS Appl Mater Interfaces 8(33):21585–21594. https://doi.org/10.1021/acsami.6b06903
48. Yu Y, Yang T, Sun T (2020) New insights into the synthesis, toxicity and applications of gold nanoparticles in CT imaging and treatment of cancer. Nanomedicine (Lond) 15(11):1127–1145. https://doi.org/10.2217/nnm-2019-0395
49. Guo J, Rahme K, He Y, Li LL, Holmes JD, O'Driscoll CM (2017) Gold nanoparticles enlighten the future of cancer theranostics. Int J Nanomed 12:6131–6152. https://doi.org/10.2147/IJN.S140772
50. Siddique S, Chow JCL (2020) Gold nanoparticles for drug delivery and cancer therapy. Appl Sci 10(11):3824. https://doi.org/10.3390/app10113824
51. Babaei M, Ganjalikhani M (2014) A systematic review of gold nanoparticles as novel cancer therapeutics. Nanomed J 1(4):211–219
52. Huang K, Ma H, Liu J et al (2012) Size-dependent localization and penetration of ultrasmall gold nanoparticles in cancer cells, multicellular spheroids, and tumors in vivo. ACS Nano 6(5):4483–4493. https://doi.org/10.1021/nn301282m
53. Ma N, Wu FG, Zhang X et al (2017) Shape-dependent radiosensitization effect of gold nanostructures in cancer radiotherapy: comparison of gold nanoparticles, nanospikes, and nanorods. ACS Appl Mater Interfaces 9(15):13037–13048

54. Li J, Li JJ, Zhang J, Wang X, Kawazoe N, Chen G (2016) Gold nanoparticle size and shape influence on osteogenesis of mesenchymal stem cells. Nanoscale 8(15):7992–8007. https://doi.org/10.1039/C5NR08808A
55. Kumar R, Korideck H, Ngwa W, Berbeco RI, Makrigiorgos GM, Sridhar S (2013) Third generation gold nanoplatform optimized for radiation therapy. Transl Cancer Res 2(4). https://doi.org/10.3978/j.issn.2218-676X.2013.07.02
56. Caldorera-Moore M, Guimard N, Shi L, Roy K (2010) Designer nanoparticles: incorporating size, shape and triggered release into nanoscale drug carriers. Expert Opin Drug Deliv 7(4):479–495. https://doi.org/10.1517/17425240903579971
57. Morozov KV, Kolyvanova MA, Kartseva ME et al (2020) Radiosensitization by gold nanoparticles: impact of the size, dose rate, and photon energy. Nanomaterials (Basel) 10(5):952. https://doi.org/10.3390/nano10050952
58. Chen Y et al (2020) Gold nanoparticles as radiosensitizers in cancer radiotherapy. Int J Nanomed 15:9407
59. Jain S, Coulter JA, Butterworth KT et al (2014) Gold nanoparticle cellular uptake, toxicity and radiosensitisation in hypoxic conditions. Radiother Oncol 110(2):342–347. https://doi.org/10.1016/j.radonc.2013.12.013
60. Gilles M, Brun E, Sicard-Roselli C (2014) Gold nanoparticles functionalization notably decreases radiosensitization through hydroxyl radical production under ionizing radiation. Colloids Surf B Biointerfaces 123:770–777. https://doi.org/10.1016/j.colsurfb.2014.10.028
61. Schuemann J, Berbeco R, Chithrani DB et al (2016) Roadmap to clinical use of gold nanoparticles for radiation sensitization. Int J Radiat Oncol Biol Phys 94(1):189–205. https://doi.org/10.1016/j.ijrobp.2015.09.032
62. Coulter JA, Jain S, Butterworth KT et al (2012)Cell type-dependent uptake, localization, and cytotoxicity of 1.9 nm gold nanoparticles. Int J Nanomed 7:2673–2685. https://doi.org/10.2147/IJN.S31751
63. Hainfeld JF, Slatkin DN, Smilowitz HM (2004) The use of gold nanoparticles to enhance radiotherapy in mice. Phys Med Biol 49(18):N309–N315. https://doi.org/10.1088/0031-9155/49/18/N03
64. Hainfeld JF, Dilmanian FA, Zhong Z, Slatkin DN, Kalef-Ezra JA, Smilowitz HM (2010) Gold nanoparticles enhance the radiation therapy of a murine squamous cell carcinoma. Phys Med Biol 55(11):3045–3059
65. Hainfeld JF, Smilowitz HM, O'Connor MJ, Dilmanian FA, Slatkin DN (2013) Gold nanoparticle imaging and radiotherapy of brain tumors in mice. Nanomedicine (Lond) 8(10):1601–1609. https://doi.org/10.2217/nnm.12.165
66. Cifter G, Chin J, Cifter F et al (2015) Targeted radiotherapy enhancement during electronic brachytherapy of accelerated partial breast irradiation (APBI) using controlled release of gold nanoparticles. Phys Med 31(8):1070–1074
67. Joh DY, Sun L, Stangl M et al (2013) Selective targeting of brain tumors with gold nanoparticle-induced radiosensitization. PLoS ONE 8(4):e62425. https://doi.org/10.1371/journal.pone.0062425
68. Hau H, Khanal D, Rogers L et al (2016) Dose enhancement and cytotoxicity of gold nanoparticles in colon cancer cells when irradiated with kilo- and mega-voltage radiation. Bioeng Transl Med 1(1):94–102. https://doi.org/10.1002/btm2.10007
69. Abbasian M, Baharlouei A, Arab-Bafrani Z, Lightfoot DA (2019) Combination of gold nanoparticles with low-LET irradiation: an approach to enhance DNA DSB induction in HT29 colorectal cancer stem-like cells. J Cancer Res Clin Oncol 145(1):97–107. https://doi.org/10.1007/s00432-018-2769-3
70. MousavieAnijdan SH, Mahdavi SR, Shirazi A, Zarrinfard MA, Hajati J (2013) Megavoltage X-ray dose enhancement with gold nanoparticles in tumor bearing mice. Int J Mol Cell Med 2(3):118–123
71. Kanavi MR, Asadi S, Balagholi S, Alikarami F, Nosrati H, Ahmadieh H (2018) Gamma irradiation of ocular melanoma and lymphoma cells in the presence of gold nanoparticles: in vitro study. J Appl Clin Med Phys 19(3):268–275. https://doi.org/10.1002/acm2.12336

72. Liu Y, Liu X, Jin X et al (2015) The dependence of radiation enhancement effect on the concentration of gold nanoparticles exposed to low- and high-LET radiations. Phys Med 31(3):210–218. https://doi.org/10.1016/j.ejmp.2015.01.006
73. Jain S, Hirst DG, O'Sullivan JM (2012) Gold nanoparticles as novel agents for cancer therapy. Br J Radiol 85(1010):101–113. https://doi.org/10.1259/bjr/59448833
74. Chithrani DB, Jelveh S, Jalali F et al (2010) Gold nanoparticles as radiation sensitizers in cancer therapy. Radiat Res 173(6):719–728
75. Perrault SD, Walkey C, Jennings T, Fischer HC, Chan WC (2009) Mediating tumor targeting efficiency of nanoparticles through design. Nano Lett 9(5):1909–1915. https://doi.org/10.1021/nl900031y
76. Lechtman E, Chattopadhyay N, Cai Z, Mashouf S, Reilly R, Pignol JP (2011) Implications on clinical scenario of gold nanoparticle radiosensitization in regards to photon energy, nanoparticle size, concentration and location. Phys Med Biol 56(15):4631–4647. https://doi.org/10.1088/0031-9155/56/15/001
77. Chithrani DB (2010) Nanoparticles for improved therapeutics and imaging in cancer therapy. Recent Pat Nanotechnol 4(3):171–180. https://doi.org/10.2174/187221010792483726
78. Wong C, Stylianopoulos T, Cui J et al (2011) Multistage nanoparticle delivery system for deep penetration into tumor tissue. Proc Natl Acad Sci USA 108(6):2426–2431. https://doi.org/10.1073/pnas.1018382108
79. Cedervall T, Lynch I, Lindman S et al (2007) Understanding the nanoparticle-protein corona using methods to quantify exchange rates and affinities of proteins for nanoparticles. Proc Natl Acad Sci USA 104(7):2050–2055. https://doi.org/10.1073/pnas.0608582104
80. Choi CH, Alabi CA, Webster P, Davis ME (2010) Mechanism of active targeting in solid tumors with transferrin-containing gold nanoparticles. Proc Natl Acad Sci USA 107(3):1235–1240
81. Yasui H, Takeuchi R, Nagane M et al (2014) Radiosensitization of tumor cells through endoplasmic reticulum stress induced by PEGylated nanogel containing gold nanoparticles. Cancer Lett 347(1):151–158
82. Alalaiwe A, Roberts G, Carpinone P, Munson J, Roberts S (2017) Influence of PEG coating on the oral bioavailability of gold nanoparticles in rats. Drug Deliv 24(1):591–598. https://doi.org/10.1080/10717544.2017.1282554
83. Hinkley GK, Carpinone P, Munson JW, Powers KW, Roberts SM (2015) Oral absorption of PEG-coated versus uncoated gold nanospheres: does agglomeration matter? Part Fibre Toxicol 12:9. https://doi.org/10.1186/s12989-015-0085-5
84. Cao-Milán R, Liz-Marzán LM (2014) Gold nanoparticle conjugates: recent advances toward clinical applications. Expert Opin Drug Deliv 11(5):741–752
85. Charbgoo F, Nejabat M, Abnous K et al (2018) Gold nanoparticle should understand protein corona for being a clinical nanomaterial. J Control Release 272:39–53. https://doi.org/10.1016/j.jconrel.2018.01.002
86. Jokerst JV, Lobovkina T, Zare RN, Gambhir SS (2011) Nanoparticle PEGylation for imaging and therapy. Nanomedicine (Lond) 6(4):715–728. https://doi.org/10.2217/nnm.11.19
87. Liu S, Lämmerhofer M (2019) Functionalized gold nanoparticles for sample preparation: a review. Electrophoresis 40:2438–2461
88. Mousavi M, Nedaei HA, Khoei S et al (2017) Enhancement of radiosensitivity of melanoma cells by pegylated gold nanoparticles under irradiation of megavoltage electrons. Int J Radiat Biol 93(2):214–221. https://doi.org/10.1080/09553002.2017.1231944
89. Muxika A, Etxabide A, Uranga J, Guerrero P, de la Caba K (2017) Chitosan as a bioactive polymer: processing, properties and applications. Int J Biol Macromol 105:1358–1368. https://doi.org/10.1016/j.ijbiomac.2017.07.087
90. Fathy MM, Mohamed FS, Elbialy N, Elshemey WM (2018) Multifunctional chitosan-capped gold nanoparticles for enhanced cancer chemo-radiotherapy: an in vitro study. Phys Med 48:76–83. https://doi.org/10.1016/j.ejmp.2018.04.002
91. Cheung RC, Ng TB, Wong JH, Chan WY (2015) Chitosan: an update on potential biomedical and pharmaceutical applications. Mar Drugs 13(8):5156–5186. https://doi.org/10.3390/md13085156

92. Dong L, Li M, Zhang S et al (2015) Cytotoxicity of BSA-stabilized gold nanoclusters: in vitro and in vivo study. Small 11(21):2571–2581. https://doi.org/10.1002/smll.201403481
93. Al-Jawad SMH, Taha AA, Al-Halbosiy MMF, Al-Barram LFA (2018) Synthesis and characterization of small-sized gold nanoparticles coated by bovine serum albumin (BSA) for cancer photothermal therapy. Photodiagn Photodyn Ther 21:201–210. https://doi.org/10.1016/j.pdpdt.2017.12.004
94. Murawala P, Tirmale A, Shiras A, Prasad BL (2014) In situ synthesized BSA capped gold nanoparticles: effective carrier of anticancer drug methotrexate to MCF-7 breast cancer cells. Mater Sci Eng C Mater Biol Appl 34:158–167. https://doi.org/10.1016/j.msec.2013.09.004
95. Kefayat A, Ghahremani F, Motaghi H, Mehrgardi MA (2019) Investigation of different targeting decorations effect on the radiosensitizing efficacy of albumin-stabilized gold nanoparticles for breast cancer radiation therapy. Eur J Pharm Sci 130:225–233. https://doi.org/10.1016/j.ejps.2019.01.037
96. Zhou B, Song J, Wang M et al (2018) BSA-bioinspired gold nanorods loaded with immunoadjuvant for the treatment of melanoma by combined photothermal therapy and immunotherapy. Nanoscale 10(46):21640–21647
97. Koutcher L, Sherman E, Fury M et al (2011) Concurrent cisplatin and radiation versus cetuximab and radiation for locally advanced head-and-neck cancer. Int J Radiat Oncol Biol Phys 81(4):915–922. https://doi.org/10.1016/j.ijrobp.2010.07.008
98. Dasari S, Tchounwou PB (2014) Cisplatin in cancer therapy: molecular mechanisms of action. Eur J Pharm 740:364–378. https://doi.org/10.1016/j.ejphar.2014.07.025
99. Rosenblum D, Joshi N, Tao W, Karp JM, Peer D (2018) Progress and challenges towards targeted delivery of cancer therapeutics. Nat Commun 9(1):1410. https://doi.org/10.1038/s41467-018-03705-y
100. Sinha R, Kim GJ, Nie S, Shin DM (2006) Nanotechnology in cancer therapeutics: bioconjugated nanoparticles for drug delivery. Mol Cancer Ther 5(8):1909–1917. https://doi.org/10.1158/1535-7163.MCT-06-0141
101. Khoshgard K, Hashemi B, Arbabi A, Rasaee MJ, Soleimani M (2014) Radiosensitization effect of folate-conjugated gold nanoparticles on HeLa cancer cells under orthovoltage superficial radiotherapy techniques. Phys Med Biol 59(9):2249–2263. https://doi.org/10.1088/0031-9155/59/9/2249
102. Samadian H, Hosseini-Nami S, Kamrava SK, Ghaznavi H, Shakeri- ZA (2016) Folate-conjugated gold nanoparticle as a new nanoplatform for targeted cancer therapy. J Cancer Res Clin Oncol 142(11):2217–2229. https://doi.org/10.1007/s00432-016-2179-3
103. Cruz E, Kayser V (2019) Synthesis and enhanced cellular uptake in vitro of anti-HER2 multifunctional gold nanoparticles. Cancers 11(6):870. https://doi.org/10.3390/cancers11060870
104. Jóźwiak P, Lipińska A (2012) The role of glucose transporter 1 (GLUT1) in the diagnosis and therapy of tumors. PostepyHig Med Dosw 66:165–174 (Article in Polish)
105. Barron CC, Bilan PJ, Tsakiridis T, Tsiani E (2016) Facilitative glucose transporters: implications for cancer detection, prognosis and treatment. Metabolism 65(2):124–139. https://doi.org/10.1016/j.metabol.2015.10.007
106. Calvo MB, Figueroa A, Pulido EG, Campelo RG, Aparicio LA (2010) Potential role of sugar transporters in cancer and their relationship with anticancer therapy. Int J Endocrinol 2010:205357. https://doi.org/10.1155/2010/205357
107. Kong T, Zeng J, Wang X et al (2008) Enhancement of radiation cytotoxicity in breast-cancer cells by localized attachment of gold nanoparticles. Small 4(9):1537–1543. https://doi.org/10.1002/smll.200700794
108. Roa W, Zhang X, Guo L et al (2009) Gold nanoparticle sensitize radiotherapy of prostate cancer cells by regulation of the cell cycle. Nanotechnology 20(37):375101. https://doi.org/10.1088/0957-4484/20/37/375101
109. Wang C, Jiang Y, Li X, Hu L (2015) Thioglucose-bound gold nanoparticles increase the radiosensitivity of a triple-negative breast cancer cell line (MDA-MB-231). Breast Cancer 22(4):413–420. https://doi.org/10.1007/s12282-013-0496-9

110. Soleymanifard S, Rostami A, Aledavood SA, Matin MM, Sazgarnia A (2017) Increased radiotoxicity in two cancerous cell lines irradiated by low and high energy photons in the presence of thio-glucose bound gold nanoparticles. Int J Radiat Biol 93(4):407–415. https://doi.org/10.1080/09553002.2017.1268282
111. Kunjachan S, Pola R, Gremse F et al (2014) Passive versus active tumor targeting using RGD- and NGR-modified polymeric nanomedicines. Nano Lett 14(2):972–981. https://doi.org/10.1021/nl404391r
112. Ruoslahti E, Pierschbacher MD (1987) New perspectives in cell adhesion: RGD and integrins. Science 238(4826):491–497. https://doi.org/10.1126/science.2821619
113. Wu PH, Onodera Y, Ichikawa Y et al (2017) Targeting integrins with RGD-conjugated gold nanoparticles in radiotherapy decreases the invasive activity of breast cancer cells. Int J Nanomed 12:5069–5085. https://doi.org/10.2147/IJN.S137833
114. Özcelik S, Pratx G (2020) Nuclear-targeted gold nanoparticles enhance cancer cell radiosensitization. Nanotechnology 31(41):415102. https://doi.org/10.1088/1361-6528/aba02b
115. Kunjachan S, Detappe A, Kumar R et al (2015) Nanoparticle mediated tumor vascular disruption: a novel strategy in radiation therapy. Nano Lett 15(11):7488–7496. https://doi.org/10.1021/acs.nanolett.5b03073
116. Banerjee D, Harfouche R, Sengupta S (2011) Nanotechnology-mediated targeting of tumor angiogenesis. Vasc Cell. 3(1):3. https://doi.org/10.1186/2045-824X-3-3
117. Yang C, Neshatian M, van Prooijen M (2014) Cancer nanotechnology: enhanced therapeutic response using peptide-modified gold nanoparticles. J Nanosci Nanotechnol 14(7):4813–4819. https://doi.org/10.1166/jnn.2014.9280
118. Yang C, Uertz J, Yohan D, Chithrani BD (2014) Peptide modified gold nanoparticles for improved cellular uptake, nuclear transport, and intracellular retention. Nanoscale 6(20):12026–12033. https://doi.org/10.1039/C4NR02535K
119. Mackey MA, El-Sayed MA (2014) Chemosensitization of cancer cells via gold nanoparticle-induced cell cycle regulation. PhotochemPhotobiol 90(2):306–312. https://doi.org/10.1111/php.12226
120. Luo CH, Huang CT, Su CH, Yeh CS (2016) Bacteria-mediated hypoxia-specific delivery of nanoparticles for tumors imaging and therapy. Nano Lett 16(6):3493–3499. https://doi.org/10.1021/acs.nanolett.6b00262
121. Roma-Rodrigues C, Pombo I, Raposo L, Pedrosa P, Fernandes AR, Baptista PV (2019) Nanotheranostics targeting the tumor microenvironment. Front Bioeng Biotechnol 7:197
122. Zhao Z, Xu H, Li S et al (2019) Hypoxic radiosensitizer-lipid coated gold nanoparticles enhance the effects of radiation therapy on tumor growth. J Biomed Nanotechnol 15(9):1982–1993. https://doi.org/10.1166/jbn.2019.2830
123. Wang Y, Shang W, Niu M, Tian J, Xu K (2019) Hypoxia-active nanoparticles used in tumortheranostic. Int J Nanomed 14:3705–3722. https://doi.org/10.2147/IJN.S196959
124. Chen Q, Feng L, Liu J et al (2016) Intelligent albumin-MnO2 nanoparticles as pH-/H_2O_2—responsive dissociable nanocarriers to modulate tumor hypoxia for effective combination therapy. Adv Mater 28(33):7129–7136. https://doi.org/10.1002/adma.201601902
125. Ding Y, Sun Z, Tong Z et al (2020) Tumor microenvironment-responsive multifunctional peptide coated ultrasmall gold nanoparticles and their application in cancer radiotherapy. Theranostics 10(12):5195–5208. https://doi.org/10.7150/thno.45017
126. Zhang Y, Yang L, Yang C, Liu J (2020) Recent advances of smart acid-responsive gold nanoparticles in tumor therapy. Wiley Interdiscip Rev Nanomed Nanobiotechnol 12(4):e1619. https://doi.org/10.1002/wnan.1619
127. Yao L, Daniels J, Moshnikova A et al (2013) PHLIP peptide targets nanogold particles to tumors. Proc Natl Acad Sci USA 110(2):465–470. https://doi.org/10.1073/pnas.1219665110
128. Daniels JL, Crawford TM, Andreev OA, Reshetnyak YK (2017) Synthesis and characterization of pHLIP coated gold nanoparticles. BiochemBiophys Rep 10:62–69
129. Antosh MP, Wijesinghe DD, Shrestha S et al (2015) Enhancement of radiation effect on cancer cells by gold-pHLIP. Proc Natl Acad Sci USA 112(17):5372–5376. https://doi.org/10.1073/pnas.1501628112

130. Astolfo A, Schültke E, Menk RH, Kirch RD, Juurlink BH, Hall C, Harsan LA, Stebel M, Barbetta D, Tromba G, Arfelli F (2013) In vivo visualization of gold-loaded cells in mice using x-ray computed tomography. Nanomed Nanotechnol Biol Med 9(2):284–292
131. Li K, Liu G, Wu Y, Hao P, Zhou W, Zhang Z (2014) Gold nanoparticle amplified optical microfiber evanescent wave absorption biosensor for cancer biomarker detection in serum. Talanta 120:419–424
132. Daraee H, Pourhassanmoghadam M, Akbarzadeh A, Zarghami N, Rahmati-Yamchi M (2016) Gold nanoparticle–oligonucleotide conjugate to detect the sequence of lung cancer biomarker. Artif Cells Nanomed Biotechnol 44(6):1417–1423
133. Sengupta M, Bag A, Das S, Shukla A, Konathala LNS, Naidu CA, Bordoloi A (2016) Reaction and mechanistic studies of heterogeneous hydroamination over support-stabilized gold nanoparticles. ChemCatChem 8(19):3121–3130
134. Zhang J, Lakshmipriya T, Gopinath SC (2020) Electroanalysis on an interdigitated electrode for high-affinity cardiac troponin I biomarker detection by aptamer-gold conjugates. ACS Omega 5(40):25899–25905
135. Ahammad AS, Choi YH, Koh K, Kim JH, Lee JJ, Lee M (2011) Electrochemical detection of cardiac biomarker troponin I at gold nanoparticle-modified ITO electrode by using open circuit potential. Int J Electrochem Sci 6(6):1906–1916
136. Mohan RR, Tovey JC, Sharma A, Tandon A (2012) Gene therapy in the cornea: 2005–present. Prog Retin Eye Res 31(1):43–64
137. Rahaman W, Bag A, Pal S (2022) Influence of linker orientation and regulative factor(s) in liposomal gene delivery: a molecular level investigation. J Phys Chem A 126(11):1816–1822
138. Adijanto J, Naash MI (2015) Nanoparticle-based technologies for retinal gene therapy. Eur J Pharm Biopharm 95:353–367
139. Gao Q, Zhang J, Gao J, Zhang Z, Zhu H, Wang D (2021) Gold nanoparticles in cancer theranostics. Front Bioeng Biotechnol 9:647905
140. Dreifuss T, Betzer O, Shilo M, Popovtzer A, Motiei M, Popovtzer R (2015) A challenge for theranostics: is the optimal particle for therapy also optimal for diagnostics? Nanoscale 7(37):15175–15184
141. Singh M, Harris-Birtill DC, Markar SR, Hanna GB, Elson DS (2015) Application of gold nanoparticles for gastrointestinal cancer theranostics: a systematic review. Nanomed Nanotechnol Biol Med 11(8):2083–2098
142. Boulos SP, Davis TA, Yang JA, Lohse SE, Alkilany AM, Holland LA, Murphy CJ (2013) Nanoparticle–protein interactions: a thermodynamic and kinetic study of the adsorption of bovine serum albumin to gold nanoparticle surfaces. Langmuir 29(48):14984–14996
143. Chakraborty S, Joshi P, Shanker V, Ansari ZA, Singh SP, Chakrabarti P (2011) Contrasting effect of gold nanoparticles and nanorods with different surface modifications on the structure and activity of bovine serum albumin. Langmuir 27(12):7722–7731
144. Seredych M, Mikhalovska L, Mikhalovsky S, Gogotsi Y (2018) Adsorption of bovine serum albumin on carbon-based materials. C J Carbon Res 4(1):3
145. Devanabanda M, Sana SS, Vadde R, Madduri R, Venkatesan R, Eldesoky GE, Kim SC (2023) Ex vivo fluorescence imaging for the identification of rhodamine-labeled bovine serum albumin and chitosan-coated gold and silver nanoparticles. J Biophotonics 16(9):e202300110
146. Begines B, Ortiz T, Pérez-Aranda M, Martínez G, Merinero M, Argüelles-Arias F, Alcudia A (2020) Polymeric nanoparticles for drug delivery: recent developments and future prospects. Nanomaterials 10(7):1403
147. Nagarwal RC, Kant S, Singh PN, Maiti P, Pandit JK (2009) Polymeric nanoparticulate system: a potential approach for ocular drug delivery. J Control Release 136(1):2–13
148. Quinteros DA, Ferreira LM, Schaffazick SR, Palma SD, Allemandi DA, Cruz L (2016) Novel polymeric nanoparticles intended for ophthalmic administration of acetazolamide. J Pharm Sci 105(10):3183–3190
149. Boddu HSS (2012) Polymeric nanoparticles for ophthalmic drug delivery: an update on research and patenting activity. Recent Pat Nanomed 2(2):96–112

150. Mittal N, Kaur G (2019) Investigations on polymeric nanoparticles for ocular delivery. Adv Polym Technol 2019:1316249
151. Patel A, Patel M, Yang X, Mitra AK (2014) Recent advances in protein and peptide drug delivery: a special emphasis on polymeric nanoparticles. Protein Pept Lett 21(11):1102–1120
152. Mohsen AM (2022) Cationic polymeric nanoparticles for improved ocular delivery and antimycotic activity of terconazole. J Pharm Sci 111(2):458–468
153. Zhou Y, Fang A, Wang F, Li H, Jin Q, Huang L, Fu C, Zeng J, Jin Z, Song X (2020) Core-shell lipid-polymer nanoparticles as a promising ocular drug delivery system to treat glaucoma. Chin Chem Lett 31(2):494–500
154. Romeo A, Musumeci T, Carbone C, Bonaccorso A, Corvo S, Lupo G, Anfuso CD, Puglisi G, Pignatello R (2021) Ferulic acid-loaded polymeric nanoparticles for potential ocular delivery. Pharmaceutics 13(5):687
155. Bejjani RA, Behar-Cohen F, Benezra D, Gurny R, Delie F (2005) Polymeric nanoparticles for drug delivery to the posterior segment of the eye. Chimia 59(6):344–344
156. Kumari A, Yadav SK, Yadav SC (2010) Biodegradable polymeric nanoparticles based drug delivery systems. Colloids Surf, B 75(1):1–18

Fundamentals of Nano-Based Drug Delivery Systems

Yamini Shiva, Pranavi Vadla, Shreya Rodda, Divya S. Parimi, Bhagyasree Paila, Veda V. Dasari, and Anil K. Suresh

Abstract The field of drug delivery has undergone a transformative paradigm shift with the advent of nanoscience and nanotechnology. This chapter explores the fundamentals of nano-based drug delivery systems, shedding insights into the innovative approaches and key principles that underpin their design and application in the realm of biomedicine. Beginning with a comprehensive overview of nanotechnology, this chapter delves into the unique properties of nanomaterials that make them ideal for drug theragnostics. Various forms of nanocarriers, including plasmonic nanoparticles, polymeric liposomes, micelles, vesicles, and dendrimers, are discussed, emphasizing their structures, fabrication methods, and the inherent advantages they offer in optimizing imaging, diagnostics, and drug delivery. The chapter also addresses the critical role of surface modifications and targeting strategies employed to enhance the specificity and efficacy of nano-based drug carriers. Ultimately, this chapter serves as a comprehensive guide for researchers and clinicians, illustrating the clinically approved nano-based formulations for clinical purposes, fostering the realization of more effective and targeted therapeutic interventions.

Chapter Overview

This chapter highlights the background of the evolution of drug delivery systems, along with the pivotal role played by nanotechnology in revolutionizing biomedicine. Overall the chapter includes; the basics of nanotechnology defining the key principles of nanotechnology. The unique properties of nanomaterials relevant to drug delivery using examples of various nanocarriers (liposomes, micelles, nanoparticles, dendrimers) are illustrated. Alongside their structural characteristics, fabrication methods, and applications of each nanocarrier. Finally, spotlight on the targeted drug delivery systems to improve specificity and therapeutic outcomes for clinical validation and commercial purposes.

Y. Shiva · P. Vadla · S. Rodda · D. S. Parimi · B. Paila · V. V. Dasari · A. K. Suresh (✉)
Bionanotechnology and Sustainable Laboratory, Department of Biological Sciences, School of Engineering and Applied Sciences, SRM University-AP, Amaravati, India
e-mail: anil.s@srmap.edu.in

Md. M. Ansari et al. (eds.), *Emergence of Sustainable Biomaterials in Tackling Inflammatory Diseases*, Smart Nanomaterials Technology,
https://doi.org/10.1007/978-981-96-2112-5_4

1 Introduction to Nano-Cargoes

Nanoparticles are a type of matter in nanotechnology that consists of compounds with particle sizes ranging from 10 to 100 nm. Their size distributions can affect the physiochemical visualization (optical properties) of the nanomaterial such as *gold (Au), platinum (Pt), silver (Ag), and palladium (Pd).* The colours of nanoparticles are in the order of wine red, yellowish-grey, black, dark black and so forth. Nanoparticles are made up of three structural elements, to begin with a surface layer, which may be functionalized by a range of small molecules, metal ions, polymers, and surfactants. Then, the shell layer, which is a chemically varied material in contrast to the centre of this structure, in addition, the core, which is essentially the central portion of the nanoparticle and is commonly referred to as the nanoparticle itself. Drug delivery, chemical and biological sensing, gas sensing, CO_2 capture, and other related uses are only a few of the health-related uses for which NPs employed in the biomedical system [1].

Nanomaterials, with their distinctive properties that differ from their bulk counterparts, have captured the attention of science and technology for over a decade. These unique characteristics, such as exceptional optical, magnetic, and electrical properties, hold significant potential for various applications in electronics, spintronics, medicine, and other devices. While some nanomaterials occur naturally, most are synthesized through engineering methods to meet specific industrial requirements. Nanomaterials have been intentionally designed for use in various industries, including cell phones, CD-ROMs, magnetic recordings, cosmetics, drug delivery in medicine, imaging, and water splitting. Nanomaterials are deliberately engineered through specialized methods to maximize their unique properties resulting from their small size and distinct characteristics that are not observed in their bulk counterparts. The primary reasons for these differences are a large surface area and confinement effects. With a higher surface area to volume ratio, nanomaterials exhibit greater chemical reactivity, which can impact lattice pressure. The advancement in technology has allowed for the manipulation of nanostructured materials to meet specific application needs through the use of physical chemistry techniques that enable the customization of low-dimensional material properties [2].

Nanomaterials can take on various morphologies, including zero-dimensional quantum dots, one-dimensional nanowires, two-dimensional nanofilms, and three-dimensional nanoparticles. These materials can exist in different forms such as isolated, merged, accumulated, or agglomerated shapes with spherical, cylindrical, or irregular geometries [2].

2 What is Drug Delivery?

Drug delivery is the process of delivering a pharmaceutical ingredient to humans or animals in order to have a therapeutic effect. Various techniques, formulations, technologies, and systems may be used in this process in order to facilitate the drug's distribution to the intended place.

The process of giving medication to a patient in a way that increases the concentration of the drug in certain body parts relative to other body parts is known as drug delivery. Any delivery system's ultimate objective is to target, confine, and extend the medication in the diseased tissue while maintaining a safe interaction [3]. Each dosage form consists of an excipient or additive—a non-drug component—and a drug, or active pharmaceutical ingredients (APIs). The actual chemical ingredients used to treat illnesses are known as *APIs*. According to the FDA, an *API* (active pharmaceutical ingredient) is a substance that is officially recognized in pharmaceuticals and is intended for use in the diagnosis, cure, mitigation, treatment, or prevention of disease [4].

3 Nano-Based Drug Delivery Systems and Underlying Mechanisms

Due to the numerous drawbacks of conventional DDSs, nanocarriers have gained immense attention. Nanocarriers like polymeric nanoparticles, mesoporous nanoparticles, nanomaterials, carbon nanotubes, dendrimers, liposomes, metallic nanoparticles, and engineered nanomaterials are used as carriage systems for targeted delivery at specific sites of affected areas in the body [5]. Nanomedicine has rapidly grown to treat certain diseases like brain cancer, lung cancer, breast cancer, cardiovascular diseases, and many others. These nanomedicines can improve drug bioavailability and drug absorption time, reduce release time, eliminate drug aggregation, and enhance drug solubility in the blood [6, 7].

Nano-based systems can be used to transport active medicinal products to specific parts of the body. Most challenges with drug delivery, such as low water solubility and poor bioavailability, can be solved using nanotechnology. In addition, nanoparticles can overcome various physiological obstacles to increase load distribution to desired sites. Nanoparticles can carry a load of medication or therapeutic agent, such as a DNA-related substance, to enhance distribution time and deliver the drug to the target site in either a nonspecific (through enhanced permeability and retention (EPR)) or specific (through binding specific target receptors) manner [6–8].

Several hybrid NPs have recently been investigated for use in the treatment of inflammation. Current research has indicated that one or more materials are synthesized into nanostructures, completely using each material's features, for improved biocompatibility and targeting capabilities. Every structure has a unique characteristic that makes it possible to insert NPs in the right tissues. To create an NPs-based drug delivery system, the produced NPs are loaded with anti-inflammatory medications and utilized as carriers [9–11].

Pharmacokinetics: The link between a patient's dosage of a medication and the subsequent change in their physiological state—or their body's reaction to the drug—is the fundamental concept of pharmacology. A molecule's pharmacokinetics are determined by its absorption, distribution, metabolism, and excretion; many of these processes are governed by the physicochemical characteristics of the molecule [5].

Absorption: The transfer of a medication from the point of administration to the body's circulatory system is referred to as the "absorption step". There are five different ways that drugs might enter the systemic circulation: transcellular, paracellular, transport protein-mediated, pinocytosis, and endocytosis. The primary impediments to medication absorption are low permeability and low aqueous solubility. A nanocarrier can be adsorbed by entering an organism's interior or by passing across biocarriers [5].

Distribution: Drugs travel from the systemic circulation to the tissue during the distribution stage. Target delivery and blood circulation prolongation are facilitated by nanocarriers. The blood–brain barrier is one of the main obstacles to medication delivery, however, it can be overcome via transcytosis, intranasal administration, or direct injection of the drug into the central nervous system [5].

Metabolism: Drugs are broken down by enzymes during the metabolism process, which might take place in the skin, lungs, kidneys, gastrointestinal tract, or liver. A drug's metabolism happens in two stages (modification and conjugation) albeit occasionally both phases happen at the same time to complete the metabolism. Nanoparticle formulations may protect lipophilic drugs from metabolizing enzymes in the liver. Thus, while creating a medicine, these considerations must be taken into account [5].

Elimination: The body expels nanocarriers through structural deterioration. The hepatic and renal systems are the two primary routes by which nanoparticles (NPs) are eliminated from the body. To identify and phagocytose NPs, the hepatic route depends on elements of the mononuclear phagocyte system (MPS), including Kupffer cells. Kidney elimination involves glomerular filtration, active tubular secretion and tubular reabsorption. The size, charge, and form of the nanocarriers determine whether or not they are eliminated [5].

Several techniques can be used to achieve the targeted distribution of a nano drug, including conjugating with other materials like peptides, regulating particle size, increasing permeability and retention, and coating with hydrophilic (unhygienic) components. These techniques make it easier for medications to be absorbed, metabolized, distributed, and excreted. It also increases effectiveness and lessens cytotoxicity toward normal cells.

Regarding biodistribution, degradation, and drug release behaviour, nanocarriers have distinct pharmacokinetic profiles. These differences also impact the pharmacodynamic profile. Two mechanisms allow the nano-drug to concentrate in target sites: first, the drug-loaded nanocarrier diffuses into the target cell, and second, the drug-loaded nanocarrier travels to the target location and releases the drug [5, 11].

4 Roles of Nanocarriers in Drug Delivery

The cutting-edge method of treating infections, cancer, and neurological conditions is nanotechnology. Both organic and inorganic nanocarriers can be used to transport drugs. Solid–liquid nanoparticles, liposomes, dendrimers, polymeric nanoparticles, polymeric micelles, and virus-based nanoparticles are examples of organic nanocarriers. Mesoporous silica nanoparticles and carbon nanotubes are examples of inorganic nanoparticles [12]. The use of various polymeric carbon-based nano-vectors in diverse therapeutic applications is given in Fig. 1 and Table 1[13].

4.1 Solid Lipid Nanoparticles

Solid lipid nanoparticles (SLNs) are made by melting solid lipids in water and then adding an emulsifier to create a stable solution. The size of *SLNs* ranges from 50 to 1000 nm. SLNs can be used to administer heat-sensitive medications, have a low pharmacokinetic profile, and are not well physicochemical compatible. Some of the problems associated with focused and sustained medication delivery can be removed

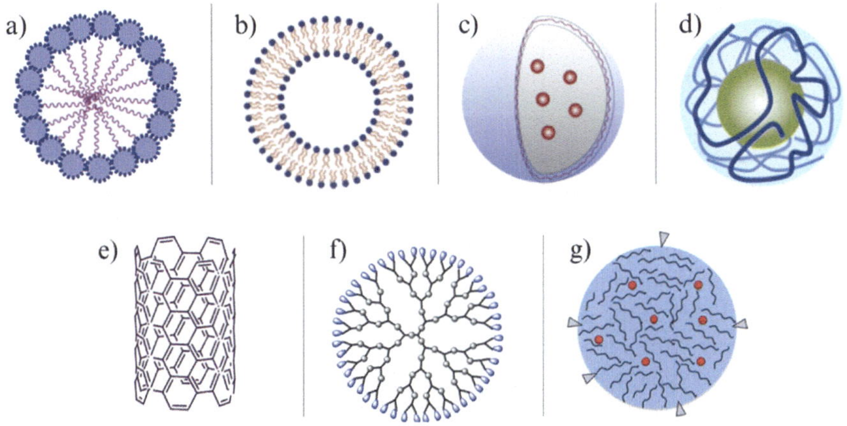

Fig. 1 Schematic representation of various polymeric carbon-based nano-vectors used in various therapeutic applications copyright@ Refs. [4, 15–18]

Table 1 List of various nanocarriers and their applications in theragnostics [22]

Nanocarrier	Applications
Nanocapsules	In the areas of vaccinations, transdermal drug delivery, peptide and protein delivery, enhancing the bioavailability and dissolving rates of poorly water-soluble medicines, and targeted medication delivery
Polymeric micelles	For the purpose of delivering photosensitizers for photodynamic treatment, targeted medication administration, gene delivery, oral drug delivery, and as pH-responsive particle emulsifiers
Nanogels	MRI contrast agent, injectable contrast agent for bioimaging, gene therapy, drug delivery, ophthalmologic applications, cancer therapy, neurodegenerative disorders, acute pulmonary inflammation treatment, bone loss disorder, vaccines, controlled release, nanoglucometer, insulin delivery, as scaffold, and the incorporation of nanomaterial into nanogel for use in nanomedicine
Nanofibers	For example, in controlled release, dressings for wound healing, scaffolds for tissue engineering, dental applications, medical implants, targeted medication administration, and the preservation of active biological molecules
Nanospheres	Hollow gold nanospheres for photothermal tumour therapy, pH-responsive magnetic nanospheres for the reversibly selective capture and release of glycoproteins
Dendrimers	In vitro diagnostics, boron neutron capture therapy, targeted drug delivery, gene delivery, oral drug delivery, transdermal drug delivery, as contrast agents, as vaginal virucide, as solublizing agents, as therapeutic agents, many uses, as hydrogels
Liposomes	In gene therapy, tailored medication administration, and the removal of generic drug toxicity

by *SLNs*. Using emulsion solvent evaporation to encapsulate *Amphotericin B* and *Paromomycin* in *SLN* can increase the bioavailability of oral medication delivery [12].

4.2 Liposomes

Liposomes are spherical vesicles with an aqueous core surrounded by a lipid bilayer that can transport bioactive substances like medications. Two microfluidic devices may be used to manufacture *liposomes*, load drugs, and remove unwanted components in a continuous process. While liposomes work well for medication delivery, they don't release material on demand. For this reason, *verteporfin* and gold nanoparticles are included in the design of X-ray-triggerable *liposomes*. According to this study, an X-ray with a 6 MeV energy stimulates the photosensitive *verteporfin* to create singlet oxygen, which in turn causes the liposomal membrane to destabilize and release its contents. use of thermosensitive liposomes, which release the medicine upon increasing the tumour's temperature by ultrasonic technology, might improve the drug's effectiveness, distribution, and concentration at the intended spot [14].

4.3 Dendrimers

Several arms that emerge from the centre of the *dendrimer* are made of sugar, nucleotides, and amino acids. Dendrimers are a useful cancer treatment because they reduce the amount of bioavailable copper in tumours. Using the advantageous properties of dendrimers in medication distribution has been the subject of several investigations. *Dendrimer* combined with N-acetyl-L-cysteine (DNAC) to reduce the risk of premature delivery. DNAS has shown the ability to decrease CD3+ T-cell recruitment to the placenta 10 mg/kg of DNAC was more effective than 100 mg/kg of N-acetyl-L-cysteine in reducing the preterm birth rate *PEGylated dendrimers* in a study demonstrated reduced cytotoxicity on cells [5].

4.4 Polymeric Nanoparticles

Colloidal particles made of biodegradable polymer, *polymeric nanoparticles* can have two forms: *nanospheres* and *nanocapsules*. The medicine is dissolved or dispersed in a liquid core (oil or water) and then enclosed in a polymeric membrane to create a *nanocapsule*. A medication is encapsulated in a polymeric matrix to create a *nanosphere*. The procedure of creating nanocapsules is intricate; thus, a one-step self-assembly technique was created to address this difficulty [5, 19].

4.5 Polymeric Micelles

Polymeric micelles consist of a spherical shell that self-assembles in aqueous conditions with the use of amphiphilic di- or tri-block copolymers. In one work, amphiphilic deblock copolymer micelles with a hydrophobic core and a hydrophilic shell were created using controlled radical polymerization. The degree of polymerization was shown to affect the doxorubicin release rate during nine hours, causing it to drop from 11.3% to 6.12%. In a study, AS1411 aptamer was covalently linked to the polymeric micelle core by including camptothecin. The results indicated 93% tumour inhibition [20].

4.6 Virus-Based Nanoparticles

Strong protein cages made by self-assembly, virus-based nanoparticles can encapsulate medicines, peptides, proteins, and nucleic acids. Because of their significantly reduced size, these protein cages have trouble functionalizing with virus-binding moieties. DNA origami triangle tiles were created with the capacity to self-assemble into icosahedral shells to solve this issue. Virus-based nanoparticles are essential for the precise delivery of hydrophobic medications to tumour cells for therapeutic purposes. A scalable and promising nanoparticle platform technology for disease imaging or detection is virus-based nanoparticles. Such drug delivery can be utilized to ensure continuous drug release and prevent off-target distribution [5].

4.7 Carbon Nanotubes

A graphene sheet is folded into a tube-like structure with a diameter that is nanoscale and a length that is a thousand times the diameter to create carbon nanotubes (CNTs). CNTs are biocompatible, flexible, and have electrical conductivity. It can have one or more walls; for example, DOX forms a stronger association with CNTs with multiple walls and improves the release of medication from CNTs with one wall. In a study, carbon nanotubes were created using the bacterium *Magneto spirillum magneticum*. This scaffold's capacitance was enhanced to six times that of pure carbon nanotubes [6].

4.8 Mesoporous Silica Nanoparticles

Mesoporous silica nanoparticles, or MSNs, are silica particles that resemble honeycombs in structure and may be produced in a range of sizes by including varied pore diameters. The most popular technique for creating MSNs is the sol–gel procedure, which includes hydrolysing and condensing precursor silicon alkoxide. One of the most prevalent problems with medication delivery is that it often reaches its target before it does. However, by employing MSNs, this problem may be solved since exposure to either internal or external stimuli will cause the pore to open and release the drug [20, 21].

4.9 Nanocapsules

Nanocapsules offers several advantages, including protection against chemical and biological degradation for peptides, hormones, proteins, enzymes, medicines, and metabolites. Incorporated drugs release slowly and continuously, show site specificity, and enhance stability and bioavailability. Modifications to the pH or the addition of polymers that are sensitive to changes in temperature, redox, or pH can alter the permeability of the capsule. Targeted intercellular protein delivery for the treatment of tumour cells is also made easier by these intelligent polymeric nanocapsules due to their diverse biological performance. Overall, nanocapsules offer broad advantages in targeted drug administration of intracellular degradable active compounds [5, 21].

4.10 Nanogels

Nanogels, three-dimensional hydrogel structures created through the cross-linking of manufactured or naturally occurring polymers, have gained popularity as a carrier system for various types of drug molecules due to their ability to trap multiple kinds of drugs such as small molecules, fluorophores, peptides, proteins, and inorganic nanoparticles. Nanogels offer benefits for polymer-based drug delivery systems due to their robust internal network that facilitates biomolecule integration and high surface area for multivalent bioconjugation [16]. Nanogels can retain biological molecules through covalent bonds with polymer chains or interactions such as electrostatic, van der Waals, hydrophobic, and hydrophilic interactions. Hydrophilic polymer chains emerge at the surface during the drug-nanogel complex collapse, enveloping the nanogel in a layer of defence. The diverse features of nanogels, including swelling, softness, stimulus response behaviour, and site specificity, make them suitable for delivering various types of drugs such as oligonucleotides, combination drugs, and proteins [18].

4.11 Nanofibers

Polymeric nanofibers with diameters less than 100 nm possess unique properties due to their high surface-to-weight ratio, making them a promising material for various applications. Electrospinning is a preferred method for large-scale production, and triaxial electrospinning has been developed for medicated nanofibers. Natural polymers such as collagen, gelatine, chitosan, hyaluronic acid, and silk fiber have been employed in tissue engineering applications. Nanofibers exhibit enhanced encapsulation efficiency, increased drug loading capacity, and simultaneous administration of multiple treatments. Their cost-effectiveness and simplicity of preparation make them a better option for medication delivery. The versatility and unique properties of nanofibers make them a promising material for various applications in different fields, including biomedical, environmental, and electronics [7].

4.12 Nanospheres

Nanospheres, with diameters ranging from 10 to 200 nm, are spherical polymeric matrices that protect drugs from degradation and provide controlled release.

Biodegradable nanospheres, such as those made of poly-lactic acid (PLA), poly-D-L-glycolide (PLG), poly-D-L-lactide-co-glycolide (PLGA), albumin, polypropylene dextran, gelatine, modified starch, and poly-cyanoacrylate (PCA), are commonly used. Magnetic nanospheres respond to magnetic forces. Nanospheres can avoid clearance by phagocytes and pass through cell gaps and tissues. The rate of drug release depends on the polymer matrix's composition and absorption capacity [13, 21].

Drug distribution by nanocarriers improves bioavailability, increases permeability, lowers dose, inhibits degradation, targets both actively and passively, lowers the likelihood of aggregation-related adverse effects, and accelerates the healing of illness. Despite the many benefits of nanocarriers, their use is restricted because of their toxicity, difficulties in scaling up the procedure, and poor effectiveness of drug loading [22].

5 Modes of Drug Delivery

Efficient drug delivery can be achieved through a rational design of nanosystems based on a comprehensive understanding of their interactions with the biological environment. This encompasses considerations such as the target cell population, specific cell-surface receptors, alterations in cell receptors associated with disease progression, the mechanism and site of drug action, drug retention, multiple drug administration, molecular mechanisms, and the pathobiology of the targeted disease. It is equally crucial to comprehend the barriers to drug delivery, such as the stability of therapeutic agents within the living cell environment. A pictorial representation of nano-cargos mediated various modes of systemic drug delivery is illustrated in Fig. 2 [23].

Various factors can contribute to reduced drug efficacy, including instability of the drug inside the cell, unavailability due to multiple targeting or chemical properties of the delivering molecules, changes in the genetic makeup of cell-surface receptors, over-expression of efflux pumps, alterations in signalling pathways with disease progression, or drug degradation. For instance, the escalation of DNA methylation during cancer progression can lead to the failure of several anti-neoplastic agents like doxorubicin and cisplatin. A more profound understanding of uptake mechanisms, intracellular trafficking, retention, and protection from degradation within a cell is essential for enhancing the efficacy of encapsulated therapeutic agents [23].

Nanomaterials have been successfully employed in the formulation of various anti-cancer drugs, including paclitaxel, doxorubicin, 5-fluorouracil, and dexamethasone. Specifically, nanoparticles utilizing polylactic/glycolic acid (PLGA) and polylactic acid (PLA) have been designed to encapsulate dexamethasone, a glucocorticoid renowned for its intracellular action. Dexamethasone, serving as a chemotherapeutic agent, exhibits both anti-proliferative and anti-inflammatory effects. Following its binding to cytoplasmic receptors, the resultant drug-receptor complex is transported

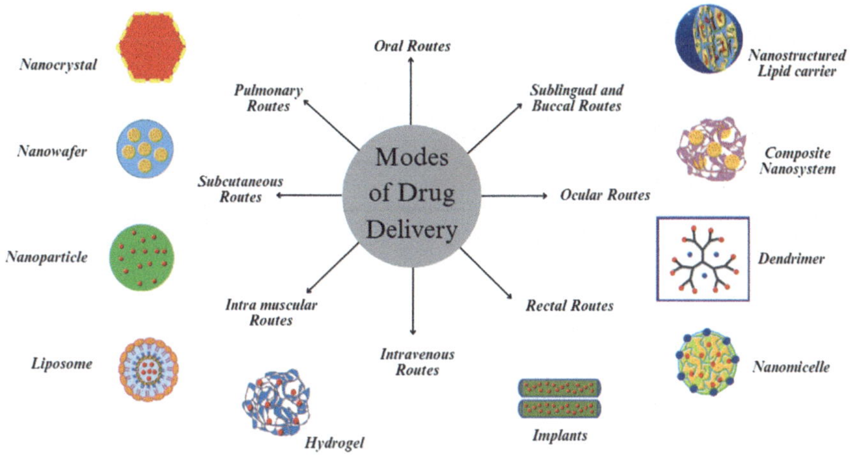

Fig. 2 Modes of various nano-based drug delivery systems and the participating nanoparticles Copyright@ Ref. [23]

to the nucleus, prompting the expression of specific genes regulating cell proliferation. These formulations of nanoparticles loaded with the drug, releasing higher concentrations over an extended period, have effectively halted the proliferation of vascular smooth muscle cells [23].

6 Nanoparticle-Based Drug Delivery

6.1 Targeting Cancer Cells with Nanoparticles

Cancer remains one of the most formidable challenges in contemporary medicine, with brain cancer posing a particularly intricate challenge for detection and treatment. The primary obstacle lies in the formidable blood–brain barrier, which hampers the passage of imaging and therapeutic agents into the brain. Apolipoprotein E has been proposed as a facilitator for drug transport across this barrier. For instance, loperamide, traditionally unable to traverse the blood–brain barrier but demonstrating antinociceptive effects upon direct injection into the brain was encapsulated into human serum albumin nanoparticles and linked to apolipoprotein. Administering this complex intravenously to mice induced antinociceptive effects in the tail-flick test, with the effectiveness of this drug delivery system contingent on the recognition of lipoprotein receptors [10].

Uniquely, Probes Encapsulated by Biologically Localized Embedding (PEBBLE), serve multiple functions while carrying various unique agents on their surface. With one target molecule guiding the PEBBLE to a tumour, another aiding in visualizing the target via magnetic resonance imaging, and a third delivering a potent dose of

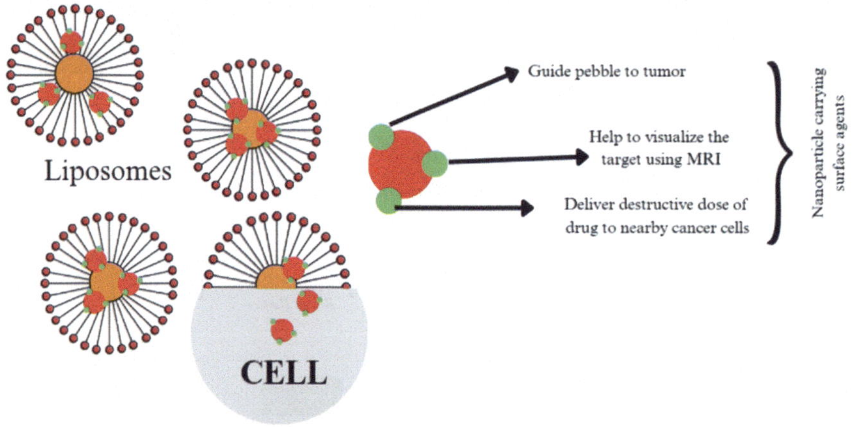

Fig. 3 PEBBLE consisting of nanoparticles carrying three surface agents for combinatorial theragnostics

drug or toxin to nearby cancer cells, these polymer spheres emerge as powerful tools against cancer. In another example, the anti-cancer drug doxorubicin, when bound to polysorbate-coated nanoparticles, effectively traverses the intact blood–brain barrier and releases therapeutic concentrations in the brain. superparamagnetic iron oxide particle conjugates can precisely target and locate brain tumours [10]. The multimodality functional capability of PEBBLE for diverse surface antigens for targeting cancer cells is depicted in Fig. 3.

6.2 *Targeting Angiogenesis with Nanoparticles*

The aggressive growth of tumours is facilitated by angiogenesis, making it imperative to inhibit this process as a means of starving tumour cells. Recent evidence underscores the significant regulatory roles played by integrin αvβ3 and vascular endothelial growth factors (VEGFs) in angiogenesis. Consequently, selectively targeting αvβ3 integrin and VEGFs emerges as a novel anti-angiogenesis strategy applicable to a wide range of solid tumours. One method involves coating nanoparticles with peptides designed to bind specifically to αvβ3 integrin and the VEGF receptor [11].

A synthetic peptide containing the Arg-Gly-Asp (RGD) sequence is known for its specific binding to αvβ3 integrin expressed on endothelial cells within angiogenic blood vessels. This interaction holds the potential to effectively inhibit tumour growth and proliferation. By undergoing hydrophobic modifications, glycol chitosan can form self-aggregated nanotubes and has been utilized as a carrier for the RGD peptide, labelled with fluorescein isothiocyanate (FITC-GRGDS). These nanotubes loaded with FITC-GRGDS could prove valuable for monitoring or disrupting the angiogenic tissue and blood vessels surrounding tumour tissue [11].

Rosette nanotubes (RGDSK-RNT), self-assembling from RGDSK, represent a unique class of *nanotubes* inspired by biological structures, naturally water-soluble upon synthesis. Constructed from a guanine-cytosine motif, these nanotubes possess the distinctive ability to accept various functional groups at the G/C motif, providing functional versatility for specific medical or biological applications. Consequently, RNTs have the potential to be modified for in vivo targeting of various therapeutic molecules, offering promising avenues for the treatment of cancer and inflammatory diseases [10].

7 Nanosystems in Tacking Inflammation

7.1 Targeting Macrophages to Control Inflammation

Leveraging the swift recognition and clearance abilities of macrophages for foreign particles provides a rational basis for directing nanoparticles to macrophages. Macrophages, integral in regulating inflammation across various diseases, emerge as potential targets for pharmaceutical interventions in both human and animal diseases. Despite their proficiency in eliminating most microbes, certain microorganisms like *Toxoplasma gondii, Leishmania sp., Mycobacterium tuberculosis,* and *Listeria monocytogenes* have developed strategies to resist macrophage phagocytosis. These pathogens circumvent the molecular machinery designed by macrophages for their destruction, taking refuge in modified lysosomes. Hence, nanoparticle-mediated delivery of antimicrobial agents into pathogen-containing intracellular vacuoles within macrophages stands as a promising strategy for eradicating cellular reservoirs. This method aims to achieve therapeutic drug concentrations in the vacuoles of infected macrophages, mitigating side effects associated with drug administration and the release of pro-inflammatory cytokines [23].

Polyalkylcyanoacrylates (PACA) nanoparticles have been harnessed as carriers for targeting antileishmanial drugs into macrophages. This nanomaterial demonstrates the advantage of not inducing interleukin-1 release by macrophages, suggesting its potential in addressing macrophage infections in chronic diseases. The antifungal and anti-leishmanial agent amphotericin B (AmB) has been intricately complexed with lipid-based nanotubes to create a less toxic formulation of AmB. The resultant AmB in trilaurin-based nanosized lipid particles (emulsomes), stabilized by soya phosphatidylcholine, introduces a novel intravenous drug delivery system for macrophage targeting [11].

The nanocarrier-mediated delivery of macrophage toxins proves effective in eliminating undesired macrophages in gene therapy and various clinically relevant scenarios, encompassing autoimmune blood disorders, T-cell-mediated autoimmune diabetes, rheumatoid arthritis, spinal cord injury, sciatic nerve injury, and restenosis following angioplasty. Alternatively, the exploration of nanoparticles with macrophage-lethal properties remains a viable avenue. The strategic utilization of diverse macrophage cell receptors as therapeutic targets may offer a more potent approach for antigen delivery and targeting, employing particulate nanocarriers [23].

7.2 Targeting Inflammatory Molecules

In the past two decades, the identification of numerous cell adhesion molecules has advanced significantly. These glycoproteins, situated on the cell surface, function as receptors for both cell-to-cell and cell-to-extracellular matrix adhesion. Categorized into four classes—integrins, cadherins, selectins, and the immunoglobulin superfamily—these molecules are pivotal for the effective migration of inflammatory cells like neutrophils and monocytes into inflamed organs, contributing to the host's response to infections. However, there is substantial evidence suggesting that an excessive migration of neutrophils into inflamed lungs can lead to heightened tissue damage and mortality. Consequently, substantial efforts are being made to finely regulate neutrophil migration into inflamed organs. The recent advancements in understanding cell adhesion molecules have significantly influenced the design and development of drugs, including peptides and proteins, for potential use in treating conditions such as cancer, heart diseases, and autoimmune diseases. These molecules play crucial roles in various diseases, including cancer, thrombosis, and autoimmune conditions like type-1 diabetes [10].

RGD peptides have been employed to target integrins $\alpha v\beta 3$ and $\alpha v\beta 5$, while peptides derived from intercellular adhesion molecule-1 (ICAM-1) target the $\alpha v\beta 2$ integrin. Peptides derived from $\alpha v\beta 2$ can effectively target ICAM-1 expressing cells. Cyclic RGD peptides have been conjugated to paclitaxel (PTX-RGD) and doxorubicin (Dox-RGD4C) to enhance the specific delivery of these drugs to tumour cells. In mice bearing human breast carcinoma cells (MDA-MB-435), treatment with Dox-RGD4C resulted in survival, while all untreated control mice succumbed to the disease. This conjugate specifically targets $\alpha v\beta 3$ and $\alpha v\beta 5$ integrins on tumour vasculature during angiogenesis [11].

8 Why Nanoparticles for Drug Delivery?

Traditional molecules have been used by humans for millennia to treat a wide range of illnesses. This is primarily due to the special benefits that come with natural medications, which include reduced toxicity and adverse effects, affordability, and promising therapeutic potential. As a result of this, they are good leads for the development of new medications [7]. However, using natural medications presents a great challenge due to concerns associated with the biocompatibility and toxicity of natural compounds. Therefore, one way to potentially address these important problems is to use novel drug delivery systems that allow medications to be targeted at particular body parts. As a result, nanotechnology is crucial to advanced medicine and drug formulations, focusing on the targeted area and successfully delivering controlled drug release [24].

Nanomedicine is the area of medicine that applies the science of nanotechnology to the prevention and treatment of different diseases. In comparison to conventional drugs, nanoparticles possess improved properties like auto-reactive stability and self-reassembly, which make them adaptable and capable of being altered to acquire desired properties like large surface areas. These characteristics enable them to specifically target fatigued cells while avoiding healthy ones, which can boost medication efficacy and lessen adverse effects [24].

Applications of nanotechnology and nanobiotechnology will become more significant in medical technology over the next ten years. The standard of living of patients could be greatly enhanced by nanomedicine. The future of nano therapy involves treating patients at the molecular level as soon as the disease is in its development stage using specially designed medications, or "personalized medicine." Here, the creation of nanodrugs and the many approaches to the targeted transportation of active ingredients (drug delivery) will be crucial. These could make it feasible to create well-tolerated, efficient treatments for illnesses that were previously incurable. Theragnostics is a new specialty medical field where therapy and diagnostics are increasingly combined because drugs and nanotechnology techniques can be used for both therapeutic and diagnostic purposes [24].

Recently, theragnostic applications have shown effectiveness in modulating chimeric antigen receptors (CARs) on-switch for the treatment of chemotherapy-resistant B cell cancers. In a draconic engineered T cell therapeutic approach, a switch-on-switch-off mechanism could be successfully established by introducing small molecules, thereby avoiding the harmful effects of the cell killer function [22].

Nanobiotechnology presents new opportunities, above all, in the field of regenerative medicine. The process known as "tissue engineering" allows for the artificial production or regeneration of damaged or absent tissue, including hair, cartilage, and bone, muscles, organs, and nerve cells. This is accomplished through stimulation and targeted control of cell growth. Nanoporous carrier materials are already being used as matrices for controlled cell growth in plastic surgery and wound healing. Successful targeted growth of nerve cells could lead to the development of new treatments for neurological diseases like Parkinson's, Alzheimer's, multiple sclerosis, and epilepsy, which were previously incurable [22, 25].

Gene therapy is the field where nanomedicine is expected to grow in significance in the upcoming years. A great deal of research is being done to determine whether and how different mechanisms of targeted transport of active substances can be used to introduce individual genes, DNA fragments, and nucleic acids using non-viral nanoparticles into tissue and cells. A further procedure is magnetofection, in which negatively charged nucleic acids are loaded onto positively charged iron-containing nanoparticles, similar to active MDT. These nanoparticles are then transported to the target cells using a magnetic field and, because of their surface-structure preparation, sluiced into them. There, the particles enter the cell's iron metabolism, and the nucleic acids are released. This technique, also known as "magnetic gene targeting" (MGT), has the benefit of not harming the target cells, making it particularly useful for treating neurodegenerative illnesses, cancer, and myocardial infarction [26].

Hence, nanoparticles are essential due to their ability to readily enter target cells and build up in subcellular structures to alter cellular functions, which could be advantageous in the management of chronic illnesses like diabetes, cancer, and kidney disorders. Because of this, the Federal Drug Administration in the US has already approved a large number of these nanoparticles for clinical use, which are discussed in the next section [26].

9 Clinically Approved Nano-Based Drugs and& Drug Delivery Platforms

In recent years, both the broadening of nanocarrier typology and the increase in the complexity of particles and materials employed have inspired examinations into novel nanodelivery technologies produced a range of goods and led to a large number of biotechnology and nanomedicine clinical trials and applications. Regulatory bodies must first validate nanocarriers through a variety of preclinical and clinical procedures before granting premarket authorization. Currently, the approval process for nanomedicines in humans is regulated by the Food and Drug Administration (FDA) in the USA and is essentially the same as it would be for any other biologic, device, or regulated medicine [22, 26].

When a medication is approved by the FDA, it means that CDER has assessed the effect data of the medication and determined that, for the intended population, the medication's benefits outweigh any known or prospective hazards. An estimated $1 billion and ten to fifteen years are needed for the full procedure for each new medicine. Animal studies are typically conducted during the pre-clinical stage of testing once the material has been discovered or invented in order to show its safety, efficacy, toxicity profile, and acceptable dose ranges. While polymeric, liposomal, and nanocrystal formulations account for the majority of FDA-approved materials, there is a trend toward the development of more complex materials that include protein-based NPs, micelles, and the emergence of a range of inorganic and metallic particles in clinical trials (Fig. 4) [5].

Inorganic nanoparticles have also been approved by the FDA for clinical use. Some examples of inorganic nanoparticles that have received FDA approval include gold nanoparticles for cancer imaging and phototherapy, iron oxide nanoparticles for magnetic resonance imaging (MRI) contrast enhancement, and silica nanoparticles for drug delivery. There are several types of organic nanoparticles that have been approved by the FDA for clinical use. Some examples include liposomes, polymeric nanoparticles, and dendrimers [12].

Liposomes are spherical structures made of lipid bilayers that can encapsulate drugs, genes, or other therapeutic agents for targeted delivery. Polymeric nanoparticles can be designed to release drugs in a controlled manner, making them ideal

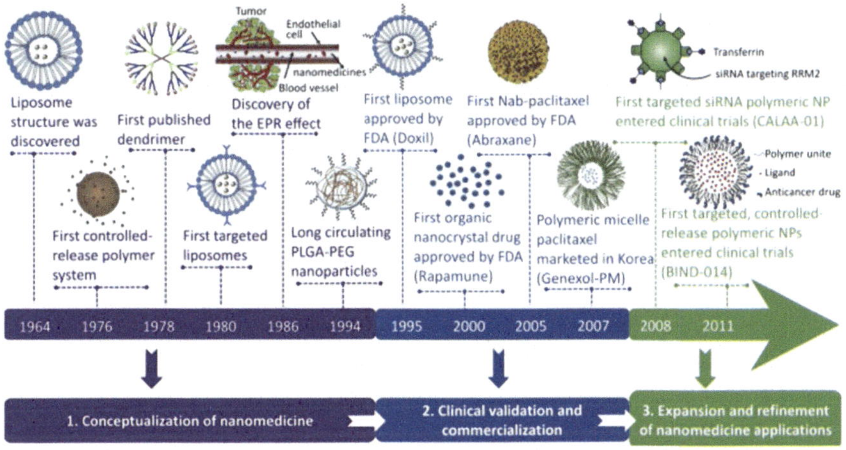

Fig. 4 Timeline advent and milestones of nano-based formulations in nanomedicine towards clinical approach Copyright@Ref. [5]

for sustained-release therapies. Dendrimers are highly branched, tree-like structures that can be functionalized with various targeting moieties and therapeutic agents. These organic nanoparticles have been approved by the FDA for a variety of medical applications, such as cancer treatment, gene therapy, and vaccine delivery. As with inorganic nanoparticles, the safety and biocompatibility of organic nanoparticles are carefully evaluated by the FDA before approval to ensure that they do not pose any potential risks to human health [5, 25] (Table 2).

10 Limitations of Nanoparticles in Drug Delivery

While nanoparticles hold great promise in drug delivery due to their small size, high surface area, and ability to target specific cells and tissues, several important limitations must be considered. Nanoparticles in medicine face challenges in safety and toxicity, high cost, stability and biocompatibility, regulatory environment, and clinical translation. Safety concerns arise from the unpredictable interaction of nanoparticles with biological systems, and more research is needed to ensure safe use [23]. The high cost of production and purification limits commercial viability. Stability and biocompatibility can be affected by factors such as pH, temperature, and storage conditions. Regulatory hurdles add time and cost to development, and manufacturing challenges limit the ability to produce nanoparticles with desired properties. Few successful clinical translations have occurred due to safety concerns, regulatory hurdles, and technical challenges in scaling up production and delivering nanoparticles to patients [13, 19].

Table 2 Inorganic and metallic nano-based drugs approved by the food and drug administration [5]

Sl. no	Drug name and company	Nanoparticle type	Nanoparticle advantage	Applications	Approval
1	NFeD (sanofi avertis), dexIron/ dexferrum (sanofi avertis)	Iron dextran (low MW) Iron dextran (high MW)	Permits a higher dosage	Iron deficiency in chronic kidney disease (CKD)	1957
2	Feridex/Endorem (AMAG pharmaceuticals) Magnetite (MagnaChem) GastroMARK; umirem (AMAG Pharmaceuticals)	SPION coated with dextran Iron oxide SPION coated with silicone	Superparamagnetic character Magnetic hyperthermia Superparamagnetic character	Imaging	1996 2008 2001
3	Ferrlecit (sanofi avertis)	Sodium ferric gluconate	Enables a higher dosage	In chronic kidney disease (CKD), iron deficiency	1999
4	Venofer (luitpold pharmaceuticals)	Iron sucrose	Allows increased dose	Iron deficiency in chronic kidney disease (CKD)	2000
5	Feraheme/ ferumoxytol (AMAG pharmaceuticals	Ferumoxytol SPION with polyglucose sorbitol carboxymethylether	Magnetite suspension allows for prolonged steady release, decreasing number of doses	Deficiency anemia iron deficiency in chronic kidney disease (CKD)	2009
6	Nanotherm (MagForce)	Iron oxide	Longer half life	Glioblastoma	2010

10.1 Limitations of Various Nanocarriers in Drug Delivery

Nano-cargo category-wise limitations of various nano-cargos for their use in therapeutic applications are given below [6, 7, 14, 19, 20, 23, 24].

Liposomes:

- Might crystallize if stored for a long time;
- Inadequate management of the drug release rate;
- Inability to overcome biological obstacles;
- Adequate loading of medications whose pH and ion gradients are not applicable;
- Fusion and leakage of loaded medication;
- Phospholipids can oxidize as well as hydrolysis.

Dendrimers:

- The intricate process of synthesis;
- Potential for partial reactions involving terminal groups;
- Steric hindrance to dendrons and the core molecule prevents high-generation dendrimer formation.

Exosomes:

- Quick removal from the blood
- Currently used techniques have poor drug loading and retention. Massive extraction and purification processes are challenging

Metal nanoparticles:

- Cytotoxicity and poor biocompatibility could arise from RES uptake;
- Amorphous nature of nanoparticles;

Silica nanoparticles:

- Further research on cytotoxicity is required;
- Hemolysis is caused by interactions between high-surface silanol groups and the phospholipids in red blood cell membranes.

Carbon nanotubes:

- Increased cytotoxicity, carcinogenicity, and immunogenicity;
- Not biodegradable;
- Insufficient water solubility and absorption.

Nano capsules/nanospheres:

- Non-biodegradable polymers build up in tissues;
- The pathways for in vivo metabolism and elimination are unclear.

Quantum dots:

- Reactive oxygen species (ROS) are produced when heavy metals like cadmium leach as a result of quantum dot degradation and elevated cytotoxicity.

Nanofibers:

- Scalability is a problem;
- Inadequate management of nanofiber dimensions;
- In the electrospinning process, the solvent system for each polymer must be optimized.

10.2 Limitations Based on Modes of Drug Delivery

Oral Drug Delivery System: This mode of drug delivery presents challenges for pediatric, geriatric, and mentally challenged patients. Administering the drug can be difficult, and its stability may be compromised in the gastrointestinal tract environment, resulting in diminished bioavailability. Moreover, there is a risk of unnecessary drug accumulation [23].

Buccal Drug Delivery System: The drug has a limited residence time, making it susceptible to scattering. Some drugs may be unstable in the buccal pH, and those requiring high dosages cannot be effectively administered [11].

Rectal Drug Delivery System: This approach may induce rectal irritation, and absorption can be disrupted by defecation [11].

Intravenous Drug Delivery System: Potential undesirable immune reactions, especially with proteins and peptides, are a concern. Thrombophlebitis, tissue necrosis, and air embolism are also possible [23].

Subcutaneous Drug Delivery System: Absorption is gradual, painful and irritating at the injection site and accumulation of unabsorbed drug can lead to tissue damage [23].

Intramuscular Drug Delivery System: Pain at the injection site is a potential drawback. The quantity injected depends on muscle mass, and there is a risk of hematoma. Accidental rupture of blood vessels may result in the incorporation of the drug into the bloodstream [23].

Pulmonary Drug Delivery System: Aerodynamic filters may impede efficient deposition, and the mucus lining may transport deposited agents toward the throat [23].

Transdermal Drug Delivery System: The skin is impermeable to hydrophilic components, and there is a risk of enzymatic deterioration. Incomplete utilization of the drug may occur [23].

11 Conclusion

In conclusion, the fundamentals of nano-based drug delivery systems represent a ground-breaking avenue in pharmaceutical research and development. The remarkable progress in nanotechnology has paved the way for designing innovative strategies to enhance drug delivery, providing unprecedented precision and efficiency. Through the manipulation of nanoscale materials, researchers have achieved targeted delivery, prolonged drug release, and improved therapeutic efficacy while minimizing adverse effects. Nano-based drug delivery systems have demonstrated their potential to revolutionize the treatment landscape for various diseases, including cancer, inflammatory diseases, and chronic conditions. The ability to overcome biological barriers, optimize pharmacokinetics, and tailor drug formulations to specific patient needs holds great promise for personalized medicine. In the future, advancements in nanotechnology and drug delivery research are likely to lead to the development of increasingly sophisticated nano-based systems. These systems may incorporate smart materials, smart-systemic self-navigating and self-endured robots, responsive drug release mechanisms, and real-time monitoring capabilities. As we navigate this exciting frontier, ongoing research efforts and a commitment to ethical and responsible innovation will be crucial to unlocking the full potential of nano-based drug delivery systems and improving patient outcomes in the realm of modern medicine.

Conflict of Interest The author declares no competing interest.

References

1. Sahu T et al (2021) Nanotechnology based drug delivery system: current strategies and emerging therapeutic potential for medical science. J Drug Deliv Sci Technol 63:102487
2. Hong NH (2019) Introduction to nanomaterials: basic properties, synthesis, and characterization. In: Nano-sized multifunctional materials. Elsevier, pp 1–19
3. Mitra AK, Kwatra D, Vadlapudi AD (2014) Drug delivery
4. Vinothini K, Rajan M (2019) Mechanism for the nano-based drug delivery system. In: Characterization and biology of nanomaterials for drug delivery. Elsevier, pp 219–263
5. Sultana A et al (2022) Nano-based drug delivery systems: conventional drug delivery routes, recent developments and future prospects. Med Drug Discov 15:100134
6. Yusuf A et al (2023) Nanoparticles as drug delivery systems: a review of the implication of nanoparticles' physicochemical properties on responses in biological systems. Polymers 15(7):1596
7. Patra JK et al (2018) Nano-based drug delivery systems: recent developments and future prospects. J Nanobiotechnol 16(1):1–33
8. Obaid A et al (2022) Nanoparticles in drug delivery: from history to therapeutic applications. Nanomaterials 12(24):4494
9. Abdellatif AAH, Alsowinea AF (2021) Approved and marketed nanoparticles for disease targeting and applications in COVID-19. Nanotechnol Rev 10(1):1941–1977
10. Wang H et al (2021) Update on nanoparticle-based drug delivery system for anti-inflammatory treatment. Front Bioeng Biotechnol 9:630352
11. Afshari AR et al (2022) Nanoparticle-based drug delivery systems in cancer: A focus on inflammatory pathways. In: Seminars in cancer biology, vol 86. Academic Press
12. Patra JK et al (2018) Nano based drug delivery systems: recent developments and future prospects. J Nanobiotechnol 16(1):1–33
13. Alshawwa SZ et al (2022) Nanocarrier drug delivery systems: characterization, limitations, future perspectives and implementation of artificial intelligence. Pharmaceutics 14(4):883
14. Nsairat H et al (2022) Liposomes: structure, composition, types, and clinical applications. Heliyon
15. Oti VB (2020) Nanoparticles and its implications in HIV/AIDS therapy. Curr Drug Discov Technol 17(4):448–456
16. Chan M, Almutairi A (2016) Nanogels as imaging agents for modalities spanning the electromagnetic spectrum. Mater Horiz 3(1):21–40
17. Fadeel B, Pietroiusti A, Shvedova AA (eds) (2017) Adverse effects of engineered nanomaterials: exposure, toxicology, and impact on human health. Academic Press
18. García MC et al (2018) Self-assembled nanomaterials. In: Nanobiomaterials. Woodhead Publishing, pp 41–94
19. Zielińska A et al (2020) Polymeric nanoparticles: production, characterization, toxicology and ecotoxicology. Molecules (Basel, Switzerland) 25(16):3731

20. Bose A et al (2021) Nanomicelles: types, properties and applications in drug delivery. IET Nanobiotechnol 15(1):19–27
21. Mendes M et al (2018) Clinical applications of nanostructured drug delivery systems: From basic research to translational medicine. In: Core-shell nanostructures for drug delivery and theranostics. Woodhead Publishing, pp 43–116
22. Hoseinnejad M et al (2018) Inorganic and metal nanoparticles and their antimicrobial activity in food packaging applications. Crit Rev Microbiol 44(2):161–181
23. Suri SS, Fenniri H, Singh B (2007) Nanotechnology-based drug delivery systems. J Occup Med Toxicol 2:1–6
24. Krukemeyer MG et al (2015) History and possible uses of nanomedicine based on nanoparticles and nanotechnological progress. J Nanomed Nanotechnol 6(6):336
25. Li T, Mattei A (eds) (2018) Pharmaceutical crystals
26. Zhang C, Yan L, Wang X, Zhu S, Chen C, Gu Z, Zhao Y (2020) Progress, challenges, and future of nanomedicine. Nano Today 35:101008

Surface Modification, Functionalization and Bioconjugation of Nanomaterials for Targeted Drug Delivery

Anas Ahmad, Rakesh Kumar Mishra, Ajay Kumar, Akshay Vyawahare, Kanika, Aneesh Ali, Md. Meraj Ansari, and Rehan Khan

Abstract Nanomaterials (NMs) are ultrafine materials made of different organic and/or inorganic fabrics that have a size range of 10 to 300 nm. Nanomaterials are widely applicable in many fields, such as energy and the environment, health and medicine, and disease detection, treatment, and prevention. At the beginning of their research and exploration, naïve nanomaterials were investigated for their potential applications in all of these fields. However, as time went on and research methodologies, instruments, and techniques advanced, a significant amount of changes and adjustments to their structural, morphological, and physicochemical properties became apparent. Due to this, their capacities and properties such as targetability, drug entrapment efficiency, cellular uptake, preferential accumulation at the desired site and specific interactions with the desired receptors have improved significantly. Additionally, their therapeutic efficacy has improved, leading to a notable increase in the likelihood that they will be useful in clinical settings. The functionalization of nanomaterials in general and nanoparticles in particular to alter their surfaces involves a process that seeks to significantly enhance and add some beneficial qualities for the usage of nanoparticles in biomedical applications. Various kinds of nanomaterials can be employed in the initial stages of functionalization because of their unique chemical

---

A. Ahmad
Department of Microbiology, Immunology and Infectious Diseases, Cumming School of Medicine, Julia McFarlane Diabetes Research Centre (JMDRC), University of Calgary, Foothills Medical Centre, Snyder Institute for Chronic Diseases, Hotchkiss Brain Institute, Calgary, Canada

R. K. Mishra · A. Kumar · A. Vyawahare · Kanika · A. Ali · R. Khan (✉)
Institute of Nano Science and Technology (INST), Knowledge City, Sahibzada Ajit Singh Nagar, Punjab, India
e-mail: rehankhan@inst.ac.in

Md. M. Ansari
Department of Animal Biotechnology, Faculty of Biotechnology, College of Applied Life Sciences, Jeju National University, Jeju Special Self-Governing Province, Jeju-si, South Korea

Interdisciplinary Graduate Program in Advanced Convergence Technology and Science, Jeju National University, Jeju Special Self-Governing Province, Jeju-si, South Korea

Md. M. Ansari et al. (eds.), *Emergence of Sustainable Biomaterials in Tackling Inflammatory Diseases*, Smart Nanomaterials Technology,
https://doi.org/10.1007/978-981-96-2112-5_5

characteristics and exposed functional groups. In this chapter, we discussed surface modification, functionalization and bioconjugation of nanomaterials for targeted drug delivery.

Keywords Nanomaterials · Surface modification · Functionalization · Bioconjugation · Targeted drug delivery

1 Introduction

Nanomaterials (NMs) are ultrafine materials with a size range between 10 and 300 nm and composed of various organic and/or inorganic fabrics. Nanomaterials have found a wide range of employability in various arenas including energy and environment, health and medicine and diagnosis, treatment and prevention of diseases [1, 2].

Earlier during the initial phases of their research and exploration, naïve nanomaterials were being studied for their applications in all these areas; however, with the passage of time and advancement in research methodologies and tools and techniques, a great degree of alterations and modifications in their structural, morphological and physicochemical characteristics have come into the picture [3, 4]. This has led to some significant improvements in their capacities and properties like targetability, drug entrapment efficiency, cellular uptake, preferential accumulation at the desired site, specific interactions with the desired receptors, significantly reduced toxicological outcomes, improved therapeutic efficacy and as a result quite an appreciable improvement in enhancing chances of their applicability in the clinical settings as well. As far as their applications in nanomedicine and biomedical areas are concerned, these nanomaterials have been studied to great extents due to their unique characteristics and possess suitable applications in a large number of areas including biomolecule detection, vaccines, regenerative medicine, and tissue engineering, gene and drug delivery, cancer therapy, high accuracy diagnosis and theranostics [5–8].

Surface alterations by functionalization of nanomaterials in general and nanoparticles, in particular, implicate a method that aims to appreciably improve as well as add some useful properties for the applications of nanoparticles in their biomedical applications. Different types of nanomaterials have characteristic chemical properties and functional groups exposed on their surface to be used in the first steps of functionalization [9–11]. Surface biofunctionalization of these nanoparticles by exploitation of their surface characteristics as well as by using ligands that possess a strong affinity and specificity for many of the disease-specific receptors overexpressed on the diseased body cells or expressed proteins in disease-specific microenvironment. Firstly passive and then active targeting are the phenomenon that often occur side by side, and one may not hamper the other in their peculiar characteristics [12, 13].

For example, in the case of antitumor therapy, there should always be an ample drug concentration in the blood circulation so as to permit for an optimum and effective dose at the tumor locality. The drug target should be powerfully inhibited, with the purpose that is quite essential for cancer cell viability. The desired drug in question must have a preferential higher toxicity for cancer or a favourable and desirable therapeutic range. Many researchers have provided a wide spectrum of drugs like hydrophilic and hydrophobic pharmacological agents, proteins, enzymes and vaccines, as well as other biological macromolecular compounds employing nanoparticles as some of the very efficient and targeted carriers [14, 15].

Multifunctionalised nanoparticle-mediated drug delivery paradigms could effectively manipulate the targeting, diagnosis, and therapeutic potential of these nanoparticles under study. There have already been many astonishing numbers of forthcoming applications of these agents. This purpose either takes the advantage of the unique and specific characteristics of these nanomaterials as some drugs or their components are uniquely and specifically designed for some of the newer advances so as to regulate and control drug release, amplify and improve drug targetability, and improvement in overall properties of drugs which exhibit low bioavailability [16, 17]. Surface functionalization of nanoparticles and nanomaterials also aids in the stabilization of these NPs and allows them to possess specific shapes as well as size and maximize their receptor affinity. It also prevents aggregation as well as agglomeration and settling down of the colloidal nanosuspensions and creates various sites on their nano-surfaces for the attachments which are vitally necessary for all these ligands in order to offer targeting potential to these nanoparticles [18–21].

It is only because of the possibility of altering their physicochemical characteristics of nanomaterials in general and nanoparticles in particular, that it has become possible to improve nanoparticles' biocompatibility and cellular uptake efficiency by functionalization of nanoparticles' surface. In this review, we discuss various recent studies in which nanoparticles' surface modification improves their biocompatibility and enhance cellular uptake. Furthermore, many of the most widely used techniques employed to evaluate nanoparticles' biocompatibility and cellular uptake as well as other physicochemical and pharmacological properties have also been reported [22, 23].

2 Types of Nanomaterial-Based Platforms for Targeted Drug Delivery

a. **Polymeric nanoparticles**

Nanomaterials have been categorized into two chief classifications: Nano-structured as well as nanocrystalline. Nano-structured formulations can be further divided into polymeric nanosytems, non-polymeric nanoparticles, and lipid-based nanocarriers. Polymeric nanoformulations include nanoparticles, micelles, nanogels, protein nanoparticles, as well as drug conjugates. Non-polymeric nanosystems contain silica-containing nanoformulations, metallic nanoformulations, carbon nanotubes, nanodiamonds, quantum dots, and. Lipid-containing nanoformulations can further be categorized into liposomes as well as solid lipid nanoformulations. So far, almost all of these nanoparticulate formulations that have been clinically approved for therapeutic purposes are either polymeric nanosystems or lipid-based formulations [24].

The surface modifying capability of nanomaterials with long-chain biocompatible and biodegradable polymeric systems like polyethylene glycol (PEG) has been exhibited to reduce the non-specific protein absorption on these nanoparticulate surfaces. Because of their inherent physicochemical characteristics, polyethylene glycol acts as a favorable polymer for the formulation of therapeutic nanoparticles, which reduces the uptake of nanoparticles by phagocytosis and minimizes their concentrations in off-target organ systems [25]. Many of the factors including but not limited to length, surface morphological features, size, shape, as well as density of poly ethylene glycol chains regulate the surface hydrophilicity and therefore, must be taken into preformulation studies before coating of nanoparticles for their therapeutic potential. Conjugation of the targeted ligand onto the surface of these nanoformulations can appreciably enhance the target specificity and controlled drug delivery through these nanoparticles; however, it has also been shown to affect the biodistribution of these nanoparticulate materials [26].

Because of the possession of higher biocompatibilities and biodegradabilities, many polymers of synthetic nature which include, e.g., polyvinyl alcohol, poly-l-lactic acid, polyethylene glycol, and poly(lactic-co-glycolic acid), as well as some other natural polymers, like alginates and chitosan based nanomaterials, have been extensively employed for the fabrication of various types of nanoformulations. Polymeric nanomedicinal systems can also be divided into nanospheres and nanocapsules where both of these can act as excellent drug delivery nanocarrier formulations [27].

b. **Magnetic nanoparticles**

The concept of magnetic nanoparticles mediated drug delivery was first introduced in the 1980s but in recent years the enthusiasm for magnetic NPs-based targeting enhanced because of the development of more powerful magnets and highly sophisticated magnetic probes having a wide variety of functions, i.e., theranostic probes. Such probes permit the combining of diagnosis (like magnetic resonance based imaging (MRI) or magnetic nanoparticle imaging), and their therapeutic potential, tha can include hyperthermia as well as the release of therapeutic payload and hence targeted delivery of the desired drug (e.g. with the application of applied alternating magnetic field (AMF)). Magnetic nanoparticles can be formulated to surpass two of the major concerns that are inherent with non-magnetic nanoparticles: reticuloendothelial system (RES) entrapment and reduced site specificity [28, 29].

One approach is the development of porous or semi-porous nanoparticles from magnetic-based spinels ferrite. The higher magnetic potential of these will mean that nanoparticles can be easily manipulated by a magnet within the vascular system and, more specifically, remain in the target organ capillaries. Researchers employed the hollow nanoparticle (NP) along with some mesoporous kinds of shells which can create a high surface area and can offer a larger cavity in which the drug of choice could be entrapped in both of these mesoporous cavities. Additionally, these can also develop more heat under the microwave radiations which will allow easy release of loaded drug payload. However, doping of the Fe-oxides may cause a slight decrement in saturation of the magnetization potential that will diminish nanoparticles' capability for the magnetic targeting of the disease [30, 31].

Iron oxide particles having core–shell nanostructures are some of the most widely employed as sources of magnetic material in drug delivery. Iron oxide has many polymorphs which are called hematite, maghemite, magnetite as well as some other amorphous forms. However, only maghemite and magnetite have the greatest interest in their applications as drug carriers. Also, carbonyl iron is a well-known material which has some unique forms of iron due to its smaller nanoparticle size and is also exploited for its magnetic core [14, 32]. In other scientific reports, some other metals with high purity, including Fe and Co are selected as magnetic materials due to various advantages as compared to iron oxide, viz. having better magnetic potential, higher saturation of their magnetization potential, as well as higher specific loss in magnetic power. One group of researchers formulated iron-incorporated porous carbon materials having excellent magnetic properties by some facile nano scaling method and used it for carrying and delivering tetracycline hydrochloride. Furthermore, iron and cobalt exhibit poorer oxidative stability, and exhibit toxicity as compared to only iron oxides. Functionalization of these magnetic nanoparticles with either the amino functional group or silica, coating with some biocompatible and biodegradable polymers, as well as many surfactants and various other organic compounds often provides better achievement of their physical as well as chemical characteristics [33, 34]. Moreover, these core–shell nanostructures of magnetic nanoparticles possess various advantages like better dispersibility, higher stabilities against their oxidation, and significantly enhanced concentrations of drugs that could be entrapped in their polymeric shells. Further, many of the functional groups from these polymers on their surfaces may also be employed for extensive functionalization to obtain other desired physicochemical characteristics. It can thus be said that that these magnetic nanoparticles will retain enough hydrophilicity with polymer coating, and will not surpass 100 nm of their particle size in order to avoid quick removal by RES.

c. **Liposomes**

Liposomes are one of the most promising drug delivery carriers which are employed for targeting particular cells by attaching some specialized and specific components like antibodies, protein molecules or appropriate agents which target certain specific cellular receptors. Examples of these approaches include polymeric cationic liposomal carriers, which have been recently applied for delivering the mutated RAF genes. These liposomes for RAF0 genes were systemically administered into mice models which caused the apoptosis of cancer cells. Moreover, liposomal formulations also enhance the time of drug exposure to target cells and act as sustained and regulated drug release reservoirs for various drug moieties, which were entrapped into these. Conversely, these liposomal agents can also offer protection to patients against various adverse effects of anti-cancer compounds. Dose limitation of these anti-tumor drugs, like doxorubicin, which can cause irreversible damage of cardiac muscles. Liposomal encapsulation of all these drugs can appreciably decrease the exposure of these heart muscles to doxorubicin, thus significantly reducing the toxicological outcomes. As these liposomal carriers can solubilize various lipophilic agents, like paclitaxel, etc. as well as some of their derivatives of cisplatin also, this capability can be exploited for delivering the drugs that possess poor aqueous solubility profiles. Taking into account all these multidimensional applicabilities of liposomal nanocarriers, it is very obvious that these can provide highly flexible drug carrier modalities having various promising aspects for the delivery of anti-cancer agents [35, 36].

More recently various liposomes conjugated with polysaccharides are formulated and acting as carriers that are furnished with various targeting ligands, and which coating of these polysaccharide agents offers stability to the overall system in both the in vitro as well as in-vivo conditions whereas the firmly fixed recognizing ligands impart target specificity. Polysaccharide-conjugated liposomal agents are studied widely as highly targetable and quite stable carriers of drugs in cancer therapy, particularly for the introduction of chemotherapeutics in cancer cells [37]. Experiments have been performed by employing caner-recognizing ligands like 1-amino lactose, attached to these liposomal carriers which contain Adriamycin and the results showed that on the attachment of cell recognizing agents, anti-tumor agents, these drug carriers are customized for active targeting of these cancer cells. Therefore, carbohydrate-based ligands could offer directional capability to these liposomal agents modulating the delivery of drugs quite precisely controlled as well as quite targeted in its approach, and paradigms like these could also be exploited for the delivery of anti-tumor agents to the breast cancer cell lines [38].

d. **Nano Lipid carriers (NLCs)**

Nanostructured lipid carriers (NLCs) are basically lipid-derived nanoformulations that are generally researched as nanosized drug delivery platforms. They possess a solid matrix at room temperatures and are believed to be superior in comparison to other conventional lipid-based nanosystems like nanoemulsion, liposomal formulations and solid lipid nanocarriers because of enhanced physical stabilities, improved drug-loading capacities, and bio- as well as cytocompatibility. Nonetheless, SLNs are the drug carriers that are limited by lower cargo loading efficiencies and enhanced in the risk of expulsions of payload from formulations upon their storage because of the polymorphic transition in these [39]. To overcome these restrictions and limitations, Müller and co-workers [39] formulated nanostructured lipid carriers by using mixture of the solid lipid and liquid lipid which forms amorphous solidified matrices at both the body as well as room temperature. Incorporations of liquid lipids in matrices are among the fundamental steps since they appreciably increase the physicochemical properties of nanoformulation in comparison to the solid lipid nanoparticles. Liquid lipid contributes to the formation of an amorphous lattice with much imperfectness in the crystalline solid matrices and through that the more drugs can further be incorporated as in contrast to perfectly crystalline solids matrices created in solid lipid nanoparticles which limit the concentration of encapsulated drugs and cause its expulsion because of the restricted spatial availabilities.

Further, NLCs can widen the spectra of drug delivery platforms and surmount many of the other restrictions linked with other traditional nano-based drug delivery carriers. For example, possession of solid state by the NLCs at room temperatures can increase their physical stabilities which are major concerns with emulsion-based formulae for the preparation of these lipid-based platforms [40]. The availability of some optimized methods for larger scale formulations of the NLC can resolve expensive technology considerations for large-scale preparation of liposomal delivery agents. Additionally, NLCs offer a biocompatible paradigm as distinguished by rigid morphological features which can contribute to the distinct features in comparison to various other lipid carrier systems [41].

Nanostructured lipid carriers have many potential applications in various arenas. These applications are categorized into two broad classifications which cover their therapeutical applicabilities including many routes of the drug administrations in these drug delivery paradigms while the second class describes their applicabilities in many other areas which include cosmeceuticals, nutraceutical agents, foods, anti-cancer chemotherapy platforms, and gene delivery [42].

e. **Dendrimers**

Dendrimers have travelled a far and long path within the past 30 years after their origination. Generally formulated as the wonder molecules in chemical sciences, dendrimers are in the fourth class of the polymer category. Dr. Donald Tomalia first wrote about one of his premium and the seminal workings of Poly Amido Amine or PAMAM dendrimer somewhere around 1985. Employability of dendrimer agents as the drug delivery platform was started in the later parts of the 1990s. Dendrimer-based drug delivery strategies undertake mainly two advances: (a) Firstly, formulation and then (b) Secondly nano-construction. In the earlier one viz. formulation strategy, medicinal compounds can be physically encapsulated in the dendrimers by employing non-covalent interaction forces, while these medicinal agents can be covalently joined or linked on the dendrimer carrier in second nano-construction paradigms [43].

Drug encapsulation efficiency as well as the release of drugs from the dendrimer-based nanocarriers is regulated by modification of the dendrimers' physical surface and then further its various subsequent generations. PAMAM dendrimers can further increase the percutaneous penetration of drugs and pharmacological agents and can also be employed for their specific and distinct drug-targeting capabilities. Dendrimer-based drug delivery platforms can be manipulated so as to conjugate targeting ligands and several other imaging moieties for the creation of nanodevices. Dendrimer nanotechnology, because of its multifunctional abilities, possesses great potential for creation of the future generations' nanoscale devices [44].

Functional groups on the surfaces of dendrimers may either be positive or negative and also sometimes neutral, which proves significant for the probing of the desired dendrimer agents as the optimized drug delivery platforms. Dendrimers bearing the positive functionalized group generally lead to the destabilizing of the anionic cellular membranes and further may cause lysis of the target cells, which further results in lower biocompatibilities, while anionic or neutral charge-bearing dendrimer agents comparatively cause lesser haemolysis. To surmount this problem, surface modification is brought about like PEGylation which can change the zeta potential, retention time in circulation, and sometimes even in vivo distribution [45]. Gene therapy being a very promising paradigm for various diseases; still faces many obstacles in its inability to target the cells in vivo. Cationic dendrimer bearing agents electrostatically attract polynucleotides possessing a negative charge for the formation of more stable dendriplexes. These dendriplexes can then be released via some "proton sponges" effects in the endosomal compartments after their uptake inside the cells thereby facilitating the expression of genes in those particular specialised cells. As significant carriers for delivering the genes, dendrimers are able to protect the gene moieties from biological degradation, and can further cause facilitation of the nucleic acid to penetrate more inside the cells, so as to maintain the pharmacological activities of the genes [46].

3 Characterization of Nanoparticles

Nanoparticles were found to be the most emerging growing class that can be used in huge applications. A polydispersion of nanoparticles with a broad distribution of size, shapes and defects is always yielded by the synthesis of nanoparticles. Different characterization techniques are based on different physical properties, therefore only providing a partial picture of the nanoparticle characteristics. Nanoparticle characterization is a broad and complex discipline. Nanoparticles exist in various chemical compositions ranging from micelles to metal (oxides), from synthetic polymers to large biomolecules. Each of these materials features a completely different chemistry, which can be analysed by a variety of methods including optical spectroscopy, X-ray fluorescence and absorbance, Raman spectroscopy, etc. [47, 48].

However, often the behaviour of nanoparticles is largely governed by their nanometre dimensions. As such, throughout nanoparticle characterization, the investigation of size, shape, surface charge and porosity is a fundamental step for fully understanding and predicting their behaviour. The size, crystal structure, composition of elements and various different other physical properties of nanoparticles can be analysed. Two of the principal criteria for characterizing NPs are size and type, and distribution of size may also be calculated, degree of aggregation, surface charge and surface area, and to some degree, surface chemistry can also be analysed. Scale, size distribution and organic ligands present on the surface of the particles can affect other properties and potential applications of nanoparticles [49, 50].

There are various techniques that help in analysing the major characteristic of nanoparticles. The information regarding the size, morphology and crystal structure are provided by the various stated microscopy techniques which are TEM, HRTEM, and AFM whereas other techniques that are specially meant for magnetic properties like SQUID, VSM, FMR, and XMCD. Furthermore, the techniques that give information about optical properties, structure information, composition of elements and various other physical properties of the nanoparticle are spectroscopy and scattering techniques.

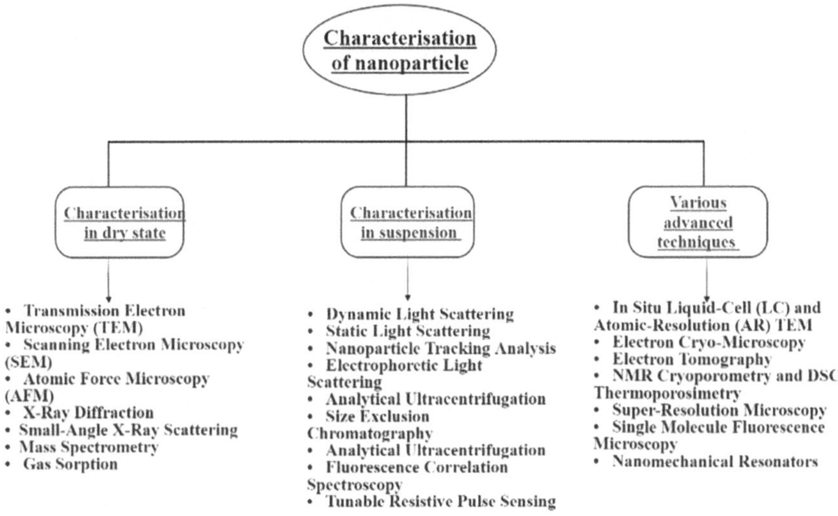

(a) **X-Ray diffraction**

The most extensively used technique provides information about various characteristics like lattice parameters, nature of the phase and regarding crystalline structure. The major benefit of XRD techniques is that it usually performed in samples that are in powdered form. Extended X-ray absorption fine structure (EXAFS) and X-ray absorption near edge structure XANES come under X-ray absorption spectroscopy (XAS). EXFAS was found to be the most suitable way to detect the chemical state of the species that may even occur in very low concentrations [51–53].

(b) **UV–Vis spectroscopy**

A technique that is very often used to measure the intensity of light that is used to reflect from the sample and compares to the intensity that is reflected back from a reference material. UV- Vis has become a crucial tool because of the optical properties of nanomaterials that makes them sensitive to various features like shape, concentration and refractive index that are present near the Nanoparticle surface. UV–Vis spectroscopy is mainly used to determine the size and concentration of Ag nanostructures [54, 55].

(c) **Nuclear magnetic resonance (NMR)**

It is used to determine the quantitative and structural characteristics of nanoscale particles and is majorly used to study the interaction or coordination between the ligand and the surface of diamagnetic or antiferromagnetic nanoparticles. NMR spectrum is used to analyse the size of nanoparticles via the analysis of particle diffusion. Hence, NMR acts as a complement to various standard nanoparticle sizing techniques and is found to be the most convenient technique in finding out the information about size, shape, chemical structure, binding and exchange processes and intermolecular molecular interactions in complex nano-systems [56, 57].

(d) **Dynamic Light Scattering (DLS)**

This technique is mainly used to determine the size of nanoparticles that are present in the colloidal suspensions and are in continuous Brownian motion. The crucial thing for DLS is a low concentration of nanoparticles hence multiple scattering can be avoided. DLS is found to be highly sensitive and reproducible for monodisperse and homogenous samples. Hence, DLS is used to measure the hydrodynamic radius of nanoparticles accurately but the major limitation is that it lacks the resolution to detect the small aggregates. A stable colloidal solution is formed by the high repulsion of positively and negatively charged particles that show only minor trends to agglomerate. The zeta-potential of the sample is found to be the key indicator of the stability of colloidal dispersions. The concentration of the suspension and solvent composition affects the zeta-potential. Hence, it can be said that DLS when combined with the zeta-potential measurements provides a more complete characterization of nanoparticles [35, 58–60].

(e) **Mass spectrometry**

It was found to be the most reliable way for the analytical characterisation of a nanoparticle. The composition, structure and chemical state of nanoparticles and their bioconjugation to target biomolecules are offered by mass spectrometry. Ahead it can also be used for quantification of bioconjugation. It can be furthermore coupled with other separation techniques to obtain real-time information [61, 62].

(f) **Transmission electron microscopy (TEM) and HR-TEM**

It determines the interaction between a thin sample and the uniform current density electron beam. This technique was found to be the most accurate and common that aids in analysing the homogeneity, shape and size of the nanoparticle. Furthermore, various factors affect the magnitude of the interaction are the density of the sample, the size of the sample and its elemental composition. To measure the larger number of particles and limit the requirement of large sample, it is usually combined with other techniques like UV–Vis and DLS. The mode uses phase-contrast imaging where the image is formed by the combination of both the scattered and transmitted electrons. It gives information about the structure of nanoparticles whereas another conventional electron microscope gives a statistical assessment of nanoparticle morphology [63, 64].

(g) **Scanning transmission electron microscopy**

This was found to be the most powerful technique for the characterization of the 3D morphology of nanoparticles. The major limitation of STEM is the time that is needed for taking the full tomographs and that might exclude various samples that are electron beam sensitive [65].

(h) **Atomic force microscopy**

This technique has the capability to create a highly magnified three-dimensional image. It measures the interaction between the between the sample and the fine probe. Tapping mode is the most common for the characterization of nanoparticles. The major advantage of using the AFM is that it does not require any coating or any surface modification before imaging hence, gives the topological analysis of the nanoparticle and also helps in the characterization of the height of the nanoparticle. The other factors that make it best compared with other electron microscope techniques are its cost is much less and less space requirement hence, provides comparable results when analysing the size of nanoparticles [66, 67].

(i) **Electrophoretic mobility (EPM)**

The surface charge of nanomaterials can be measured by EPM. The formation of large aggregates resulted in low EPM values whereas high EPM values were observed in the case of stable nanoparticles. These values also aid in obtaining the zeta potential values [68, 69].

(j) **Gel permeation chromatography (GPC)**

The highly valuable tool also known as size exclusion chromatography helps in separating the molecules based on their hydrodynamic volume or size. Various other characteristics like information about polymers such as molecular weight distribution, average molecular mass can be obtained when coupled with advanced detection systems [70–72].

(k) **Differential scanning calorimetry (DSC)**

Phase transitions of C18SH-derivatized Au NPs can be detected by DSC and these phase transitions were found to be related to reversible disordering of alkyl chains. Hence, the technique in which a difference in the amount of heat is required to raise the temperature of the sample and reference, and then it is measured [73].

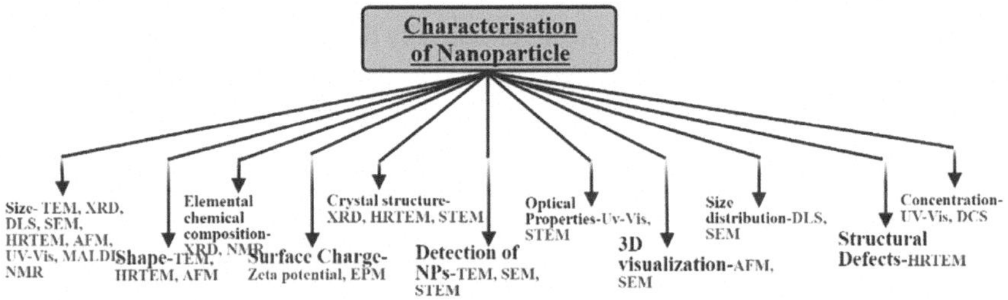

The above topic disseminates the overall picture of features that are related to the nanoparticles that entail the usage of various techniques and sometimes even one single property is analysed by more than one technique. The above-mentioned role of each technique will be helpful in the choice of the most suitable technique for characterization and with their capability to use in an accurate manner.

4 Need for the Surface Modification, Functionalization and Bioconjugation of Nanomaterials

Nanotechnology is a multidisciplinary branch of science that covers a wide range of research and development in the production of nanoparticles. Nanoparticles are the end products of nanotechnology that refer to a variety of interactions, such as physical, chemical, quantum, and biological interactions, and these are measured in the range of 1–100 nm according to the FDA. Due to their versatile nature nanoparticles are commonly used in many fields of research. Surface modification, bio-conjugation, and modification in surface functionality have resulted in the development of thousands of new nanoparticle-based products [74]. The therapeutic efficacy of a drug molecule is determined by the availability of the drug's concentration at the target site and the required duration for the onset of action. We need to monitor the drug's unspecific binding near the target site to reduce the drug's adverse effects. The nature of drug small molecules limits their solubility; it is estimated that about 70% of recently discovered molecules are hydrophobic, and the majority of these have a very short half-life, less residence time, and rapid clearance time [75, 76]. Nano-drug delivery systems have been used for improving the short half-life of drug molecules along with drug solubility. Several nanocarriers such as micelles, dendrimers, lipid nanoparticles, solid lipid nanoparticles, polymeric conjugates, and inorganic nanoparticles are used for nanoparticle-based drug delivery [77].

Characteristics of nanocarriers play a key role in the successful delivery of therapeutic agents to the active site when these are designed. The main objective of surface modification is nanoparticles should be non-toxic, biocompatible, immunologically inactive, and cause minor side effects. Surface modification is getting popular in drug delivery because it enhances the therapeutic effect of drug molecules. There are polymers of both natural and synthetic origin that are used to synthesize nanoparticles, and these have similar properties to their native form. Surface modification is carried out to achieve particular goals that are determined by the drug and its target. Since the surface is the first point of contact for a cell's outer members, body fluids, and organs, it's more important to investigate the surface's properties before modifying it for the desired therapeutic effect [78]. Since the majority of synthetic polymers and the nanoparticles made from these are hydrophobic, the human body perceives them as foreign bodies and activates the immune system to combat these. These foreign particles are quickly identified by the body's immune system's mononuclear phagocytic system (MPS), which causes opsonization and phagocytic destruction of nanoparticles. Surface modification's main objective is to render particles unrecognizable to MPS and direct them to the desired targeted site [79].

Functionalization is known as the surface modification of nanoparticles such as the bio-conjugation of drugs for improved biological effects with minimal toxic effects [80]. Surface functionalization improves the characteristics and properties of nanoparticles by modifying their surface, allowing them to play an important role in therapy. Surface functionalization not only aids in the improvement of surface properties in drug delivery, but it aids in the improvement of surface properties in medical imaging too. It aids in the exploration of nanoparticle interactions as functionalization helps in targeting the particular cells as well as the reduction of false and background signals at the molecular level in various therapies [81].

a. **Surface functionalization for improvement in diagnosis**

NP functionalization can enhance imaging thus the development of improved bioimaging/biosensors for diagnostic agents is the major focus of recent researchers. Secret et al. reported that porphyrin is used for enhancing fluorescence imaging which enables photodynamic therapy to be carried out and allows TPE, therefore imaging at a greater depth is possible [82]. Tasso et al. also developed the Zwitter ion ligands attached to quantum dots for improved fluorescence imaging by improving the fluorescence of particles by passivation of their surfaces [83]. Ratzinger et al. developed MRI imaging with a very high positive contrast imaging in which Gd (III) ions are used [84]. Cole et al., reported for computed tomography, that bi phosphate attached to Au improves contrast, allowing the identification of hydroxyl particles to increase appetite far beyond the limit of detection [85]. Iodine [86] was used by Yang et al., for PET scanning radioisotope which has an enhanced lifetime [87].

b. **Surface functionalization for improving theranostic effect**

NP functionalization is a tool that helps in enhancing imaging and therapy jointly for a theranostic purpose. The theranostic agent keeps improving with time and tumors and other diseases are the main objective of recent researches. Xin et al. reported that PEG–Polycaprolactone (PCL) and Angiopep-2 conjugated nanoparticles would selectively accumulate in glioma tissue due to the EPReffect, as nanoparticles efficiently bypass the BBB in larger proportion than non-functionalized nanoparticles [88]. A multimodal magnetic resonance imaging nanoplatforms for cancer theranostics was reported by Benyettou et al., which combines magnetic aiming to the therapeutic agents with both magnetic resonance and fluorescence imaging in a single system. It can be used as an MRIcontrast agent or conjugated with therapeutic molecules of the hydroxylmethylene bisphosphonate family [89]. Sun et al. reported the functionalized PLNPs for the theranostic purpose which can work as biosensors along with therapeutic effects [90]. Soleymani et al., designed Folate-targeted iron oxide nanoparticles (FA@Fe_3O_4 NPs) using surface functionalization for use in cancer theranostics with joint magnetic hyperthermia therapy (MHT) and magnetic resonance imaging (MRI) [91]. Kong et al., developed self-assembly of molecular motifs using coordination chemistry which is a new and efficient way to make functional materials for biomedical applications [92].

c. **Surface functionalization for improved therapeutic effect**

Amendoeira et al. reported multifunctional gold nanoparticles (AuNPs) with innovative targeted therapies which include selective targeting due to their unique optical properties, which enable them to interact with light of a particular wavelength and deliver treatment with incredible temporal and spatial accuracy [93]. According to Garg et al. Triazole functionalized heteroatom co-doped carbon quantum dots (TFH-CQDs) can produce a synergistic antiviral response for the treatment of human coronavirus [94]. Fu et al. recently reported Dual-Functionalized Mesenchymal stem cells (MSCs) that enhance the expression of IL-25 and CX3CR1 which shows phenomenal improved therapeutic effects in the treatment of Inflammatory Bowel Disease (IBD) [95].

5 Polymer Coating of Nanoparticles

a. **Nanomaterials and Targeted Drug Delivery**

To achieve the optimum therapeutic efficacy of a drug molecule, it is a prime requirement to deliver that particular drug at the site of action. After administration of the drug, it faces various environmental changes in the biological system where its efficacy and potency change. There are various factors that affect the efficacy of the drug after administration. Drug metabolism is also an important approach to be considered when dealing with the therapeutic efficacy of drugs. Water solubility, controlled or sustained release, stability of drugs in different biological environments, etc. are some of the important factors that deal with the need for drug delivery systems [96]. To overcome the physicochemical properties of drugs together with unspecific side effects and improve the therapeutic efficacy of drugs researchers need a targeted drug delivery system (TDD) [97, 98]. These systems directly deliver drugs or pharmaceuticals at the site of action, i.e., organ, tissues, cells, etc. through a variety of delivery systems [99]. There are several formulation approaches that effectively achieve drug targeting such as nanoparticles, liposomes, nanostructured lipid carriers, micelles, and dendrimers [100].

b. **Polymer coating of Nanoparticles**

In modern systems of medicine, polymeric nanoparticles are mostly used for encapsulation of drugs. They are of mainly two types such as nanospheres and nanocapsules. In nanocapsules, the cavity is enclosed by a shell or polymeric membrane whereas nanospheres belong to solid matrix systems [18, 101]. In TDD polymeric nanoparticles have various advantages as drug delivery vehicles they possess controlled release, protect the drug from different biological environments, improve bioavailability, etc. [102–104]. Polymeric nanoparticles have the potency to deliver vast types of biological molecules including vaccines, proteins, drugs and pharmaceuticals at the desired site. They are biodegradable, nontoxic, biocompatible and have the ability to possess prolonged blood circulation period [105]. On the basis of the physicochemical parameters of the drug, type of polymer and route of administration there are several approaches have been employed to synthesize the polymeric nanoparticles [102]. They are synthesized using the techniques like emulsification, reverse salting out, nanoprecipitation and microfluidics.

i. **Emulsification**

Emulsification techniques have been used in two ways for the formulation of polymeric nanoparticles which are solvent evaporation and solvent diffusion. In the solvent evaporation method firstly, polymer is dissolved in an organic solvent then the drug is dispersed in it. Another phase consists of an aqueous phase which contains water-soluble surfactant. Upon mixing of both the phases, an o/w emulsion has been made which is then processed for ultrasonication or high-speed homogenization for proper dispersion of droplets and control of particle size [106–109]. Solvent evaporation has been performed with magnetic stirring of emulsion at room temperature or at low pressure depending on solvents [110].

In the solvent diffusion technique, drug and polymers are mixed in water-miscible solvent and the surfactant is dissolved in water. Generally, polymer solutions are prepared in volatile solvents. After mixing both the phases, an o/w emulsion has been formed. The internal phase of this emulsion contains partially water-miscible organic solvent and is thermodynamically stable by water at room temperature [111–113]. Further excess dilution with water ensures the formation of colloidal particles via diffusion of solvent from the dispersed droplets into the external phase. This method is mainly employed for the formulation of nanospheres but in the organic phase if a small amount of oil has been taken then nanocapsules can be formulated.

ii. **Reverse salting out**

This method is the advancement of the solvent diffusion method of polymeric nanoparticle formulation. This technique consists of o/w emulsion which is prepared with water-miscible organic solvents, aqueous medium containing gel, salting-out agent and colloidal stabilizer [114, 115]. The salting-out agents used in this method are calcium chloride, magnesium chloride, magnesium acetate, sucrose, etc. [1, 116]. The reverse salting out method is mainly achieved by the separation of water-miscible solvents from aqueous solvents with the effect of reverse salting out agents. The emulsion is prepared by reducing the water miscibility of organic solvents and then the emulsion is sufficiently diluted by the use of deionized water or aqueous solvents and allowed the diffusion of organic solvents to the external phase. Then the precipitation of the polymer is performed and finally, the obtained product is nanospheres. Cross-flow filtration is performed for the separation of reverse salting-out agent and remaining solvents. The particle size in this method varies between 200 and 900 nm which is controlled by optimizing the concentration of polymers and solvents [117].

iii. **Nanoprecipitation Method**

This process consists of organic phase and aqueous phase and the formation of polymeric nanoparticles takes place after the displacement of organic solvent at the aqueous phase from the lipophilic solution. Polymer is dissolved in organic solvents of intermediate polarity and water miscibility which is added dropwise in aqueous solution under constant stirring. The spontaneous diffusion of the organic phase takes place into the aqueous phase and instantly nanoparticles have been formed. The organic solvent is easily removed from the mixture due to the polarity difference with water [118–121]. Polymers were precipitated in the form of polymeric nanoparticles after the diffusion of solvent from nanodroplets occurred. Sometimes to ensure the stability of formulation surfactant can also be used. The nanoprecipitation method is suitable for the formulation of both the nanoparticles nanospheres and nanocapsules. Nanospheres are formed when the drug is dispersed in a polymeric solution whereas nanocapsules are obtained when the drug is dispersed in oil then it is emulsified in an organic solvent and finally the internal phase is dispersed in the external phase [122, 123].

c. **Purpose of coating nanomaterials with different polymers**

In the field of drug delivery researchers are regularly exploited the rational use of drugs and pharmaceuticals to achieve optimum benefits. The physicochemical properties and route of administration govern the efficacy and potency of a particular drug. Every drug has its own therapeutic range and concentration as well as a target area in the biological system where it shows its maximum effect. The formulation approach and coating of nanomaterials with different polymers made it possible to overcome drug-related problems and with the help of delivery systems, the optimum efficacy of drugs and pharmaceuticals can be achieved. Drug delivery systems significantly affect the pharmacokinetics of drugs and thus pharmacodynamic effects of drugs [86, 124, 125].

Polymers have different structures and functional groups due to which they have various properties. In drug delivery, they provide maximum efficacy of drugs and selectivity of formulations. When polymers are coated on the surface of nanomaterials they add their functionality and specificity which are important for the biological effects and therapeutic efficacy [114, 126–128].

d. **Colon specific drug delivery system**

To deliver a therapeutic amount of drug at the target site is the major aim of drug delivery system. Colon targeting is beneficial in the treatment of colonic diseases like ulcerative colitis, Crohn's disease, polyps, etc. as well as colon cancer [129, 130]. Difference in the pH of the entire gastrointestinal tract is the major challenge of the colonic delivery systems. There are some approaches that can be employed for colon targeting such as bacterial enzymes, pH-sensitive delivery systems, time-dependent delivery, etc. [131]. In these approaches, enteric coatings are widely used for the colon-specific drug delivery systems. Eudragit, alginates, chitosan, pectin, dextran, etc. are some of the extensively used polymers to make the delivery systems colon specific [132–135].

e. **Liver-specific drug delivery system**

Liver diseases like cirrhosis, hepatitis and liver cancer are some of the commonly reported ailments with high mortality and morbidity in the world. The liver is the major organ for drug metabolism and elimination. Some of the substances used for the diagnosis and treatment of liver diseases may have inadequate effectiveness and adverse side effects in biological systems due to nonspecific uptake [136–138].

Polymeric nanoparticles and other delivery systems are developed for the treatment of hepatic diseases including liver cancer by a passive targeting mechanism. Surface coating of nanomaterials with polymers can modify their properties such as hydrophobicity, targeting capability, and biocompatibility. [139, 140]. Polyethylene glycol coated nanomaterial has been reported with high liver bioavailability. Chitosan-decorated nanoparticles reported with good cellular uptake in the tumor rather than the liver. The study also reported that positively charged coated nanoparticles are more concentrated in the liver [141]. More recently PLGA nanoparticle have been reported in the treatment of liver cancer. This polymeric nanoparticle has prolonged half-life and higher plasma and liver concentration [142].

f. **CNS-specific drug delivery system**

The diagnosis and treatment of the central nervous system appear very challenging because of the specific entry of theranostics to the brain via the highly restricted behaviour of blood–brain barrier [143, 144]. To overcome the limitations of treatment of brain diseases pharmaceutical scientists have proposed various nanocarriers which are able to directly deliver the active compounds to the brain cells successfully [145]. There are many polymers used to formulate brain-targeted drug delivery systems such as amino acids, proteins, polyethyleneimines, and polyesters. For the polymeric nanoparticles delivering the drug to the CNS have the capability to possess adsorptive mediated transcytosis and receptor-mediated transcytosis through the blood–brain barrier. The first report for polymeric nanoparticles in brain drug delivery is a poly (butylcynoacrylate) nanoparticle which is coated with polysorbate 80 that delivers dalargin to the CNS. Polysorbate coating made the nanoparticle interact with endothelial cells of blood–brain barrier and also provided reduced clearance by reticuloendothelial cells and proposed to work in a similar way like low-density lipoproteins via modulation of efflux transporters and tight junctions [146, 147]. Polyester nanoparticles also have good efficacy for the loading and delivery of active drugs to the brain [148, 149].

6 Surface Functionalization of Nanoparticles

The past two decades witnessed an evolution in the designing of nanoparticles or drug delivery strategies has been observed which actively interact with the surrounding system and help in better targeting of drugs to the intended locations. Surface functionalization is one of the concepts by which either we prevent the immunogenic reactions in response to the foreign nanoparticle or bypass the system like stealth operation of the military. There are a number of strategies by which surface functionalization of nanoparticles takes place.

a. **Functionalization of nanocarriers with drugs**

The characteristic of nanoparticles that determines the colloidal stability of nanoparticles as well as self-assembly and many other important biomedical applications is the charge. It has been observed that surface charge greatly influences the size of metal oxide nanoparticles, higher the surface charge corresponds to the smaller size of nanoparticles [150]. The addition of the positive charge over the nanoparticles determines the delivery of specific constructs like non-viral gene transfection or delivery of DNA construct to the nucleus of cancer cells. The liposomes and polymers are used for the specific purpose and deliver the cargo by endocytosis or direct penetration via the cell membrane with the help of cationic surface moiety present over the nanoparticles and glycoproteins of the cell surface [151]. Another strategy involves the delivery of DNA via lipoplex, in this, the cytoplasmic anionic lipids diffuse into the endosomes resulting in the formation of a neutral charge complex resulting release of DNA from the lipoplex-endosome complex [152].

It was observed by Bantz et al. that the stability of SiO_2-based nanomaterials differs in different conditions like at physiological salinity, the stability of negatively charged silica nanoparticles was remarkable. However, it is unstable in the presence of proteins. It was also observed that if the electrostatically stabilized nanoparticles of poly (organosiloxane) had positive surface charge, it was precipitated at physiological salinity [153]. The effect of charge at the surface of inorganic nanoparticles plays a crucial role in their activity. Abbas Zadegan et al. performed a study in which it was noted that charge present over the surface of silver nanoparticles determines its activity against Gram-positive and Gram-negative bacteria. The bactericidal activity is greatly influenced by surface charge with respect to silver nanoparticles. They found that the silver nanoparticles with positive surface charge were the most effective against the microorganism. Whereas silver nanoparticles with negative surface charge were the least effective against a variety of tested microorganisms. However, they also tested neutral surface charge silver nanoparticles and found that it was effective against most of the tested bacterial strains [154].

Pelaz et al. using polyethylene glycol for surface functionalization studied its effect on protein adsorption and cellular uptake and compared it with non-decorated nanoparticles. It was observed that the adsorption of human serum albumin (HSA) over decorated nanoparticles was relatively low in comparison to the non-decorated nanoparticles. There was a 10% reduction in cellular uptake in 3T3 fibroblasts upon treatment with pegylated nanoparticles with respect to non-decorated nanoparticles [155]. Surface functionalization of cationic polymethylmethacrylate (PMMA) nanoparticles was evaluated with sialic acid and cholesterol. It has been reported that the anticancer effect of benznidazole was significantly increased in human carcinoma cervical cancer (HeLa) and human colorectal cancer (HT-29) with surface functionalization [156].

b. **Fluorescent dyes**

Functionalization route is crucial for emission in dye nano-assemblies of gold nanorods. In this study, author has developed multifunctional fluorescent nanoprobes by the conjugation of dye-labelled DNA oligonucleotides with gold nanorods. There has been a tenfold increment in the emission reported with respect to the free dye which indicates that tip-functionalized gold nanorods were potentially more active than free dye molecules [157].

In another study, the fluorescent dye was functionalized covalently by copper catalysis and strain-promoted azide-alkyne click chemistry over calcium phosphate nanoparticles. It has been reported by the author that the introduction of azide group will help them to covalently attach all kinds of alkyne-carrying molecules. It has also been reported that a number of dyes and more than one molecule can be covalently linked with these azide groups following aggregation-induced emission properties and are stable under biological conditions making these nanoparticles easy to uptake by the cells [158].

Advancement in the fluorescent probe for bio-imaging leads to the development of various functionalized molecules, in this series Yen et al. developed near-infrared (NIR) IR-820 dye and its conjugates functionalized with amphiphilic polymer-coated over iron oxide (Fe_3O_4) magnetic nanoparticles (MNPs) for optical and magnetic resonance imaging. The fabrication is a 3 steps process, the resulting nanoparticles exhibit a wide emission range and minimal cytotoxicity promising advancement in bioimaging by direct labelling of cancerous HeLa cells via NIR fluorescence microscopy and MR imaging [159].

c. **Other polymers**

The surface of the nanoparticles was coated with appropriate material to enable them to interact with the cell or neighboring molecules within the cell. Luo R et al., coated the poly(DL-Lactide-co-glycolide) (PLGA) nanoparticles with polyelectrolyte by layer-by-layer deposition method and evaluated their physicochemical and biological parameters. The author has coated PLGA nanoparticles with two polyelectrolyte pairs: (1) poly (allylamine hydrochloride) (PAH) and poly (styrene sulfonate) (PSS) and (2) poly(L-lysine hydrobromide) (PLL) and dextran sulfate (DES). Upon determining the release behavior, it was observed that both coatings were resulted in the suppression of the burst release behavior associated with uncoated PLGA nanoparticles. Nanoparticles with modified surfaces with negatively charged groups show no hemolysis. However, positive surface charged PLGA nanoparticles show hemolytic activity. Their study concluded that surface modification of PLGA nanoparticles by layer-by-layer coating method results in controlling the physico-chemical and biological behavior of the nanoparticles [160].

The surface coating of magnetic iron oxide nanoparticles to prevent agglomeration, the use of organic compound makes the surface of iron oxide nanoparticles less reactive, thus preventing agglomeration. Hydrophobic interaction between iron oxide nanoparticles causes agglomeration resulting in clusters of nanoparticles followed by an increment in the particle size and a decrement in the biological activity. Coating of organic compound over the iron oxide nanoparticle results in the preservation of the magnetic properties of the nanoparticle and functionalization with organic properties of the coated material. The coating is usually done by layer-by-layer coating of organic molecules usually polymer or biomolecule, resulting in the formation of a core of iron oxide and the outer layer of polymer results in specific characteristics of magnetic nanoparticles with biocompatibility and biodegradability.

Coating of amphiphilic or surfactant material onto the surface of metal oxide nanoparticles divided them into the three types: oil-soluble, water-soluble and amphiphilic. The oil-soluble type represents weak attraction towards the surrounding environment and contains usually hydrophobic groups. In the water-soluble type, strong attraction was observed with the surrounding environment and they usually contain hydrophilic groups. Whereas amphiphilic type represents surface functionalization with both hydrophobic and hydrophilic groups and it makes them either water-soluble or oil-soluble. Depending on the surface functionalization stability, drug targeting is affected, and reportedly oil-soluble surface functionalized metal oxide nanoparticles are highly stable and show monodispersity with biocompatibility but it is not suitable for delivery of proteins, DNA or antibiotics into the system, therefore water-soluble surface functionalized metal oxide nanoparticles are preferred [161].

Atomic layer deposition has been performed for surface functionalization in the areas of energy, environment and so on. This technique protects the material with ultrathin films resulting in the protection against erosion by the external environment. The degradation of conformal coating films technique is employed to prevent possible degradation by the external environment mainly by humidity and oxygen preserving the property of the original material. There are mainly two types of methods employed for surface functionalization: (1) liquid phase method, in this method particles are encapsulated into compatible material to prevent degradation, creating a thick and porous layer at low cost. (2) Gas phase method, the limiting factor of this technique is the lack of solvent due to which coalescence in the particles is observed. However, by using the atomic layer deposition method, ultrathin coating of material can be deposited onto the surface. This reaction allows the deposition of atomic-level thickness of material [162].

7 Bioconjugation of Nanoparticles with Targeting Ligands

Bio-conjugation is the fastest growing field of research. Innovative methods of the conjugation of various bioactive ligands have been continued since the concept of active targeting came into the picture [163]. Drawing the idea the proteins and ligands perform their activity by binding to specific ligands; accordingly, particular ligands can be conjugated having an enormous role in the disease pathogenesis. The conjugation of bioactive moieties like proteins, peptides, DNA, RNA, etc., has been experimented so far [164, 165]. Under this heading, we are going to elucidate the magical role of various bio-conjugates in drug delivery. To address various issues related to the drug's nature while going in the systemic level there might be poor uptake and various side effects of the drug, this issue can be tackled via conjugating the nanoparticle with the peptides.

a. **Peptides**

Peptides have attracted the attention of researchers; from so many years they have been tried for the development of novel drug delivery systems to target complex disease. In the field of bio-conjugation, peptides were found to be the most suitable candidates due to their ease of conjugation, high stability at various reaction conditions [165]. Some scientists have noted the conjugation of bioactive peptide with non-biological molecules such as metal chelates, polymers, nanoparticles [166]. Scientists have prepared stealth nanoparticles based on peptides for targeted and pH-triggered release. They first synthesized the stealth lipopeptide which is highly lipophilic and able to penetrate the lipid bilayer via solid phase synthesis, and then incorporated liposomes containing 10% w/w of DSPE-PEG to incorporate stealth properties. They reported that it has highly efficient targeting properties without affecting the normal cellular architecture [167]. Peptide nanoparticle conjugates emerged as an efficient tool for various biomedical implementations. The nanoparticles have various physical, chemical properties due to their very small size. Their ultra-low size and high surface area to volume ratio are utilized for designing of coherent drug delivery system. Self-assembly is the easiest way of building the peptide nanoparticle conjugates. Various researchers carrying out work on the RGD peptide, cRGD-Gem-HSA-NPs have been developed for targeting $\alpha V\beta 3$ integrins that are highly present on the cancerous cell and can easily recognize RGD motifs, they have used previously prepared gemcitabine loaded albumin nanoparticle using the nab technology. Their in-vivo and in vitro results confirmed efficient delivery of the cargo to the targeted site and the suppression of tumor growth [168].

b. **Antibodies**

Various proteins and biomolecules are upregulated in disease conditions, these proteins with respect to antibodies can be targeted via conjugating certain antibodies as ligands. Conjugating antibodies as ligands is one of the most promising approaches because of the binding efficiency for the target and biocompatibility [169]. HER-2 gene is overexpressed in 20% of the women diagnosed with breast cancer. Various antibodies are targeted against the extracellular domain of the HER-2 receptor for example trastuzumab and biosimilars, pertuzumab or T-DM1. Polymeric nanoparticles are the most suitable carriers for drug delivery due to their tiny size and its capability for onsite reaching. NPs conjugated with monoclonal antibody trastuzumab, which selectively binds with HER-2 highly present in breast cancer patients, so these NPs selectively accumulate in the breast cancer cell only and the bioavailability of the drug will be high [170].

Various conjugation approaches were attempted for the successful binding of antibodies to the carrier. The maleimide chemistry approach has been widely used for conjugation. It involves sulfhydryl groups of antibodies which are less as compared to amines and thiols. The –SH groups are generated on the antibodies via thiolation with Traut's reagent (2-Iminothiolane), N-succinimidyl S-acetylthioacetate (SATA) are mostly followed. Carbodiimide [1-ethyl-3-(-3-dimethylaminopropyl) carbodiimide (EDC)] are cross-linking agent to be used to activate the carboxyl group. N-Hydroxysuccinimide (NHS) or sulfo N-Hydroxysuccinimide (Sulfo-NHS) are used to enhance the coupling effectiveness. Click chemistry, Ionic binding and physical adsorption are also used for the conjugation of the antibodies to nanoparticles [166, 171].

c. **Nuclear localization signals**

Nuclear localization signals (NLSs) are sequences of peptides used to deliver the proteins into the subcellular of the nuclear level. The limit of size-dependent diffusion of molecules is 200 and 310 bp. Molecules bigger than 30–60 kDa need an active targeting system to carry the cargo through the nuclear localization signal, which is delivered into the nucleus from the cytoplasm. SV40 is a basic nuclear localization signal peptide significantly used for carrying large molecules like proteins [172]. The increase in nuclear uptake of plasmid DNA is reported upon conjugation with nuclear localization signal peptide. But they also reported that DNA lost its inherent activity. A group of scientists working on nuclear localization signals demonstrated the incorporation of high affinity bis- and tris-intercalating compounds in contemplation to gain robust binding of nuclear localization signal to double-stranded DNA (dsDNA). They can be employed as emergency molecules for nurturing the nuclear localization of cellularly transfected DNA vectors [173].

d. **Nucleic acid (aptamers)**

Selective localization of the drug or molecules in targeted therapies is secret of the successful targeted therapies. Conjugation of nucleic acid or proteins has shown promising results respect to binding to targets or receptors [174]. This has emerged as a novel therapy to treat the complex disease. This strategy is highly suitable for carrying the nucleic acid based cargo. Any potential nucleic acid sequence can be tagged with the aptamer sequence outcomes as a bivalent moiety having the potential to target RNA/DNA, including siRNA, microRNA and Deoxyribozymes. There are various reports suggesting the use of aptamer sequence to target the RNA/DNA, siRNA delivery approach after testing in-vitro and in-vivo [175, 176].

e. **Biotin and avidin derivatives**

Biotin avidin interaction is a powerful non-covalent interaction that is 10^3 to 10^6 times stronger than antigen–antibody interaction; they have been extensively utilized for various applications including drug delivery and diagnosis [177]. The dexterity of building avidin-based conjugates for the targeted delivery of drugs makes them one of the suitable candidates as ligands. Avidin is a basic tetrameric glycoprotein built with four subunits, which binds with biotin with maximum specificity and affinity. The most beautiful property of these interactions is their high affinity interactions, that are strong and able to face the harsh environment of the body such as pH, temperature, denaturating agents. Streptavidin is widely used to target the delivery of siRNA in liver fibrosis. They have been extensively used for the delivery of various agents such as DNA, RNA, siRNA, DNA vectors, vaccines, peptides, and diagnostic applications [178]. Biotin is also called vitamin H, coenzyme R or vitamin B7. Biotin consists of a tetrahydrothiophene ring bound to a tetrahydroimidizalone (ureido) ring. To access the complex subcellular structure for the delivery of small synthetic molecules, streptavidin is used to enhance the transport of such molecules and potentiate localization at subcellular level [179].

8 Conclusion and Future Perspective

Physicochemical characteristics of nanoparticles have a significant role in the in vitro as well as in vivo behavior of nanoparticles mainly influencing their bio-distribution features. While the passive targetability is considerably influenced by hydrodynamic diameter, as the nanoparticles with particle size lesser than 7–10 nm are majorly retained by the filtration in the renal system and then get excreted via urine, nanoparticles with hydrodynamic diameter >100 nm are taken out from the bloodstream. Similarly, the charge on the nanoparticle surface also influences the nanoparticles' functions. For example, a positive charge on the nanoparticle is favouring the nanoparticle internalization inside the cells. Modifications in the surface characteristics of nanoparticles with polymers of either hydrophilic or hydrophobic origin allow altered circulation behaviours and half-lives and also regulate their clearance by the mononuclear phagocyte systems, thus also controlling the accumulation of nanoparticles at desired and/or target sites, as is the specified case with the PEGylation process. Several other polymers of hydrophilic nature or some surfactants, like poloxamines, dextran, hyaluronic acid, PVA, poloxamer, chitosan, PVP), are employed in the stealth phenomena by their coating of nanomaterials. Furthermore, passive targeting by the nanosystems has a negative influence on the cellular internalization inside the pertinent cells. Hence, surface functionalizing of nanomaterials with specified and particular functionalities serves as the alternative approach for surpassing this problem, and thus it results in the active targetability of drug-loaded nanocarriers. These paradigms involving the surface and morphological customization of nanomaterials with some very specific functional groups or other ligands, increase the nanomaterials' intractability with some specialised cells via the ligand-receptor recognition by those very specific receptors available on cellular surfaces and then can further lead to nanomaterials cellular uptake through the receptor-mediated internalization phenomena. Similar becomes the case with transferrin antibody or some of its portions as well as biotin, which possess frequent applicability as nanomaterials surface functionalizing ligand agents due to their receptors which become overexpressed on several of these disease-affected cell lines, thereby increasing the nanomaterials cellular uptake and amending the pharmacological therapeutic outcomes. More lately, dual drug-loaded nanomaterials have been formulated as either adjuvant therapy or as an alternative approach for singly targeting nanoformulations as these possess a synergistic outcome by containing two distinct targets or target localities. Such an approach, in addition to enhancing the nanomaterials' chemical attraction for the desired targeting site, also further pushed for the minimization of the undesired and off-target effects of drugs in the cells.

The future centre of these current nanomaterials surface modification and functionalization approaches is advanced and hi-tech improvements in the delivery of pharmacological compounds in cancer therapeutics and genetic employability. Nanomaterials will offer a more effective and better method with appreciably enhanced drugs and nanomaterials safety for RNA interference (RNAi) approaches and gene therapies. Since several conventional as well as traditional drugs are difficult to apply in clinical settings because of many issues like enzymatic degradability and minimal cellular uptake so different combinations of certain polymers and other materials are being tested to create more suitable and better carriers for future use. Irrational or misuse of nanomaterials and polymers has demonstrated higher toxicities, immunological reactions, and formulation instabilities. Cyclodextrin-associated polymeric nanomaterials are presently under investigation for their better safety and pharmacological efficacy. Metallic nanoformulations are now under study for the possible destruction of drug-mediated as well as hyperthermia-mediated selective targeting of tumors and inhibition of cancer cells. These are quite therapeutically efficacious and active and are not degraded so quickly in shorter time periods, and further adding the light or magnetic field waves these can get heated to higher temperatures. These can then destroy the tumor cells using either photothermia or hyperthermia methodology and also shall be entirely bio- and cytocompatible with human cells and tissues and thus a great extent toxicity risks shall be minimized.

References

1. Pal SL, Jana U, Manna PK, Mohanta GP, Manavalan R (2011) Nanoparticle: an overview of preparation and characterization. J Appl Pharm Sci 1(6):228–234
2. Baer DR, Engelhard MH, Johnson GE, Laskin J, Lai J, Mueller K et al (2013) Surface characterization of nanomaterials and nanoparticles: important needs and challenging opportunities. J Vac Sci Technol A 31(5). https://www.ncbi.nlm.nih.gov/pmc/articles/PMC3869349/
3. Delfi M, Ghomi M, Zarrabi A, Mohammadinejad R, Taraghdari ZB, Ashrafizadeh M et al (2020) Functionalization of polymers and nanomaterials for biomedical applications: antimicrobial platforms and drug carriers. Prosthesis 2(2):117–139
4. Ahmad A, Khan F, Mishra RK, Khan R (2019) Precision cancer nanotherapy: evolving role of multifunctional nanoparticles for cancer active targeting. J Med Chem 62(23):10475–10496
5. Ahmad A, Gupta A, Jayamurugan G, Khan R (2018) Applying synthetic lethality to nanomedicine: LCS-1 loaded magnetite and polymeric nanoparticles for the treatment of BLM and CHEK2-deficient colorectal cancer cells. Cancer Med, Wiley:27–28
6. Gupta A, Ahmad A, Singh H, Kaur S, Neethu KM, Ansari MM et al (2019) Nanocarrier composed of magnetite core coated with three polymeric shells mediates LCS-1 delivery for synthetic lethal therapy of BLM-defective colorectal cancer cells. Biomacromol 20(12):4623–4623
7. Attia MF, Anton N, Wallyn J, Omran Z, Vandamme TF (2019) An overview of active and passive targeting strategies to improve the nanocarriers efficiency to tumour sites. J Pharm Pharmacol 71(8):1185–1198

8. Din FU, Aman W, Ullah I, Qureshi OS, Mustapha O, Shafique S et al (2017) Effective use of nanocarriers as drug delivery systems for the treatment of selected tumors. Int J Nanomed 12:7291–7309
9. Jeevanandam J, Barhoum A, Chan YS, Dufresne A, Danquah MK (2018) Review on nanoparticles and nanostructured materials: history, sources, toxicity and regulations. Beilstein J Nanotechnol 3(9):1050–1074
10. Sanità G, Carrese B, Lamberti A (2020) Nanoparticle surface functionalization: how to improve biocompatibility and cellular internalization. Front Mol Biosci 7. https://www.ncbi.nlm.nih.gov/pmc/articles/PMC7726445/
11. Chenthamara D, Subramaniam S, Ramakrishnan SG, Krishnaswamy S, Essa MM, Lin F-H et al (2019) Therapeutic efficacy of nanoparticles and routes of administration. Biomater Res 23(1):20
12. Ansari MM, Ahmad A, Kumar A, Alam P, Khan TH, Jayamurugan G et al (2021) Aminocellulose-grafted-polycaprolactone coated gelatin nanoparticles alleviate inflammation in rheumatoid arthritis: a combinational therapeutic approach. Carbohyd Polym 258:117600
13. Guerrini L, Alvarez-Puebla RA, Pazos-Perez N (2018) Surface modifications of nanoparticles for stability in biological fluids. Materials 11(7):1154
14. Ahmad A, Gupta A, Ansari MM, Vyawahare A, Jayamurugan G, Khan R (2019) Hyperbranched polymer-functionalized magnetic nanoparticle-mediated hyperthermia and niclosamide bimodal therapy of colorectal cancer cells. ACS Biomater Sci Eng 6(2):1102–1111
15. Lombardo D, Kiselev MA, Caccamo MT (2019) Smart nanoparticles for drug delivery application: development of versatile nanocarrier platforms in biotechnology and nanomedicine. J Nanomater 27(2019):e3702518
16. Mishra RK, Ahmad A, Vyawahare A, Alam P, Khan TH, Khan R (2021) Biological effects of formation of protein corona onto nanoparticle. Int J Biol Macromol
17. Schenone M, Dančík V, Wagner BK, Clemons PA (2013) Target identification and mechanism of action in chemical biology and drug discovery. Nat Chem Biol 9(4):232–240
18. Mitchell MJ, Billingsley MM, Haley RM, Wechsler ME, Peppas NA, Langer R (2021) Engineering precision nanoparticles for drug delivery. Nat Rev Drug Discov 20(2):101–124
19. Pathak YV (ed) (2019) Surface modification of nanoparticles for targeted drug delivery. Springer International Publishing. https://www.springer.com/gp/book/9783030061142
20. Minelli C (2016) Measuring nanoparticle properties: are we high and dry or all at sea? In: Nanoparticle characterisation–challenges for the community. IOP(Institute of Physics), book of abstracts
21. Gupta A, Ahmad A, Dar AI, Khan R (2018) Synthetic lethality: from research to precision cancer nanomedicine. Curr Cancer Drug Targets 18(4):337–346
22. De Jong WH, Borm PJ (2008) Drug delivery and nanoparticles: applications and hazards. Int J Nanomed 3(2):133–149
23. Khan I, Saeed K, Khan I (2019) Nanoparticles: properties, applications and toxicities. Arab J Chem 12(7):908–931
24. Bennet D, Kim S (2014) Polymer nanoparticles for smart drug delivery. In: Application of nanotechnology in drug delivery. IntechOpen. https://www.intechopen.com/books/application-of-nanotechnology-in-drug-delivery/polymer-nanoparticles-for-smart-drug-delivery
25. Yetisgin AA, Cetinel S, Zuvin M, Kosar A, Kutlu O (2020) Therapeutic nanoparticles and their targeted delivery applications. Molecules 25(9):2193
26. Patra JK, Das G, Fraceto LF, Campos EVR, del Pilar R-T, Acosta-Torres LS et al (2018) Nano based drug delivery systems: recent developments and future prospects. J Nanobiotechnol 16(1):1–33
27. Song R, Murphy M, Li C, Ting K, Soo C, Zheng Z (2018) Current development of biodegradable polymeric materials for biomedical applications. Drug Des Devel Ther. 24(12):3117–3145
28. Price PM, Mahmoud WE, Al-Ghamdi AA, Bronstein LM (2018) Magnetic drug delivery: where the field is going. Front Chem 6. https://doi.org/10.3389/fchem.2018.00619/full

29. Guo T, Lin M, Huang J, Zhou C, Tian W, Yu H et al (2018) The recent advances of magnetic nanoparticles in medicine. J Nanomater 29(2018):e7805147
30. Sensenig R, Sapir Y, MacDonald C, Cohen S, Polyak B (2012) Magnetic nanoparticle-based approaches to locally target therapy and enhance tissue regeneration in vivo. Nanomedicine (Lond) 7(9):1425–1442
31. Chen P, Cui B, Bu Y, Yang Z, Wang Y (2017) Synthesis and characterization of mesoporous and hollow-mesoporous $M_xFe_{3-x}O_4$ (M=Mg, Mn, Fe Co, Ni, Cu, Zn) microspheres for microwave-triggered controllable drug delivery. J Nanopart Res 19(12):398
32. Gupta A, Ahmad A, Singh H, Kaur S, KM N, Ansari MM et al (2018) Nanocarrier composed of magnetite core coated with three polymeric shells mediates LCS-1 delivery for synthetic lethal therapy of BLM-defective colorectal cancer cells. Biomacromolecules 19(3):803–815
33. Mou X, Ali Z, Li S, He N (2015) Applications of magnetic nanoparticles in targeted drug delivery system. J Nanosci Nanotechnol 15(1):54–62
34. Ulbrich K, Holá K, Šubr V, Bakandritsos A, Tuček J, Zbořil R (2016) Targeted drug delivery with polymers and magnetic nanoparticles: covalent and noncovalent approaches, release control, and clinical studies. Chem Rev 116(9):5338–5431
35. Mishra RK, Ahmad A, Kumar A, Vyawahare A, Raza SS, Khan R (2020) Lipid-based nanocarrier-mediated targeted delivery of celecoxib attenuate severity of ulcerative colitis. Mater Sci Eng, C 116:111103
36. Olusanya TOB, Haj Ahmad RR, Ibegbu DM, Smith JR, Elkordy AA (2018) Liposomal drug delivery systems and anticancer drugs. Molecules 23(4). https://www.ncbi.nlm.nih.gov/pmc/articles/PMC6017847/
37. Pistone S, Rykke M, Smistad G, Hiorth M (2017) Polysaccharide-coated liposomal formulations for dental targeting. Int J Pharm 516(1–2):106–115
38. Sihorkar V, Vyas SP (2001) Potential of polysaccharide anchored liposomes in drug delivery, targeting and immunization. J Pharm Pharm Sci 4(2):138–158
39. Müller RH, Radtke M, Wissing SA (2002) Nanostructured lipid matrices for improved microencapsulation of drugs. Int J Pharm 242(1–2):121–128
40. Müller RH, Shegokar R, Keck CM (2011) 20 years of lipid nanoparticles (SLN and NLC): present state of development and industrial applications. Curr Drug Discov Technol 8(3):207–227
41. Müller RH, Mäder K, Gohla S (2000) Solid lipid nanoparticles (SLN) for controlled drug delivery—a review of the state of the art. Eur J Pharm Biopharm 50(1):161–177
42. Jaiswal P, Gidwani B, Vyas A (2016) Nanostructured lipid carriers and their current application in targeted drug delivery. Artif Cells Nanomed Biotechnol 44(1):27–40
43. Chauhan AS (2018) Dendrimers for drug delivery. Molecules 23(4)
44. Bugno J, Hsu H-J, Hong S (2015) Tweaking dendrimers and dendritic nanoparticles for controlled nano-bio interactions: potential nanocarriers for improved cancer targeting. J Drug Target 23(7–8):642–650
45. Zhu Y, Liu C, Pang Z (2019) Dendrimer-based drug delivery systems for brain targeting. Biomolecules 9(12). https://www.ncbi.nlm.nih.gov/pmc/articles/PMC6995517/
46. Abedi-Gaballu F, Dehghan G, Ghaffari M, Yekta R, Abbaspour-Ravasjani S, Baradaran B et al (2018) PAMAM dendrimers as efficient drug and gene delivery nanosystems for cancer therapy. Appl Mater Today 1(12):177–190
47. Zhang XF, Liu ZG, Shen W, Gurunathan S (2016) Silver nanoparticles: synthesis, characterization, properties, applications, and therapeutic approaches. Int J Mol Sci 17(9). https://www.ncbi.nlm.nih.gov/pmc/articles/PMC5037809/
48. Iravani S, Korbekandi H, Mirmohammadi SV, Zolfaghari B (2014) Synthesis of silver nanoparticles: chemical, physical and biological methods. Res Pharm Sci 9(6):385–406
49. Ploetz E, Engelke H, Lächelt U, Wuttke S (2020) The chemistry of reticular framework nanoparticles: MOF, ZIF, and COF materials. Adv Func Mater 30(41):1909062
50. Gatoo MA, Naseem S, Arfat MY, Mahmood Dar A, Qasim K, Zubair S (2014) Physicochemical properties of nanomaterials: implication in associated toxic manifestations. Biomed Res Int 6(2014):e498420

51. Bak S-M, Shadike Z, Lin R, Yu X, Yang X-Q (2018) In situ/operando synchrotron-based X-ray techniques for lithium-ion battery research. NPG Asia Mater 10(7):563–580
52. Meneghini C, Dalconi MC, Nuzzo S, Mobilio S, Wenk RH (2003) Rietveld refinement on x-ray diffraction patterns of bioapatite in human fetal bones. Biophys J 84(3):2021–2029
53. Cametti G, Scheinost AC, Churakov SV (2020) Structural evolution of Ag-LEV zeolite upon heating: an in situ single-crystal x-ray diffraction (SC-XRD) and x-ray absorption spectroscopy (XAS) study. ACS Omega. https://doi.org/10.1021/acsomega.0c04455
54. Mourdikoudis S, Pallares RM, Thanh NTK (2018) Characterization techniques for nanoparticles: comparison and complementarity upon studying nanoparticle properties. Nanoscale 10(27):12871–12934
55. Losurdo M, Bergmair M, Bruno G, Cattelan D, Cobet C, de Martino A et al (2009) Spectroscopic ellipsometry and polarimetry for materials and systems analysis at the nanometer scale: state-of-the-art, potential, and perspectives. J Nanopart Res 11(7):1521–1554
56. Akbarzadeh A, Samiei M, Davaran S (2012) Magnetic nanoparticles: preparation, physical properties, and applications in biomedicine. Nanoscale Res Lett 7(1):144
57. Marbella LE, Millstone JE (2015) NMR techniques for noble metal nanoparticles. Chem Mater 27(8):2721–2739
58. Prakash R, Mishra RK, Ahmad A, Khan MA, Khan R, Raza SS (2021) Sivelestat-loaded nanostructured lipid carriers modulate oxidative and inflammatory stress in human dental pulp and mesenchymal stem cells subjected to oxygen-glucose deprivation. Mater Sci Eng, C 120:111700
59. Ahmad A, Ansari MM, Mishra RK, Kumar A, Vyawahare A, Verma RK et al (2021) Enteric-coated gelatin nanoparticles mediated oral delivery of 5-aminosalicylic acid alleviates severity of DSS-induced ulcerative colitis. Mater Sci Eng, C 119:111582
60. Ahmad A, Vaghasiya K, Kumar A, Alam P, Raza SS, Verma RK et al (2021) Enema based therapy using liposomal formulation of low molecular weight heparin for treatment of active ulcerative colitis: new adjunct therapeutic opportunity. Mater Sci Eng, C 121:111851
61. Stevens KG, Pukala TL (2020) Conjugating immunoassays to mass spectrometry: solutions to contemporary challenges in clinical diagnostics. TrAC, Trends Anal Chem 1(132):116064
62. Gao Y, Zhou Y, Chandrawati R (2020) Metal and metal oxide nanoparticles to enhance the performance of enzyme-linked immunosorbent assay (ELISA). ACS Appl Nano Mater 3(1):1–21
63. Murata K, Wolf M (2018) Cryo-electron microscopy for structural analysis of dynamic biological macromolecules. Biochimica et Biophysica Acta (BBA) Gen Subj 1862(2):324–334
64. Lyumkis D (2019) Challenges and opportunities in cryo-EM single-particle analysis. J Biol Chem 294(13):5181–5197
65. Mayence A (2016) Design and characterization of nanoparticles and their assemblies: transmission electron microscopy investigations from atomic to mesoscopic length scales [PhD Thesis]. Department of Materials and Environmental Chemistry (MMK), Stockholm University
66. Kaliva M, Vamvakaki M (2020) Chapter 17—nanomaterials characterization. In: Narain R (ed) Polymer science and nanotechnology. Elsevier, pp 401–33. https://www.sciencedirect.com/science/article/pii/B9780128168066000170
67. Lin P-C, Lin S, Wang PC, Sridhar R (2014) Techniques for physicochemical characterization of nanomaterials. Biotechnol Adv 32(4):711–726
68. Lowry GV, Hill RJ, Harper S, Rawle AF, Hendren CO, Klaessig F et al (2016) Guidance to improve the scientific value of zeta-potential measurements in nanoEHS. Environ Sci Nano 3(5):953–965
69. Saeb A, Alshammari AS, Al-Brahim H, Al-Rubeaan KA (2014) Production of silver nanoparticles with strong and stable antimicrobial activity against highly pathogenic and multidrug resistant bacteria. Sci World J 2014
70. Netopilík M, Podzimek S (2020) Retention mechanism of branched macromolecules in size exclusion chromatography. ACS Omega 5(24):14254–14260

71. Gaborieau M, Castignolles P (2011) Size-exclusion chromatography (SEC) of branched polymers and polysaccharides. Anal Bioanal Chem 399(4):1413–1423
72. Hong P, Koza S, Bouvier ESP (2012) A review size-exclusion chromatography for the analysis of protein biotherapeutics and their aggregates. J Liq Chromatogr Relat Technol 35(20):2923–2950
73. Leyva-Porras C, Cruz-Alcantar P, Espinosa-Solís V, Martínez-Guerra E, Piñón-Balderrama CI, Compean Martínez I et al (2019) Application of differential scanning calorimetry (DSC) and modulated differential scanning calorimetry (MDSC) in food and drug industries. Polymers (Basel) 12(1). https://www.ncbi.nlm.nih.gov/pmc/articles/PMC7023573/
74. Patel P, Hanini A, Shah A, Patel D, Patel S, Bhatt P et al (2019) Surface modification of nanoparticles for targeted drug delivery. In: Surface modification of nanoparticles for targeted drug delivery. Springer, pp 19–31
75. Markovsky E, Baabur-Cohen H, Eldar-Boock A, Omer L, Tiram G, Ferber S et al (2012) Administration, distribution, metabolism and elimination of polymer therapeutics. J Control Release 161(2):446–460
76. Sleep D, Cameron J, Evans LR (2013) Albumin as a versatile platform for drug half-life extension. Biochimica et Biophysica Acta (BBA) Gen Subj 1830(12):5526–5534
77. Byrne JD, Betancourt T, Brannon-Peppas L (2008) Active targeting schemes for nanoparticle systems in cancer therapeutics. Adv Drug Deliv Rev 60(15):1615–1626
78. Choi SW, Kim WS, Kim JH (2005) Surface-functionalized nanoparticles for controlled drug delivery. In: Rosenthal SJ, Wright DW (eds) NanoBiotechnology protocols. Humana Press, Totowa, NJ, pp 121–31 (Methods in Molecular Biology™). https://doi.org/10.1385/1-59259-901-X:121
79. Gref R, Lück M, Quellec P, Marchand M, Dellacherie E, Harnisch S et al (2000) 'Stealth' corona-core nanoparticles surface modified by polyethylene glycol (PEG): influences of the corona (PEG chain length and surface density) and of the core composition on phagocytic uptake and plasma protein adsorption. Colloids Surf, B 18(3):301–313
80. Herranz F, Pellico J, Jús Ruiz-Cabello J (2012) Covalent functionalization of magnetic nanoparticles for biomedical imaging. Resonance 4:6
81. Thiruppathi R, Mishra S, Ganapathy M, Padmanabhan P, Gulyás B (2017) Nanoparticle functionalization and its potentials for molecular imaging. Adv Sci 4(3):1600279
82. Secret E, Maynadier M, Gallud A, Chaix A, Bouffard E, Gary-Bobo M et al (2014) Two-photon excitation of porphyrin-functionalized porous silicon nanoparticles for photodynamic therapy. Adv Mater 26(45):7643–7648
83. Tasso M, Singh MK, Giovanelli E, Fragola A, Loriette V, Regairaz M et al (2015) Oriented bioconjugation of unmodified antibodies to quantum dots capped with copolymeric ligands as versatile cellular imaging tools. ACS Appl Mater Interfaces 7(48):26904–26913
84. Ratzinger G, Agrawal P, Körner W, Lonkai J, Sanders HMHF, Terreno E et al (2010) Surface modification of PLGA nanospheres with Gd-DTPA and Gd-DOTA for high-relaxivity MRI contrast agents. Biomaterials 31(33):8716–8723
85. Cole LE, Vargo-Gogola T, Roeder RK (2015) Contrast-enhanced x-ray detection of micro-calcifications in radiographically dense mammary tissue using targeted gold nanoparticles. ACS Nano 9(9):8923–8932
86. Su H, Wang Y, Liu S, Wang Y, Liu Q, Liu G et al (2019) Emerging transporter-targeted nanoparticulate drug delivery systems. Acta Pharm Sinica B 9(1):49–58
87. Yang K, Feng L, Hong H, Cai W, Liu Z (2013) Preparation and functionalization of graphene nanocomposites for biomedical applications. Nat Protoc 8(12):2392–2403
88. Xin H, Jiang X, Gu J, Sha X, Chen L, Law K et al (2011) Angiopep-conjugated poly(ethylene glycol)-co-poly(ε-caprolactone) nanoparticles as dual-targeting drug delivery system for brain glioma. Biomaterials 32(18):4293–4305
89. Benyettou F, Lalatonne Y, Chebbi I, Di Benedetto M, Serfaty J-M, Lecouvey M et al (2011) A multimodal magnetic resonance imaging nanoplatform for cancer theranostics. Phys Chem Chem Phys 13(21):10020–10027

90. Sun S-K, Wang H-F, Yan X-P (2018) Engineering persistent luminescence nanoparticles for biological applications: from biosensing/bioimaging to theranostics. Acc Chem Res 51(5):1131–1143
91. Soleymani M, Khalighfard S, Khodayari S, Khodayari H, Kalhori MR, Hadjighassem MR et al (2020) Effects of multiple injections on the efficacy and cytotoxicity of folate-targeted magnetite nanoparticles as theranostic agents for MRI detection and magnetic hyperthermia therapy of tumor cells. Sci Rep 10(1):1695
92. Kong W, Wang L, Zhao Y, Chen H, Liu Y, Han X et al (2021) Sequential module coordination-driven programmable function switch of metal-molecule nanoframeworks for cancer theranostics. Nano Today 1(38):101126
93. Amendoeira A, García LR, Fernandes AR, Baptista PV (2020) Light irradiation of gold nanoparticles toward advanced cancer therapeutics. Adv Ther 3(1):1900153
94. Garg P, Sangam S, Kochhar D, Pahari S, Kar C, Mukherjee M (2020) Exploring the role of triazole functionalized heteroatom co-doped carbon quantum dots against human coronaviruses. Nano Today 1(35):101001
95. Fu Y, Ni J, Chen J, Ma G, Zhao M, Zhu S et al (2020) Dual-functionalized MSCs that express CX3CR1 and IL-25 exhibit enhanced therapeutic effects on inflammatory bowel disease. Mol Ther 28(4):1214–1228
96. Peng Q, Wei X-Q, Yang Q, Zhang S, Zhang T, Shao X-R et al (2015) Enhanced biostability of nanoparticle-based drug delivery systems by albumin corona. Nanomedicine 10(2):205–214
97. Gao W, Zhang L (2015) Coating nanoparticles with cell membranes for targeted drug delivery. J Drug Target 23(7–8):619–626
98. Begines B, Ortiz T, Pérez-Aranda M, Martínez G, Merinero M, Argüelles-Arias F et al (2020) Polymeric nanoparticles for drug delivery: recent developments and future prospects. Nanomaterials (Basel) 10(7). https://www.ncbi.nlm.nih.gov/pmc/articles/PMC7408012/
99. Zhang Q, Wang X, Li P-Z, Nguyen KT, Wang X-J, Luo Z et al (2014) Biocompatible, uniform, and redispersible mesoporous silica nanoparticles for cancer-targeted drug delivery in vivo. Adv Func Mater 24(17):2450–2461
100. Kumari P, Ghosh B, Biswas S (2016) Nanocarriers for cancer-targeted drug delivery. J Drug Target 24(3):179–191
101. Kumar A, Ahmad A, Vyawahare A, Khan R (2020) Membrane trafficking and subcellular drug targeting pathways. Front Pharmacol:11
102. Zielińska A, Carreiró F, Oliveira AM, Neves A, Pires B, Venkatesh DN et al (2020) Polymeric nanoparticles: production, characterization, toxicology and ecotoxicology. Molecules 25(16):3731
103. Rizvi SAA, Saleh AM (2018) Applications of nanoparticle systems in drug delivery technology. Saudi Pharm J 26(1):64–70
104. Castro KC de, Costa JM, Campos MGN (2020) Drug-loaded polymeric nanoparticles: a review. Int J Polym Mater Polym Biomater:1–13
105. Han J, Zhao D, Li D, Wang X, Jin Z, Zhao K (2018) Polymer-based nanomaterials and applications for vaccines and drugs. Polymers 10(1):31
106. Desgouilles S, Vauthier C, Bazile D, Vacus J, Grossiord J-L, Veillard M et al (2003) The design of nanoparticles obtained by solvent evaporation: a comprehensive study. Langmuir 19(22):9504–9510
107. Vieira R, Souto SB, Sánchez-López E, López Machado A, Severino P, Jose S et al (2019) Sugar-Lowering drugs for type 2 diabetes mellitus and metabolic syndrome—strategies for in vivo administration: part-II. J Clin Med 8(9):1332
108. Jose S, Sowmya S, Cinu TA, Aleykutty NA, Thomas S, Souto EB (2014) Surface modified PLGA nanoparticles for brain targeting of Bacoside-A. Eur J Pharm Sci 15(63):29–35
109. Sharma N, Madan P, Lin S (2016) Effect of process and formulation variables on the preparation of parenteral paclitaxel-loaded biodegradable polymeric nanoparticles: a co-surfactant study. Asian J Pharm Sci 11(3):404–416
110. Rao JP, Geckeler KE (2011) Polymer nanoparticles: Preparation techniques and size-control parameters. Prog Polym Sci 36(7):887–913

111. Kumar S, Dilbaghi N, Saharan R, Bhanjana G (2012) Nanotechnology as emerging tool for enhancing solubility of poorly water-soluble drugs. BioNanoSci 2(4):227–250
112. Souto EB, Severino P, Santana MHA (2012) Preparação de nanopartículas poliméricas a partir da polimerização de monômeros: parte I. Polímeros 22(1):96–100
113. Souto EB, Souto SB, Campos JR, Severino P, Pashirova TN, Zakharova LY et al (2019) Nanoparticle delivery systems in the treatment of diabetes complications. Molecules 24(23):4209
114. Wang B, Wang S, Zhang Q, Deng Y, Li X, Peng L et al (2019) Recent advances in polymer-based drug delivery systems for local anesthetics. Acta Biomater 15(96):55–67
115. Lim K, Hamid ZAA (2018) 10—polymer nanoparticle carriers in drug delivery systems: Research trend. In: Inamuddin, Asiri AM, Mohammad A (eds) Applications of nanocomposite materials in drug delivery. Woodhead Publishing, pp 217–237 (Woodhead Publishing Series in Biomaterials). https://www.sciencedirect.com/science/article/pii/B9780128137413000108
116. Pinto Reis C, Neufeld RJ, Ribeiro António J, Veiga F. Nanoencapsulation I (2006) Methods for preparation of drug-loaded polymeric nanoparticles. Nanomed Nanotechnol Biol Med 2(1):8–21
117. Crucho CIC, Barros MT (2017) Polymeric nanoparticles: a study on the preparation variables and characterization methods. Mater Sci Eng, C 1(80):771–784
118. Krishnamoorthy K, Mahalingam M (2015) Selection of a suitable method for the preparation of polymeric nanoparticles: multi-criteria decision making approach. Adv Pharm Bull 5(1):57–67
119. Cañadas C, Alvarado H, Calpena AC, Silva AM, Souto EB, García ML et al (2016) In vitro, ex vivo and in vivo characterization of PLGA nanoparticles loading pranoprofen for ocular administration. Int J Pharm 511(2):719–727
120. Sánchez-López E, Egea MA, Cano A, Espina M, Calpena AC, Ettcheto M et al (2016) PEGylated PLGA nanospheres optimized by design of experiments for ocular administration of dexibuprofen—in vitro, ex vivo and in vivo characterization. Colloids Surf, B 1(145):241–250
121. Sánchez-López E, Egea MA, Davis BM, Guo L, Espina M, Silva AM et al (2018) Memantine-loaded PEGylated biodegradable nanoparticles for the treatment of glaucoma. Small 14(2):1701808
122. Salatin S, Barar J, Barzegar-Jalali M, Adibkia K, Kiafar F, Jelvehgari M (2017) Development of a nanoprecipitation method for the entrapment of a very water soluble drug into Eudragit RL nanoparticles. Res Pharm Sci 12(1):1–14
123. Martínez Rivas CJ, Tarhini M, Badri W, Miladi K, Greige-Gerges H, Nazari QA et al (2017) Nanoprecipitation process: from encapsulation to drug delivery. Int J Pharm 532(1):66–81
124. Tibbitt MW, Dahlman JE, Langer R (2016) Emerging frontiers in drug delivery. J Am Chem Soc 138(3):704–717
125. Kohrs NJ, Liyanage T, Venkatesan N, Najarzadeh A, Puleo DA (2019) Drug delivery systems and controlled release
126. Venditti I (2019) Morphologies and functionalities of polymeric nanocarriers as chemical tools for drug delivery: a review. J King Saud Univ Sci 31(3):398–411
127. George A, Shah PA, Shrivastav PS (2019) Natural biodegradable polymers based nano-formulations for drug delivery: a review. Int J Pharm 20(561):244–264
128. Hajebi S, Rabiee N, Bagherzadeh M, Ahmadi S, Rabiee M, Roghani-Mamaqani H et al (2019) Stimulus-responsive polymeric nanogels as smart drug delivery systems. Acta Biomater 1(92):1–18
129. Hua S, Marks E, Schneider JJ, Keely S (2015) Advances in oral nano-delivery systems for colon targeted drug delivery in inflammatory bowel disease: selective targeting to diseased versus healthy tissue. Nanomed Nanotechnol Biol Med 11(5):1117–1132
130. Joseph SK, Sabitha M, Nair SC (2020) Stimuli-responsive polymeric nanosystem for colon specific drug delivery. Adv Pharm Bull 10(1):1–12
131. Jain A, Jain A, Jain A, Jain A (2016) Quasi emulsion spherical crystallization technique based environmentally responsive Tulsion® (pH dependent) microspheres for colon specific delivery. J Appl Biomed 14(2):147–155

132. Sunoqrot S, Abujamous L (2019) PH-sensitive polymeric nanoparticles of quercetin as a potential colon cancer-targeted nanomedicine. J Drug Deliv Sci Technol 1(52):670–676
133. Mishra RK, Ahmad A, Kumar A, Vyawahare A, Raza SS, Khan R (2020) Lipid-based nanocarrier-mediated targeted delivery of celecoxib attenuate severity of ulcerative colitis. Mater Sci Eng, C 1(116):111103
134. Lee SH, Back S-Y, Song JG, Han H-K (2020) Enhanced oral delivery of insulin via the colon-targeted nanocomposite system of organoclay/glycol chitosan/Eudragit®S100. J Nanobiotechnol 18(1):104
135. Vaidya A, Jain S, Agrawal RK, Jain SK (2015) Pectin–metronidazole prodrug bearing microspheres for colon targeting. J Saudi Chem Soc 19(3):257–264
136. Hamel F, Grondin M, Denizeau F, Averill-Bates DA, Sarhan F (2006) Wheat extracts as an efficient cryoprotective agent for primary cultures of rat hepatocytes. Biotechnol Bioeng 95(4):661–670
137. Li L, Wang H, Ong ZY, Xu K, Ee PLR, Zheng S et al (2010) Polymer- and lipid-based nanoparticle therapeutics for the treatment of liver diseases. Nano Today 5(4):296–312
138. Wang H, Thorling CA, Liang X, Bridle KR, Grice JE, Zhu Y et al (2015) Diagnostic imaging and therapeutic application of nanoparticles targeting the liver. J Mater Chem B 3(6):939–958
139. Wei Y, Quan L, Zhou C, Zhan Q (2018) Factors relating to the biodistribution & clearance of nanoparticles & their effects on in vivo application. Nanomedicine 13(12):1495–1512
140. Xiao K, Li Y, Luo J, Lee JS, Xiao W, Gonik AM et al (2011) The effect of surface charge on in vivo biodistribution of PEG-oligocholic acid based micellar nanoparticles. Biomaterials 32(13):3435–3446
141. He C, Hu Y, Yin L, Tang C, Yin C (2010) Effects of particle size and surface charge on cellular uptake and biodistribution of polymeric nanoparticles. Biomaterials 31(13):3657–3666
142. Mandal D, Shaw TK, Dey G, Pal MM, Mukherjee B, Bandyopadhyay AK et al (2018) Preferential hepatic uptake of paclitaxel-loaded poly-(d-l-lactide-co-glycolide) nanoparticles—a possibility for hepatic drug targeting: pharmacokinetics and biodistribution. Int J Biol Macromol 1(112):818–830
143. Upadhyay RK (2014) Drug delivery systems, CNS protection, and the blood brain barrier. Biomed Res Int. https://www.ncbi.nlm.nih.gov/pmc/articles/PMC4127280/
144. Kuittinen O, Siniluoto T, Isokangas M, Turpeenniemi-Hujanen T, Peltonen J, Alahuhta S, Sonkajärvi E (2013) Chemotherapy in conjunction with blood brain barrier disruption in the treatment of primary central nervous system lymphoma. Duodecim 129(15):1563–1570
145. Patel T, Zhou J, Piepmeier JM, Saltzman WM (2012) Polymeric nanoparticles for drug delivery to the central nervous system. Adv Drug Deliv Rev 64(7):701–705
146. Kreuter J (2004) Influence of the surface properties on nanoparticle-mediated transport of drugs to the brain. J Nanosci Nanotechnol 4(5):484–488
147. Kreuter J, Shamenkov D, Petrov V, Ramge P, Cychutek K, Koch-Brandt C et al (2002) Apolipoprotein-mediated transport of nanoparticle-bound drugs across the blood-brain barrier. J Drug Target 10(4):317–325
148. Ferreira NN, Granja S, Boni FI, Prezotti FG, Ferreira LMB, Cury BSF et al (2020) Modulating chitosan-PLGA nanoparticle properties to design a co-delivery platform for glioblastoma therapy intended for nose-to-brain route. Drug Deliv Transl Res 10(6):1729–1747
149. Cai Q, Wang L, Deng G, Liu J, Chen Q, Chen Z (2016) Systemic delivery to central nervous system by engineered PLGA nanoparticles. Am J Transl Res. 8(2):749–764
150. Jolivet J-P, Froidefond C, Pottier A, Chanéac C, Cassaignon S, Tronc E et al (2004) Size tailoring of oxide nanoparticles by precipitation in aqueous medium a semi-quantitative modelling. J Mater Chem 14(21):3281–3288
151. El Ouahabi A, Thiry M, Pector V, Fuks R, Ruysschaert JM, Vandenbranden M (1997) The role of endosome destabilizing activity in the gene transfer process mediated by cationic lipids. FEBS Lett 414(2):187–192
152. Xu Y, Szoka FC (1996) Mechanism of DNA release from cationic liposome/DNA complexes used in cell transfection. Biochemistry 35(18):5616–5623

153. Bantz C, Koshkina O, Lang T, Galla H-J, Kirkpatrick CJ, Stauber RH et al (2014) The surface properties of nanoparticles determine the agglomeration state and the size of the particles under physiological conditions. Beilstein J Nanotechnol 5(1):1774–1786
154. Abbaszadegan A, Ghahramani Y, Gholami A, Hemmateenejad B, Dorostkar S, Nabavizadeh M et al (2015) The effect of charge at the surface of silver nanoparticles on antimicrobial activity against gram-positive and gram-negative bacteria: a preliminary study. J Nanomater 2015
155. Pelaz B, del Pino P, Maffre P, Hartmann R, Gallego M, Rivera-Fernández S et al (2015) Surface functionalization of nanoparticles with polyethylene glycol: effects on protein adsorption and cellular uptake. ACS Nano 9(7):6996–7008
156. dos Santos-Silva AM, de Caland LB, do Nascimento EG, Oliveira ALC de SL, de Araújo-Júnior RF, Cornélio AM et al (2019) Self-assembled benznidazole-loaded cationic nanoparticles containing cholesterol/sialic acid: physicochemical properties, in vitro drug release and in vitro anticancer efficacy. Int J Mol Sci 20(9):2350
157. Botequim D, Silva IIR, Serra SG, Melo EP, Prazeres DMF, Costa SMB et al (2020) Fluorescent dye nano-assemblies by thiol attachment directed to the tips of gold nanorods for effective emission enhancement. Nanoscale 12(11):6334–6345
158. Rojas-Sánchez L, Sokolova V, Riebe S, Voskuhl J, Epple M (2019) Covalent surface functionalization of calcium phosphate nanoparticles with fluorescent dyes by copper-catalysed and by strain-promoted azide-alkyne click chemistry. ChemNanoMat 5(4):436–446
159. Yen SK, Jańczewski D, Lakshmi JL, Dolmanan SB, Tripathy S, Ho VHB et al (2013) Design and synthesis of polymer-functionalized NIR fluorescent dyes-magnetic nanoparticles for bioimaging. ACS Nano 7(8):6796–6805
160. Luo R, Neu B, Venkatraman SS (2012) Surface functionalization of nanoparticles to control cell interactions and drug release. Small 8(16):2585–2594
161. Wu W, He Q, Jiang C (2008) Magnetic iron oxide nanoparticles: synthesis and surface functionalization strategies. Nanoscale Res Lett 3(11):397
162. Cao K, Cai J, Shan B, Chen R (2020) Surface functionalization on nanoparticles via atomic layer deposition. Sci Bull 65(8):678–688
163. Jeong W, Bu J, Kubiatowicz LJ, Chen SS, Kim YS, Hong S (218) Peptide–nanoparticle conjugates: a next generation of diagnostic and therapeutic platforms? Nano Converg 5(1):1–18
164. Curr KJ, Chem O (2012) Curr Org Chem 29(2):997–1003
165. Rizvi SAA, Saleh AM (2018) Applications of nanoparticle systems in drug delivery technology. Saudi Pharm J Elsevier B.V.:64–70
166. Gauthier MA, Klok HA (2008) Peptide/protein-polymer conjugates: synthetic strategies and design concepts. Chem Commun 23:2591–2611
167. Ranalli A, Santi M, Capriotti L, Voliani V, Porciani D, Beltram F et al (2017) Peptide-based stealth nanoparticles for targeted and pH-triggered delivery. Bioconjug Chem 28(2):627–635
168. Hawryłkiewicz A, Ptaszyńska N (2021) Gemcitabine peptide-based conjugates and their application in targeted tumor therapy. Molecules (Basel, Switzerland) 26(2)
169. Shu JY, Panganiban B, Xu T (2013) Peptide-polymer conjugates: from fundamental science to application. Annu Rev Phys Chem 64:631–657
170. Juan A, Cimas FJ, Bravo I, Pandiella A, Ocaña A, Alonso-Moreno C (2020) An overview of antibody conjugated polymeric nanoparticles for breast cancer therapy. Pharmaceutics 12(9):1–20
171. Farkaš P, Bystrický S (2010) Chemical conjugation of biomacromolecules: a mini-review. Chem Pap 64(6):683–695
172. Lanford RE, Butel JS (1984) Construction and characterization of an SV40 mutant defective in nuclear transport of T antigen. Cell 37(3):801–813
173. Shiraishi T, Hamzavi R, Nielsen PE (2005) Targeted delivery of plasmid DNA into the nucleus of cells via nuclear localization signal peptide conjugated to DNA intercalating bis- and trisacridines. Bioconjug Chem 16(5):1112–1116
174. Wang H, Yang R, Yang L, Tan W (2009) J Phys Chem C 3(9):19021–19027

175. Panigaj M, Reiser J (2016) Aptamer guided delivery of nucleic acid-based nanoparticles. DNA RNA Nanotechnol 2(1):42–52
176. Aravind A, Yoshida Y, Maekawa T, Kumar DS (2012) Aptamer-conjugated polymeric nanoparticles for targeted cancer therapy. Drug Deliv Transl Res 2(6):418–436
177. Zhang C, He T, Vedadghavami A, Bajpayee AG (2020) Avidin-biotin technology to synthesize multi-arm nano-construct for drug delivery. MethodsX 7:100882
178. Qi K, Ma Q, Remsen EE, Clark CG, Wooley KL (2004) Determination of the bioavailability of biotin conjugated onto shell cross-linked (SCK) nanoparticles. J Am Chem Soc 126(21):6599–6607
179. Ludtke JJ, Zhang G, Sebestyén MG, Wolff JA (1999) A nuclear localization signal can enhance both the nuclear transport and expression of 1 kb DNA. J Cell Sci 112(12):2033–2041

Nanomaterials and Management of Inflammatory Disorders

Shambhu Kumar, Pranav Kanti, and Md. Meraj Ansari

Abstract Inflammation is a natural reaction to pathogenic infections or tissue injury that helps restore homeostasis by protecting the host from external pathogens and healing damaged tissues. It has three phases: induction, pro-inflammatory, and resolution. Unchecked inflammation can lead to the development of many inflammatory diseases including sepsis, obesity, cancer, cardiovascular, neurological, and immunological diseases. Dysregulated localized or chronic inflammation can be caused by an imbalance between the immune system's innate and adaptive components. Anti-inflammatory drugs are critical for treating inflammatory diseases, but they have limitations such as poor tissue selectivity, systemic adverse effects, and inefficiency in overcoming biological barriers. Nanotechnology has grown quickly in medicine since the FDA-approved PEGylated liposomal doxorubicin (Doxil®) in 1995. Nanoparticles (NPs) in the 1–100 nm size range have received attention for their novel applications in biomedicine.

Keywords Nanomaterials · Inflammation · Innate immune response · Inflammatory disorders

S. Kumar (✉)
Department of Zoology, Ram Bahadur Singh College, Andaur, Samastipur 848502, Bihar, India
e-mail: shambhu@tezu.ac.in; sumit.2910@gmail.com

S. Kumar · P. Kanti
Department of Zoology, L. N. M. University, Darbhanga, Bihar, India

P. Kanti
Department of Zoology, A. N. D. College, Shahpur Patori, Samastipur 848504, Bihar, India

Md. M. Ansari
Department of Animal Biotechnology, Faculty of Biotechnology, College of Applied Life Sciences, Jeju National University, Jeju Special Self-Governing Province, Jeju-Si 63243, South Korea

Interdisciplinary Graduate Program in Advanced Convergence Technology and Science, Jeju National University, Jeju Special Self-Governing Province, Jeju-Si 63243, South Korea

Md. M. Ansari et al. (eds.), *Emergence of Sustainable Biomaterials in Tackling Inflammatory Diseases*, Smart Nanomaterials Technology,
https://doi.org/10.1007/978-981-96-2112-5_6

1 Introduction

Inflammation is the physiological and normal reaction to signals resulting from pathogenic infection or tissue injury [64]. Inflammation aids in the restoration of homeostasis following trauma or infection by shielding the host from external pathogens and mending damaged tissues [120]. The process of inflammation consists of three different sequential phases, namely induction phase, pro-inflammatory phase, and resolution phase. Inflammation begins with a quick induction phase that initiates and stimulates a pro-inflammatory response, and ends with a resolution phase [126]. As a result, inflammation is necessary for the healing of wounds. On the other hand, unchecked or extended inflammation can cause several inflammatory illnesses, including sepsis, obesity, cancer, cardiovascular, neurological, and autoimmune disorders, to form and worsen [38]. There are many phases of dysregulated localized or chronic inflammation. It begins with the activation of local immune cells by damage- associated molecular patterns (DAMPs) or pathogen-associated molecular patterns (PAMPs) signals that are, respectively, released from damaged cells/ tissue or from pathogens. These activated resident immune cells produce cytokines, chemokines, growth factors, protease, and oxygen free radicals [156]. After that, circulating immune cells use endothelial cell adhesion to settle at the inflammation site, enhancing pro-inflammatory reactions [79]. This coordinated immunological response eventually restores homeostasis. Nevertheless, late-stage inflammation may result from an inadequate or overreaction that upsets the equilibrium between the immune system's innate and adaptive arms. Damage to local or systemic tissues and an overabundance of danger signals are two hallmarks of a catastrophic cascade that might result from such dysregulated immune response. The change of an inflammatory response from acute to chronic may lead to a breakdown in the immunological tolerance which can advance an inflammatory illness and perhaps cause mortality [112]. Anti-inflammatory medications are the mainstay of treatment for inflammatory illnesses nowadays, since they may assist in achieving a balance between the immune system and inflammation [32]. The challenges that the existing anti-inflammatory drugs face include poor tissue selectivity, systemic side effects and inefficiency in overcoming biological barriers that demand new and more effective anti-inflammatory techniques. Over the past 10 years, nanotechnology has advanced quickly, particularly in the area of medicine. Since the FDA (US Food and Drug Administration) approval of PEGylated liposomal doxorubicin (Doxil®) as the first nano-drug in 1995, nanotechnology has attracted a lot of attention for the creation of innovative theranostic strategies [10]. Nanoparticles (NPs) in the 1–100 nm size range have gained recognition in recent years, particularly in the biomedical field, for their special applications. This is because of their clever combination of notable physical and chemical properties and appropriate interaction with the building blocks of living organisms at the nanoscale [110].

2 Mechanisms of Inflammation

Inflammation begins with the identification of infection or injury. Thereupon, soluble immune signaling molecules known as cytokines orchestrate a carefully regulated cascade of immunological, physiological, and behavioural events. The identification of infection or injury is respectively performed by pathogen-associated molecular patterns (PAMPs) or damage-associated molecular patterns (DAMPs). PAMPs are the surface molecules of the pathogens that have general motifs and are crucial for the survival of the pathogens whereas, DAMPs (alarmins) are naturally occurring molecules that alert the immune system to damage or necrosis. Detecting these signals becomes advantageous in minimizing the inadvertent targeting of host cells and tissues. Germ-line encoded receptors including leucine-rich repeat (LRR)-containing and intracellular nucleotide-binding domain and transmembrane Toll-like receptors (TLRs) identify numerous damage signals [70, 109, 118]. TLRs, upon ligand identification, initiate shared signalling pathways that ultimately activate a transcription factor NF-κB. This transcription factor is present in almost all cell types but remains inactive due to its association with an inhibitor protein IκB [45]. The event of signal transduction releases NF-κB from its inhibitor IκB and makes it active. The active NF-κB then migrates to the nucleus and binds to the promoters of the genes of proinflammatory cytokines, including TNF-α, IL-6, interleukin-1-beta (IL-1β), etc. resulting in the inducible expression of the genes. Inducible expression of theses soluble proteins begins the third phase of inflammation. These soluble proteins, in combination with chemokines (attractants) and other costimulatory molecules, help to recruit effector cells, such as neutrophils and monocytes, to the site of disruption. Neutrophil releases noxious substances from the cytoplasmic granules (a process known as degranulation) to produce a cytotoxic environment. The respiratory burst, marked by the simultaneous consumption of oxygen and glucose, is necessary for the rapid release of toxic compounds, for example, highly reactive radicals of oxygen and nitrogen (ROS and RNS, respectively) and different proteinases. These compounds are toxic to both hosts and infections, and they induce the liquefaction of surrounding tissue to prevent the spread of microorganisms [91]. Therefore, these effector mechanisms play a significant role in host collateral damage. Resolution, the final stage of inflammation, is essential for minimizing collateral host damage [128]. Following the initial hours of inflammation, tissue-resident and recruited macrophages initiate a concerted resolution activity by producing proinflammatory prostaglandins and leukotrienes during acute inflammation, but they quickly convert to lipoxins, which prevent the recruitment of more neutrophils and instead support increased monocyte infiltration, which is crucial for wound healing. The combined result of these interactions is the classic set of symptoms associated with localized inflammation, which include redness, pain, swelling, heat, and loss of function. The adaptive immune system further modulates the activities of effectors of inflammation.

This coordinated immunological response of innate and adaptive immune response eventually restores homeostasis. Nevertheless, late-stage inflammation may result from an inadequate or overreaction that upsets the equilibrium between the immune system's innate and adaptive arms. Damage to local or systemic tissues and an overabundance of danger signals are two hallmarks of a catastrophic cascade that might result from such dysregulated immune response. The change of an inflammatory response from acute to chronic may lead to a breakdown in the immunological tolerance which can advance an inflammatory illness and perhaps cause mortality [112].

Treating inflammatory illnesses requires targeting damaged areas [138]. High pharmacological doses may be necessary to effectively treat inflammatory cells, but this can sometimes have negative effects on tissues that are healthier. Medicines having anti-inflammatory effects are classified as either steroidal (betamethasone, prednisone, and dexamethasone) or nonsteroidal also called non-sterioidal-anti-inflammatory drugs (NSAIDs) include aspirin, ibuprofen, and naproxen [3]. Low-dose NSAIDs are available over-the-counter to treat inflammation-related pain, whereas greater doses and steroidal medicines require a prescription. Although, steroidal and non-steroidal anti-inflammatory drugs are used to treat both acute and chronic inflammatory disorders their long-term use has been found to be associated with many side effects. For example, adrenal atrophy, fluid retention, thinning of bones which can lead to osteoporosis, and increased risk of infection or injury can be caused by steroidal medicines and peptic ulcers, allergies, high blood pressure, or liver and renal disorders are caused by the long-term usage of NSAIDs. This warrants the search for new anti-inflammatory principles with the purpose of getting improved safety, better efficacy, and a more economical way to treat inflammation.

Nanomedicine is quickly expanding to address the need for better treatments for inflammatory, developmental, viral, and degenerative nervous system illnesses [44]. The rising number of affected individuals, the severity of the condition, and the associated financial burden make these disorders a considerable burden. Thus, there is a pressing need for better diagnosis and care. In the past ten years, nanotechnology has been the subject of extensive research and has shown to be a promising tactic for the medical industry and healthcare system. This cutting-edge method provides a wealth of nanomaterials and instruments that have the potential to accurately identify and treat a wide range of inflammatory diseases and disorders. This chapter focuses on nanomedicine's existing and future potential to treat inflammatory illnesses.

3 Nanoparticles (NPs)

Material scientists have spent the last fifty years investigating the various applications of nanoparticles and their derivatives, called nanostructured materials, in the healthcare industries [43]. The particles of matter with a diameter of 1 to 100 nm are typically referred to as "NPs," however larger particles with a diameter of up to 500 nm can be referred to by different labels. As an illustration, nanorods, nanowires, and nanofibers are nanoparticles with one dimension outside the nanoscale dimension with a diameter between 1 and 100 nm [13]. Nanostructured materials have a single dimension of nanoscale range (less than 100 nm), are formed of one or more materials, and have interconnected portions at the nanoscale [57]. Simple materials (such as metal, carbon, or polymer) [11], composite materials (such as polymer-metal, silica-metal, or graphene-metal), or core–shell materials [15, 30, 49, 108] can all be used to create nanoparticles and nanostructured materials. Comparing nanomaterials to their bulk counterparts, several intriguing characteristics are evident, including a large surface area, variable porosity, greater mechanical performance, the potential for surface functionalization [1, 2, 84]. Due to these exceptional qualities, nanomaterials become the ideal choice in the biomedical industry for the development of tissue-engineered scaffolds, drug delivery systems, chemical sensors [11, 108], biosensors [30, 49], and dressings for wounds [113, 114]. Surprisingly, many studies indicate that as early as 2500 BC, ancient civilizations in Egypt, China, and India may have employed metallic gold, or nanotechnology, for medical purposes [132]. Dimensionality divides nanomaterials primarily into four categories [43]:0D, in which width, height, and length are fixed at one location; One-dimensional systems (like carbon nanotubes) have only one parameter; two-dimensional systems (like graphene) have length and width; and three-dimensional systems (like graphene) have length, height, and breadth [67]. Nanoparticles are classified as organic, inorganic, carbon-based, and composite nanostructures based on their chemical composition. Organic nanomaterials include fullerenes, carbon nanofibers, graphene, polymerosomes, dendrimers, hydrogels, and nanoconjugates. Inorganic nanomaterials include metals, metal oxides, and ceramic nanomaterials [43, 125]. Nanomaterials are characterized as porous or non-porous based on their porosity [57]. Because of having comparatively less dense density, porous materials allow airflow or the absorption of atoms, ions, and molecules whereas due to higher density, non-porous materials not a good absorber and have limited ventilation. Nanomaterials having pore-size between 2 and 50 nm in diameter are called mesoporous (super-nanoporous) nanomaterials [12, 14]. Because of having few excellent characteristics such as high porosity with small pore size, and surface functionality, mesoporous nanomaterials can have regulated release and efficient drug loading at the target site. Current researches are focussed on the development of suitable mesoporous nanomaterials for the targeted delivery of therapeutic drugs to the tumour cells without minimal leakage to the healthy cells in the vicinity or en-route [125].

4 The Interaction of NPs with the Immune System

Immune cells, immune organs and immunological molecules are the basic components of the immune system. Currently, the primary way that NP formulations affect the immune system is by utilizing particular interactions with different types of immunological components specifically immune cells and molecules. Nanoparticles, based on their unique characteristics, can be designed to target specific cellular and molecular components to impart accurate modulating functions and this becomes the basis for the development of NP-based therapy. In this section, we are highlighting the development of NP techniques that interplay with different components of the immune system with a particular emphasis on how they are utilized to treat inflammatory disorders.

4.1 The Interaction of NPs with Monocytes/Macrophages

Macrophages are crucial innate immune system cells. Although macrophages have a complicated evolutionary history recent pieces of evidence have shown that macrophages in the liver, kidney, lung, brain, and heart originate from the yolk sac or foetal liver and are maintained in an adult state of homeostasis without the help of monocytic precursors in circulation while circulating monocytes in the dermis, gut, and heart contribute to macrophage populations [46]. According to the current view, tissue macrophages in the human body may have two different sources (local, self-sustaining, or generated from monocytes), as demonstrated by heart macrophages [9]. Additionally, based on their distinct modes of activation and functional properties, macrophages can be further divided into conventionally activated (also known as inflammatory) macrophages (M1) and alternatively activated (also known as wound-healing) macrophages (M2) [37]. Pro-inflammatory M1 cells are polarized by lipopolysaccharide alone or in combination with Th1 cytokines, for example, interferon-γ (IFN-γ) and granulocyte–macrophage colony-stimulating factor (GM-CSF) and this polarized M1 cells can produce cytokines, for example, interleukin (IL)-1β, IL-6, IL-12, IL-23, and tumor necrosis factor α (TNF-α) that orchestrate the stage of inflammation. [129]. On the contrary, M2 cells are immunoregulatory and release anti-inflammatory cytokines, for example, IL-10 and transforming growth factor-β (TGF-β) upon stimulation with IL-4 and IL-13 (Th2 cytokines) [37, 90, 129]. Although its exact mechanism is not known the homeostasis of pro- and anti-inflammatory responses of the human body is regulated by the M1/M2 polarization of macrophages. This activation mechanism is important in several physiological and pathological processes such as inflammatory diseases, infection, cancer and pathological pregnancy [78, 129]. A range of nanoparticle formulations have been developed to interact with monocytes and macrophages, by taking into account their pivotal involvement in many pathogenic processes [42]. To accomplish these dual goals of selectively targeting macrophages, nanoparticles must be surface-adorned with various ligands, such as peptides, monoclonal antibodies, oligomers, or small molecules such as mannose and legumain. [106]. These ligands bind with specific overexpressed surface receptors of macrophages. Nanoparticles have a variety of impacts on macrophages and monocytes, including modifying polarization regulation, microvascular permeability augmentation, activation of bone marrow, mobilization, and recruitment of monocytes. The inherent characteristics of monocytes/macrophages can also be exploited by the nanoparticles to improve targeting efficiency while avoiding clearance by the immune system (Fig. 1).

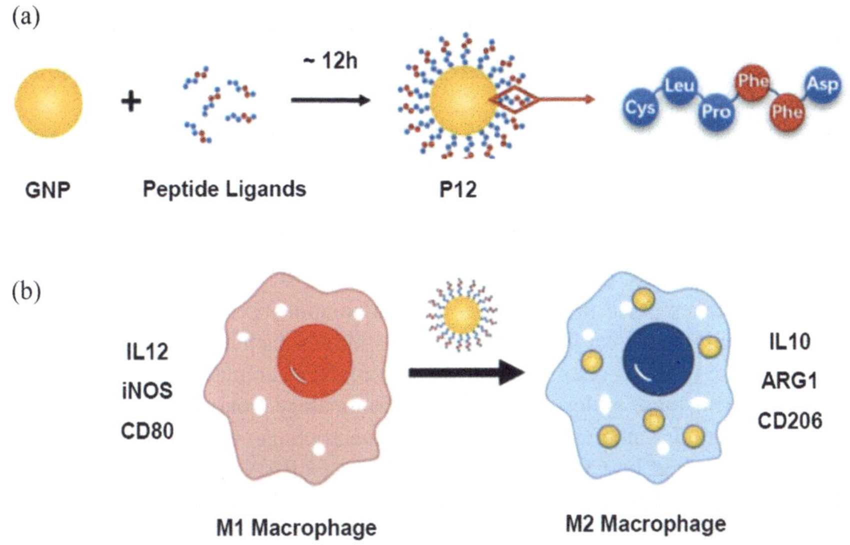

Fig. 1 Macrophage polarization by gold nanoparticles (GNP). **a** Surface-coating of GNP with peptide (5-mer), **b** Modulation of macrophage polarization from pro-inflammatory (M1) to anti-inflammatory (M2) by gold nanoparticles (GNP) P12. Image created with Biorender.com

4.1.1 Inhibition of the Monocyte Generation and Egression

Bone marrow activation, one of the main characteristics of inflammatory diseases, leads to the increase in the number of hematopoietic stem cells (HSCs) because of the high levels of inflammatory cytokines, Toll-like receptors (TLR) agonists, and noradrenaline that results in the overproduction of inflammatory monocytes which accumulates in the inflamed tissues [53, 86]. In addition to this, the interaction of chemokine C–C motif receptor 2 (CCR2) and chemokine C–C motif ligand 2 (CCL2) as well as dilated blood vessels, would boost the egression of inflammatory monocytes in the bone marrow environment [130]. Consequently, to diagnose and treat inflammatory disorders, NPs may be engineered to target distinct phases of the generation and egression of inflammatory monocytes. To control the response of bone marrow to inflammation, for instance, corticosteroids [97] and CCR2 small interfering RNA (siRNA) [73] loaded NPs have been developed. Such intervention inhibited the development of inflammatory disorders such as myocardial infarction, atherosclerosis, and pancreatic islet transplantation in diabetes, as well as dramatically lowered the number of monocytes accumulated in inflamed sites [73]. Moreover, measures to improve the accumulation of nanoparticles in the bone marrow have been developed. When conjugated to various nanoparticle forms, bisphosphonates (such as alendronate) can improve their propensity for binding to hydroxyapatite, the primary mineral that makes up bone [95]. Synthetic compounds, such as the anionic amphiphilic L-glutamic acid N-(3-carboxy-1-oxopropyl)-1,5-hexadecyl ester, can both boost bone marrow liposome accumulation and concurrently block hepatic absorption [137].

4.1.2 Modulation of Monocyte/Macrophage Mobilization and Recruitment

NPs can be engineered to control the recruitment and mobilization of monocytes and macrophages. During chronic inflammation, accumulation of monocytes/macrophages takes place in the region of inflammation because of the mobilization and recruitment effect of the various chemokines such as CCL2, CCL7, and angiotensin II (Ag II) receptor [65, 80]. Therefore, NPs can play anti-inflammatory roles by targeting these molecules to inhibit inflammatory monocytes/macrophages. In a finding, cellular atherosclerotic model of macrophage, CCR2 expression was successfully targeted by CCR2 short hairpin RNA (shRNA)-loaded EGFP-EGF1-conjugated poly (lactic-co-glycolic acid) (PLGA) nanoparticles (ENPs) to specifically knock-down the expression of CCR2 gene [147]. A similar approach using lipid polymer NPs loaded with MCP-1 siRNA was successful in preventing the mobilization of inflammatory monocytes from the hematopoietic niche to the damaged heart (myocardial infarction) of a mouse model [66]. The main step of the monocyte recruitment process, monocyte-endothelial adhesion, is the starting step of the development of inflammatory vascular diseases like atherosclerosis [59] and a variety of adhesion molecules involved in this process have been identified. Accordingly, it has been demonstrated that administering siRNA-loaded NPs to knock down the expression of 5 different cell adhesion molecules significantly reduces the mobilization of monocytes into atherosclerotic lesions and decreases the activity of plaque protease that degrades matrix following myocardial infarction in mice.

As a result, local inflammation of the infarcted myocardium was decreased and the recovery following ischemia was enhanced [122]. An Inadequate macrophage response is connected to the process behind the poor repair of tissue damage resulting from inflammatory disorders such as diabetes and myocardial infarction, which carries a significant public health cost [40, 54, 60]. The α-gal epitope, a carbohydrate antigen, can attach to its naturally occurring antibody in the human body and can create immunological complexes that attract antigen-presenting cells (APCs) such as macrophages [39]. α-gal nanoparticles created from the RBC cell membrane of rabbits was able to recruit pro-reparative macrophages in the infarcted region in a mouse model of myocardial infarction. When compared to the control group, the treated animals showed a significant remission after 28 days of therapy in terms of infarct size (about tenfold less), restoration of normal myocardial structure, and contractile function [40]. This approach has also been used with positive results in treating diabetic wound healing [60] (Fig. 2).

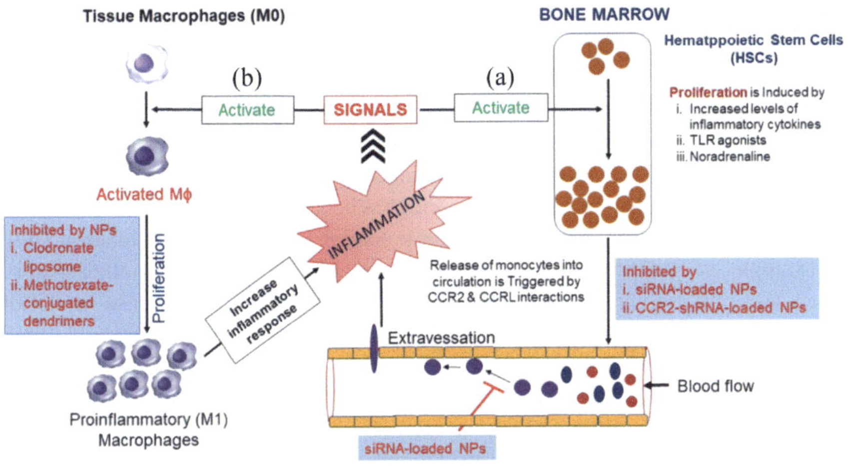

Fig. 2 Schematic Illustration of modulation of **a** Monocyte egression from bone marrow, their mobilization and recruitment **b** Depletion of hyperactive macrophages. Image created with Biorender.com

4.1.3 Reducing Hyperactive Macrophage Populations

Lesional inflammation is triggered by both circulating monocytes and the proliferation of local macrophage populations [51, 58, 119], therefore, in an established inflammatory condition it would be a wise decision to block both the process, i.e., monocyte recruitment and local macrophage proliferation [75, 119]. In this direction, many products of nanoparticles were developed to block the proliferation of macrophages or directly reduce their number to lower the burden of inflammation. For example, clodronate liposome has been used for macrophage depletion in the treatment of a variety of diseases [8, 83] and in different studies, Clodronate liposome was used for the depletion of macrophages in mouse models of endometriosis [8] and rheumatoid arthritis (RA) [116, 153]. In another study, methotrexate-conjugated dendrimer was found to alleviate inflammation in the rat model of RA by specifically targeting folate receptor overexpressing and inflammation-activated macrophages [56]. Being a folate inhibitor, methotrexate imparts anti-proliferative and anti-inflammatory activities [111].

4.1.4 Modulation of Macrophage Polarization

As previously mentioned, macrophage polarization is essential to the inflammatory process, particularly in the pathogenesis of many inflammatory disorders.

Macrophages can differentiate into the wound-healing type (M2) or the inflammatory type (M1) in response to distinct inputs from the inflammatory milieu. Pro-inflammatory cytokines and chemokines generated by M1-type cells draw in several other cell types to the inflammatory location, worsening the condition of the inflamed tissue in the process [87]. In contrast, M2 cells work oppositely, releasing anti-inflammatory cytokines that reduce inflammation. Based on such theories, nanoparticle formulations have been designed to increase M2 macrophage polarization while limiting the M1 macrophage population [4, 20]. Methods include the plasmid DNA/microRNA delivery, introduction of cytokines, metal nanoparticles, etc. Targeted delivery of cytokines and metal nanoparticles has been shown to activate macrophage M2 polarization in vivo, highlighting the potential for cytokine modulation [69]. For example, mice exposed to gold nanoparticles (AuNP) exhibit anti-inflammatory effects and possess various beneficial biochemical features. In a mouse model of bacterial sepsis using caecal ligation and puncture, researchers administered AuNP as an adjuvant to antibiotics and as a result, they observed a reduction in the M1 macrophage population while an increase in the M2 macrophage population in the spleens of sepsis mice [139, 142]. In addition, a variety of synthetic immunomodulatory nanoparticles have been developed for the repolarization of macrophages with microRNA or plasmid DNA. For example, the use of multiple hyaluronic acid-poly (ethyleneimine) NPs (HA-PEI) encapsulated IL-4/IL-10 expressing plasmid DNA, or microRNA-223, have been found to modulate the reprogramming of macrophages from the M1 to M2 subtype [141, 142]. Additionally, particles of graphene oxide (GO) loaded with a plasmid that expresses IL-4 can also provide comparable polarization-modifying effects [50]. These findings suggested that nucleic acid delivery may be used to modify macrophage polarization (Fig. 3).

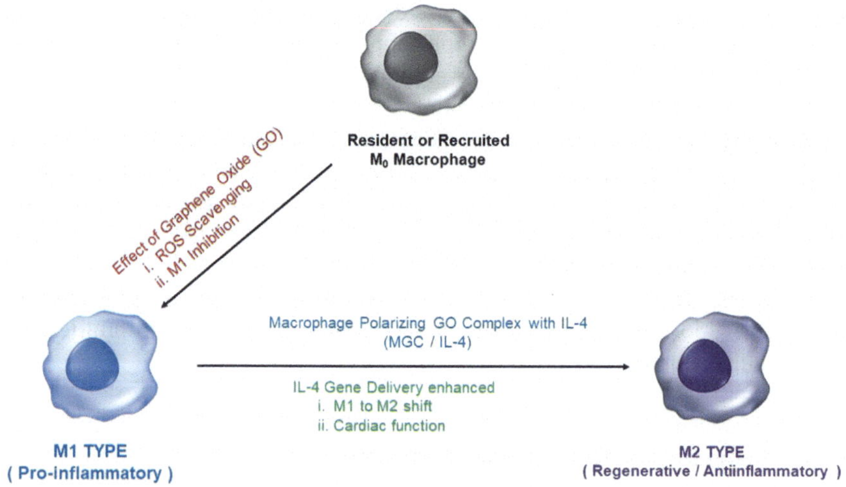

Fig. 3 Schematic illustration of Modulation of Macrophage Polarization

4.1.5 Macrophage Biomimetic Techniques

To attain maximum therapeutic efficiency, nanoparticles must bypass the immune system, traverse biological barriers, and reach target tissues [102]. To give nanoparticles enhanced bio-interfacing capabilities, currently, researchers use naturally derived cell membranes from monocytes or macrophages being an important source [35]. For example, silicon NPs (nanoparticles coated with cellular membranes isolated from leucocytes) [102] can avoid opsonization and lessen immune system phagocytosis, establish interactions with endothelial cells by receptor-ligand interactions and transport and release their payload across an inflamed reconstructed endothelium. Thus, these hybrid nanoparticles may find their use in the treatment of tumours and inflammatory disorders. In another study, proteolipid vesicles were generated from lipid nanoparticles and protein from the macrophage plasma membrane, which decreased phlogosis in a localized model of inflammation and preferentially targeted inflammatory vasculature, allowing dexamethasone to be delivered to inflamed tissues efficiently and selectively [89]. Apart from this, an alternate strategy of cellular hitchhiking or backpacking is accomplished by creating nanoparticles that are connected to the surface of or incorporated in monocytes or macrophages [6, 105]. A similar mixture, which has been used to treat inflammation-related illnesses such as viral infection and Parkinson's disease, makes use the innate capacity of the circulatory cells to target inflammatory areas without being cleared by the immune system [21, 33].

4.2 The Interaction of NPs and Neutrophils

Neutrophils, which make up 50–70% of all leucocytes in humans, and are distinguished by having a segmented nucleus, numerous granules and secretory vesicles-rich cytoplasm, that functionally play a vital role in acute inflammation to clear extracellular pathogens [74, 82]. Neutrophils originate from myeloid progenitor cells in the bone marrow and before becoming mature and entering into the circulation they have to pass through three cellular pools: the stem cell pool, the mitotic pool, and the postmitotic pool [18]. Neutrophils, during inflammation, are attracted by endothelial cell adhesion molecules such as P- and E-selectins and aggregate at the inflammatory site after a series of recruitment processes (tethering, rolling, adhesion, crawling, and transmigration) [63]. Once activated and recruited, these cellular components can eradicate infections through several mechanisms such as degranulation, phagocytosis, and the release of neutrophil extracellular traps [81]. Because of the roles of neutrophils in the pathophysiology of many inflammatory disorders, scientists have created a variety of nanoparticles that specifically target them to modify the inflammatory milieu to reduce inflammation in disease models. The way by which NPs exert their effects includes, modulation of neutrophil migration, reduction of the number of neutrophils associated with inflammation, drug delivery based on neutrophils, neutrophil biomimetic approaches, etc.

4.2.1 Modulation of Neutrophil Migration

Adherence to endothelial cells and migration of neutrophils into the lesional tissue during inflammation becomes the basis of the development of many inflammatory disorders [74]. Therefore, blocking the migratory pathway to reduce inflammation is a viable approach. Researchers revealed that some drug-free nanoparticle formulations can limit the migration of neutrophils involved in inflammation. For example, reprogramming of the immune response linked to neutrophils was made possible in traumatic primary spinal cord injury (SCI) of mice. PLG nanoparticle administration has resulted in a four-times reduction of the total amount of neutrophils accumulated at the injury site, as well as a three-fold reduction in fibrotic and gliotic scarring when compared to the sham group [101]. Additionally, scientists discovered a significant increase in the number of neutrophils gathered in the spleen, indicating a change in the migratory pathway of innate immune cells. Similar approaches have been used in models of experimental autoimmune encephalomyelitis [123] and acute lung damage [36]. The illness severity was dramatically reduced, with more neutrophils moving to the spleen rather than the inflammatory lesion. When combined, these findings demonstrated that artificial nanoparticles could influence neutrophil migration patterns in inflammatory environments and might potentially be used to treat inflammatory disorders in the future (Fig. 4).

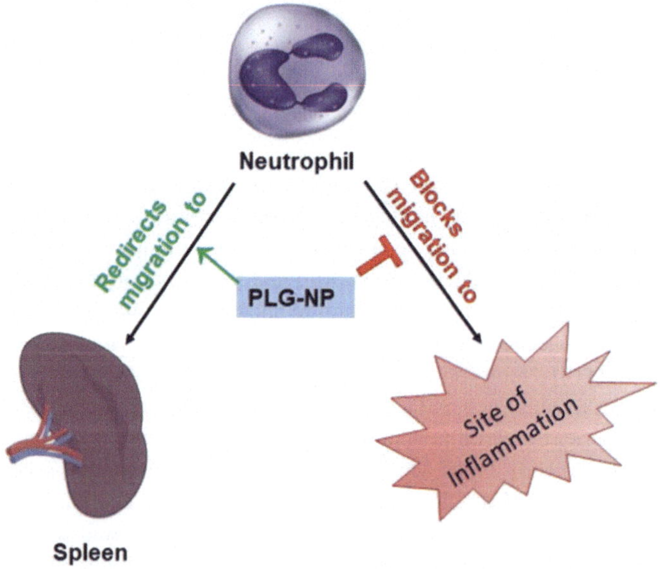

Fig. 4 Schematic illustration of modulation of neutrophil migration

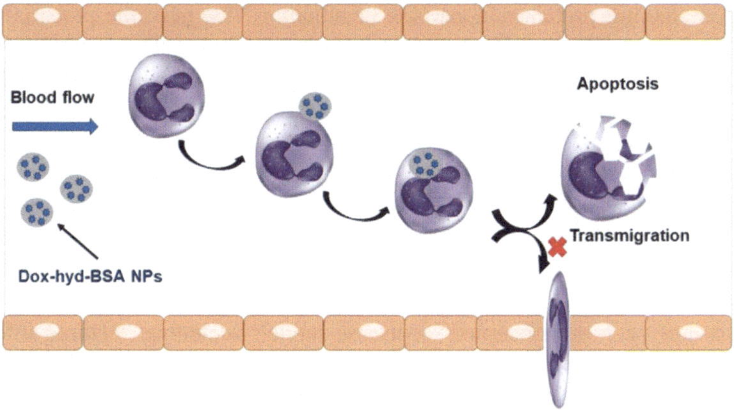

Fig. 5 Schematic illustration of reduction of inflammation-associated neutrophils

4.2.2 Reducing Inflammation-Associated Neutrophils

In acute immune reactions, neutrophils accumulate at the site of inflammation and these highly activated neutrophils are crucial for the development of inflammatory disorders. Various tactics have been devised to mitigate the intensity of regional inflammation, either by reducing the number of local neutrophils or by suppressing their proliferation and functions. It has been shown that doxorubicin (DOX) linked pH-responsive albumin nanoparticles may selectively target and deplete active neutrophils [151] where nanoparticles and DOX were joined by a pH-labile hydrazone link. Following the intravenous delivery of these nanoparticles, neutrophils absorbed them in response to infections or tissue damage. The acidic environment of active neutrophils would then dissolve the hydrazone link, releasing DOX, which in turn causes apoptosis of neutrophils. For example, a more recent study concentrated on the coronavirus-induced cytokine storms and inflammation linked to neutrophils. COVID-19 patients have increased levels of proinflammatory cytokines and cell-free DNA (cfDNA) as the disease worsens, which ultimately leads to multiple organ failure [29, 85]. Studies found that exogenously delivered polydopamine-poly (ethylene glycol) nanoparticles coated with DNase-1 might inhibit the neutrophil activities mediated by the coronavirus and the ensuing cytokine storm [72]. Mouse models of sepsis treated with DNase-1 formulations showed a substantial decrease in neutrophil cell numbers and levels of several inflammatory cytokines, along with an improvement in overall survival when compared to untreated groups (Fig. 5).

4.2.3 Strategies for Delivering Drugs Relevant to Neutrophils

Neutrophils may be utilized as transporters of medicinal drugs, much like the hitchhiking techniques associated with macrophages and monocytes. According to a study, denatured albumin protein nanoparticle or albumin NPs, have the potential to be therapeutic drug delivery vehicles since neutrophils may internalize them [145]. Unconjugated dyes are seen to spread in these cells upon internalization of such dye-coated nanoparticles by neutrophils [26]. Interestingly, further studies revealed that active neutrophils attached to endothelium were only able to take up albumin NPs and then these neutrophils carried the NPs to the inflamed tissues by making a way through the blood vessel barrier [27]. Subsequent research validated the association between the FcγRIII receptors on neutrophil surfaces and this specificity. These findings demonstrated the potential for using medication delivery strategies that target neutrophils in the treatment of inflammatory illnesses. The effectiveness of these strategies has been demonstrated in disease models including bacterial infection and acute lung inflammation [27, 145]. For example, a mouse model of acute lung inflammation treated with albumin NPs loaded with TPCA-1, an NF-κB inhibitor. In the treatment group, there was a substantial decrease in inflammatory markers such as neutrophil infiltration and cytokine release. Additionally, NPs enhanced lung integrity to avoid pulmonary edema [27]. When evaluated in *P. aeruginosa*-infected mice, albumin NPs loaded with cefoperazone acid (Cefo-A) significantly reduced bacterial multiplication by three times as compared to control [27].

4.2.4 Neutrophil Membrane-Coated NPs

NPs coated with neutrophil membranes maintain some characteristics of the native neutrophils such as the ability to evade immune monitoring, eliminate harmful molecules, and specifically target inflamed tissue. Consequently, it has been thought that neutrophil-mimicking nanoparticles offer a viable therapeutic platform for the management of inflammatory disorders [35, 152]. For example, nanoparticles coated with neutrophil membranes were created by fusing neutrophil membranes with polymeric cores [152]. Since these nanoparticles carry membrane-associated exterior antigenic functions of the neutrophil cells, they are able to neutralize proinflammatory cytokines, quench synovial inflammation, target depth of cartilage matrix, and provide strong chondroprotection against joint damage in a mouse model of collagen-induced arthritis and a human transgenic mouse model of arthritis [152]. Moreover, tailored medication administration can make use of nanovesicles derived from neutrophil membranes. For example, when TPCA-1 was injected into nanovesicles derived from HL-60 cells [26] and the same nanovesicles were then administered in a model organism of acute lung inflammation generated by lipopolysaccharide (LPS) the neutrophil infiltration and levels of TNF-α and IL-6 significantly decreased in the treatment group [41], indicating that novel drug carriers targeting inflamed tissue may be found in cell membrane-derived nanovesicles.

4.3 The Interaction of NPs with Lymphocytes

Lymphocytes, (T and B lymphocytes) the principal components and major effectors of the adaptive immune system, originate from hematopoietic stem cells (HSCs) through the common lymphoid progenitor (CLP) cells in the bone marrow [96]. Upon migration of multipotent progenitor cells from bone marrow to thymus, cells undergo many changes, such as gene rearrangement of T cell surface receptor and expression of the molecules present on cell surface (CD3, CD4, CD8, etc.), and this entire process is known as T cell maturation that leads to the development of subset of mature T lymphocytes such as effector T (T_E), regulatory T (T_{reg}), and memory T (T_M) cells. Based on their surface markers mature T cells are of two types, $CD4^+$ T cells and $CD8^+$ T cells [19]. $CD8^+$ T lymphocytes, also known as cytotoxic T lymphocytes (CTL) perform the main effector functions of cellular immunity whereas the $CD4^+$ T lymphocytes, also known as helper T (Th) cells, can regulate various immunological responses through the release of cytokines. On the contrary, T_{reg} cells function as negative regulators and suppress the activity of effector cells [154]. T_M cells are the main carriers of immunological memory and another crucial component of the immune response [92]. The bone marrow initiates B cell maturation directly. In the peripheral lymphoid tissues, when B cells encounter antigens directly or through T cells, they can differentiate into plasma cells and memory cells. Plasma B cells can secrete multiple antibodies and initiate humoral immunity whereas memory B cells memorize previous infection and react quickly when exposed to the same antigen again. [127, 144]. To modify immune response by targeting lymphocytes, various nanoparticle formulation has been developed. NPs thus developed can also be used to treat several inflammatory disorders. The mechanisms of treatment involve the induction of immunological tolerance linked to T cells, the depletion of auto-reactive T lymphocytes, and the modulation of immune responses related to B cells.

4.3.1 Establishing Immunological Tolerance Linked to T Cells

One of the main issues with the pathophysiology of many inflammatory illnesses is the auto-reactive immune response. Thus, one of the main objectives of treating such illnesses is to restore immunological tolerance. Immune tolerance has been thought to be reliably induced by relevant antigen administration. In a study, when airway inflammation was treated with poly(lactide-co-glycolide) NPs carrying encapsulated antigen [PLG(Ag)], both as a preventative measure and therapeutic intervention, surprisingly, it was observed that these antigens were well tolerated and efficiently suppressed ovalbumin (OVA)-induced Th2 responses and airway inflammation [134]. A similar approach was also implemented for the treatment of autoimmune diseases caused by Th1/Th17 cells [56]. Additionally, a subset of Treg cells $FoxP3^+$ Treg cells Additionally, $FoxP3^+$ T_{reg} cells, a subset of T_{reg} cells, mediate immune tolerance by interacting with various cellular components, hence researchers attempted to create nanoparticle platforms that target the expansion of T_{reg} cells in autoimmune disease models [150]. For example, treatment with nanoparticles coated with autoimmune disease-relevant peptides attached to major histocompatibility complex class II (pMHCII) molecules in various disease models such as non-obese diabetes (NOD), experimental autoimmune encephalomyelitis (EAE), and collagen-induced arthritis (CIA) stimulated the differentiation and expansion of antigen-specific regulatory $CD4^+$ T cell type 1 (T_R1)-like cells [28]. Such administration resulted in the restoration of established autoimmune phenomena. Additionally, they presented a brand-new, straightforward production technique that could reliably produce very stable, powerful nanomedicines based on metal oxide that activates or enlarges T cells [133]. To treat NOD mice, another group created gold nanoparticles that contained the β cell antigen proinsulin ($NP_{ITE+Ins}$) and the aryl hydrocarbon receptor (AhR) ligand [150]. This prevented Type 1 diabetes (T1D) from developing spontaneously. It was revealed that $NP_{ITE+Ins}$ worked by causing dendritic cells (DCs) to take on a tolerogenic phenotype and by promoting $FoxP3^+$ T_{reg} cell development in a way that was reliant on Socs2/NF-κB [150]. In addition to these recipes, other approaches involve the direct introduction of antigens into the liver, a natural tolerogenic setting [55], where antigen presentation fosters tolerance to both internal and external antigens [76]. The mouse allergen ovalbumin (OVA) was delivered to the liver using poly (lactic-co-glycolic acid) (PLGA) nanocarriers [76]. The ligands used to adorn the NPs target the liver sinusoidal endothelial cells (LSECs), which are specialized tolerogenic cells that can produce regulatory T-cells (Tregs). OVA-NPs were almost entirely transported to the liver following intravenous delivery, according to in vivo investigations, and they were able to considerably reduce airway inflammation in rats that had been exposed to and challenged with OVA. OVA-NPs suppressed anti-OVA IgE responses and promoted the development of Foxp3+ Treg cells in addition to producing large amounts of TGF-β, IL-4, and IL-10. Pharmacological development has shown interest in naturally occurring substances such as curcumin. A kind of hydrophobic polyphenol called curcumin is taken out of turmeric, a classic Indian spice [68]. Due to its anti-inflammatory properties and multi-targeted actions, it has lately been anticipated for the treatment of malignancies and inflammatory illnesses [34, 68, 93]. Nevertheless, curcumin's poor oral bioavailability has restricted its therapeutic use; thus, new delivery strategies are required to increase the rate of absorption. In this direction, an evaluation of the therapeutic effects of curcumin nanoparticles on the progression of inflammatory bowel diseases (IBD), studies demonstrated that the use of nanoparticle curcumin could reduce the disease activity index, histological colitis score, rate of body weight loss, and improve mucosal permeability in a model of colitis induced by dextran sulfate sodium (DSS) while also increasing the expansion of $CD4^+$ $Foxp3^+$ regulatory T cells [93].

4.3.2 The Reduction of Autoreactive T Lymphocytes

Rheumatism and acute inflammation are two examples of the many inflammatory disorders influenced by autoreactive T cells. Numerous platforms including nanoparticles attempted to deplete or block these pathogenic cells directly such as the administration of cytotoxic medicines, the suppression of cell receptors, and the blockage of cell signaling. Methotrexate (MTX), an antiproliferative drug, is frequently used to treat cancer and inflammatory conditions [5, 23, 94] because it can impede leukocyte activation and migration, including those of T cells, macrophages, and neutrophils, as well as limit cell proliferation [23]. It was found that the combination of gold nanoparticles and MTX exhibited greater anti-inflammatory activity than MTX alone when tested in mouse models of psoriasis induced by imiquimod [98]. In recent years, hedgehog (Hh) signaling has been linked to the activation of T lymphocytes and thus it could be a therapeutic target for inflammatory illnesses [135]. A Hh inhibitor, Eggmanone (Egm), has been specifically delivered to $CD4^+$ T cells using polymeric NPs decorated with anti-CD4 F(ab′) fragments of antibody and found that the new delivery method was able to target $CD4^+$ T cells specifically and significantly limit their responses, as seen by a decrease in cytokine release and activation of helper T cells [52]. Asthma etiology has been linked to group 2 innate lymphoid cells (ILC2s) which is a subset of type 2 (T2) inflammatory cells [104] therefore, targeting ICL2 receptor, suppression of tumorigenicity 2 (ST2) with NPs has been the subject of research [146]. Superparamagnetic iron oxide (SPIO) NPs coupled with anti-ST2 antibodies were used to treat OVA-induced mice and found that OVA-induced lung inflammation was markedly attenuated. Additional research revealed that anti-ST2-conjugated NPs inhibited the differentiation and proliferation of $CD4^+$ T cells by lowering the ILC2 capacity to generate IL-5 and IL-13 [146]. The activation of T cells is facilitated by these studies, and it has also been shown that delivery platforms of nanoparticles may enhance the effect of monoclonal antibodies (mAbs), which might eventually be used as strategic drug delivery methods in clinical settings.

4.3.3 Immune Regulation by B Lymphocytes

The B lymphocytes, that execute adaptive immune response, develop into plasma cells and release antibodies in response to matching antigens on the B cell receptor (BCR) [149]. Autoantibodies, produced by the autoreactive B cells have been found to be linked with the etiology of many inflammatory disorders. Therefore, depletion of autoreactive B cells by specifically targeting these cells is the primary focus of the NP techniques. For instance, anti-citrullinated protein autoantibodies (ACPA), one of the defining characteristics of rheumatoid arthritis (RA), are produced by the memory citrulline-specific B cells [31]. To treat RA, researchers created Siglec-engaging Tolerance-inducing Antigenic Liposomes (STALs) that co-display synthetic citrullinated antigen (CCP STALs) and CD22, an immunological inhibitor [16]. In a mouse model, findings indicated that CCP STALs successfully established the tolerance to citrullinated antigens and in addition to that formation of ACPA from memory B cells in RA patients was decreased [16]. In another study, poly (DL-lactic-co-glycolic acid) NPs conjugated with complement-activating lytic peptide and synthetic citrullinated peptide were able to significantly decrease the production of ACPA by B cells in peptide seropositive RA patients [107]. The B cell inhibition/depletion technique also applies to autoimmune illnesses of the central nervous system (CNS). For example, a cell-mediated drug delivery method, based on nanotechnology, was created by researchers to load anti-CD20 monoclonal antibody conjugated NPs into T cells that are specific to myelin antigen. This mixture could dramatically alleviate the disease course and pathology while drastically reducing B cells in the brain and spleen of the multiple sclerosis (MS) animal model [22].

4.4 The Impact of Nanoparticles on Molecules/Compounds Linked to Inflammation

A variety of immune chemicals such as cytokines, receptors, enzymes, reactive oxygen species (ROS), and others, in addition to the cellular components, play critical roles in the inflammatory process and are now being targeted therapeutically to treat inflammatory illnesses. These molecules play pivotal roles in the cross-talk of the cells of the immune system. Immune cells secrete effectors and have unique receptors that accept certain signals, which allow them to interact with one another and control immune responses. Therefore, modifying the concentration of various molecules linked to inflammation might be a dependable approach for the control of the immune system, and management of inflammatory conditions, in particular. As previously discussed, NPs can target source cells and modify production patterns of cytokines, which in turn affects the level of cytokines [98, 146]. Additionally, through direct interactions such as neutralizing effect, absorption, energy, etc., NPs can disrupt molecules linked to inflammation. Membrane-coated NPs can function as decoys that neutralize and absorb different types and complicated pathogenic chemicals, without any structural specificity [152]. Studies revealed that neutrophil membrane-coated polylactic acid nanoparticles (PLGA NPs) are able to bind and neutralize IL-1β and TNF-α cytokines [152], accompanied by a striking improvement in joint damage and a reduction in the overall severity of arthritis of collagen-induced and a human transgenic model of the mouse. In a sepsis model, nanoparticles coated with macrophage membranes have also been used to counteract proinflammatory cytokines and endotoxins [140]. The development of inflammatory disorders has also been linked to reactive oxygen and nitrogen species (RONS). RONS can generate oxidative stress, which can lead to protein, lipid, and DNA damage in addition to cytotoxicity [155]. To control RONS, researchers have attempted to create nanotechnology-based platforms [77, 148]. Studies, in a rat model of osteoarthritis, revealed dopamine melanin (DM) NPs as a novel RONS scavenger and they exhibited a potent capacity to eradicate an extensive variety of reactive oxygen and nitrogen species (RONS), such as superoxide, hydroxyl radicals, and peroxynitrite. In addition, (DM) NPs also suppress the synthesis of pro-inflammatory cytokines such as IL-1β with optimal anti-inflammatory and chondro-protective effects [155]. Liu et al. used a similar approach to treat ischemic stroke, showing distinctive anti-inflammatory, anti-oxidative, and biocompatible characteristics [77]. Like melanin nanomedicines, bilirubin nanomedicines have shown potent antioxidative properties in several models of inflammation-related diseases [25, 61, 71]. Bilirubin NPs, as efficient ROS scavenger, are capable of lowering the levels of inflammatory cytokines in addition to the protection of tissues from oxidative stress [61, 71]. Myeloperoxidase (MPO), a lytic enzyme produced by the activated neutrophils is involved in several inflammatory illnesses as a local mediator of tissue destruction [7], but polystyrene nanoparticles (NPs) were able to impede the activity of MPO [124]. This finding offers a potential method for controlling MPO activity during the pathological process of inflammation. SARS-CoV-2 disease is characterized by the activation of neutrophils, cytokine storm, and large amounts of cell-free DNA (cfDNA). cfDNA are encapsulated in neutrophil extracellular traps (NETs), causing severe organ failure in patients [72, 100]. NPs targeting cfDNA coated with recombinant DNase-1 have been developed and they have successfully suppressed the cytokine storm caused by SARS-CoV-2 [72]. Additionally, studies have shown that NPs can control the amounts of certain cytokines to treat inflammatory disorders, for instance, TNF-α and IL-1β levels can be suppressed by metformin hydrochloride-loaded PLGA NPs [103]. The level of pro-inflammatory cytokines (e.g., IL-1β, IL-6, IL-12, and IL-17) was reduced in a rat model of inflammatory bowel disease by NPs based on silk fibroin [47, 121]. Although not fully understood, the underlying mechanisms may be linked to the impact of NPs on immune cells rather than cytokines per se, which requires further investigations. The complement system, another crucial component of the immune system, consists of many soluble and membrane-bound proteins and is important for the clearance of pathogens, infected, dead, and damaged cells [17]. Complement system activation is a multi-step cascade based on proteolysis that primarily involves three routes: the lectin, classical, and alternative pathways [117]. The processes of all three of the currently available NP formulae' interactions with the complement system have been examined earlier [17] but more research is required because only a small number of them have been connected to the treatment of inflammatory illnesses.

4.5 Perspectives and Discussion

As previously mentioned, platforms of nanoparticles are capable of interacting with every element of the immune system giving us a new way for the cure of immune system-linked illnesses. Clinically, inflammatory illness treatment mostly depends on the administration of anti-inflammatory medications such as biologicals or glucocorticoids [115, 136]. These medications have a systemic distribution and a relatively low specificity and have significant adverse effects. By manipulating various properties such as size, shape, biodegradability, and surface charge of NPs by combining multiple functional molecules like enzymes, cell membranes, and antibodies, NPs can have numerous therapeutic properties and limitless potential to counteract inflammation. In contrast, NPs can target more precisely. We provide an overview of the numerous interactions that exist between NPs and immune system components in this article, as well as how they may be used to treat inflammatory illnesses. These investigations imply that we may create NP formulas that target the immune system's implicated cellular or molecular components and accurately interfere with the inflammatory milieu if the pathogenic process of a given disease is known. The use of NPs in immunological modulation still has many challenges. First, modulatory NPs can work by reducing or suppressing immune cells to control excessive inflammation. However, using such a tactic might cause immunodeficiency or immunological malfunction, which would raise the risk of cancer and opportunistic infections. Second, although NPs have been designed for their particular interactions with the cells and molecules of the immune system, they may also affect the healthy immune or non-immune cells which may lead to unwanted side effects and are not seen in preclinical data. Thirdly, despite the seemingly "exciting" pre-clinical outcomes, NPs need more examination in clinical trials and some may not fit to exhibit immunomodulatory effects in humans [143]. Fourthly, some investigations have suggested that NPs are unique medication carriers. They protect therapeutic compounds from the rapid clearance activity of the mononuclear phagocytic system, lengthen their half-lives and stability, and deliver medications directly to the inflammatory cells. However, a transportation method like this might potentially make the agents' adverse effects worse by allowing for a longer half-life and closer interaction with the organism. A comprehensive assessment of the biosecurity of these nanoparticles necessitates systemic clinical trials and more research into the processes pertaining to the anti-inflammatory properties of NPs. To more effectively connect pre-clinical findings to clinical impacts, we must consider and quantify physiological differences between model systems and human bodies. Additionally, studies have demonstrated that certain physical attributes like shape, size, and surface potential can have impacts on the effects of NPs [62, 131]. To this end, modulation of physical properties may be a viable strategy to maximize the therapeutic efficiency and pharmacokinetics of NPs. The global burden of disease from inflammatory disorders is increasing, making the development of sophisticated techniques for effective disease intervention imperative. The present developments in precision medicine have bestowed onto NP platforms an infinite range of applications. However, despite the constant development of innovative NP methods, clinically approved nanotherapeutics remain relatively uncommon due to the abundance of unidentified variables in this cutting-edge field of research. Numerous properties of NPs may be altered, including size, shape, charge, surface modification, and more, allowing for intelligent NP creation. For instance, NP charge plays a crucial role in both much-penetrating application and endosomal escape, and targeted surface indicators can be used to facilitate the absorption of particular cell types [88]. Numerous options are available for the implementation of multifunctional strategies thanks to these qualities.

However, researchers prefer to add additional manufacturing components to their formulae as clinical needs and design concerns get more complex, which makes their success in clinical trials more unpredictable. The design of nanoparticle and its likely interactions with human body needs careful investigations to assure the safety and specificity of the formulations in order to identify the most effective NP techniques for a given patient sub-group. In addition, to ensure batch-to-batch uniformity, industrial-scale NP production should adhere to good manufacturing practices (GMP) [24]. Notwithstanding all of the difficulties we face, it is indisputable that NPs, due to their ability of interaction with nearly all disease-causing components of the immune system, have changed the therapeutic focus of inflammatory disorders from symptomatic relief to etiological treatment of inflammation, as we have outlined in this chapter. There has always been a practical issue in translating NPs from the bench to the bedside, particularly when treating inflammatory disorders. Even if there are currently few FDA-approved NPs and those undergoing clinical trials, these successful examples teach us some common traits. The majority of NP formulae that are authorized for use in clinical trials or the market are typically polymeric NPs, liposomes, or delivery methods connected to inert metals. For example, triamcinolone acetonide can be released gradually into the synovium via the PLGA NP Zilretta®, an FDA-approved treatment for osteoarthritis [99]. Several innovative NP techniques with distinctive properties and significant application potential in the biomedical area have been presented in the last 10 years. Emerging nanotools that can make use of the sourced cells' innate capacity to affect the surroundings include, for example, cell membrane-coated nanoparticles. Nowadays, nearly any type of cell, including leukocytes, erythrocytes, stem cells, and even cancer cells, may be used to produce cell membranes [35]. Different membrane coatings provide NPs with a range of functionalities. When combined with encapsulated core compositions, these NPs form a unique platform with a broad range of application possibilities, such as drug delivery, detoxification, photoactivatable therapy, and imaging. Another illustration of a "combination strategy" using polyphenols and multivalent metal ions is the metal–polyphenol network (MPN). Polyphenol serves as an essential antioxidant and an anti-inflammatory agent, while metal offers polyvalent complexing, anti-bacteria, and tissue repair boosting ability, detoxifying, photoactivatable treatment, etc. Another illustration of a "combination strategy" using polyphenols and multivalent metal ions is the metal-polyphenol network (MPN). Polyphenols are significant antioxidants and anti-inflammatory agents, whereas metal is responsible for polyvalent complexing, anti-bacteria, and tissue repair promotion properties. Because of their noteworthy qualities, including as structural flexibility, heat stability, and pH responsiveness, MPN-based NPs have piqued the interest of researchers and have since been developed into a variety of NP platforms for the treatment of various illnesses [48]. These illustrations imply that to increase the breadth of applications and enhance current structures, future NP tactics will continue to concentrate on using biochemical synthesis techniques to incorporate various chemical or biological components. However, because of the intricacy of their structure and the fact that several crucial features have not been well investigated, these unique formulations should also be handled carefully. More thorough research and clinical assessment are still required for translation from labs to clinics.

Conflict of Interest The author declares no competing interest.

References

1. Abdel-Haleem FM, Gamal E, Rizk MS, Madbouly A, El Nashar RM, Anis B, Barhoum A et al (2021) Molecularly imprinted electrochemical sensor-based Fe(2)O(3)@MWCNTs for ivabradine drug determination in pharmaceutical formulation, serum, and urine samples. Front Bioeng Biotechnol 9:648704
2. Abdel-Haleem FM, Mahmoud S, Abdel-Ghani NET, El Nashar RM, Bechelany M, Barhoum A (2021) Polyvinyl chloride modified carbon paste electrodes for sensitive determination of levofloxacin drug in serum, urine, and pharmaceutical formulations. Sensors (Basel) 21(9)
3. Abdulkhaleq LA, Assi MA, Abdullah R, Zamri-Saad M, Taufiq-Yap YH, Hezmee MNM (2018) The crucial roles of inflammatory mediators in inflammation: a review. Vet World 11(5):627–635
4. Ahamad N, Kar A, Mehta S, Dewani M, Ravichandran V, Bhardwaj P, Banerjee R et al (2021) Immunomodulatory nanosystems for treating inflammatory diseases. Biomaterials 274:120875
5. Aletaha D, Westhovens R, Gaujoux-Viala C, Adami G, Matsumoto A, Bird P, Burmester GR et al (2021) Efficacy and safety of filgotinib in methotrexate-naive patients with rheumatoid arthritis with poor prognostic factors: post hoc analysis of FINCH 3. RMD Open 7(2):e001621
6. Anselmo AC, Mitragotri S (2014) Cell-mediated delivery of nanoparticles: taking advantage of circulatory cells to target nanoparticles. J Control Release 190:531–541
7. Aratani Y (2018) Myeloperoxidase: Its role for host defense, inflammation, and neutrophil function. Arch Biochem Biophys 640:47–52
8. Bacci M, Capobianco A, Monno A, Cottone L, Di Puppo F, Camisa B, Rovere-Querini P et al (2009) Macrophages are alternatively activated in patients with endometriosis and required for growth and vascularization of lesions in a mouse model of disease. Am J Pathol 175(2):547–556
9. Bajpai G, Schneider C, Wong N, Bredemeyer A, Hulsmans M, Nahrendorf M, Lavine KJ et al (2018) The human heart contains distinct macrophage subsets with divergent origins and functions. Nat Med 24(8):1234–1245
10. Barenholz Y (2012) Doxil®–the first FDA-approved nano-drug: lessons learned. J Control Release 160(2):117–134
11. Barhoum A, El-Maghrabi HH, Nada AA, Sayegh S, Roualdes S, Renard A, Bechelany M et al (2021) Simultaneous hydrogen and oxygen evolution reactions using free-standing nitrogen-doped-carbon–Co/CoOx nanofiber electrodes decorated with palladium nanoparticles. J Mater Chem A 9(33):17724–17739
12. Barhoum A, Melcher J, Van Assche G, Rahier H, Bechelany M, Fleisch M, Bahnemann D (2017) Synthesis, growth mechanism, and photocatalytic activity of Zinc oxide nanostructures: porous microparticles versus nonporous nanoparticles. J Mater Sci 52(5):2746–2762
13. Barhoum A, Pal K, Rahier H, Uludag H, Kim IS, Bechelany M (2019) Nanofibers as new-generation materials: From spinning and nano-spinning fabrication techniques to emerging applications. Appl Mater Today 17:1–35
14. Barhoum A, Van Assche G, Rahier H, Fleisch M, Bals S, Delplancked M-P, Bahnemann D et al (2017) Sol-gel hot injection synthesis of ZnO nanoparticles into a porous silica matrix and reaction mechanism. Mater Des 119:270–276

15. Barhoum A, Van Lokeren L, Rahier H, Dufresne A, Van Assche G (2015) Roles of in situ surface modification in controlling the growth and crystallization of $CaCO_3$ nanoparticles, and their dispersion in polymeric materials. J Mater Sci 50(24):7908–7918
16. Bednar KJ, Nycholat CM, Rao TS, Paulson JC, Fung-Leung WP, Macauley MS (2019) Exploiting CD22 to selectively tolerize autoantibody producing B-cells in rheumatoid arthritis. ACS Chem Biol 14(4):644–654
17. Boraschi D, Italiani P, Palomba R, Decuzzi P, Duschl A, Fadeel B, Moghimi SM (2017) Nanoparticles and innate immunity: new perspectives on host defence. Semin Immunol 34:33–51
18. Borregaard N (2010) Neutrophils, from marrow to microbes. Immunity 33(5):657–670
19. Brandstadter JD, Maillard I (2019) Notch signalling in T cell homeostasis and differentiation. Open Biol 9(11):190187
20. Brown BN, Ratner BD, Goodman SB, Amar S, Badylak SF (2012) Macrophage polarization: an opportunity for improved outcomes in biomaterials and regenerative medicine. Biomaterials 33(15):3792–3802
21. Brynskikh AM, Zhao Y, Mosley RL, Li S, Boska MD, Klyachko NL, Batrakova EV et al (2010) Macrophage delivery of therapeutic nanozymes in a murine model of Parkinson's disease. Nanomedicine (Lond) 5(3):379–396
22. Carnasciali A, Amoriello R, Bonechi E, Mazzoni A, Ravagli C, Doumett S, Ballerini C et al (2021) T cell delivery of nanoparticles-bound Anti-CD20 monoclonal antibody: successful B cell depletion in the spinal cord during experimental autoimmune encephalomyelitis. J Neuroimmune Pharmacol 16(2):376–389
23. Chan ES, Cronstein BN (2010) Methotrexate–how does it really work? Nat Rev Rheumatol 6(3):175–178
24. Chen W, Schilperoort M, Cao Y, Shi J, Tabas I, Tao W (2022) Macrophage-targeted nanomedicine for the diagnosis and treatment of atherosclerosis. Nat Rev Cardiol 19(4):228–249
25. Chen Z, Vong CT, Gao C, Chen S, Wu X, Wang S, Wang Y (2020) Bilirubin nanomedicines for the treatment of reactive oxygen species (ROS)-mediated diseases. Mol Pharm 17(7):2260–2274
26. Chu D, Dong X, Shi X, Zhang C, Wang Z (2018) Neutrophil-based drug delivery systems. Adv Mater 30(22):e1706245
27. Chu D, Gao J, Wang Z (2015) Neutrophil-mediated delivery of therapeutic nanoparticles across blood vessel barrier for treatment of inflammation and infection. ACS Nano 9(12):11800–11811
28. Clemente-Casares X, Blanco J, Ambalavanan P, Yamanouchi J, Singha S, Fandos C, Santamaria P et al (2016) Expanding antigen-specific regulatory networks to treat autoimmunity. Nature 530(7591):434–440
29. Colón DF, Wanderley CW, Franchin M, Silva CM, Hiroki CH, Castanheira FVS, Cunha FQ et al (2019) Neutrophil extracellular traps (NETs) exacerbate severity of infant sepsis. Crit Care 23(1):113
30. Cremers V, Rampelberg G, Barhoum A, Walters P, Claes N, de Oliveira TM, Detavernier C et al (2018) Oxidation barrier of Cu and Fe powder by atomic layer deposition. Surf Coat Technol 349:1032–1041
31. Derksen V, Huizinga TWJ, van der Woude D (2017) The role of autoantibodies in the pathophysiology of rheumatoid arthritis. Semin Immunopathol 39(4):437–446
32. Dinarello CA (2010) Anti-inflammatory agents: present and future. Cell 140(6):935–950
33. Dou H, Grotepas CB, McMillan JM, Destache CJ, Chaubal M, Werling J, Gendelman HE et al (2009) Macrophage delivery of nanoformulated antiretroviral drug to the brain in a murine model of neuroAIDS. J Immunol 183(1):661–669
34. Epstein J, Sanderson IR, Macdonald TT (2010) Curcumin as a therapeutic agent: the evidence from in vitro, animal and human studies. Br J Nutr 103(11):1545–1557
35. Fang RH, Kroll AV, Gao W, Zhang L (2018) Cell membrane coating nanotechnology. Adv Mater 30(23):e1706759

36. Fromen CA, Kelley WJ, Fish MB, Adili R, Noble J, Hoenerhoff MJ, Eniola-Adefeso O et al (2017) Neutrophil-particle interactions in blood circulation drive particle clearance and alter neutrophil responses in acute inflammation. ACS Nano 11(11):10797–10807
37. Funes SC, Rios M, Escobar-Vera J, Kalergis AM (2018) Implications of macrophage polarization in autoimmunity. Immunology 154(2):186–195
38. Furman D, Campisi J, Verdin E, Carrera-Bastos P, Targ S, Franceschi C, Slavich GM et al (2019) Chronic inflammation in the etiology of disease across the life span. Nat Med 25(12):1822–1832
39. Galili U (2021) Biosynthesis of α-gal epitopes (Galα1–3Galβ1–4GlcNAc-R) and their unique potential in future α-gal therapies. Front Mol Biosci 8
40. Galili U, Zhu Z, Chen J, Goldufsky JW, Schaer GL (2021) Near complete repair after myocardial infarction in adult mice by altering the inflammatory response with intramyocardial injection of α-gal nanoparticles. Front Cardiovasc Med 8:719160
41. Gao J, Chu D, Wang Z (2016) Cell membrane-formed nanovesicles for disease-targeted delivery. J Control Release 224:208–216
42. Gaspar N, Zambito G, Löwik C, Mezzanotte L (2019) Active nano-targeting of macrophages. Curr Pharm Des 25(17):1951–1961
43. Gaur M, Misra C, Yadav AB, Swaroop S, Maolmhuaidh F, Bechelany M, Barhoum A (2021) Biomedical applications of carbon nanomaterials: fullerenes, quantum dots, nanotubes, nanofibers, and graphene. Materials (Basel) 14(20)
44. Gendelman HE, Anantharam V, Bronich T, Ghaisas S, Jin H, Kanthasamy AG, Mallapragada SK et al (2015) Nanoneuromedicines for degenerative, inflammatory, and infectious nervous system diseases. Nanomedicine 11(3):751–767
45. Ghosh S, May MJ, Kopp EB (1998) NF-kappa B and Rel proteins: evolutionarily conserved mediators of immune responses. Annu Rev Immunol 16:225–260
46. Ginhoux F, Guilliams M (2016) Tissue-resident macrophage ontogeny and homeostasis. Immunity 44(3):439–449
47. Gou S, Huang Y, Sung J, Xiao B, Merlin D (2019) Silk fibroin-based nanotherapeutics: application in the treatment of colonic diseases. Nanomedicine (Lond) 14(17):2373–2378
48. Guo Y, Sun Q, Wu FG, Dai Y, Chen X (2021) Polyphenol-containing nanoparticles: synthesis, properties, and therapeutic delivery. Adv Mater 33(22):e2007356
49. Hammani S, Moulai-Mostefa N, Samyn P, Bechelany M, Dufresne A, Barhoum A (2020) Morphology, rheology and crystallization in relation to the viscosity ratio of polystyrene/polypropylene polymer blends. Materials (Basel) 13(4)
50. Han J, Kim YS, Lim MY, Kim HY, Kong S, Kang M, Kim BS et al (2018) Dual roles of graphene oxide to attenuate inflammation and elicit timely polarization of macrophage phenotypes for cardiac repair. ACS Nano 12(2):1959–1977
51. Hashimoto D, Chow A, Noizat C, Teo P, Beasley MB, Leboeuf M, Merad M et al (2013) Tissue-resident macrophages self-maintain locally throughout adult life with minimal contribution from circulating monocytes. Immunity 38(4):792–804
52. Haycook CP, Balsamo JA, Glass EB, Williams CH, Hong CC, Major AS, Giorgio TD (2020) PEGylated PLGA nanoparticle delivery of eggmanone for T cell modulation: applications in rheumatic autoimmunity. Int J Nanomed 15:1215–1228
53. Heidt T, Sager HB, Courties G, Dutta P, Iwamoto Y, Zaltsman A, Nahrendorf M et al (2014) Chronic variable stress activates hematopoietic stem cells. Nat Med 20(7):754–758
54. Hesketh M, Sahin KB, West ZE, Murray RZ (2017) Macrophage phenotypes regulate scar formation and chronic wound healing. Int J Mol Sci 18(7)
55. Heymann F, Tacke F (2016) Immunology in the liver–from homeostasis to disease. Nat Rev Gastroenterol Hepatol 13(2):88–110
56. Hunter Z, McCarthy DP, Yap WT, Harp CT, Getts DR, Shea LD, Miller SD (2014) A biodegradable nanoparticle platform for the induction of antigen-specific immune tolerance for treatment of autoimmune disease. ACS Nano 8(3):2148–2160
57. Jeevanandam J, Barhoum A, Chan YS, Dufresne A, Danquah MK (2018) Review on nanoparticles and nanostructured materials: history, sources, toxicity and regulations. Beilstein J Nanotechnol 9:1050–1074

58. Jenkins SJ, Ruckerl D, Cook PC, Jones LH, Finkelman FD, van Rooijen N, Allen JE et al (2011) Local macrophage proliferation, rather than recruitment from the blood, is a signature of TH2 inflammation. Science 332(6035):1284–1288
59. Ji H, Qiu R, Gao X, Zhang R, Li X, Hei Z, Yuan D (2019) Propofol attenuates monocyte-endothelial adhesion via modulating connexin43 expression in monocytes. Life Sci 232:116624
60. Kaymakcalan OE, Abadeer A, Goldufsky JW, Galili U, Karinja SJ, Dong X, Spector JA et al (2020) Topical α-gal nanoparticles accelerate diabetic wound healing. Exp Dermatol 29(4):404–413
61. Kim MJ, Lee Y, Jon S, Lee DY (2017) PEGylated bilirubin nanoparticle as an anti-oxidative and anti-inflammatory demulcent in pancreatic islet xenotransplantation. Biomaterials 133:242–252
62. Kinnear C, Moore TL, Rodriguez-Lorenzo L, Rothen-Rutishauser B, Petri-Fink A (2017) Form follows function: nanoparticle shape and its implications for nanomedicine. Chem Rev 117(17):11476–11521
63. Kolaczkowska E, Kubes P (2013) Neutrophil recruitment and function in health and inflammation. Nat Rev Immunol 13(3):159–175
64. Kotas ME, Medzhitov R (2015) Homeostasis, inflammation, and disease susceptibility. Cell 160(5):816–827
65. Kratofil RM, Kubes P, Deniset JF (2017) Monocyte conversion during inflammation and injury. Arterioscler Thromb Vasc Biol 37(1):35–42
66. Krohn-Grimberghe M, Mitchell MJ, Schloss MJ, Khan OF, Courties G, Guimaraes PPG, Nahrendorf M et al (2020) Nanoparticle-encapsulated siRNAs for gene silencing in the haematopoietic stem-cell niche. Nat Biomed Eng 4(11):1076–1089
67. Kumar S, Bhushan P, Bhattacharya S (2018) Fabrication of nanostructures with bottom-up approach and their utility in diagnostics, therapeutics, and others. In: Bhattacharya S, Agarwal AK, Chanda N, Pandey A, Sen AK (eds) Environmental, chemical and medical sensors. Springer, Singapore, pp 167–198
68. Kunnumakkara AB, Bordoloi D, Padmavathi G, Monisha J, Roy NK, Prasad S, Aggarwal BB (2017) Curcumin, the golden nutraceutical: multitargeting for multiple chronic diseases. Br J Pharmacol 174(11):1325–1348
69. Kwon D, Cha BG, Cho Y, Min J, Park EB, Kang SJ, Kim J (2017) Extra-large pore mesoporous silica nanoparticles for directing in vivo M2 macrophage polarization by delivering IL-4. Nano Lett 17(5):2747–2756
70. Lange C, Hemmrich G, Klostermeier UC, López-Quintero JA, Miller DJ, Rahn T, Rosenstiel P et al (2011) Defining the origins of the NOD-like receptor system at the base of animal evolution. Mol Biol Evol 28(5):1687–1702
71. Lee Y, Kim H, Kang S, Lee J, Park J, Jon S (2016) Bilirubin nanoparticles as a nanomedicine for anti-inflammation therapy. Angew Chem Int Ed Engl 55(26):7460–7463
72. Lee YY, Park HH, Park W, Kim H, Jang JG, Hong KS, Park CG et al (2021) Long-acting nanoparticulate DNase-1 for effective suppression of SARS-CoV-2-mediated neutrophil activities and cytokine storm. Biomaterials 267:120389
73. Leuschner F, Dutta P, Gorbatov R, Novobrantseva TI, Donahoe JS, Courties G, Nahrendorf M et al (2011) Therapeutic siRNA silencing in inflammatory monocytes in mice. Nat Biotechnol 29(11):1005–1010
74. Liew PX, Kubes P (2019) The neutrophil's role during health and disease. Physiol Rev 99(2):1223–1248
75. Lindau A, Härdtner C, Hergeth SP, Blanz KD, Dufner B, Hoppe N, Hilgendorf I et al (2016) Atheroprotection through SYK inhibition fails in established disease when local macrophage proliferation dominates lesion progression. Basic Res Cardiol 111(2):20
76. Liu Q, Wang X, Liu X, Kumar S, Gochman G, Ji Y, Nel AE et al (2019) Use of polymeric nanoparticle platform targeting the liver to induce treg-mediated antigen-specific immune tolerance in a pulmonary allergen sensitization model. ACS Nano 13(4):4778–4794

77. Liu Y, Ai K, Ji X, Askhatova D, Du R, Lu L, Shi J (2017) Comprehensive insights into the multi-antioxidative mechanisms of melanin nanoparticles and their application to protect brain from injury in ischemic stroke. J Am Chem Soc 139(2):856–862
78. Locati M, Curtale G, Mantovani A (2020) Diversity, mechanisms, and significance of macrophage plasticity. Annu Rev Pathol 15:123–147
79. Luster AD, Alon R, von Andrian UH (2005) Immune cell migration in inflammation: present and future therapeutic targets. Nat Immunol 6(12):1182–1190
80. Mack M (2018) Inflammation and fibrosis. Matrix Biol 68–69:106–121
81. Manfredi AA, Ramirez GA, Rovere-Querini P, Maugeri N (2018) The neutrophil's choice: phagocytose vs make neutrophil extracellular traps. Front Immunol 9:288
82. Mantovani A, Cassatella MA, Costantini C, Jaillon S (2011) Neutrophils in the activation and regulation of innate and adaptive immunity. Nat Rev Immunol 11(8):519–531
83. McCloskey E, Paterson AH, Powles T, Kanis JA (2021) Clodronate. Bone 143:115715
84. Mehrotra P (2016) Biosensors and their applications—a review. J Oral Biol Craniofac Res 6(2):153–159
85. Mehta P, McAuley DF, Brown M, Sanchez E, Tattersall RS, Manson JJ (2020) COVID-19: consider cytokine storm syndromes and immunosuppression. Lancet 395(10229):1033–1034
86. Mendelson A, Frenette PS (2014) Hematopoietic stem cell niche maintenance during homeostasis and regeneration. Nat Med 20(8):833–846
87. Miao X, Leng X, Zhang Q (2017) The current state of nanoparticle-induced macrophage polarization and reprogramming research. Int J Mol Sci 18(2)
88. Mitchell MJ, Billingsley MM, Haley RM, Wechsler ME, Peppas NA, Langer R (2021) Engineering precision nanoparticles for drug delivery. Nat Rev Drug Discov 20(2):101–124
89. Molinaro R, Corbo C, Martinez JO, Taraballi F, Evangelopoulos M, Minardi S, Tasciotti E et al (2016) Biomimetic proteolipid vesicles for targeting inflamed tissues. Nat Mater 15(9):1037–1046
90. Murray PJ (2017) Macrophage polarization. Annu Rev Physiol 79:541–566
91. Nathan C (2002) Points of control in inflammation. Nature 420(6917):846–852
92. Natoli G, Ostuni R (2019) Adaptation and memory in immune responses. Nat Immunol 20(7):783–792
93. Ohno M, Nishida A, Sugitani Y, Nishino K, Inatomi O, Sugimoto M, Andoh A et al (2017) Nanoparticle curcumin ameliorates experimental colitis via modulation of gut microbiota and induction of regulatory T cells. PLoS ONE 12(10):e0185999
94. Ong SY, de Mel S, Grigoropoulos NF, Chen Y, Tan YC, Tan MSY, Nagarajan C et al (2021) High-dose methotrexate is effective for prevention of isolated CNS relapse in diffuse large B cell lymphoma. Blood Cancer J 11(8):143
95. Ordikhani F, Zandi N, Mazaheri M, Luther GA, Ghovvati M, Akbarzadeh A, Annabi N (2021) Targeted nanomedicines for the treatment of bone disease and regeneration. Med Res Rev 41(3):1221–1254
96. Orkin SH, Zon LI (2008) Hematopoiesis: an evolving paradigm for stem cell biology. Cell 132(4):631–644
97. Ozbakir B, Crielaard BJ, Metselaar JM, Storm G, Lammers T (2014) Liposomal corticosteroids for the treatment of inflammatory disorders and cancer. J Control Release 190:624–636
98. Özcan A, Sahin D, Impellizzieri D, Nguyen TT, Hafner J, Yawalkar N, Kolios AGA et al (2020) Nanoparticle-coupled topical methotrexate can normalize immune responses and induce tissue remodeling in psoriasis. J Invest Dermatol 140(5):1003-1014.e1008
99. Paik J, Duggan ST, Keam SJ (2019) Triamcinolone acetonide extended-release: a review in osteoarthritis pain of the knee. Drugs 79(4):455–462
100. Papayannopoulos V (2018) Neutrophil extracellular traps in immunity and disease. Nat Rev Immunol 18(2):134–147
101. Park J, Zhang Y, Saito E, Gurczynski SJ, Moore BB, Cummings BJ, Shea LD et al (2019) Intravascular innate immune cells reprogrammed via intravenous nanoparticles to promote functional recovery after spinal cord injury. Proc Natl Acad Sci USA 116(30):14947–14954

102. Parodi A, Quattrocchi N, van de Ven AL, Chiappini C, Evangelopoulos M, Martinez JO, Tasciotti E et al (2013) Synthetic nanoparticles functionalized with biomimetic leukocyte membranes possess cell-like functions. Nat Nanotechnol 8(1):61–68
103. Pereira A, Brito GAC, Lima MLS, Silva Júnior AAD, Silva EDS, de Rezende AA, Araújo AA et al (2018) Metformin hydrochloride-loaded PLGA nanoparticle in periodontal disease experimental model using diabetic rats. Int J Mol Sci 19(11)
104. Peters MC, Ringel L, Dyjack N, Herrin R, Woodruff PG, Rios C, Seibold MA et al (2019) A transcriptomic method to determine airway immune dysfunction in T2-high and T2-low asthma. Am J Respir Crit Care Med 199(4):465–477
105. Polak R, Lim RM, Beppu MM, Pitombo RN, Cohen RE, Rubner MF (2015) Liposome-loaded cell backpacks. Adv Healthc Mater 4(18):2832–2841
106. Poupot R, Goursat C, Fruchon S (2018) Multivalent nanosystems: targeting monocytes/macrophages. Int J Nanomed 13:5511–5521
107. Pozsgay J, Babos F, Uray K, Magyar A, Gyulai G, Kiss É, Sármay G et al (2016) In vitro eradication of citrullinated protein specific B-lymphocytes of rheumatoid arthritis patients by targeted bifunctional nanoparticles. Arthritis Res Ther 18:15
108. Prasad S, Kumar V, Kirubanandam S, Barhoum A (2018) Chapter 11—engineered nanomaterials: nanofabrication and surface functionalization. In: Barhoum A, Makhlouf ASH (eds) Emerging applications of nanoparticles and architecture nanostructures. Elsevier, pp 305–340
109. Proell M, Riedl SJ, Fritz JH, Rojas AM, Schwarzenbacher R (2008) The nod-like receptor (NLR) family: a tale of similarities and differences. PLoS ONE 3(4):e2119
110. Prosperi D, Colombo M, Zanoni I, Granucci F (2017) Drug nanocarriers to treat autoimmunity and chronic inflammatory diseases. Semin Immunol 34:61–67
111. Qi R, Majoros I, Misra AC, Koch AE, Campbell P, Marotte H, Baker JR Jr et al (2015) Folate receptor-targeted dendrimer-methotrexate conjugate for inflammatory arthritis. J Biomed Nanotechnol 11(8):1431–1441
112. Rajendran P, Chen YF, Chen YF, Chung LC, Tamilselvi S, Shen CY, Huang CY et al (2018) The multifaceted link between inflammation and human diseases. J Cell Physiol 233(9):6458–6471
113. Rasouli R, Barhoum A, Bechelany M, Dufresne A (2019) Nanofibers for biomedical and healthcare applications. Macromol Biosci 19(2):e1800256
114. Rasouli R, Barhoum A, Uludag H (2018) A review of nanostructured surfaces and materials for dental implants: surface coating, patterning and functionalization for improved performance. Biomater Sci 6(6):1312–1338
115. Rendon A, Schäkel K (2019) Psoriasis pathogenesis and treatment. Int J Mol Sci 20(6)
116. Richards PJ, Williams AS, Goodfellow RM, Williams BD (1999) Liposomal clodronate eliminates synovial macrophages, reduces inflammation and ameliorates joint destruction in antigen-induced arthritis. Rheumatology (Oxford) 38(9):818–825
117. Ricklin D, Hajishengallis G, Yang K, Lambris JD (2010) Complement: a key system for immune surveillance and homeostasis. Nat Immunol 11(9):785–797
118. Roach JC, Glusman G, Rowen L, Kaur A, Purcell MK, Smith KD, Aderem A et al (2005) The evolution of vertebrate Toll-like receptors. Proc Natl Acad Sci USA 102(27):9577–9582
119. Robbins CS, Hilgendorf I, Weber GF, Theurl I, Iwamoto Y, Figueiredo JL, Swirski FK et al (2013) Local proliferation dominates lesional macrophage accumulation in atherosclerosis. Nat Med 19(9):1166–1172
120. Rock KL, Latz E, Ontiveros F, Kono H (2010) The sterile inflammatory response. Annu Rev Immunol 28:321–342
121. Rodriguez-Nogales A, Lozano-Pérez AA, Aznar-Cervantes SD, Algieri F, Garrido-Mesa J, Garrido-Mesa N, Gálvez J et al (2016) Effect of aqueous and particulate silk fibroin in a rat model of experimental colitis. Int J Pharm 511(1):1–9
122. Sager HB, Dutta P, Dahlman JE, Hulsmans M, Courties G, Sun Y, Nahrendorf M et al (2016) RNAi targeting multiple cell adhesion molecules reduces immune cell recruitment and vascular inflammation after myocardial infarction. Sci Transl Med 8(342):342ra380
123. Saito E, Kuo R, Pearson RM, Gohel N, Cheung B, King NJC, Shea LD et al (2019) Designing drug-free biodegradable nanoparticles to modulate inflammatory monocytes and neutrophils for ameliorating inflammation. J Control Release 300:185–196

124. Sanfins E, Correia A, Gunnarsson SB, Vilanova M, Cedervall T (2018) Nanoparticle effect on neutrophil produced myeloperoxidase. PLoS One 13(1):e0191445
125. Sawy AM, Barhoum A, Abdel Gaber SA, El-Hallouty SM, Shousha WG, Maarouf AA, Khalil ASG (2021) Insights of doxorubicin loaded graphene quantum dots: synthesis, DFT drug interactions, and cytotoxicity. Mater Sci Eng C Mater Biol Appl 122:111921
126. Schett G, Neurath MF (2018) Resolution of chronic inflammatory disease: universal and tissue-specific concepts. Nat Commun 9(1):3261
127. Seifert M, Küppers R (2016) Human memory B cells. Leukemia 30(12):2283–2292
128. Serhan CN, Savill J (2005) Resolution of inflammation: the beginning programs the end. Nat Immunol 6(12):1191–1197
129. Shapouri-Moghaddam A, Mohammadian S, Vazini H, Taghadosi M, Esmaeili SA, Mardani F, Sahebkar A et al (2018) Macrophage plasticity, polarization, and function in health and disease. J Cell Physiol 233(9):6425–6440
130. Shi C, Jia T, Mendez-Ferrer S, Hohl TM, Serbina NV, Lipuma L, Pamer EG et al (2011) Bone marrow mesenchymal stem and progenitor cells induce monocyte emigration in response to circulating toll-like receptor ligands. Immunity 34(4):590–601
131. Silva AL, Peres C, Conniot J, Matos AI, Moura L, Carreira B, Florindo HF et al (2017) Nanoparticle impact on innate immune cell pattern-recognition receptors and inflammasomes activation. Semin Immunol 34:3–24
132. Singh KRB, Nayak V, Singh J, Singh AK, Singh RP (2021) Potentialities of bioinspired metal and metal oxide nanoparticles in biomedical sciences. RSC Adv 11(40):24722–24746
133. Singha S, Shao K, Yang Y, Clemente-Casares X, Solé P, Clemente A, Santamaria P et al (2017) Peptide–MHC-based nanomedicines for autoimmunity function as T-cell receptor microclustering devices. Nat Nanotechnol 12(7):701–710
134. Smarr CB, Yap WT, Neef TP, Pearson RM, Hunter ZN, Ifergan I, Miller SD et al (2016) Biodegradable antigen-associated PLG nanoparticles tolerize Th2-mediated allergic airway inflammation pre- and postsensitization. Proc Natl Acad Sci USA 113(18):5059–5064
135. Smelkinson MG (2017) The hedgehog signaling pathway emerges as a pathogenic target. J Dev Biol 5(4)
136. Smolen JS, Aletaha D, McInnes IB (2016) Rheumatoid arthritis. Lancet 388(10055):2023–2038
137. Sou K, Goins B, Takeoka S, Tsuchida E, Phillips WT (2007) Selective uptake of surface-modified phospholipid vesicles by bone marrow macrophages in vivo. Biomaterials 28(16):2655–2666
138. Sun D, Zhuang X, Xiang X, Liu Y, Zhang S, Liu C, Zhang HG et al (2010) A novel nanoparticle drug delivery system: the anti-inflammatory activity of curcumin is enhanced when encapsulated in exosomes. Mol Ther 18(9):1606–1614
139. Taratummarat S, Sangphech N, Vu CTB, Palaga T, Ondee T, Surawut S, Leelahavanichkul A et al (2018) Gold nanoparticles attenuates bacterial sepsis in cecal ligation and puncture mouse model through the induction of M2 macrophage polarization. BMC Microbiol 18(1):85
140. Thamphiwatana S, Angsantikul P, Escajadillo T, Zhang Q, Olson J, Luk BT, Zhang L et al (2017) Macrophage-like nanoparticles concurrently absorbing endotoxins and proinflammatory cytokines for sepsis management. Proc Natl Acad Sci USA 114(43):11488–11493
141. Tran T-H, Krishnan S, Amiji MM (2016) MicroRNA-223 induced repolarization of peritoneal macrophages using CD44 targeting hyaluronic acid nanoparticles for anti-inflammatory effects. PLoS ONE 11(5):e0152024
142. Tran T-H, Rastogi R, Shelke J, Amiji MM (2015) Modulation of macrophage functional polarity towards anti-inflammatory phenotype with plasmid DNA delivery in CD44 targeting hyaluronic acid nanoparticles. Sci Rep 5(1):16632
143. van der Valk FM, van Wijk DF, Lobatto ME, Verberne HJ, Storm G, Willems MC, Stroes ES et al (2015) Prednisolone-containing liposomes accumulate in human atherosclerotic macrophages upon intravenous administration. Nanomedicine 11(5):1039–1046
144. Wang Y, Liu J, Burrows PD, Wang JY (2020) B cell development and maturation. Adv Exp Med Biol 1254:1–22

145. Wang Z, Li J, Cho J, Malik AB (2014) Prevention of vascular inflammation by nanoparticle targeting of adherent neutrophils. Nat Nanotechnol 9(3):204–210
146. Wu Y, Shi W, Wang H, Yue J, Mao Y, Zhou W, Wang Y et al (2020) Anti-ST2 nanoparticle alleviates lung inflammation by targeting ILC2s-CD4(+)T response. Int J Nanomed 15:9745–9758
147. Wu Z, Chen C, Luo J, Davis JRJ, Zhang B, Tang L, Liao D et al (2020) EGFP-EGF1-conjugated poly (lactic-co-glycolic acid) nanoparticles as a carrier for the delivery of CCR2- shRNA to atherosclerotic macrophage in vitro. Sci Rep 10(1):19636
148. Yang B, Chen Y, Shi J (2019) Reactive oxygen species (ROS)-based nanomedicine. Chem Rev 119(8):4881–4985
149. Yang J, Reth M (2016) Receptor dissociation and B-cell activation. Curr Top Microbiol Immunol 393:27–43
150. Yeste A, Takenaka MC, Mascanfroni ID, Nadeau M, Kenison JE, Patel B, Quintana FJ et al (2016) Tolerogenic nanoparticles inhibit T cell-mediated autoimmunity through SOCS2. Sci Signal 9(433):ra61
151. Zhang CY, Dong X, Gao J, Lin W, Liu Z, Wang Z (2019) Nanoparticle-induced neutrophil apoptosis increases survival in sepsis and alleviates neurological damage in stroke. Sci Adv 5(11):eaax7964
152. Zhang Q, Dehaini D, Zhang Y, Zhou J, Chen X, Zhang L, Zhang L et al (2018) Neutrophil membrane-coated nanoparticles inhibit synovial inflammation and alleviate joint damage in inflammatory arthritis. Nat Nanotechnol 13(12):1182–1190
153. Zhang Q, Yuan R, Li C, Wei W, Shen W, Cui Y, Yuan X (2019) Macrophage depletion with clodronate-containing liposomes affects the incidence and development of rheumatoid arthritis. Z Rheumatol 78(10):996–1003
154. Zhao X, Shan Q, Xue HH (2022) TCF1 in T cell immunity: a broadened frontier. Nat Rev Immunol 22(3):147–157
155. Zhong G, Yang X, Jiang X, Kumar A, Long H, Xie J, Zhao J et al (2019) Dopamine-melanin nanoparticles scavenge reactive oxygen and nitrogen species and activate autophagy for osteoarthritis therapy. Nanoscale 11(24):11605–11616
156. Zindel J, Kubes P (2020) DAMPs, PAMPs, and LAMPs in immunity and sterile inflammation. Annu Rev Pathol 15:493–518

Nanomaterial-Based Drug Delivery System in Targeting Inflammatory Disorders

Shambhu Kumar, Arun Ghuge, and Indarchand Gupta

Abstract Inflammatory diseases can occur due to many factors including environmental exposure, infection and as the part of immune response. However, the prolonged and dysregulated inflammation could be harmful. It could persist over time and hence it could damage a particular organ or a system. To manage the medical conditions developed during the inflammation, there is a need to make use of anti-inflammatory drugs. Many types of drugs are already available to prevent the organ-specific and system-specific damage. However, there are some limitations in the treatment. The target-specific therapy is the major challenge in delivering the anti-inflammatory drug at the anticipated site. To overcome this challenge, nanomaterials could be a better alternative. Nanomaterials have many advantages over other classical drug delivery agents, due to which they are preferred for the delivery of anti-inflammatory drugs for the treatment of inflammatory diseases. Hence, the present chapter is focused on a brief discussion of the treatment of inflammation, followed by a detailed discussion of the various nanomaterials that can be used as a delivery agent for drugs for the treatment of inflammation. Moreover, the nanotoxicity as the major challenge in the said therapy is also discussed with the various options to avoid the nanotoxic effects.

Keywords Anti-inflammatory drugs · Liposomes · Solid lipid nanoparticles · Membrane-coated nanoparticles

---

S. Kumar
Department Zoology, Ram Bahadur Singh College, Andaur, Samastipur, Bihar, India

Department Zoology, L. N. Mithila University, Kameshwar Nagar, Darbhanga, Bihar, India

A. Ghuge
Department of Forensic Biology, Government Institute of Forensic Science, Civil Lines, Nagpur, Maharashtra, India

I. Gupta (✉)
Department of Biotechnology, Government Institute of Science, Chhatrapati Sambhajinagar, Maharashtra, India
e-mail: indarchandgupta@gmail.com; indar.gupta@mah.gov.in

Md. M. Ansari et al. (eds.), *Emergence of Sustainable Biomaterials in Tackling Inflammatory Diseases*, Smart Nanomaterials Technology,
https://doi.org/10.1007/978-981-96-2112-5_7

1 Introduction

Inflammation is the physiological and normal reaction to signals resulting from pathogenic infection or tissue injury [37]. Inflammation aids in the restoration of homeostasis following trauma or infection by shielding host from pathogens and mending damaged tissues [54]. The inflammation process consists of a quick induction phase that initiates a pro-inflammatory response, succeeded gradually by a resolution phase [58]. As a result, inflammation is necessary for the healing of wounds. On the other hand, unchecked or extended inflammation can cause several inflammatory illnesses, including sepsis, obesity, cancer, cardiovascular, neurological, and autoimmune disorders, to form and worsen [24].

There are many phases of dysregulated localized or chronic inflammation. Early on, resident immune cells are activated to produce various cytokines, chemokines, and other factors by danger signals viz. damage-associated molecular patterns (DAMPs) or pathogen-associated molecular patterns (PAMPs), which are released from damaged host cells or dead pathogens [76]. After that, circulating immune cells use endothelial cell adhesion to settle at the inflammation site, enhancing pro-inflammatory reactions [42]. This synchronized immunological response eventually restores homeostasis. Nevertheless, late-stage inflammation may result from an inadequate or overreaction that upsets the equilibrium between the immune system's innate and adaptive arms. Damage to local or systemic tissues and an overabundance of danger signals are two hallmarks of a catastrophic cascade that might result from such dysregulated immune response. A breakdown in immunological tolerance may also be brought on by changes in the acute to chronic inflammatory response, which can advance an inflammatory illness and perhaps cause mortality [52]. Anti-inflammatory medications are the mainstay of treatment for inflammatory illnesses nowadays, since they may assist achieve a balance between the immune system and inflammation [21]. The necessity for new and more effective anti-inflammatory techniques is suggested by the challenges that existing anti-inflammatory drugs face, including poor tissue selectivity, systemic side effects, and inefficiency in overcoming biological barriers.

Over the past 10 years, nanotechnology has advanced quickly, particularly in the area of medicine. Since the US Food and Drug Administration (FDA) approved Doxil®, a PEGylated liposomal doxorubicin, in 1995 as the first nanodrug, nanotechnology has attracted a lot of attention for the creation of innovative theranostic strategies [7]. Nanoparticles (NPs) in the 1–100 nm size range have gained recognition in recent years, particularly in the biomedical field, for their special applications. This is because of their clever combination of notable physical and chemical properties and appropriate interaction with the building blocks of living organisms at the nanoscale [50]. As some NPs may find it difficult to pass through the cell membrane, therefore, to overcome these challenges, delivery mechanisms are needed. Consequently, a variety of NP delivery methods have been created, some of which include chitosan, liposomes, micelles, and synthetic dendrimers [29, 49, 57, 69].

2 Challenges and Opportunities of Nano-based Delivery

The abundance of lipids with distinct compositions and characteristics makes it simple to screen for and select the best lipid mixture for maximum delivery efficiency. Lipidic nanoparticle formulation is facilitated by this [3]. For instance, such an optimized lipid formulation plays a role in the efficacy of the COVID-19 mRNA vaccine [36]. The encapsulation of tiny hydrophobic medicines and the functionalization of lipid nanoparticles are still difficult issues, albeit they may be resolved by changing the lipid composition. Because of their high compositional and structural variety and molecular weight, polymeric nanoparticles can encapsulate any medication. In addition, polymeric nanoparticles can have targeting, stimulus–response, and biodegradability added to them. Nevertheless, differences in the molecular weight and content of several batches of the polymeric carrier could provide challenges in clinical translation. Due to the wide molecular weight range, polydispersity in the size as well as the surface potential of nanoparticles influences performance and encourages aggregation. Since their size, composition, and structure are more consistent, inorganic nanoparticles have electrical, magnetic, and electrical capabilities that can be used for imaging or external control release. Mesoporous silica nanoparticles, for example, are porous inorganic nanoparticles that provide highly effective drug molecule loading into their interior pores [59, 60]. Drugs that are restricted to surface immobilization will have less drug loading and be more vulnerable to surface-exposed drug breakdown. To increase nanocarrier delivery efficiency, a variety of strategies are needed [43, 45]. For instance, creating nanocarriers that react to inflammatory chemicals can lead to more precise targeting of inflammation. To treat inflammation at different phases, it would also be beneficial to transport several cargos with a single vehicle.

To increase the anti-inflammatory effect for the late stage of inflammation, the delivery system might, for instance, discharge scavengers first to eliminate cfDNA or reduce RONS, and then drugs that can block leucocyte-endothelial interactions. In addition to inflammatory chemicals, it is possible to investigate how cells and cellular communication work together to draw leucocytes to inflammatory cells. For instance, various leucocyte populations and arterial beds express various combinations of adhesion molecules, that might be taken into consideration when designing tissue-specific targeted drug delivery platforms [30]. Connexin is a membrane protein that is essential to gap junctions and controls the transmission of signals between adjacent cells by facilitating the quick passage of metabolites, second messengers, and ions. It has been discovered that leucocyte and epithelial cells in the lungs and intestine communicate through gap junction connexin 43 channels [72]. The artificial peptide ACT1 imitates the carboxyl terminus of connexion 43, stabilizing gap junctions and decreasing the infiltration of inflammatory neutrophils into wounds [26]. The communication between leucocytes and epithelial or endothelial cells could be modulated by targeting these connexins [25]. Exosomes are also involved in cell signalling, transporting lipids, proteins, microRNA, and mRNA between cells [35]. In response to a variety of stimuli, including inflammation, different cell types, such as leucocytes and endothelial cells, produce exosomes. Exosomes have the potential to function as drug-delivery vehicles by expressing specific surface molecules and delivering them to targeted targets [75]

The abundance of lipids with distinct compositions and characteristics makes it simple to screen for and select the best lipid mixture for maximum delivery efficiency. Lipidic nanoparticle formulation is facilitated by this [3]. For instance, such an optimized lipid formulation plays a role in the efficacy of the COVID-19 mRNA vaccine [36]. The encapsulation of tiny hydrophobic medicines and the functionalization of lipid nanoparticles are still difficult issues, albeit they may be resolved by changing the lipid composition.

Because of their high compositional and structural variety and molecular weight, polymeric nanoparticles can encapsulate any medication. In addition, polymeric nanoparticles can have targeting, stimulus–response, and biodegradability added to them. Nevertheless, differences in the molecular weight and content of several batches of the polymeric carrier could provide challenges for clinical translation. For instance, due to the wide molecular weight range, polydispersity in the size and surface potential of nanoparticles influences performance and encourages aggregation. Since their size, composition, and structure are more consistent, inorganic nanoparticles have electrical, magnetic, and electrical capabilities that can be used for imaging or external control release. Mesoporous silica nanoparticles, for example, are porous inorganic nanoparticles that provide highly effective drug molecule loading into their interior pores [59, 60]. Drugs that are restricted to surface immobilization will have less drug loading and be more vulnerable to surface-exposed drug breakdown.

To increase nanocarrier delivery efficiency, a variety of strategies are needed [43, 45]. For instance, creating nanocarriers that react to inflammatory chemicals can lead to more precise targeting of inflammation. To treat inflammation at different phases, it would also be beneficial to transport several cargos with a single vehicle.

To increase the anti-inflammatory efficacy at the late stage of inflammation, the delivery system might, for instance, release scavengers first to remove cfDNA or reduce RONS, and then drugs that can block leucocyte-endothelial interactions.

In addition to inflammatory chemicals, it is possible to investigate how cells and cellular communication work together to draw leucocytes to inflammatory cells. For instance, various leucocyte populations and arterial beds express different combinations of adhesion molecules, which might be taken into consideration when designing tissue-specific targeted drug delivery platforms [30].

Connexin is a membrane protein that is essential to gap junctions and controls the transmission of signals between adjacent cells by facilitating the quick passage of metabolites, second messengers, and ions. It has been discovered that leucocyte and epithelial cells in the lungs and intestine communicate through gap junction connexin 43 channels [72]. The artificial peptide ACT1 imitates the carboxyl terminus of connexin43, stabilizing gap junctions and decreasing the infiltration of inflammatory neutrophils into wounds [26]. The communication between leucocytes and epithelial or endothelial cells could be modulated by targeting these connexins [25]. Exosomes are also involved in cell signalling, transporting lipids, proteins, microRNA, and mRNA between cells [35].

In response to a variety of stimuli, including inflammation, different cell types, such as leucocytes and endothelial cells, produce exosomes. Exosomes have the potential to function as drug-delivery vehicles by expressing specific surface molecules and delivering them to targeted targets [75].

3 Therapies for the Treatment of Inflammatory Diseases

Countless therapies are available for treating inflammatory diseases, such as Physical, Biological, Immunosuppressants, Surgery, Corticosteroids, and Nonsteroidal Anti-Inflammatory Drugs etc. These are a few examples of therapies that can be used to treat or control inflammatory disorders having some pros and cons.

Biological Therapies: Biological drugs are target-oriented molecules used to treat inflammatory diseases. Biological therapies like anti-tumour necrosis factor (TNF), anti-interleukin, and JAK inhibitors antibodies are extensively used as an active treatment for immune-mediated inflammatory diseases, such as psoriasis, inflammatory bowel diseases, psoriatic arthritis, rheumatoid arthritis [11], ulcerative colitis (UC) and Crohn's disease (CD). Some of the inflammatory disease conditions are responsible for progressive disability, pain, and loss of working ability if not cured [23, 65, 73]. Moreover, chronic inflammatory diseases can be responsible for depression and anxiety [9].

Immunosuppressants: Azathioprine, mycophenolate mofetil, and cyclosporine are the immunosuppressor drugs used to suppress the natural immune system and thus reduce inflammation [32]. They are generally used in autoimmune disease conditions where the immune system incorrectly attacks the self-body tissues, such as systemic lupus erythematosus, multiple sclerosis ankylosing spondylitis, psoriatic arthritis, and inflammatory bowel disease [20].

In case of transplantation, the immunosuppressive treatments are determined by a number of parameters, such as the length of time since transplantation, the disease's aetiology, and/or the state of the graft. Individualized immunosuppressive anti-inflammatory treatment requires different protocols and preparations. Triple-drug therapy, which consists of corticosteroids, antiproliferative medications, and calcineurin inhibitors, is the mainstay of maintenance treatment for recipients of renal transplants [64]. The use of immunosuppressive medications includes a risk of various adverse effects in addition to the intended impact. In order to minimize side effects and maximize efficacy while reducing toxicity, new immunosuppressive anti-inflammatory drugs and modes of treatments, are being searched for. It will not only increase the opportunity to have effective medication, but it will also help to lower the morbidity and mortality.

Surgery: In a few complex conditions, surgery is the prime therapy to treat the complications of inflammatory diseases, for example, intestinal strictures in Crohn's disease or joint damage in rheumatoid arthritis. Crohn's disease is inflammation all over the layers of the bowel, Recurrence of this disease happens periodically, resulting in a chronic condition and leading to various complications. Laparoscopic surgery is the major therapy for the first-line approach to Crohn's disease [40, 55].

Physical Therapy: Physical therapy can benefit in controlling or remedial Inflammatory conditions affecting the musculoskeletal system Exercises, stretches, and yoga can help improve flexibility, strength, and range of motion while reducing pain and inflammation.

Corticosteroids: Corticosteroids are alternatively well-known as steroids, glucocorticosteroids or glucocorticoids. They are frequently used drugs and are effective in many inflammatory diseases. The inhaled corticosteroids are the first-line therapy. Moreover, they are effective in the anti-inflammatory treatment in asthmatic conditions [53]. Prednisone and cortisone mimic the effects of body hormones produced naturally and are potent anti-inflammatory drugs. They are generally recommended for rheumatoid arthritis, asthma, and inflammatory bowel disease because of their quick anti-inflammatory property [8].

Nonsteroidal Anti-Inflammatory Drugs (NSAIDs): NSAIDs are mostly consumed drugs either over-the-counter or medical practitioner prescription. Diclofenac, Diflunisal, Etodolac, Fenoprofen, Flurbiprofen, Ibuprofen, Mefenamic acid, Meloxicam, Naproxen, Oxaprozin, Piroxicam, Sulindac, Celecoxib, Rofecoxib, Valdecoxib are the non-selective and selective NSAIDs generally used to help reduce inflammation and relieve pain [10, 47].

Other than above recommended therapies for inflammatory diseases lifestyle modification such as regular exercise, maintaining a healthy weight, a healthy diet plan, and smoking cessation can be helpful in the reduction of inflammation. Individuals with inflammatory diseases need to work closely with their healthcare providers to develop a treatment plan personalized to their specific requirements and medical history. Treatment may involve a combination of therapies to effectively manage symptoms and prevent disease progression.

3.1 Passive Targeting

Transport of drug formulations across the wall of leaky fenestrated blood capillaries through passive diffusion or convection is known as passive targeting. Biomolecules can enhance the targeting ability via physical and chemical interactions as well as physical factors of anti-inflammatory therapeutics to target cells, tissues, and organs to accomplish specific drug delivery and anti-inflammatory efficacy and to reduce toxicity [5]. Increased permeability of tissue space and other microenvironment biochemical properties such as reactive oxygen species (ROS), pH, and enzymes are useful for improving the passive targeting of drugs [16].

Many biologically active molecules are not able to cross the biological barriers that separate the site of drug administration from the site of action. The biological systems are compartmentalized and divided into several sections such as epithelium, endothelium, cellular membrane, and mechanical or physicochemical barriers. Some molecules are not able to diffuse naturally into the cell whereas their therapeutic target is with intracellular localization. The targeted delivery of therapeutic agents to an organ, a tissue, or a type of cell is the major concern and challenge in treating various human diseases [12].

Many researchers have reported issues like drugs poor water solubility, unexpected side effects and biological degradation. To solve such issues there is a need to search for a novel drug carrier system to control inflammatory disorders.

Passive targeting is a novel solution for targeted drug delivery to achieve the goal of inflammatory disease control. The drug targeting by nanoparticles highly relies on their physicochemical properties, surface functionalization and pathophysiological characteristics.

3.2 Active Targeting

Active targeting comprises fragments that specifically bind inflammatory molecules and cells to the surface of biomaterial-based drugs. Pathogens, enzymes, inflamed vasculature, immune cells, or inflammatory products are the key inflammatory targets. Inflamed vasculature is a common symptom in the pathogenesis of atherosclerosis [62]. Circulating leukocyte and endothelial cell interaction results in vascular inflammation that is usually mediated by endothelial cell adhesion molecules (CAMs). Managing vascular inflammation by therapeutic targeting endothelial CAMs is a promising approach [70]. In such conditions targeted drug delivery can be achieved by coating nanoparticles with leukocyte plasma membranes which can specifically bind endothelial cell receptors, allowing targeted delivery to inflamed tissues [46].

4 Nanomaterial-Based Drug Delivery Strategies

Nanomaterial-based drug delivery strategies have more advantages over available conventional methods.

Improved Bioavailability: Inflammatory disease drug degradation can be avoided in the body through nanoparticles. Drugs can be released in a controlled manner by designing or modifying the nanoparticles. Nanoparticles play an important role in increasing their bioavailability and ensuring a more sustained release profile which can improve patient compliance and reduce the frequency of dosing. Mesoporous silica nanoparticles have a higher drug-loading capacity. They also have diverse controlled release profiles due to their porous structure as compared to other polymer-based carriers. These nanoparticles are exceptional hydrophobic anti-inflammatory drug delivery nanocarriers [14].

Biological Barriers Crossing: Nanoparticles can traverse biological barriers such as the blood–brain barrier, allowing for the delivery of drugs to previously inaccessible sites. The drug transport across the tight junctions of a human epithelial colorectal adenocarcinoma cell monolayer (Caco-2) was enhanced by resveratrol drug encapsulation in silica nanoparticles [74]. Nanomaterials can enhance the solubility of many drugs for better absorption and distribution to the target site.

Chitosan nanoparticles consist of randomly dispersed β-(1–4)-linked D-glucosamine and N-acetyl-D-glucosamine units. Protonated amino groups have a positive charge, which is why they are capable of regulating pH and operate as a water-soluble cationic polyelectrolyte that can interact with negatively charged molecules. Through the encapsulation of dexamethasone (DEX) sodium phosphate for topical ocular administration, chitosan nanoparticles improved drug permeability and reduced drug residence time in the cornea and thus promising the treatment of ocular inflammation [71].

Targeted Delivery: By using nanomaterials as a foundation, nanotechnology can be used to develop medication or chemical delivery systems or even to help tissue regeneration. Food and Drug Administration (FDA) approved nanoparticle therapy since 1989 and it changed the face of medical science for targeted drug delivery. The use of NPs in various applications including drug delivery, gene delivery, tissue engineering, photodynamic therapy, imaging, and wound healing highly depends on their physical and chemical properties. Many studies have reported that the size and surface of the nanoparticles frequently amended to decrease particle elimination, escape the biological defensive system, increase circulation time, and prolong drug presence at the target sites. Targeted drug delivery helps to reduce systemic exposure and minimizes the side effects on healthy tissues.

The NPs can be used for targeted drug delivery mainly because of some reasons. It is well known that signals from biomaterials' micro- and nano-environments can change the interactions between cells and biomaterials. The NP-delivered medications can reach the target location without causing systemic side effects and at lower doses thanks to the enormous potential that nanoparticles have demonstrated as delivery platforms. Thus, by targeting immunomodulation with nanomaterials, it has been observed that nanomaterials can have immunosuppressive or immune-stimulating effects [41].

Anti-inflammatory medications and inorganic carriers are frequently combined to create inorganic nanoparticles used to target the treatment of inflammation. As they are more stable, metal oxide nanoparticles have a wide range of applications in nanomedicine. Since zinc is essential for the transfer of genetic signals, zinc oxide nanoparticles, or ZnO NPs, have outstanding biological qualities [1]. These ZnO nanoparticles have demonstrated bacteriostatic qualities, and zinc oxide is a potent antibacterial agent. Research has been conducted to determine the mechanism, and one theory is that the bacterial cell's integrity was broken by the electrostatic contact

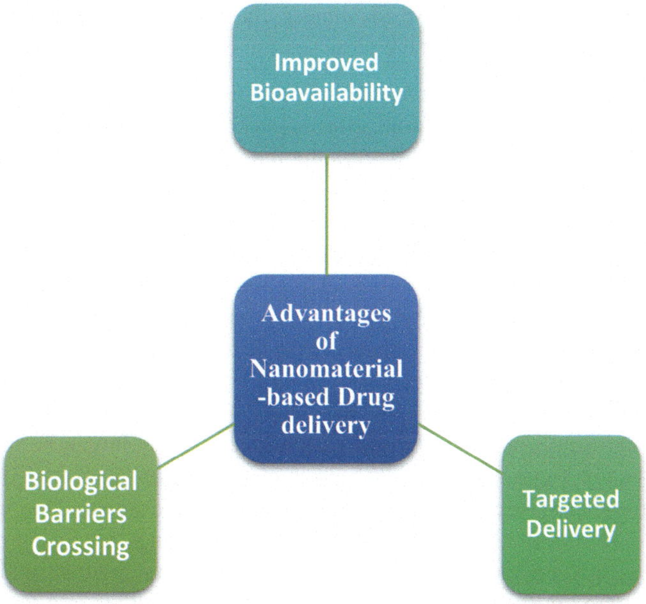

Fig. 1 Advantages of nanomaterial-based drug delivery

between ZnO and the cell wall [13]. However, one study showed the ZnO NPs also showed the targeted release of folic acid-modified PEG-ZnO NPs for anti-inflammatory and anti-tumor effects.

Other than the above-discussed points nanomaterials have various advantages and applications like nanoparticle combination therapy, imaging, and diagnostic properties of some nanomaterials, revolutionizing the field of medicine by improving the efficacy, safety, and specificity of drug therapies across various disease conditions. The advantages of the use of nanomaterial-based drug delivery are summarized in Fig. 1

4.1 Liposome-Based Nano-platforms for Targeted Delivery of Anti-Inflammatory Small-Molecular Drugs

Liposomes are the spherical bodies containing the lipid bilayers. The core of the liposome is aqueous and it can accommodate the hydrophilic drugs. Whereas the membrane of the liposome can house hydrophobic drugs [2]. Thus, liposomes are versatile carriers for both hydrophilic and hydrophobic drugs. Their surface can thus be modified for targeted delivery. The liposomes are one of the potential candidates for drug delivery as they prolong the drug circulation time, augment the drug stability, target delivery to specific cells or tissues and assist in the controlled discharge of drugs at the target site [39]. Due to these advantages, the liposome-based nano-platforms are specifically well-matched for the purpose of anti-inflammatory drug delivery.

Delivery of Morin: Morin is an anti-inflammatory, naturally available flavonoid. It is used for the treatment of Rheumatoid arthritis [44]. The therapeutic effects of morin can be enhanced with the help of liposome-based targeted delivery [67]. Under physiological conditions, naked morin is prone to enzymatic degradation, which limits its therapeutic effect. It can be encapsulated in the liposome aqueous core, which thus protects it from getting degraded and oxidised [68], thereby increasing its stability and preserving its bioactivity. Moreover, the liposomal encapsulation of morin increases its water solubility, subsequently increasing its bioavailability [67]. For targeted delivery of morin to inflamed cells or tissues, the liposome surface can be conjugated to the targeting ligand which can precisely bind to the receptors or overexpressed biomarkers on it [61]. Thus, through conjugation of targeting ligand to the drug-carrying liposome, the therapeutic effect of morin can be maximised. Liposomes can be engineered to release morin in a controlled manner [63]. It can be achieved through passive diffusion or a triggered release mechanism. This process lets the local and persistent release of morin at the inflammation site, which prolongs the therapeutic action and decreases the dose frequency. Due to target-specific delivery of morin through liposome-based nano-platform, the exposure to non-targeted cells and tissues is curtailed. It also lessens the adverse side effects accompanying the naked or free morin.

4.1.1 Delivery of Quercetin

Quercetin is another naturally available, anti-inflammatory flavonoid. However, it also has antioxidant and anticancer properties [6]. Quercetin is hydrophobic and hence it could be accommodated in the phospholipid bilayer of the liposome [66]. Liposome helps to improve the cellular uptake of quercetin by intended cells or tissues. It helps in the direct delivery of quercetin by getting fused with the target cell membrane. This mode of quercetin delivery is advantageous as it helps in intensifying its accumulation cell cytoplasm, thereby enhancing its pharmacological effects [4]. In addition to quercetin embedded in the liposomal lipid bilayer, the ligand such as antibodies or peptides can be exploited which can help for their specific delivery [19]. It minimises the adverse side effects on non-targeted cells and tissues and helps in their controlled release. Likewise, liposomes can be additionally loaded with other drugs which have antioxidant or anticancer activity. In such a scenario, this technology can have excellent therapeutic effects.

4.1.2 Delivery of Resveratrol

It is naturally found anti-inflammatory polyphenol present in peanuts, grapes and berries [56]. As far as liposome-based delivery of resveratrol is concerned, there is a lack of sufficient studies. However, it is known that it has poor aqueous solubility and stability [18]. Therefore, if used in free form, its bioavailabity and therapeutic efficacy are affected. To solve this issue, resveratrol can be encapsulated in the nano-based liposome [31]. In this case, the curcumin was linked to the hydrophobic acyl chain area of liposomes, whereas resveratrol was located in the polar hydrophilic region. The physical and chemical characteristics of the liposomes were also examined in the study [31]. The encapsulation rate and antioxidant activity peaked at a 5:1 curcumin-to-resveratrol ratio. Comparably, utilizing liposomes as the nanocarrier, co-loading quercetin and resveratrol improved cellular absorption and ROS scavenging capacity compared to single drug loading.

The liposomal encapsulation improves its stability and protects it from degradation. Subsequently, liposome promotes its uptake by the target cells. As mentioned earlier in the case of morin and quercetin, liposome-carrying resveratrol could be surface modified for target-specific drug delivery to inflamed tissue [17]. However, on the contrary, it is also known that Liposomes have short-life and can be oxidized and hydrolysed easily. It is reported that liposome-based intravenous delivery of Resveratrol caused a brief half-life, ranging from 7.8 to 33 min [15].

4.2 Membrane-Coated Nanoparticles (MCNPs)

Membrane-coated nanoparticles (MCNPs) are another way for the targeted delivery of anti-inflammatory drugs [34]. They contain nanoparticles coated with cell membranes, which are generally derived from the cells involved in inflammation or the immune cells [38]. Therefore, the cell membranes herein hold the source cell surface receptors and proteins. This phenomenon helps MCNPs to mimic the targeting characteristics of the original immune cells or cells involved in inflammation. It empowers the MCNPs to reach the targeted inflamed cells or tissues, aiding in target-specific drug delivery at the site of inflammation. As of liposomes, MCNPs are biocompatible [33]. This is due to the presence of host cell membrane coatings. They have reduced immunogenicity too thus minimizing the chances of getting cleared by the host immune system [74].

MCNPs can also be designed to act as per the signals present in the inflammatory microenvironment, including the redox potential, enzymatic activity and change in the pH. It allows for the release of the drug during specific stages of inflammation. Additionally, the MCNPs can also be equipped with anti-inflammatory drugs which are involved in cytokine inhibition or modify the activity of immune cells involved in inflammation.

4.3 Solid Lipid Nanoparticles

Solid lipid nanoparticles (SLNs) are colloidal drug delivery systems. Usually, they are in the size range of 50 to 1000 nm with an average diameter of 100–200 nm. They consist of biocompatible and biodegradable lipids like fatty acids, triglycerides, waxes, etc. They are solid at both room and body temperature and thus they provide structural integrity to the nanoparticles [48]. They have the capacity to encapsulate both the hydrophilic and hydrophobic molecules. The lipid matrix of SLNs protects the drug, prevents its degradation and improves its bioavailability.

SLNs are able to encapsulate anti-inflammatory drugs such as ibuprofen, diclofenac, resveratrol, curcumin, and corticosteroids.

5 Challenges and Opportunities of Nano-based Delivery

As mentioned earlier, various types of nanoparticles have been shown to have potential in drug delivery for the treatment of various diseases. However, with the rise in applications of nanoparticles, the reports showing their harmful effects are also increasing [28]. To address the toxic effects of nanoparticles, the term Nanotoxicology has been coined. It is the branch of nanotechnology dealing with the harmful effects of nanoparticles on the human and environment. Many studies have considered some factors such as size, shape, surface charge and dissolution to contribute to the harmful effects of nanoparticles. The smaller size nanoparticles are more toxic as compared to their larger counterparts. They can be synthesised in various shapes such as sphere, triangle, and rod [27]. Each shape of nanoparticle shows a particular level of toxic effects. Many nanoparticles either get functionalised during their synthesis or need to be functionalised afterward by a suitable capping agent. Therefore, different type of capping agent plays a role in their toxicity. In such cases, the surface capping makes a significant impact on the nanoparticle activity [51]. Moreover, many studies have shown that nanoparticles have the property of dissolution in the media in which they form the suspension. The ions released via the process of dissolution pose a risk to the cell membrane and cytoplasmic contents [22]. Therefore, these factors need to be considered for applications of any type of nanoparticles for applications such as drug delivery. However, to solve these issues, it has been suggested to make use of different capping agents. The capping agents such as certain peptides, polyethylene glycol, and SDS can be used as capping agents. They can help reduce the toxicity of nanoparticles [51]. The capping agents could also be used to anchor the drug to be used for the treating a particular inflammatory disease. Hence, these are the factors that must be considered and discussed before choosing any nanoparticle for drug delivery.

6 Conclusion

Inflammation is an important mechanism that develops in the body in response to infection or injury. It is thus required for healing of the wounds. Whereas, prolonged inflammatory response could be harmful. It can happen in the situation of dysregulation of inflammatory response which may be inadequate or over-reactive. It can result in many inflammatory diseases causing damage to local or systemic tissues. In order to control the aforesaid damage, anti-inflammatory drugs have been designed and are available in the market. However, drug delivery has the challenge of reaching the targeted tissue or organ to show the intended effect. Many types of carriers are being exploited. However, the challenge of site-specific delivery remains unsolved. Hence, there is a need to search for a better alternative for drug delivery agents. Nanomaterials are coming up to be a better solution for this purpose. A rapid change has been occurring in the field of nanotechnology. Due to the unique properties of nanomaterials, they are being used for many types of uses in the field of biomedical science. Drug delivery with the help of nanoparticles is among those applications. Nanotechnology is being extensively used in the delivery of drugs for the treatment of inflammatory diseases. Nano-based platforms such as Liposome, MCNPs, SLNs have shown promising results for the treatment of inflammatory diseases. Although they are found to assist in the delivery of the intended drug to the target site, still there is a need to reduce the lacunae associated with their drug delivery properties and mode of action. Moreover, with the applications of nanomaterials, their toxicity should also need to be considered. There is a need to perform an in-depth study that will make the nano-based anti-inflammatory drug delivery extremely specific, effective and safe for the treatment of inflammatory diseases.

References

1. Agarwal H, Shanmugam V (2020) A review on anti-inflammatory activity of green synthesized zinc oxide nanoparticle: mechanism-based approach. Bioorg Chem 94:103423
2. Akbarzadeh A, Rezaei-Sadabady R, Davaran S, Joo SW, Zarghami N, Hanifehpour Y, Samiei M, Kouhi M, Nejati-Koshki K (2013) Liposome: classification, preparation, and applications. Nanoscale Res Lett 8(1):102
3. Akinc A, Zumbuehl A, Goldberg M, Leshchiner ES, Busini V, Hossain N, Bacallado SA, Nguyen DN, Fuller J, Alvarez R, Borodovsky A, Borland T, Constien R, de Fougerolles A, Dorkin JR, Narayanannair JayaprakashK, Jayaraman M, John M, Koteliansky V, Manoharan M, Nechev L, Qin J, Racie T, Raitcheva D, Rajeev KG, Sah DWY, Soutschek J, Toudjarska I, Vornlocher H-P, Zimmermann TS, Langer R, Anderson DG (2008) A combinatorial library of lipid-like materials for delivery of RNAi therapeutics. Nat Biotechnol 26(5):561–569
4. Alyautdin R, Khalin I, Nafeeza MI, Haron MH, Kuznetsov D (2014) Nanoscale drug delivery systems and the blood-brain barrier. Int J Nanomed 9:795–811
5. Attia MF, Anton N, Wallyn J, Omran Z, Vandamme TF (2019) An overview of active and passive targeting strategies to improve the nanocarriers efficiency to tumour sites. J Pharm Pharmacol 71(8):1185–1198

6. Azeem M, Hanif M, Mahmood K, Ameer N, Chughtai FRS, Abid U (2023) An insight into anticancer, antioxidant, antimicrobial, antidiabetic and anti-inflammatory effects of quercetin: a review. Polym Bull (Berl) 80(1):241–262
7. Barenholz Y (2012) Doxil®–the first FDA-approved nano-drug: lessons learned. J Control Release 160(2):117–134
8. Barnes PJ (2006) How corticosteroids control inflammation: quintiles prize lecture 2005. Br J Pharmacol 148(3):245–254
9. Baumgart DC, Misery L, Naeyaert S, Taylor PC (2019) Biological therapies in immune-mediated inflammatory diseases: can biosimilars reduce access inequities? Front Pharmacol 10
10. Bindu S, Mazumder S, Bandyopadhyay U (2020) Non-steroidal anti-inflammatory drugs (NSAIDs) and organ damage: a current perspective. Biochem Pharmacol 180:114147
11. Boehncke WH (2015) Etiology and pathogenesis of psoriasis. Rheum Dis Clin North Am 41(4):665–675
12. Boyd BJ, Bergström CAS, Vinarov Z, Kuentz M, Brouwers J, Augustijns P, Brandl M, Bernkop-Schnürch A, Shrestha N, Préat V, Müllertz A, Bauer-Brandl A, Jannin V (2019) Successful oral delivery of poorly water-soluble drugs both depends on the intraluminal behavior of drugs and of appropriate advanced drug delivery systems. Eur J Pharm Sci 137:104967
13. Brayner R, Ferrari-Iliou R, Brivois N, Djediat S, Benedetti MF, Fiévet F (2006) Toxicological impact studies based on escherichia coli bacteria in ultrafine ZnO nanoparticles colloidal medium. Nano Lett 6(4):866–870
14. Brown TD, Whitehead KA, Mitragotri S (2020) Materials for oral delivery of proteins and peptides. Nat Rev Mater 5(2):127–148
15. Caddeo C, Pucci L, Gabriele M, Carbone C, Fernàndez-Busquets X, Valenti D, Pons R, Vassallo A, Fadda AM, Manconi M (2018) Stability, biocompatibility and antioxidant activity of PEG-modified liposomes containing resveratrol. Int J Pharm 538(1–2):40–47
16. Chiorean EG, Coveler AL (2015) Pancreatic cancer: optimizing treatment options, new, and emerging targeted therapies. Drug Des Dev Ther 9:3529–3545
17. Das SS, Verma PRP, Singh SK (2020) Quercetin-loaded nanomedicine as nutritional application. In: Rahman M, Beg S, Kumar V, Ahmad FJ (eds) Nanomedicine for bioactives: healthcare applications. Springer, Singapore, pp 259–301
18. Davidov-Pardo G, McClements DJ (2014) Resveratrol encapsulation: designing delivery systems to overcome solubility, stability and bioavailability issues. Trends Food Sci Technol 38(2):88–103
19. De Leo V, Milano F, Agostiano A, Catucci L (2021) Recent advancements in polymer/liposome assembly for drug delivery: from surface modifications to hybrid vesicles. Polymers 13(7):1027
20. Desai RJ, Bateman BT, Huybrechts KF, Patorno E, Hernandez-Diaz S, Park Y, Dejene SZ, Cohen J, Mogun H, Kim SC (2017) Risk of serious infections associated with use of immunosuppressive agents in pregnant women with autoimmune inflammatory conditions: cohort study. BMJ 356:j895
21. Dinarello CA (2010) Anti-inflammatory agents: present and future. Cell 140(6):935–950
22. Dos Santos CA, Seckler MM, Ingle AP, Gupta I, Galdiero S, Galdiero M, Gade A, Rai M (2014) Silver nanoparticles: therapeutical uses, toxicity, and safety issues. J Pharm Sci 103(7):1931–1944
23. Enns MW, Bernstein CN, Kroeker K, Graff L, Walker JR, Lix LM, Hitchon CA, El-Gabalawy R, Fisk JD, Marrie RA (2018) The association of fatigue, pain, depression and anxiety with work and activity impairment in immune mediated inflammatory diseases. PLoS ONE 13(6):e0198975
24. Furman D, Campisi J, Verdin E, Carrera-Bastos P, Targ S, Franceschi C, Ferrucci L, Gilroy DW, Fasano A, Miller GW, Miller AH, Mantovani A, Weyand CM, Barzilai N, Goronzy JJ, Rando TA, Effros RB, Lucia A, Kleinstreuer N, Slavich GM (2019) Chronic inflammation in the etiology of disease across the life span. Nat Med 25(12):1822–1832
25. Ghatnekar GS, Grek CL, Armstrong DG, Desai SC, Gourdie RG (2015) The effect of a connexin43-based Peptide on the healing of chronic venous leg ulcers: a multicenter, randomized trial. J Invest Dermatol 135(1):289–298

26. Ghatnekar GS, O'Quinn MP, Jourdan LJ, Gurjarpadhye AA, Draughn RL, Gourdie RG (2009) Connexin43 carboxyl-terminal peptides reduce scar progenitor and promote regenerative healing following skin wounding. Regen Med 4(2):205–223
27. Golinska P, Rathod D, Wypij M, Gupta I, Składanowski M, Paralikar P, Dahm H, Rai M (2017) Mycoendophytes as efficient synthesizers of bionanoparticles: nanoantimicrobials, mechanism, and cytotoxicity. Crit Rev Biotechnol 37(6):765–778
28. Gupta I, Duran N, Rai M (2012) Nano-silver toxicity: emerging concerns and consequences in human health. In: Cioffi N, Rai M (eds) Nano-antimicrobials: progress and prospects. Springer, Berlin, Heidelberg, pp 525–548.
29. Hasan M, Belhaj N, Benachour H, Barberi-Heyob M, Kahn CJ, Jabbari E, Linder M, Arab-Tehrany E (2014) Liposome encapsulation of curcumin: physico-chemical characterizations and effects on MCF7 cancer cell proliferation. Int J Pharm 461(1–2):519–528
30. He W, Holtkamp S, Hergenhan SM, Kraus K, de Juan A, Weber J, Bradfield P, Grenier JMP, Pelletier J, Druzd D, Chen CS, Ince LM, Bierschenk S, Pick R, Sperandio M, Aurrand-Lions M, Scheiermann C (2018) Circadian expression of migratory factors establishes lineage-specific signatures that guide the homing of leukocyte subsets to tissues. Immunity 49(6):1175-1190.e1177
31. Huang M, Liang C, Tan C, Huang S, Ying R, Wang Y, Wang Z, Zhang Y (2019) Liposome co-encapsulation as a strategy for the delivery of curcumin and resveratrol. Food Funct 10(10):6447–6458
32. Hussain Y, Khan H (2022). Immunosuppressive drugs. In: Rezaei N (ed) Encyclopedia of infection and immunity. Elsevier, Oxford, pp 726–740
33. Jiménez-Jiménez C, Manzano M, Vallet-Regí M (2020) Nanoparticles coated with cell membranes for biomedical applications. Biology 9(11):406
34. Joshi S, Allabun S, Ojo S, Alqahtani MS, Shukla PK, Abbas M, Wechtaisong C, Almohiy HM (2023) Enhanced drug delivery system using mesenchymal stem cells and membrane-coated nanoparticles. Molecules 28(5)
35. Kalluri R, LeBleu VS (2020) The biology, function, and biomedical applications of exosomes. Science 367(6478)
36. Khurana A, Allawadhi P, Khurana I, Allwadhi S, Weiskirchen R, Banothu AK, Chhabra D, Joshi K, Bharani KK (2021) Role of nanotechnology behind the success of mRNA vaccines for COVID-19. Nano Today 38:101142
37. Kotas ME, Medzhitov R (2015) Homeostasis, inflammation, and disease susceptibility. Cell 160(5):816–827
38. Krishnan N, Fang RH, Zhang L (2023) Cell membrane-coated nanoparticles for the treatment of cancer. Clin Transl Med 13(6):e1285
39. Liu P, Chen G, Zhang J (2022) A review of liposomes as a drug delivery system: current status of approved products, regulatory environments, and future perspectives. Molecules 27(4)
40. Liu W, Zhou W (2023) Minimally invasive surgery in Crohn's disease: state-of-the-art review. Front Surg 10:1216014
41. Lukin I, Erezuma I, Desimone MF, Zhang YS, Dolatshahi-Pirouz A, Orive G (2023) Nanomaterial-based drug delivery of immunomodulatory factors for bone and cartilage tissue engineering. Biomater Adv 154:213637
42. Luster AD, Alon R, von Andrian UH (2005) Immune cell migration in inflammation: present and future therapeutic targets. Nat Immunol 6(12):1182–1190
43. Manzari MT, Shamay Y, Kiguchi H, Rosen N, Scaltriti M, Heller DA (2021) Targeted drug delivery strategies for precision medicines. Nat Rev Mater 6(4):351–370
44. Miao Y, Wu X, Xue X, Ma X, Yang L, Zeng X, Hu Y, Dai Y, Wei Z (2023) Morin, the PPARγ agonist, inhibits Th17 differentiation by limiting fatty acid synthesis in collagen-induced arthritis. Cell Biol Toxicol 39(4):1433–1452
45. Mitchell MJ, Billingsley MM, Haley RM, Wechsler ME, Peppas NA, Langer R (2021) Engineering precision nanoparticles for drug delivery. Nat Rev Drug Discov 20(2):101–124
46. Molinaro R, Corbo C, Martinez JO, Taraballi F, Evangelopoulos M, Minardi S, Yazdi IK, Zhao P, De Rosa E, Sherman MB, De Vita A, Toledano Furman NE, Wang X, Parodi A,

Tasciotti E (2016) Biomimetic proteolipid vesicles for targeting inflamed tissues. Nat Mater 15(9):1037–1046
47. Newman O, Christian A, David Darko O, Aaron Opoku A (2017) Mechanism of action of nonsteroidal anti-inflammatory drugs. In: Ali Gamal Ahmed Ak (ed) Nonsteroidal anti-inflammatory drugs. Rijeka, IntechOpen (Ch. 2)
48. Pandey S, Shaikh F, Gupta A, Tripathi P, Yadav JS (2022) A recent update: solid lipid nanoparticles for effective drug delivery. Adv Pharm Bull 12(1):17–33
49. Park JH, von Maltzahn G, Ruoslahti E, Bhatia SN, Sailor MJ (2008) Micellar hybrid nanoparticles for simultaneous magnetofluorescent imaging and drug delivery. Angew Chem Int Ed Engl 47(38):7284–7288
50. Prosperi D, Colombo M, Zanoni I, Granucci F (2017) Drug nanocarriers to treat autoimmunity and chronic inflammatory diseases. Semin Immunol 34:61–67
51. Rai M, Gupta I (2024) Nanotoxicology for agricultural and environmental applications. Elsevier Science
52. Rajendran P, Chen YF, Chen YF, Chung LC, Tamilselvi S, Shen CY, Day CH, Chen RJ, Viswanadha VP, Kuo WW, Huang CY (2018) The multifaceted link between inflammation and human diseases. J Cell Physiol 233(9):6458–6471
53. Ricciardolo FLM, Blasi F, Centanni S, Rogliani P (2015) Therapeutic novelties of inhaled corticosteroids and bronchodilators in asthma. Pulm Pharmacol Ther 33:1–10
54. Rock KL, Latz E, Ontiveros F, Kono H (2010) The sterile inflammatory response. Annu Rev Immunol 28:321–342
55. Ruf (2005) Chirurgische Therapie der Komplikationen der chronisch entzündlichen Darmerkrankungen. Praxis 94(43):1681–1685
56. Sales JM, Resurreccion AV (2014) Resveratrol in peanuts. Crit Rev Food Sci Nutr 54(6):734–770
57. Sanyakamdhorn S, Agudelo D, Tajmir-Riahi HA (2013) Encapsulation of antitumor drug doxorubicin and its analogue by chitosan nanoparticles. Biomacromol 14(2):557–563
58. Schett G, Neurath MF (2018) Resolution of chronic inflammatory disease: universal and tissue-specific concepts. Nat Commun 9(1):3261
59. Shao D, Li M, Wang Z, Zheng X, Lao YH, Chang Z, Zhang F, Lu M, Yue J, Hu H, Yan H, Chen L, Dong WF, Leong KW (2018) Bioinspired diselenide-bridged mesoporous silica nanoparticles for dual-responsive protein delivery. Adv Mater:e1801198
60. Shao D, Zhang F, Chen F, Zheng X, Hu H, Yang C, Tu Z, Wang Z, Chang Z, Lu J, Li T, Zhang Y, Chen L, Leong KW, Dong WF (2020) Biomimetic diselenide-bridged mesoporous organosilica nanoparticles as an x-ray-responsive biodegradable carrier for chemo-immunotherapy. Adv Mater 32(50):e2004385
61. Siddique S, Chow JCL (2020) Gold Nanoparticles for drug delivery and cancer therapy. Appl Sci 10(11):3824
62. Siti HN, Kamisah Y, Kamsiah J (2015) The role of oxidative stress, antioxidants and vascular inflammation in cardiovascular disease (a review). Vascul Pharmacol 71:40–56
63. Sultana F, Neog MK, Rasool M (2017) Targeted delivery of morin, a dietary bioflavanol encapsulated mannosylated liposomes to the macrophages of adjuvant-induced arthritis rats inhibits inflammatory immune response and osteoclastogenesis. Eur J Pharm Biopharm 115:229–242
64. Szumilas K, Wilk A, Wiśniewski P, Gimpel A, Dziedziejko V, Kipp M, Pawlik A (2023) Current status regarding immunosuppressive treatment in patients after renal transplantation. Int J Mol Sci 24(12)
65. Takeshita J, Grewal S, Langan SM, Mehta NN, Ogdie A, Van Voorhees AS, Gelfand JM (2017) Psoriasis and comorbid diseases: epidemiology. J Am Acad Dermatol 76(3):377–390
66. Tomou E-M, Papakyriakopoulou P, Saitani E-M, Valsami G, Pippa N, Skaltsa H (2023) Recent Advances in nanoformulations for quercetin delivery. Pharmaceutics 15(6):1656
67. Tran H-M, Yang C-Y, Wu T-H, Yen F-L (2022) Liposomes encapsulating morin: investigation of physicochemical properties, dermal absorption improvement and anti-aging activity in pm-induced keratinocytes. Antioxidants 11(6):1183

68. Turjeman K, Bavli Y, Kizelsztein P, Schilt Y, Allon N, Katzir TB, Sasson E, Raviv U, Ovadia H, Barenholz Y (2015) Nano-drugs based on nano sterically stabilized liposomes for the treatment of inflammatory neurodegenerative diseases. PLoS ONE 10(7):e0130442
69. Wang X, Cai X, Hu J, Shao N, Wang F, Zhang Q, Xiao J, Cheng Y (2013) Glutathione-triggered "off-on" release of anticancer drugs from dendrimer-encapsulated gold nanoparticles. J Am Chem Soc 135(26):9805–9810
70. Weber C, Fraemohs L, Dejana E (2007) The role of junctional adhesion molecules in vascular inflammation. Nat Rev Immunol 7(6):467–477
71. Wei J, Mu J, Tang Y, Qin D, Duan J, Wu A (2023) Next-generation nanomaterials: advancing ocular anti-inflammatory drug therapy. J Nanobiotechnol 21(1):282
72. Westphalen K, Gusarova GA, Islam MN, Subramanian M, Cohen TS, Prince AS, Bhattacharya J (2014) Sessile alveolar macrophages communicate with alveolar epithelium to modulate immunity. Nature 506(7489):503–506
73. Wu Y, Wan S, Yang S, Hu H, Zhang C, Lai J, Zhou J, Chen W, Tang X, Luo J, Zhou X, Yu L, Wang L, Wu A, Fan Q, Wu J (2022) Macrophage cell membrane-based nanoparticles: a new promising biomimetic platform for targeted delivery and treatment. J Nanobiotechnol 20(1):542
74. Yan H, Shao D, Lao YH, Li M, Hu H, Leong KW (2019) Engineering cell membrane-based nanotherapeutics to target inflammation. Adv Sci (Weinh) 6(15):1900605
75. Yang Z, Shi J, Xie J, Wang Y, Sun J, Liu T, Zhao Y, Zhao X, Wang X, Ma Y, Malkoc V, Chiang C, Deng W, Chen Y, Fu Y, Kwak KJ, Fan Y, Kang C, Yin C, Rhee J, Bertani P, Otero J, Lu W, Yun K, Lee AS, Jiang W, Teng L, Kim BYS, Lee LJ (2020) Large-scale generation of functional mRNA-encapsulating exosomes via cellular nanoporation. Nat Biomed Eng 4(1):69–83
76. Zindel J, Kubes P (2020) DAMPs, PAMPs, and LAMPs in Immunity and Sterile Inflammation. Annu Rev Pathol 15:493–518

Nanotechnology-Based Management of Cardiovascular Disease

Shubhrat Maheshwari, Amita Verma, Aditya Singh, Vaseem A. Ansari, Aamir Anwar, Pavan Kumar, Bhupendra G. Prajapati, Nemat Ali, and Abdullah F. AlAsmari

Abstract Cardiovascular disease (CVD) continues to be a predominant cause of global mortality and morbidity, presenting an urgent need for advanced therapeutic and diagnostic strategies. Nanotechnology has emerged as a groundbreaking field with the potential to revolutionize CVD management through its unique capabilities in targeted drug delivery, improved diagnostic techniques, and novel therapeutic interventions. This book chapter delves into the current advancements and applications of nanotechnology in CVD management, highlighting the use of nanoparticles, liposomes, dendrimers, and nanosensors. These nanocarriers offer significant benefits, including enhanced drug bioavailability, reduced systemic toxicity, and precise delivery to specific diseased tissues, thereby improving therapeutic efficacy and patient outcomes. Furthermore, the integration of nanosensors in diagnostic processes allows for early and accurate detection of cardiovascular conditions, facilitating timely intervention. The chapter also discusses the challenges and future perspectives in the clinical translation of nanotechnology-based therapies for CVD,

S. Maheshwari · A. Verma
Bioorganic and Medicinal Chemistry Research Laboratory, Department of Pharmaceutical Sciences, Sam Higginbottom University of Agriculture, Technology and Sciences, Prayagraj, Uttar Pradesh, India

A. Singh · V. A. Ansari · A. Anwar
Department of Pharmacy, Faculty of Pharmacy, Integral University, Uttar Pradesh, Lucknow, India

P. Kumar
Maharishi School of Pharmaceutical Sciences, MUIT, Uttar Pradesh, Lucknow, India

B. G. Prajapati (✉)
Shree S. K. Patel College of Pharmaceutical Education and Research, Ganpat University, Gujarat, India
e-mail: bhupen27@gmail.com

N. Ali · A. F. AlAsmari
Department of Pharmacology and Toxicology, College of Pharmacy, King Saud University, Riyadh, Saudi Arabia

B. G. Prajapati
Faculty of Pharmacy, Silpakorn University, Nakhon Pathom 73000, Thailand

Md. M. Ansari et al. (eds.), *Emergence of Sustainable Biomaterials in Tackling Inflammatory Diseases*, Smart Nanomaterials Technology,
https://doi.org/10.1007/978-981-96-2112-5_8

aiming to provide a comprehensive understanding of this innovative approach in the fight against cardiovascular diseases.

Keywords Cardiovascular disease · Nanotechnology · Targeted drug delivery · Nanosensors

1 Introduction

Cardiovascular diseases (CVDs) represent significant sources of both morbidity and mortality worldwide. CVDs are responsible for an estimated 17.9 million deaths annually, contributing to 32% of all global mortality, with 85% of these attributed to heart attack or stroke [1]. The predominant origin of CVDs is characterized by the harmful buildup of lipids and fibrins in the arterial wall, coupled with an inappropriate immune response and dysregulation of cholesterol metabolism, initiating the development of atherosclerotic plaques. The left and right coronary arteries, branching off the aortic root, provide oxygen-rich blood to the myocardium. Atherosclerotic plaques in the coronary arteries can lead to coronary artery disease (CAD) [2]. Early detection is essential in preventing CAD, alongside the effective management of cardiovascular risk factors. Technological advancements and the identification of relevant biomarkers have significantly improved diagnostic approaches in clinical settings, including electrocardiography (ECG) and imaging methods for assessing abnormal cardiovascular conditions [3]. Medications play a crucial role in enhancing long-term outcomes and averting acute cardiovascular incidents [4]. Interventions such as stent placement or conventional recanalization surgeries like coronary artery bypass grafting can rescue ischemia induced by thrombotic occlusion. Despite the growing understanding of CAD, the persistently high mortality rates and resultant heart failure underscore the significant challenges still faced by current diagnostic and therapeutic approaches [5]. Nanomaterials, in comparison to bulk materials, exhibit a high surface area to volume ratio and possess adjustable optical, electronic, magnetic, mechanical, catalytic, thermal, and biological characteristics [6]. These materials can be customized in terms of size, shape, chemical composition, surface alterations, and hollow or solid structures. Such attributes empower researchers to devise innovative diagnostic and therapeutic platforms at the molecular level that surpass conventional methods [7]. Early diagnosis of CAD heightens the likelihood of successful treatment and potential recovery, thereby enhancing patient outcomes and prolonging survival periods. Biomarkers linked to CAD, such as those released into the bloodstream upon heart damage, are crucial for early detection [8]. A potential strategy for the early detection of CAD involves the development of accurate, specific, uncomplicated, stable, and rapid blood tests for such substances. The utilization of mass spectrometry is common in the identification of possible CAD biomarkers [9], albeit hindered by sensitivity and specificity issues due to the low levels of biomarkers in human plasma. Given these factors, the integration of nanotechnologies with biosensors

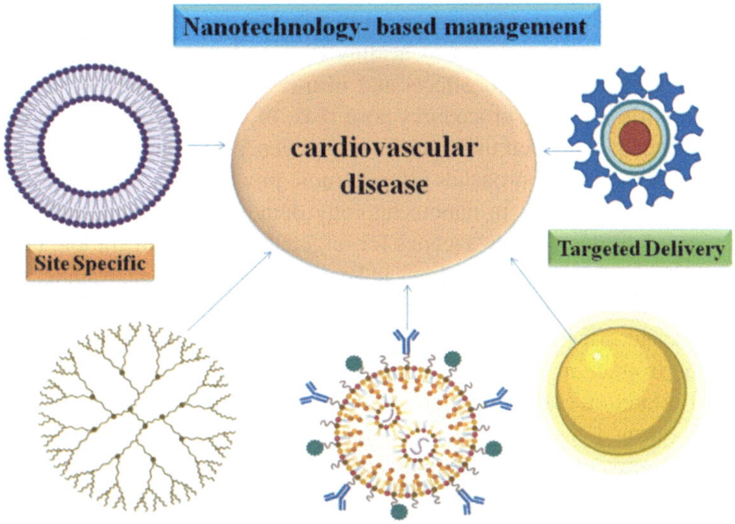

Fig. 1 Nanotechnologies can strongly bind targeted substances and reduce nonspecific adsorption

could offer a promising resolution for the early diagnosis of CAD [10]. Nanotechnologies can establish strong binding to targeted substances and reduce nonspecific adsorption through surface modifications or structural enhancements. Biosensors consist of a biological sensing component for target recognition and a transducer for data conversion (Fig. 1) [11].

Cardiovascular diseases (CVDs) are a significant cause of global morbidity and mortality, accounting for approximately 31% of total deaths worldwide. These conditions encompass a variety of disorders, including coronary artery disease (CAD), heart failure, arrhythmias, and strokes, among others [12]. Atherosclerosis, characterized by lipid accumulation, deposition of fibrous elements, and infiltration of immune cells in arterial walls, underpins many of these conditions by forming atheromatous plaques that may rupture, leading to thrombosis and acute coronary events like myocardial infarction (MI) or ischemic strokes. CAD, the primary CVD, involves the accumulation of atheromatous plaques [13]. Key risk factors for atherosclerosis and CAD include hyperlipidemia, hypertension, smoking, diabetes, obesity, and a sedentary lifestyle [14]. MI, commonly known as a heart attack, results from prolonged blockage of blood flow to a segment of the heart, causing tissue damage and presenting symptoms like chest pain, breathlessness, nausea, and sweating [15]. The current therapeutic landscape for CVDs encompasses lifestyle modifications, pharmacological treatments, and surgical interventions, which, though beneficial, have limitations and fall short of addressing the complexities of these conditions. Traditional therapies typically involve medications such as statins, beta-blockers, and anticoagulants, as well as procedures like angioplasty and coronary artery bypass grafting. However, these approaches may be associated with adverse effects, limited efficacy in advanced disease stages, and difficulties in targeting specific pathological sites while minimizing harm to healthy tissues. For instance, systemic drug delivery can lead to undesired effects and insufficient drug concentrations at the disease site, while invasive procedures pose risks of complications and prolonged recovery periods [16]. Given the significant prevalence and impact of CVDs, there is a need for therapeutic approaches offering more precise, effective, and minimally invasive options. Advances in nanotechnology demonstrate considerable promise in this regard, introducing new avenues for targeted drug administration, enhanced diagnostic imaging, tissue regeneration, and regenerative medicine. Nanotechnology can complement current therapies by improving outcomes, thereby enabling a more personalized and effective management of CVDs [17]. Nanotechnology presents a promising frontier in addressing cardiovascular diseases (CVDs), particularly myocardial infarction (MI) and atherosclerosis, with its distinctive attributes facilitating precise drug delivery, enhancing treatment outcomes, and minimizing adverse effects [18]. Nanotechnology functions at the sub-molecular scale, encompassing structures within the range of 0.1 to 100 nm. These nanoscale materials showcase distinctive properties. Within the medical realm, nanoparticles (NPs) play a crucial role in enhancing the imaging of early-stage diseases, navigating complex anatomical locations (such as the blood–brain barrier), and delivering bioactive compounds to specific cells (e.g., targeting cancerous cells while preserving healthy tissues). Furthermore, nanomaterials are widely utilized in regenerative medicine for the replacement of damaged tissues. In the context of cardiovascular diseases (CVDs) [19], nanotechnology fulfills four primary functions: diagnosis, imaging, precise drug administration, as well as tissue engineering and regeneration. This adaptability stems from their ability to facilitate the delivery of therapeutic substances to diseased sites [20].

2 Nanotechnology Used for CVDs Management

Nanotechnology lies in its theranostic methodology, which integrates diagnostic imaging with therapeutic interventions to pinpoint and address atheromatous plaques at specific locations, thereby enhancing treatment outcomes [21]. Nanotechnology exhibits its potential by enabling sustained treatment within atherosclerotic regions, thereby improving efficacy and minimizing systemic side effects [22]. Biomimetic nanoparticles, cloaked with membranes from macrophages or platelets, imitate natural cells, contributing to plaque stabilization and targeted drug delivery [23]. Nanotechnology, encompassing nano-liposomes and nanoparticle-based drug delivery systems, supports the restoration of functions and tissues in cardiovascular disorders like myocardial infarction (MI). Additionally, nanotechnology enhances the therapeutic potential of substances such as small-molecule drugs, proteins, nucleic acids, stem cells, and gas-signaling molecules in managing cardiovascular complications, thereby enhancing localized effects and sustained drug release [24]. Injectable hydrogels embedding therapeutic agents and nanocomplexes elevate drug delivery to specific target sites. Nanoparticles surmount biological impediments, refining the targeting of desired tissues or cells, bolstering the effectiveness of therapeutic delivery [25], and mitigating systemic adverse reactions. Scientists have investigated the use of precise strategies that empower interventions at the cellular and molecular levels, facilitating early intervention and treatment of CVDs (Table 1). Nanoparticles have demonstrated promise in non-invasive imaging modalities, monitoring disease progression, and therapeutic applications in clinical scenarios such as atherosclerosis [26]. Furthermore, micro and nanoparticles encapsulating therapeutic agents using biomaterials offer prolonged release, thereby enhancing treatment outcomes [27]. Nanotechnology improves thrombus treatment by enhancing targeting precision, prolonging drug half-life, and reducing the side effects associated with thrombolytic medications [28]. Combined therapies involving nanoparticles are currently under scrutiny, including polymeric nanoparticles engineered to alleviate oxidative stress and diminish muscle impairment in cardiac myocytes. Nanotechnology provides precise carriers for drug delivery to eliminate abnormal cardiac muscle cells, manage arrhythmias, and address post-operative atrial fibrillation [29]. Nanoparticles with sirolimus have proven effective in reducing restenosis rates and inflammatory responses in the arteries of the lower extremities following balloon injury. Stents coated with poly(lactic-co-glycolic acid) (PLGA) encapsulating anti-proliferative agents enhance endothelialization while minimizing the risk of restenosis post-implantation [2].

Table 1 Types of nanomaterials used in the management of CVDs Management

Biomarkers	Nanomaterials	Recognition elements	Methods	Linear range	References
Cardiac troponins(cTns)	AuNPs Fe3O4/PDDA/ Au@Pt	Aptamer	EC	0.05–100 ng/ mL	[30]
Cardiac troponins(cTns)	QDs	MIP	EIS	0.01–5.00 ng/ mL	[4]
Cardiac troponins (cTns)	AuNP	Antibody	ECL	0.1 fg/mL ~ 100 pg/mL	[31]
Myoglobin	AuNPs	Aptamer	EC	0.1–100 μg/ mL	[3]
Myoglobin	UCNPs Pt-staining AuNPs	Antibody	colorimetry	5.74–150 ng/ mL	[32]
Multitargets	TiO2 nanofibrous	Antibody (cTnI, CRP)	ELISA	10 pg/mL ~ 100 ng/mL 1 pg/mL ~ 100 ng/mL	[33]
miRNAs	Hollow Ag	complementary strands	CHA	1 fM ~ 10 nM	[34]

3 Mechanisms of Nanotechnology in Cardiovascular Disease

Nanotechnology, a multidisciplinary field involving the manipulation of matter at the atomic and molecular scale, has revolutionized various aspects of medicine, particularly in the management of cardiovascular diseases (CVDs). The mechanisms by which nanotechnology impacts cardiovascular disease are multifaceted and can be broadly categorized into diagnostics, therapeutics, and regenerative medicine [4]. These advancements are pivotal in addressing the growing global burden of CVDs, which remain the leading cause of morbidity and mortality worldwide. Nanoparticles are engineered to target specific biomarkers associated with cardiovascular conditions [11]. For instance, gold nanoparticles conjugated with biomarkers such as troponin, which is indicative of myocardial infarction, can be detected with high sensitivity using techniques such as surface-enhanced Raman spectroscopy (SERS) or fluorescence imaging. This enables early detection of cardiac events even at low biomarker concentrations [35]. Magnetic nanoparticles (MNPs) have also shown promise in cardiovascular diagnostics. When functionalized with targeting ligands, MNPs can home in on atherosclerotic plaques or areas of vascular inflammation. Once localized, these particles can be visualized using magnetic resonance imaging (MRI), providing detailed images of plaque morphology and composition. This capability is crucial for assessing the risk of plaque rupture, which can lead to acute coronary syndromes [36]. Furthermore, quantum dots (QDs), known for their unique optical properties, are being utilized for the multiplexed detection of several cardiovascular biomarkers simultaneously. The ability to detect multiple markers in a single assay enhances diagnostic accuracy and provides a comprehensive assessment of cardiovascular health status. Nanotechnology-based therapeutic strategies for cardiovascular diseases encompass drug delivery systems, gene therapy, and novel treatment modalities such as photothermal therapy and magnetic hyperthermia [37]. Lipid-based, metallic, and polymeric nanoparticles are commonly used as drug-delivery vehicles. These nanoparticles can encapsulate therapeutic agents, protecting them from degradation and enhancing their bioavailability. By using these nanoparticles, drugs can be selectively delivered to diseased tissues, such as atherosclerotic plaques or ischemic myocardium. For instance, peptides that bind to integrins in inflamed or injured vascular tissues can selectively deliver anti-inflammatory or anti-atherosclerotic drugs to these sites [38]. In gene therapy, nanoparticles offer a promising platform to modulate gene expression in cardiovascular diseases [39]. Nanoparticles can protect genetic materials from nuclease degradation and facilitate their uptake by target cells. This approach has been investigated for silencing pro-inflammatory genes in atherosclerosis or for delivering genes that promote angiogenesis in ischemic heart disease [8]. Nanotechnology also enables innovative therapeutic techniques such as photothermal therapy (PTT) and magnetic hyperthermia. In PTT, nanoparticles, particularly gold nanorods or nanoshells, are administered systemically and accumulate in the target tissue. Upon exposure to near-infrared light, these nanoparticles generate localized heat, effectively ablating the diseased tissue

[40]. This technique can be used to selectively target and destroy pathological tissues within the cardiovascular system. The regenerative potential of nanotechnology in cardiovascular disease lies in its ability to facilitate tissue engineering and repair. Nanomaterials can create scaffolds that promote cellular proliferation. These scaffolds can be seeded with stem cells or other progenitor cells to regenerate damaged myocardial or vascular tissues. Nanofibers, created using techniques such as electrospinning, have shown great promise in cardiac tissue engineering. These fibers can be functionalized with bioactive molecules, such as those promoting cellular adhesion, migration, and differentiation [41]. When implanted into the heart, these nanofiber scaffolds can promote the regeneration of myocardial tissue following a heart attack. Nanoparticles can also deliver regenerative factors. For instance, nanoparticles loaded with vascular endothelial growth factor (VEGF) can be injected into ischemic tissues to promote angiogenesis and restore blood flow. Similarly, nanoparticles carrying anti-apoptotic agents can prevent cell death in the peri-infarct region, preserving cardiac function [42]. Nanotechnology is also advancing personalized medicine and precision therapeutics in cardiovascular disease management. By enabling the precise characterization of individual patients' disease profiles through advanced diagnostic tools, treatments can be tailored to each patient. This approach not only improves therapeutic outcomes but also minimizes adverse effects. Nanoparticles can be engineered to deliver personalized drug regimens based on the genetic and molecular makeup of the patient's disease. For example, theranostic nanoparticles, which combine therapeutic and diagnostic functions, can provide real-time feedback on the delivery and efficacy of the treatment, allowing for dynamic adjustments in the therapeutic approach. This personalized strategy ensures that patients receive the most effective treatments while avoiding unnecessary side effects [43].

3.1 Types of Nanomaterials

Nanomaterials, defined as materials with structural components smaller than 100 nm, have gained substantial attention due to their enhanced catalytic characteristics. The broad range of applications for nanomaterials spans various fields, including medicine, electronics, energy, environmental protection, and agriculture, demonstrating their transformative potential [26, 44]. Nanomaterials are categorized based on their dimensionality. Zero-dimensional (0D) nanomaterials, such as quantum dots, possess all three dimensions at the nanoscale, while three-dimensional (3D) nanomaterials, like nanocomposites, have all three dimensions outside the nanoscale but contain nanoscale structures within them.

Nanomaterials hold promising potential in medicine. These nanoparticles enhance bioavailability, allowing for increased efficacy and reduced side effects of drugs. For instance, liposomal formulations, such as Doxil (liposomal doxorubicin), have shown improved therapeutic indices and reduced toxicity compared to conventional chemotherapy [11].

Nanomaterials also play a role in air purification. Nanofiber-based filters can efficiently capture particulate matter, improving indoor and outdoor air quality [45]. Additionally, nanomaterials are used in soil remediation to remove or neutralize contaminants. Zerovalent iron nanoparticles, for example, can reduce chlorinated organic compounds and heavy metals in contaminated soils, offering a more efficient and cost-effective approach to soil remediation compared to traditional methods [4].

Future research in nanomaterials is likely to focus on developing multifunctional and hybrid nanomaterials that combine the desirable properties of different materials. This holds promise for the development of smart systems that can adapt and respond to changing conditions in real time. For example, smart sensors incorporating nanomaterials could be used for continuous monitoring of environmental pollutants, providing valuable data for environmental protection efforts.

Nanomaterials are also extensively used in medical diagnostics for bioimaging and biosensing applications. They can be conjugated with biomolecules to target specific biomarkers, providing high sensitivity and specificity in the detection of diseases [11]. Gold nanoparticles are used in various diagnostic assays, including lateral flow immunoassays, to detect pathogens and biomarkers with high precision.

Nanotechnology enables simultaneous disease treatment and monitoring. For example, magnetic nanoparticles can be used for both magnetic resonance imaging (MRI) and hyperthermia therapy, where localized heating induced by an alternating magnetic field is used to ablate cancerous tissues. The unique properties of various nanomaterials make them ideal candidates for a range of applications in cardiovascular medicine [46].

Liposomes, made of phospholipid bilayers, can encapsulate both hydrophilic and hydrophobic drugs. Their biocompatibility makes them highly suitable for cardiovascular applications. Liposomes can be engineered for targeted delivery to specific tissues or cells, such as atherosclerotic plaques or ischemic myocardium [47]. By functionalizing their surface with ligands or small molecules, these carriers can selectively bind to disease sites, enhancing therapeutic efficacy while minimizing side effects. For instance, folic acid-coated liposomes can target folate receptor-overexpressing cells in atherosclerotic plaques, delivering drugs directly to the diseased area.

Liposomes can provide controlled and sustained drug release, which is particularly beneficial for chronic cardiovascular conditions requiring prolonged therapy. The lipid composition and size of liposomes can be manipulated to achieve the desired release profile [48]. Liposomes offer the capability to co-encapsulate multiple drugs, allowing for combination therapy, which is advantageous in treating complex cardiovascular diseases involving multiple pathological pathways. For example, liposomes can carry both anti-inflammatory and lipid-lowering agents, addressing inflammation and hyperlipidemia simultaneously in atherosclerosis treatment. Encapsulation of drugs in liposomes can protect non-target tissues from the toxic effects of potent drugs. This property is particularly useful in delivering chemotherapeutic agents or other drugs with significant side effects, improving the overall safety profile of the treatment [49].

Polymeric nanoparticles, crafted from materials such as PLGA (polylactic-co-glycolic acid), offer versatile platforms for drug delivery in cardiovascular diseases [50]. These nanoparticles provide a controlled release and the ability to functionalize their surface for targeted delivery. Their design allows for a steady therapeutic effect. For instance, PLGA nanoparticles have been utilized to deliver statins, providing prolonged lipid-lowering effects crucial for managing hyperlipidemia [51].

Polymeric nanoparticles play a pivotal role in gene therapy by delivering DNA, RNA, or CRISPR/Cas9 components to target cells. This approach shows promise in treating cardiovascular diseases by silencing pro-inflammatory genes or promoting angiogenesis in ischemic tissues. For example, nanoparticles delivering siRNA targeting pro-inflammatory cytokines can significantly reduce inflammation in atherosclerotic plaques. These nanoparticles are also employed in anti-restenosis therapy to prevent the re-narrowing of blood vessels after angioplasty. Drug-eluting stents coated with polymeric nanoparticles containing antiproliferative agents like paclitaxel or sirolimus inhibit smooth muscle cell proliferation, effectively reducing restenosis rates. Functionalizing polymeric nanoparticles with ligands such as antibodies or peptides can enhance their targeting capabilities, ensuring therapeutic agents reach specific sites of disease, such as inflamed vascular tissues, maximizing their therapeutic efficacy. Metallic nanoparticles, including those made from gold, silver, and iron oxide, possess unique optical, electronic, and magnetic properties that make them suitable for various biomedical applications in cardiovascular diseases [52]. Gold nanoparticles are extensively used in imaging techniques like computed tomography (CT) and photoacoustic imaging to visualize atherosclerotic plaques and monitor disease progression. Their high atomic number and strong absorption properties make them excellent contrast agents, providing high-resolution images of cardiovascular tissues. Gold nanoparticles can be utilized for photothermal therapy, where they generate localized heat upon exposure to near-infrared light, effectively ablating atherosclerotic plaques or neoplastic tissues. Magnetic nanoparticles, such as iron oxide, are employed in magnetic resonance imaging (MRI) to detect and characterize cardiovascular lesions, aiding in accurate diagnosis. Silver nanoparticles, known for their antimicrobial properties, can be used to coat cardiovascular implants such as stents and catheters to prevent infections. These antimicrobial coatings reduce the risk of post-operative infections, improving the overall success rate of cardiovascular interventions. Metallic nanoparticles can also serve as drug delivery vehicles, delivering drugs directly to diseased tissues and enhancing therapeutic efficacy while minimizing toxicity [53, 54]. Dendrimers are highly branched, tree-like polymers with a well-defined, monodisperse structure, making them excellent carriers for drug delivery and gene therapy in cardiovascular diseases. Dendrimers can encapsulate therapeutic agents within their core or attach drugs to their surface functional groups [55]. This versatility allows for the delivery of a wide range of drugs, including hydrophobic and hydrophilic molecules. Dendrimers have been used to deliver anti-inflammatory drugs, reducing vascular inflammation in atherosclerosis. Dendrimers are highly efficient in delivering nucleic acids, such as siRNA or plasmid DNA, to target cells, facilitating gene therapy for cardiovascular diseases. Their high transfection efficiency and low cytotoxicity make them ideal vectors for genetic interventions aimed at silencing pathogenic genes or promoting therapeutic gene expression [56]. Functionalizing dendrimers with targeting ligands can enhance their selectivity for diseased tissues. For example, dendrimers conjugated with peptides targeting endothelial cells in inflamed vasculature can deliver anti-atherosclerotic drugs directly to the site of disease, improving therapeutic outcomes and minimizing off-target effects. The structure of dendrimers allows for the controlled release of encapsulated drugs. By modifying the generation number and surface functional groups, the release profile can be tailored to suit specific therapeutic needs, ensuring sustained and controlled drug delivery [54].

Quantum dots (QDs) are semiconductor nanocrystals known for their unique optical and electronic properties, including size-tunable fluorescence emission and high photostability. These properties make them valuable tools for imaging and diagnostic applications in cardiovascular diseases. Quantum dots are used for high-resolution imaging of cardiovascular tissues and cells. Their size-tunable fluorescence allows for multiplexed imaging, enabling the simultaneous detection of multiple biomarkers [52]. This capability is particularly useful in studying the complex molecular mechanisms underlying cardiovascular diseases. Quantum dots can be conjugated with biomolecules to create sensitive biosensors for detecting cardiovascular biomarkers, such as troponins, natriuretic peptides, and inflammatory cytokines. These biosensors provide rapid and accurate diagnosis of cardiovascular conditions, facilitating early intervention and treatment [57]. Quantum dots can be integrated into theranostic platforms that combine diagnostic imaging and therapy. For example, QD-based nanoparticles can be used for fluorescence imaging of atherosclerotic plaques, followed by photothermal therapy to ablate the plaques, offering a comprehensive approach to cardiovascular disease management. Functionalized quantum dots, combining imaging and treatment in a single platform, enhance the precision and effectiveness of cardiovascular therapies [58].

4 Nanotechnology-Based Drug Delivery

The development of systems for targeted drug delivery (TDDS) directly to the disease site enhances treatment efficacy. These systems can target specific cell types or tissues. One of the primary mechanisms for targeting is the use of ligands on the surface of nanocarriers that bind to receptors overexpressed on diseased cells, such as cancer cells or atherosclerotic plaques in cardiovascular diseases. This ligand-receptor interaction facilitates the selective uptake of the drug-loaded nanocarriers by the target cells, ensuring that the drugs are present in the diseased environment, further enhancing their specificity and efficacy. Controlled release mechanisms are crucial for maintaining therapeutic drug levels over an extended period. This approach ensures that drugs are released at a sustained rate. Nanocarriers can be engineered to release their payload through various mechanisms, including diffusion, degradation, swelling, and external triggers.

For example, polymeric nanoparticles can be formulated to release their contents slowly in the body, providing a steady release of the drug. They can also be engineered to release their contents in response to changes in the local pH or temperature. Furthermore, controlled release can be triggered externally, ensuring that drugs are released at the desired site and time. These controlled-release strategies are particularly beneficial in treating chronic conditions such as cardiovascular diseases, where maintaining consistent drug levels is critical for effective management. This necessitates the development of more effective therapeutic strategies. Nanotechnology offers promising solutions for the treatment of heart diseases through targeted and controlled drug delivery.

One case study involves systems to treat atherosclerosis. Researchers have developed nanoparticles with targeting ligands that bind to inflamed endothelial cells and atherosclerotic plaques. These nanoparticles can deliver anti-inflammatory drugs, statins, or gene therapy agents directly to the plaques, reducing inflammation and preventing plaque rupture. Another example is the use of liposomal formulations to deliver drugs for myocardial infarction. Liposomes can be loaded with cardioprotective agents and targeted to the heart tissue following an ischemic event. By delivering drugs specifically to the affected area, liposomes can reduce infarct size and improve cardiac function. Moreover, nanotechnology has facilitated the development of advanced drug-delivery systems for heart failure. Polymeric nanoparticles can encapsulate drugs such as beta-blockers or angiotensin-converting enzyme inhibitors, ensuring sustained therapeutic levels and improving patient outcomes. These case studies underscore the potential of nanotechnology to transform the treatment of cardiovascular diseases. By enabling targeted and controlled drug delivery, nanotechnology-based systems can enhance the efficacy of existing therapies, reduce side effects, and ultimately improve patient quality of life.

The early detection of coronary artery diseases (CADs) enhances the probability of successful therapy and potential cure, thus leading to improved prognoses and extended survival durations for patients. Biomarkers associated with CADs, such as C-reactive protein (CRP) and various miRNAs, are discharged into the bloodstream following heart damage or stress. Consequently, a promising strategy for the timely identification of CADs involves the development of accurate, targeted, uncomplicated, stable, and rapid blood analyses for these molecules. While mass spectrometry is commonly utilized for the identification of potential CAD biomarkers, it is hindered by sensitivity and specificity issues due to the low concentrations of biomarkers in human plasma. Consequently, integrating nanotechnologies with biosensors could present a viable solution for the early detection of CADs. Nanotechnologies offer high-affinity binding to specific molecules and reduce non-specific adsorption through surface modification or structural optimization, while biosensors comprise a biological sensing component for target recognition and a transducer for data conversion into electrical signals. Initial detection of protein targets is achieved using recognition molecules like antibodies, aptamers, or molecularly imprinted polymers, followed by quantitative detection through various methods such as electrochemistry (EC), electrochemiluminescence (ECL), fluorescence (FL), colorimetry, surface-enhanced Raman scattering (SERS), and surface plasmon resonance technology (SPR). The utilization of nanomaterials with exceptional optoelectronic properties significantly enhances detection sensitivity by several orders of magnitude. For instance, Singh et al. developed a label-free EC biosensor for TnT detection using advanced nanomaterials.

5 Challenges and Future Directions

Despite the remarkable progress in nanotechnology for cardiovascular disease, several substantial challenges must be overcome for successful clinical translation. Central among these are concerns regarding the biocompatibility and long-term safety of nanomaterials. The interaction of nanoparticles with biological systems can provoke immune responses or cause toxicity, which necessitates comprehensive studies to understand the potential adverse effects of nanoparticle accumulation in the body. Given that nanomaterials often exhibit unique surface chemistries and high surface area-to-volume ratios, their behavior in vivo can differ significantly from that of larger particles or conventional materials, raising concerns about unforeseen biological interactions [59]. Therefore, a thorough investigation into their potential for inducing inflammatory responses or other immune reactions is imperative. Additionally, the large-scale production of uniform, well-characterized nanoparticles remains a formidable challenge. Ensuring that nanoparticles are consistently produced with precise size, shape, and surface properties is crucial for achieving reliable and reproducible clinical outcomes. This challenge is exacerbated by the need for robust quality control measures and standardized protocols for nanoparticle characterization to meet regulatory requirements. The development and implementation of these protocols are essential for gaining regulatory approval and ensuring patient safety, as discrepancies in nanoparticle specifications can lead to variability in performance and safety profiles. Furthermore, the scalability and reproducibility of nanomaterial synthesis and functionalization processes are critical for their successful integration into clinical settings [60]. Advances in manufacturing techniques, such as self-assembly and 3D printing, offer promising solutions for achieving precise control over nanomaterial properties. Self-assembly techniques leverage the inherent properties of nanomaterials to spontaneously form organized structures, which can enhance the efficiency and uniformity of nanoparticle production. 3D printing, on the other hand, enables the creation of complex nanostructures with high precision, potentially facilitating the development of multifunctional nanomaterials. Future research is likely to focus on the development of such multifunctional nanoparticles capable of simultaneously diagnosing, treating, and monitoring disease progression. By combining diagnostic, therapeutic, and monitoring functions in a single platform, these advanced nanomaterials could provide more comprehensive and personalized approaches to cardiovascular disease management. Integrating nanotechnology with other emerging fields, such as genomics and personalized medicine, represents another crucial direction for future research. The convergence of nanotechnology with genomics could enable the development of nanoparticle-based systems tailored to individual genetic profiles, allowing for more precise and personalized treatment strategies. This integration has the potential to revolutionize cardiovascular disease management by enhancing the specificity and efficacy of treatments while minimizing side effects. As the field progresses, addressing these challenges and leveraging technological innovations will be essential for translating the promising capabilities of nanotechnology into practical, clinically effective solutions for cardiovascular diseases. The ongoing development of advanced nanofabrication techniques, coupled with rigorous safety assessments and regulatory advancements, will pave the way for the successful implementation of nanotechnology-based therapies, ultimately improving patient outcomes and transforming the landscape of cardiovascular medicine [61].

6 Conclusion

Nanotechnology is a powerful toolkit for revolutionizing the management of cardiovascular diseases. Through targeted drug delivery, advanced diagnostic techniques, and regenerative medicine, it can significantly improve cardiovascular outcomes. As research continues to address current challenges, nanotechnology is poised to play a pivotal role in the fight against cardiovascular diseases. Nanomaterials, such as liposomes, metallic nanoparticles, and dendrimers, offer innovative solutions for cardiovascular diseases. Their unique properties and versatile applications hold promise for improving cardiovascular health outcomes and transforming the landscape of cardiovascular medicine. By enabling precise targeting of diseased tissues, reducing systemic side effects, and enhancing therapeutic efficacy, these nanomaterials provide significant advancements over conventional treatments. The future looks promising with the integration of nanotechnology. Continued research and development will be essential to overcoming current challenges, such as ensuring biocompatibility, optimizing delivery systems, and scaling up production for clinical use. These efforts will be crucial for translating technological innovations into practical medical solutions. Ultimately, the widespread adoption of nanotechnology-based approaches could lead to earlier detection of cardiovascular conditions, more effective and personalized treatments, and improved patient outcomes.

References

1. Chandarana M, Curtis A, Hoskins C (2018) The use of nanotechnology in cardiovascular disease. Appl Nanosci 8:1607–1619
2. Ullah A, Ullah M, Lim SI (2024) Recent advancement in nanotechnology based drug delivery for the management of cardiovascular disease. Curr Probl Cardiol 22:102396
3. Hu Q, Fang Z, Ge J, Li H (2022) Nanotechnology for cardiovascular diseases. The Innovation 3(2)

4. Wang DK, Rahimi M, Filgueira CS (2021) Nanotechnology applications for cardiovascular disease treatment: current and future perspectives. Nanomed Nanotechnol Biol Med 34:102387
5. Rhee JW, Wu JC (2013) Advances in nanotechnology for the management of coronary artery disease. Trends Cardiovasc Med 23(2):39–45
6. Karimi M, Zare H, Bakhshian Nik A, Yazdani N, Hamrang M, Mohamed E, Sahandi Zangabad P, Moosavi Basri SM, Bakhtiari L, Hamblin MR (2016) Nanotechnology in diagnosis and treatment of coronary artery disease. Nanomedicine 11(5):513–530
7. Zheng Z, Zhu S, Lv M, Gu Z, Hu H (2022) Harnessing nanotechnology for cardiovascular disease applications-a comprehensive review based on bibliometric analysis. Nano Today 1(44):101453
8. S Behera S, Pramanik K, K Nayak M (2015) Recent advancement in the treatment of cardiovascular diseases: conventional therapy to nanotechnology. Curr Pharm Des 21(30):4479–4497
9. Guo J, Wang H, Li Y, Zhu S, Hu H, Gu Z (2023) Nanotechnology in coronary heart disease. Acta Biomaterialia
10. Li F, Shao H, Zhou G, Wang B, Xu Y, Liang W, Chen L (2023) The recent applications of nanotechnology in the diagnosis and treatment of common cardiovascular diseases. Vascul Pharmacol 25:107200
11. Mohamed NA, Marei I, Crovella S, Abou-Saleh H (2022) Recent developments in nanomaterials-based drug delivery and upgrading treatment of cardiovascular diseases. Int J Mol Sci 23(3):1404
12. Ho KJ (2018) Cardiovascular diseases. In: Nutritional aspects of aging. CRC Press, pp 75–100; Laakso M (1999) Hyperglycemia and cardiovascular disease in type 2 diabetes. Diabetes 48(5):937–942
13. Yun JS, Ko SH (2021) Current trends in epidemiology of cardiovascular disease and cardiovascular risk management in type 2 diabetes. Metabolism 1(123):154838
14. FitzGerald GA (2004) Coxibs and cardiovascular disease. N Engl J Med 351(17):1709–1711
15. Laakso M (2010) Cardiovascular disease in type 2 diabetes from population to man to mechanisms: the Kelly West Award Lecture 2008. Diabetes Care 33(2):442–449
16. Katsiki N, Mikhailidis DP, Banach M (2018) Leptin, cardiovascular diseases and type 2 diabetes mellitus. Acta Pharmacol Sin 39(7):1176–1188
17. Gupta K, Rawlley B, Meloche C, Minhas AM, Hermel M, Slipczuk L, Sheikh S, Khoja A, Vaughan EM, Dalakoti M, Virani SS (2024) Highlights of cardiovascular disease prevention studies presented at the 2024 American college of cardiology conference. Curr Atheroscler Rep 3:1–5
18. Banach M, Surma S, Toth PP (2023) The year in cardiovascular disease–the year of new and prospective lipid lowering therapies: can we render dyslipidemia a rare disease by 2024? Arch Med Sci AMS 19(6):1602
19. Martin SS, Aday AW, Almarzooq ZI, Anderson CA, Arora P, Avery CL, Baker-Smith CM, Barone Gibbs B, Beaton AZ, Boehme AK, Commodore-Mensah Y (2024) Heart disease and stroke statistics: a report of US and global data from the American Heart Association. Circulation 149(8):e347-913
20. Conrad N, Molenberghs G, Verbeke G, Zaccardi F, Lawson C, Friday JM, Su H, Jhund PS, Sattar N, Rahimi K, Cleland JG (2024) Trends in cardiovascular disease incidence among 22 million people in the UK over 20 years: population based study. bmj 385
21. Li M, Yang F (2023) Cardiovascular nanotechnology. In: Nanomedicine, pp 439–468. Springer Nature, Singapore
22. Gupta P, Garcia E, Sarkar A, Kapoor S, Rafiq K, Chand HS, Jayant RD (2019) Nanoparticle based treatment for cardiovascular diseases. Cardiovasc Haematol Disord-Drug Targets (Formerly Current Drug Targets-Cardiovascular & Hematological Disorders) 19(1):33–44
23. Cicha I, Garlichs CD, Alexiou C (2014) Cardiovascular therapy through nanotechnology–how far are we still from bedside? Eur J Nanomed 6(2):63–87
24. Lankala CR, Yasir M, Ishak A, Mekhail M, Kalyankar P, Gupta K (2023) Application of nanotechnology for diagnosis and drug delivery in atherosclerosis: a new horizon of treatment. Curr Probl Cardiol 48(6):101671

25. Chauvierre C, Letourneur D (2015) The European project nanoathero to fight cardiovascular diseases using nanotechnologies. Nanomedicine 10(22):3391–3400
26. Syahputra RA, Dalimunthe A, Utari ZD, Halim P, Sukarno MA, Zainalabidin S, Salim E, Gunawan M, Nurkolis F, Park MN, Luckanagul JA (2024) Nanotechnology and flavonoids: current research and future perspectives on cardiovascular health. J Funct Foods 1(120):106355
27. Vaidya B, Gupta V (2015) Editorial (thematic issue: novel therapeutic strategies for cardiovascular disease treatment: from molecular level to nanotechnology). Curr Pharm Des 21(30):4367–4369
28. Pala R, Anju VT, Dyavaiah M, Busi S, Nauli SM (2020) Nanoparticle-mediated drug delivery for the treatment of cardiovascular diseases. Int J Nanomed 27:3741–3769
29. Almas T, Haider R, Malik J, Mehmood A, Alvi A, Naz H, Satti DI, Zaidi SM, AlSubai AK, AlNajdi S, Alsufyani R (2022) Nanotechnology in interventional cardiology: a state-of-the-art review. IJC Heart Vasc 1(43):101149
30. Tralhão A (2024) Using high-sensitivity troponin for enhancing risk prediction after non-cardiac surgery-are we there yet? Arq Bras Cardiol 17(121):e20240140
31. Li Z, Zhong A (2024) Electrochemical detection strategies for cardiovascular disease biomarkers: applications in troponin I and troponin T assays. Int J Electrochem Sci 29:100675
32. Zhong S, Chen L, Shi X, Chen G, Sun D, Zhang L (2023) Recent advances in electrochemical aptasensors for detecting cardiac biomarkers: a review. Microchem J 8:109063
33. Dong H, Cao L, Tan Z, Liu Q, Zhou J, Zhao P, Wang P, Li Y, Ma W, Dong Y (2019) A signal amplification strategy of CuPtRh CNB-embedded ammoniated Ti3C2 MXene for detecting cardiac troponin I by a sandwich-type electrochemical immunosensor. ACS Appl Bio Mater 3(1):377–384
34. Campu A, Muresan I, Craciun AM, Cainap S, Astilean S, Focsan M (2022) Cardiac troponin biosensor designs: current developments and remaining challenges. Int J Mol Sci 23(14):7728
35. Prajnamitra RP, Chen HC, Lin CJ, Chen LL, Hsieh PC (2019) Nanotechnology approaches in tackling cardiovascular diseases. Molecules 24(10):2017
36. Khan J, Yadav S, Alam MA (2024) Recent developments in nanomaterial drug delivery and upgrading treatment of cardiovascular disease. Curr Pharm Des
37. Abid J, Khalil FM, Saeed S, Khan SU, Iqbal I, Anthony S (2024) Nano revolution in cardiovascular health: nanoparticles (NPs) as tiny titans for diagnosis and therapeutics. Curr Probl Cardiol 16:102466
38. Li T, Liang W, Xiao X, Qian Y (2018) Nanotechnology, an alternative with promising prospects and advantages for the treatment of cardiovascular diseases. Int J Nanomed 9:7349–7362
39. Mossman BT, Borm PJ, Castranova V, Costa DL, Donaldson K, Kleeberger SR (2007) Mechanisms of action of inhaled fibers, particles and nanoparticles in lung and cardiovascular diseases. Part Fibre Toxicol 4:1
40. Binsalamah ZM, Paul A, Prakash S, Shum-Tim D (2012) Nanomedicine in cardiovascular therapy: recent advancements. Expert Rev Cardiovasc Ther 10(6):805–815
41. Saxer T, Holme MN (2016) Challenges in cardiovascular treatments using nanotechnology-based approaches. Nanosci Nanotechnol Hum Health 28:51–70
42. De Mel A, Bolvin C, Edirisinghe M, Hamilton G, Seifalian AM (2008) Development of cardiovascular bypass grafts: endothelialization and applications of nanotechnology. Expert Rev Cardiovasc Ther 6(9):1259–1277
43. Ruiz-Esparza GU, Flores-Arredondo JH, Segura-Ibarra V, Torre-Amione G, Ferrari M, Blanco E, Serda RE (2013) The physiology of cardiovascular disease and innovative liposomal platforms for therapy. Int J Nanomed 9:629–640
44. Tesoro L, Hernandez I, Saura M, Badimón L, Zaragoza C (2024) Novel cutting edge nano-strategies to address old long-standing complications in cardiovascular diseases: a comprehensive review. Eur J Clin Invest 54(7):e14208
45. Spyropoulos-Antonakakis N, Sarantopoulou E, Trohopoulos PN, Stefi AL, Kollia Z, Gavriil VE, Bourkoula A, Petrou PS, Kakabakos S, Semashko VV, Nizamutdinov AS (2015) Selective aggregation of PAMAM dendrimer nanocarriers and PAMAM/ZnPc nanodrugs on human atheromatous carotid tissues: a photodynamic therapy for atherosclerosis. Nanoscale Res Lett 10:1–9

46. Eniola-Adefeso O, Heslinga MJ, Porter TM (2012) Design of nano vectors for therapy and imaging of cardiovascular diseases. Methodist Debakey Cardiovasc J 8(1):13
47. Nazeer N, Gurjar V, Ratre P, Dewangan R, Zaidi K, Tiwari R, Soni N, Bhargava A, Mishra PK (2024) Cardiovascular disease risk assessment through sensing the circulating microbiome with perovskite quantum dots leveraging deep learning models for bacterial species selection. Microchim Acta 191(5):1–1
48. Arumugasamy SK, Chellasamy G, Yun K, Hyun J (2024) Bio-quantum dots for electrochemical sensing of cardiac biomarkers of acute myocardial infarction. J Ind Eng Chem 25(129):488–498
49. Sabir F, Barani M, Mukhtar M, Rahdar A, Cucchiarini M, Zafar MN, Behl T, Bungau S (2021) Nanodiagnosis and nanotreatment of cardiovascular diseases: an overview. Chemosensors 9(4):67
50. Scafa Udrişte A, Burduşel AC, Niculescu AG, Rădulescu M, Grumezescu AM (2024) Metal-based nanoparticles for cardiovascular diseases. Int J Mol Sci 25(2):1001
51. Bose RJ, Ha K, McCarthy JR (2021) Bio-inspired nanomaterials as novel options for the treatment of cardiovascular disease. Drug Discov Today 26(5):1200–1211
52. Shand H, Dutta S, Patra S, Jain H, Mondal R, Mandal AK, Ghorai S (2024) Nanoparticle-based intervention to cardiovascular diseases (CVDS). Appl Nanosci 14(1):57–67
53. Iravani S, Varma RS (2022) Advanced drug delivery micro-and nanosystems for cardiovascular diseases. Molecules 27(18):5843
54. Udrişte AS, Burduşel AC, Niculescu AG, Rădulescu M, Balaure PC, Grumezescu AM (2024) Organic nanoparticles in progressing cardiovascular disease treatment and diagnosis. Polymers 16(10):1421
55. Nenna A, Nappi F, Larobina D, Verghi E, Chello M, Ambrosio L (2021) Polymers and nanoparticles for statin delivery: current use and future perspectives in cardiovascular disease. Polymers 13(5):711
56. Shi W, Fuad AR, Li Y, Wang Y, Huang J, Du R, Wang G, Wang Y, Yin T (2023) Biodegradable polymeric nanoparticles increase risk of cardiovascular diseases by inducing endothelium dysfunction and inflammation. J Nanobiotechnol 21(1):65
57. Salmanpour Z, Abolmaali SS, Farahavr G, Salmanpour M, Tamaddon AM (2024) Magnetic nanoparticles and their hybrid biomaterials for drug delivery and theranostic applications in cardiovascular diseases. J Drug Deliv Sci Technol 16:105676
58. Vazquez-Prada KX, Lam J, Kamato D, Xu ZP, Little PJ, Ta HT (2021) Targeted molecular imaging of cardiovascular diseases by iron oxide nanoparticles. Arterioscler Thromb Vasc Biol 41(2):601–613
59. Lv Y, Li W, Liao W, Jiang H, Liu Y, Cao J, Lu W, Feng Y (2024) Nano-drug delivery systems based on natural products. Int J Nanomed 31:541–569
60. Ebrahimi Hosseini SS, Hoseinnzhad LE (2024) Advancements and potential future applications of medication delivery systems using nanotechnology. Adv Ther J 6(18):26–33
61. Pereira GC (2024) A novel degradable PEG superparamagnetic iron oxide capsule coupled with a polyphenolic nano-enzymatic conjugate (PSPM-NE), to treat ROS-driven cardiovascular-diseases, tested in atherosclerosis as a model disease, and hypothesizing autoimmunity as an atheroma's trigger. Front Cardiovasc Med 2(11):1125571
62. Levchenko TS, Hartner WC, Torchilin VP (2012) Liposomes for cardiovascular targeting. Ther Deliv 3(4):501–514

Nanomaterials Application for Diagnosis and Treatment of Common Inflammatory Disorders

Kumar Utkarsh, Sachin Kumar, Ahmad Reza Khan, Mohammad Zubair Alam, and Azhar Khan

Abstract Nanomaterials are gaining popularity in the treatment of common inflammatory disorders because of their distinct physical and chemical properties. Different inflammatory disorders and conditions may be diagnosed and treated with the help of a wide range of nanomaterials and tools made possible by this technology (carbon-based, metal-based, dendrimers, and nanocomposites). Nanoparticles have been the subject of extensive in-vivo and in-vitro research to demonstrate their effectiveness in treating inflammation and its related diseases. Boost in the release and production of proinflammatory cytokines (IL-8, IL-6, and IL-10) and deterioration of bone and articular cartilage are characteristics of inflammatory disorders which are linked with the beginnings of cardiovascular diseases like stroke and heart attack. Despite numerous imaging devices for diagnosis and monitoring and ongoing studies to increase their accuracy and sensitivity, proper assessment of inflammatory disorders, especially in their early stages, remains difficult. Nanoparticles made a significant impact because they could be customized to overcome metabolic barriers and directed to their required target without damaging surrounding cells and tissue. These can be used to reduce the severity of drugs, systemic side effects, and toxicity by targeting specific tissues. Thus, the aim of this chapter is to give a quick update on nanomaterials as a possible therapy method for inflammatory disorders. Also briefly explored are the potential uses of nanomaterials in the treatment of inflammatory disorders.

Keywords Nanomaterials · Inflammatory diseases · Cytokines · Biomarkers

---

K. Utkarsh · S. Kumar · A. R. Khan · A. Khan (✉)
Faculty of Biotechnology, Shoolini University of Biotechnology and Management Sciences, Solan, Himachal Pradesh, India
e-mail: mk.azhar1@gmail.com

M. Z. Alam
Pre-Clinical Research Unit, King Fahd Medical Research Center, King Abdulaziz University, Jeddah, Saudi Arabia

Md. M. Ansari et al. (eds.), *Emergence of Sustainable Biomaterials in Tackling Inflammatory Diseases*, Smart Nanomaterials Technology,
https://doi.org/10.1007/978-981-96-2112-5_9

1 Introduction

The past several decades have seen a significant increase in interest in a new, highly potent technology known as nanotechnology [1]. Nanotechnology is a multidisciplinary idea that brings together physics, chemistry, biology, electronics, biomedical research, and materials science [2, 3]. Traditionally, it is based on the development, characterization, production and uses of materials at the nanoscale (1–100 nm) [4]. Nanomaterials have revolutionized various fields, including medicine, due to their unique properties at the nanoscale. They have the potential to transform the diagnosis and treatment of inflammatory disorders by enabling targeted medication administration, sensitive detection of inflammatory biomarkers, and non-invasive imaging of inflamed tissues. The advent of nanotechnology provides a new avenue for addressing the challenges associated with the diagnosis and treatment of inflammatory disorders. The immune system's response to potentially harmful stimuli, such as infections, damaged cells, toxic substances, or radiation, is inflammation [5]. It simultaneously eliminates damaging impulses from entering the body and initiates the healing process [6]. As a result, inflammation is a protective mechanism that is vital for health [7]. Usually, during acute inflammatory reactions, cellular and molecular activity and interactions efficiently lower the risk of injury or infection. This mitigating procedure assists in the recovery of tissue homeostasis and the reduction of acute inflammation. Acute inflammation that is not treated, however, can turn chronic and contribute to a number of chronic inflammatory disorders [8]. These inflammatory disorders are categorized as a wide range of illnesses marked by atypical immune responses that cause chronic inflammation, tissue damage, and impaired function. These inflammatory disorders, such as inflammatory bowel disease, rheumatoid arthritis, and asthma, pose substantial diagnostic and therapy hurdles. This chapter investigates the potentially beneficial applications of nanomaterials in the diagnosis and treatment of inflammatory diseases. It dives into the advantages and promise of nanotechnology-based techniques such as tailored drug delivery systems, therapeutic medication release control, and sensitive detection of inflammatory biomarkers. It also emphasizes the significance of nanomaterials in non-invasive imaging for early diagnosis and disease progression monitoring. However, it is crucial to address the challenges associated with nanomaterials, including their safety profile, regulatory considerations, and ethical implications. While nanotechnology shows great promise, rigorous research and translational efforts are necessary to ensure the responsible and effective implementation of these nanomaterial-based strategies in clinical practice.

2 Nanomaterials for Diagnosis of Common Inflammatory Disorders

2.1 Imaging Techniques

Several imaging modalities help diagnose and evaluate inflammatory disorders, but getting an accurate read on the disease can be difficult, especially in its early stages [9]. This has led to a number of ongoing studies aimed at increasing the efficacy and precision of imaging methods for the early detection of inflammatory disorders [10]. The biomedical field is increasingly turning to nanoscale material engineering to achieve this end [11]. The following paragraphs will discuss the potential biomedical applications of imaging inflammatory arthritis using SPIO (Superparamagnetic iron oxide nanoparticle), polymeric NPs, gold, and multimodal nanomaterials (cerium and silica NPs).

In preclinical animal models as well as in human clinical trials, SPIONs have been evaluated as contrast agents in numerous disorders, including, muscular dystrophy, malformation, cancer, inflammation, coronary artery disease, transplant failure, and its associated condition [12]. Chih-Lung et al. [13] created tailored SPIONs for in-vivo MRI of T-cells in rheumatoid arthritis by coupling a monoclonal anti-CD3 antibody to carboxylated polyethylene glycol (PEG)-SPION (IOPC-CD3). Serial MRI examination of collagen-induced arthritis (CIA) model of RA in rats showed a selective decrease in the S/N ratio of IOPC-CD33-infused femoral growth plates, and immunohistochemical results showed that the target site would be the accumulation of SPIONs and T-cells.

For many years, gold nanostructures have been employed in medicine, most notably the injectable auranofin for inflammatory disease [13]. Due to its exceptional tunable surface plasmon resonance and light-scattering properties, gold nanostructures have been extensively tested in an extensive number of imaging techniques, including magnetic resonance imaging (MRI), computed tomography (CT), Raman spectroscopy (RS), fluorescence imaging (FI), and photoacoustic imaging (PAI) [14]. Because of its easy-to-obtain nature and high imaging quality, gold nanomaterials are a great substrate for inflammatory disease imaging [15]. Antibodies against tumor necrosis factor alpha (TNF-α) were used to manufacture larger gold nanorods with strong absorbance at 1064 nm by Marc et al. [16]. They demonstrated that injecting infliximab-modified and PEGylated control particles into arthritic and healthy control mice led to a longitudinal improvement of optoacoustic signal magnitudes using a fast-scanning optoacoustic imaging device based on a pulsed Nd:YAG laser and a single-centered US transducer.

NPs can also be applied in multimodal imaging modalities, that combine the advantages of many imaging processes. For instance, silica-based NPs are desirable as a multimodal imaging agent due to their capacity for multivalent binding, photostability, biocompatibility, and biodistribution [17]. Nanoceria, which are NPs based on cerium oxide, is another type of biocompatible NPs used for multimodal imaging of inflammatory disorders.

2.2 Nanomaterials for Detection of Biomarkers

Many inflammatory disorders and diseases, especially cancer, lack obvious symptoms and are notoriously hard to identify and diagnose until their advanced stages, at which point treatment options and chances of survival have significantly diminished [18]. Methods currently used in diagnostics range from noninvasive diagnosis (such as X-rays, magnetic resonance imaging, and computed tomography) to more invasive procedures like direct tissue biopsy [19]. The efficacy of such methods for early diagnosis remains low, despite recent and ongoing technological advancements.

Most diseases have numerous phases of development, and during each step, normal cells and tissues undergo changes that are connected with the creation and release of specific biomolecules of varying natures; they are now commonly referred to as biomarkers [20, 21]. In order to gain insight and aid in diagnostic and therapeutic decision-making, such biomolecular patterns can be profiled at varying biological states. Once limited to proteins, nucleic acids, and sugars, the term "biomarker" has since been expanded to include any quantifiable structure, such as vesicles (exosomes), which are known to transport a variety of possible biomarkers for a variety of disorders. Prostate-specific antigen (PSA) is a biomarker currently in use for the diagnosis of prostate cancer, while human epidermal growth factor receptor-2 (HER2) is a biomarker currently in use for the diagnosis of breast cancer, and also the target of therapeutic interventions [22]. Biomarker development and tracking are thus of paramount crucial for not just diagnostics but also for identifying possible therapeutic targets. Although biomarkers can be detected in a variety of bodily fluids (urine, feces, and spiral fluid), identification in a blood sample with a rapid blood test (liquid biopsy), has attracted a large network because of its accessibility and minimal invasiveness [23–25]. One of the fastest-growing areas of study in the biomedical sciences today is the identification of potential causes of human disease because of recent developments in high-throughput "omics" technology [26].

Biomarker analysis from biological material currently relies on a wide range of tests, methodologies, apparatus, and highly trained personnel within specialized laboratories. In some circumstances, however, the availability of such an approach and experience is minimal, requiring a simpler and speedier solution, and this is true despite the fact that some techniques still fall short of the criteria. Biosensors represent a promising new direction in the evolution of low-cost, rapid, and sensitive diagnostic tools. The identification of a specific biomarker in biological samples is made possible by immunosensors, which convert the analyte present of interest into a quantifiable readout in the form of, for example, a chemo- or photo-luminous, colorimetric, or electrochemical signal [27–30]. To improve the accuracy and specificity of immunosensors while also lowering their detection thresholds, researchers have turned to antibodies and antibody-derived structures. Following the development of the complex antigen–antibody, the most popular nanosensors use one of three different signal and transduction methods: optical, in which alternations in the surrounding optical properties of the media are generated (for example, color, luminescence, changes Hi refractive index) [31]; piezoelectric, which uses changes in mass measured by piezoelectric devices [32], and electrochemical, which uses electrical signals (current, voltage differences, resistance) [33, 34]. The majority of immunosensors commonly employ a sandwich assay format, wherein the antigen is seized by a specific capturing antibody and subsequently identified by a secondary antibody, also known as the detection antibody. The detection antibody is typically conjugated with a reporter molecule or agent for signal amplification. Recent developments in nanotechnology have enabled the production of nanoparticles (NPs) with precise physical–chemical properties and customizable surface modifications. This has led to the creation of improved immunosensors that combine the versatility and specificity of antibodies and their by-products with the unique properties of NPs. The present analysis aims to elucidate the pertinent uses of biomarker-based nanoparticles (NPs) and the crucial variables that impact their efficacy. This encompasses diverse techniques for immobilizing antibodies during device fabrication and various formats that leverage the belongings of the nanomaterials, with the main composition serving as the distinguishing factor.

2.3 *Nanosensors for Diagnosis of Inflammation*

Nanosensors are rapidly becoming essential in the diagnosis of inflammatory diseases [35]. The analytical and diagnostic communities have benefited greatly from nanosensors, which integrate nanomaterials with biosensing technologies. Different modalities of nanosensors can be built using a wide range of nanomaterials with specialized properties. These nanosensors are useful for the diagnosis of inflammatory diseases because of their enhanced analytical performances, such as sensitivity, selectivity, robustness, speed, and sample consumption. Here, we introduce several nanosensors that can be used to detect inflammatory disorders.

Chemiluminescence (CL) has been identified as a highly desirable optical technique for biochemical analysis due to its exceptional attributes such as little background noise, enhanced sensitivity, and broad dynamic range. Luminol and acridinium ester are frequently employed chemical reagents in chemiluminescence (CL) reactions. In this study, a novel method for microbial diagnostics utilizing a gold nanoplate-based chemiluminescence assay is presented [36]. For the detection of nucleases and bacteria, a CL biosensor based on mesoporous silica nanoparticles (MSNs) is described. The CL tag molecule Hemin is inserted into MSNs, and the whole thing is sealed off with a DNA gate. The DNA gate is broken down by DNA nuclease. Nuclease activity can be measured by monitoring the luminol and hydrogen peroxide (H_2O_2) CL reaction triggered by the released hemin. Bacterial lysate can also break down the encapsulated DNA, allowing for the quantification of bacteria, because DNA nuclease is present in bacteria. Target organisms include *Escherichia coli* O157:H7 and *Staphylococcus aureus*.

Various nanosensors have been developed for the purpose of diagnosing infectious diseases. A novel sensing platform utilizing localized surface plasmon resonance (LSPR) and gold nanodisks has been developed for the specific detection of *S.*

aureus [37]. The SARS-CoV-2 nucleic acid is detected using a gold nano island-based plasmonic photothermal enhanced LSPR biosensor with a limit of detection (LOD) of 0.22 pM [38]. Biosensors based on silicon microring resonators are created for *S. pneumonia* nucleic acid detection [39]. It has been reported that SARS-CoV-2 can be detected quickly and sensitively using a biosensor based on a field-effect transistor (FET) [40]. The biosensor was fabricated through the immobilization of the SARS-CoV-2 spike antibody onto the graphene sheets of the FET. The biosensor under consideration has the potential to detect the SARS-CoV-2 spike antigen, SARS-CoV-2 in cultured medium, and SARS-CoV-2 in nasopharyngeal swab specimens.SARS-CoV-2 has a LOD of 1.6 101 pfu/mL in cultured medium and 2.42 102 copies/mL in clinical samples. The utilization of thermal contrast amplification (TCA) has been implemented as a novel approach for the readout of lateral flow assays (LFA) based on gold nanoparticles (AuNPs) in order to detect pathogens. This method has demonstrated an eightfold enhancement in the detection threshold when compared to the conventional naked-eye readout technique. For the detection of *S. typhimurium*, a quartz crystal microbalance (QCM) sensor has been developed [41]. Lasers desorption/ionization mass spectrometry (LDI MS) detection and inhibition of *Escherichia coli* use a silver platform with dual functionalities [42]. For the multiplex detection of viral infections, a metal nanoparticle-based LDI MS immunoassay is developed [43].

3 Nanomaterials for the Treatment of Common Inflammatory Disorders

3.1 Nanocarriers for Drug Delivery

Colloidal nanoparticles, commonly referred to as nanocarriers, have received a great deal of interest in the realm of medicine delivery because of their ability to transport therapeutic agents and other substances to specific target sites [44]. The nanocarriers exhibit a diameter ranging from 1 to 100 nm (nm). The nanocarriers exhibit favorable biocompatibility owing to their inertness and general recognition as a safe medium. The nanocarriers are designed to have an extended circulation period and to ensure a continuous liberation of the drug, thereby conquering the endosome-lysosome-based mechanism [45]. Nanocarriers' activity can be improved with reduced side effects through manipulation of their physiochemical features, such as composition, shape, and surface [46]. It thus has broad consequences for the field of medication delivery. Despite the many different nanocarriers that have been developed, only a select number have shown extraordinary drug transport efficiency (Fig. 1) [47].

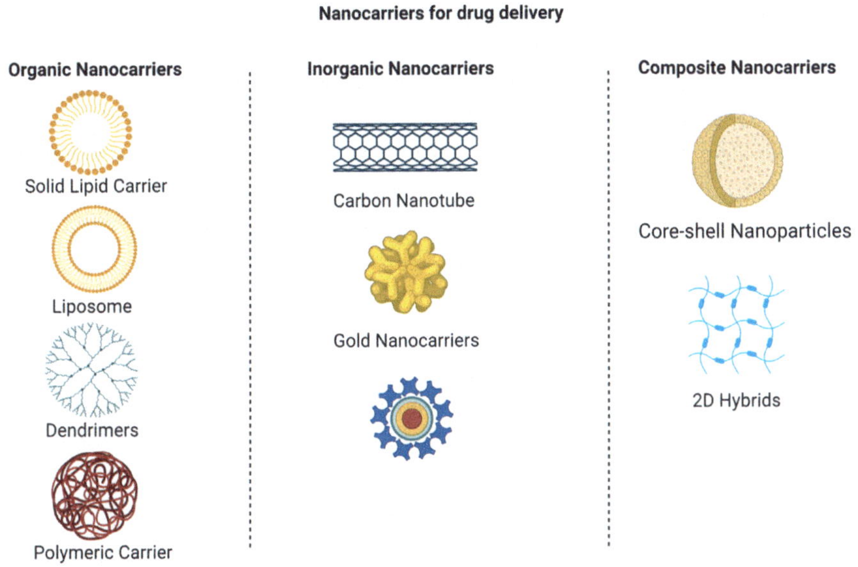

Fig. 1 Various types of nanocarriers for drug delivery

3.1.1 Organic Nanocarriers

Solid–Lipid Nanocarrier

Since the beginning of the 1990s, solid-lipid nanocarriers have been employed as an effective way of transporting lipophilic drugs. Micro-emulsification or high-pressure homogenization are two methods used to create solid lipid nanocarriers from a solution of melted solid lipids and water [48]. This solid lipid nanocarrier's adaptability allows it to circumvent the problems with standard chemotherapy. Integrating ionic as well as hydrophilic drug molecules into the typical solid lipid nanocarrier presents a problem due to its elimination by the Reticule Endothelial System and its inability to provide sustained drug release. A recent development in solid lipid nanocarrier has been employed to integrate not only lipophilic medication but also ionic and hydrophilic anticancer medicines.

Liposome

Liposomes are spherical vesicles composed of lipid bilayers that enclose an aqueous core. They have the ability to deliver drugs that are both lipophilic and hydrophilic to specific target sites. The bilayer has the capability to undergo differentiation into either a uni-lamellar vesicle, consisting of a single bilayer, or a multilamellar vesicle, consisting of more than one bilayer. The aforementioned vesicle functions as a means of conveying biologically active molecules to a particular location. The stealth liposome exhibits remarkable stability and an extended half-life in the bloodstream due to its ability to evade elimination by the Reticuloendothelial System. This results in a controlled release of the drug [49]. The incorporation of drug molecules within liposomes has been shown to enhance the pharmacokinetics and bio-distribution of the drug. The drug molecule doxorubicin, when encapsulated in stealth liposomes, exhibits reduced plasma distribution and lower concentration in healthy cells compared to the free drug in solution [50].

Dendrimers

Dendrimers are a type of macromolecule characterized by a branched structure. They are composed of a central initiator core that gives rise to multiple arms, each of which terminates in an active group [51]. Dendrimers have been observed to be synthesized from various building blocks such as nucleotides, sugar molecules, and amino acids. The molecules in question exhibit multivalency and high branching and possess a wide range of peripheral groups. Additionally, their unique molecular weight renders them a promising candidate for drug delivery applications. These dendrimers have a well-defined composition and structure which can be finely modified to encapsulate different medications, such as rifampicin, which is used to treat tuberculosis [52]. This dendrimer has many potential applications, including MRI scanning, medication administration, vaccine distribution, and gene delivery. It's also helpful for making connections to drugs. To increase its anticancer effects, the dendrimer is extensively coupled with a variety of anticancer medicines [53–55].

Polymeric Nanocarriers

Colloidal, solid nanoparticles that are biodegradable can be produced from various polymeric materials, and are commonly referred to as polymeric nanoparticles [56]. The drug molecules can be encapsulated in either a reservoir type, known as nanocapsules, where they are dissolved or dispersed within the core of a polymer, or a matrix type, known as nanospheres, where the drug molecules are entrapped within the polymer matrix. Synthetic polymers like poly(lactic-co-glycolic acid), poly(ethylene glycol), poly(glutamic acid), and poly(caprolactone) can be used to create the polymeric nanocarriers as well as natural polymers like chitosan, gelatin, albumin, collagen, alginate, etc. [57]. Advantages of this polymeric nanocarrier include increased stability, drug payload, half-life in the systemic circulation, and prolonged drug release compared to competing nanocarriers. Producing multifunctional polymeric nanocarriers that can hold numerous medicines is also feasible [58]. Recent developments in polymeric nanoparticle synthesis have shown promise for targeted drug delivery applications. Specifically, the use of stimuli-sensitive polymers, also known as smart polymers, has demonstrated significant advancements in this field. Smart polymers have the ability to release drugs in response to both internal and external environmental stimuli. These stimuli may include low pH, redox reactions, and enzymatic activity, as well as temperature, light, ultrasound, magnetic fields, and electric fields. Smart polymers present several design challenges that must be addressed for their successful implementation.

3.1.2 Inorganic Nanocarrier

Various types of inorganic nanocarriers, such as gold, magnetic nanocarriers, quantum dots, and mesoporous silica, have been identified as potential candidates for drug delivery systems. Inorganic nanocarriers are utilized due to their favorable properties. Inorganic nanocarriers have demonstrated efficacy in various applications such as biosensing, cell labeling, targeting, imaging, and diagnostics.

Carbon Nanotubes

These hollow structures have the shape of tubes and are made from sheets of graphene rolled together at precise angles [59]. Depending on how many graphene sheets are wrapped into a tube, the carbon nanotubes can be either single-walled or multi-walled. These tubes have a cross-section as small as 0.4 nm and as large as 100 nm, with a length many times that of the cross-section. Because of their unique properties, such as their nanosized needle shape, high aspect ratio, ultralight weight, and large surface area, as well as their chemical, thermal, mechanical, and electrical capabilities, carbon nanotubes find extensive use in the drug delivery process [60]. Carbon nanotubes possess unique properties that make them an ideal source for various industrial applications, in addition to their biomedical uses. Graphene, a carbon-based nanocarrier, has been identified as a significant agent for drug delivery with high efficiency.

Gold Nanocarriers

The utilization of gold, a noble metal, has had a significant impact on the field of nanotechnology. Specifically, gold nanoparticles have been developed and employed as highly effective agents for a variety of applications, including but not limited to photoacoustic imaging, chemotherapy, surface-enhanced resonance spectroscopy, and gene therapy [61]. Gold nanoparticles exhibit various anisotropic shapes, including but not limited to nanostars, nanorods, nanocages, nanoshells, and nanoprisms. The optical property of gold nanocarriers is a prominent feature that renders them appealing for use in the biomedical field. Gold nanoparticles have the capability to facilitate the attachment of various biomolecules, including but not limited to enzymes, carbohydrates, fluorophores, peptides, proteins, and genes [62]. The effective transportation of molecules within a cell is facilitated, thereby overcoming associated barriers. Analysis of computed tomography scans, positron emission tomography scans, and single photon emission computed tomography scans all make use of gold nanocarriers [63].

Quantum Dot

Quantum dots are composed of atoms belonging to either the II-VI (Se, Zn, Te, Cd) or III-V (In, As, P) element groups in the periodic table. The entities under investigation are colloidal nanocrystals that serve as energy donors [64]. The light emission of quantum dots is dependent on their size, with smaller quantum dots (~2 nm) emitting blue fluorescence and larger quantum dots (~5 nm) emitting red fluorescence across the UV-near IR regions [65]. Quantum dots have been identified as effective delivery and reporter systems. The surface modification of quantum dots was observed to be effective in attaching to the nucleolin on cancerous cells when homing tumour peptide was used. This modification also resulted in an enhancement of cellular uptake. In the field of charge transfer, quantum dots are utilised as a quencher in systems such as energy transfer and quantum dot fluorescence resonance. Acceptors of the resonant energy transmission in chemiluminescence [66].

3.2 Targeted Therapy Using Nanoparticles

The phenomenon of active targeting pertains to the binding of ligands to receptors subsequent to the delivery of nanoparticles to the intended site through systemic circulation. The interaction between a ligand and its receptor can occur solely under the condition that the two entities are situated within close proximity of each other, measuring less than 0.5 nm in distance [67]. In order to achieve active targeting in tumors, the nanoparticles must be functionalized with proteins, peptides, nucleic acid aptamers, polysaccharides, and other small molecules. Biodegradable polymers, liposomes, dendrimers, nanoshells, and nucleic acid-based NPs are only some of the forms of material that have been produced for targeted therapy. The two most important conditions for successful targeted delivery are prolonged release and localized administration. The stability of nanoparticles is a crucial factor for their sustained presence in the bloodstream and eventual accumulation in tumors [68]. The high drug-loading capacity, biocompatibility, and anion exchange capability of layered double hydroxides, a well-known kind of 2D nanocarrier, have attracted interest for their potential as a drug carrier. Layered double hydroxides (LDHs) possess the potential for versatile applications due to their high specific surface areas, which enable facile modification of their internal and external surfaces for the incorporation of targeting ligands. Drugs and macromolecules like proteins or peptides can be encapsulated inside polymeric nanoparticles, as can a wide variety of hydrophilic and hydrophobic smaller molecules [69]. The controlled release of the drug from the nanoparticles can be achieved through various mechanisms such as diffusion, swelling followed by diffusion, or bulk erosion. The release rate is dependent on time. The control of the release rate can be achieved by means of polymer modification, the creation of novel polymers, or the production of copolymers [70, 71]. The primary advantages of utilizing biodegradable polymers include the ability to sustain optimal drug concentration levels for extended periods, leading to enhanced drug efficacy and improved patient adherence.

3.3 Photothermal Therapy Using Nanoparticles

Photodynamic therapy (PTT) is a novel therapeutic approach that employs electromagnetic radiation to target and treat a range of medical conditions, including different inflammatory disorders (Fig. 2) and various forms of cancer. The process entails the targeted elevation of tissue temperature through the photoexcitation of a photosensitizer material using microwaves, radio frequencies, near-infrared (NIR), or visible light exposure at the local level [72]. Elevating the temperature within the range of 41–47 °C induces hyperthermic effects that impede anomalous tissue proliferation and pannus development. This, in turn, results in lesion ablation through mechanisms such as irreversible protein denaturation, nucleic acid impairment, and cytosol evaporation, ultimately leading to cellular apoptosis [73]. When light is absorbed, it is transformed into heat via three distinct photophysical processes. As a result of absorbing light phonons, metal nanoparticles (NPs) generate heat, which, in turn, causes charge separation particle oscillation during the first and second steps of photothermal conversion. When energy is transferred from light absorption to temperature, electron phonons relax. In the third stage, metal NPs cool down by dissipating their heat into the air. This heats the area around gold NPs (AuNPs) in arthritic tissue, leading to cell death and apoptosis [74].

PTT's great specificity and its ability to lessen the impact of invasion make it an intriguing technique [75]. The application of near-infrared (NIR) light within the range of 750–2500 nm in conjunction with a photothermal agent may result in minimal damage to adjacent healthy tissue. Upon exposure to NIR irradiation, the nanomaterials undergo photon absorption and subsequently undergo non-radiative relaxation. The utilization of laser ablation in conjunction with nanospheres, nanorods, and nano-shells has been shown to enhance the precise targeting of inflammatory cells [76]. In metallic materials, it is typical to observe the presence of both a valence band and a conduction band. This characteristic facilitates the transfer of electrons from the valence band to the conduction band. In the case of nanosized materials, it has been observed that the band gap is insignificant. As a result, the oscillation that takes place at the interface between the metal and dielectric is coherent and is commonly referred to as localized surface plasmon resonance (LSPR) [77]. The phenomenon of plasmonic nano-materials is characterized by their ability to exhibit strong light absorption when the frequency of the incident light is in resonance with the oscillation frequency of the metallic material. This results in the creation of a significantly amplified electric field on the surface of the local particle [78].

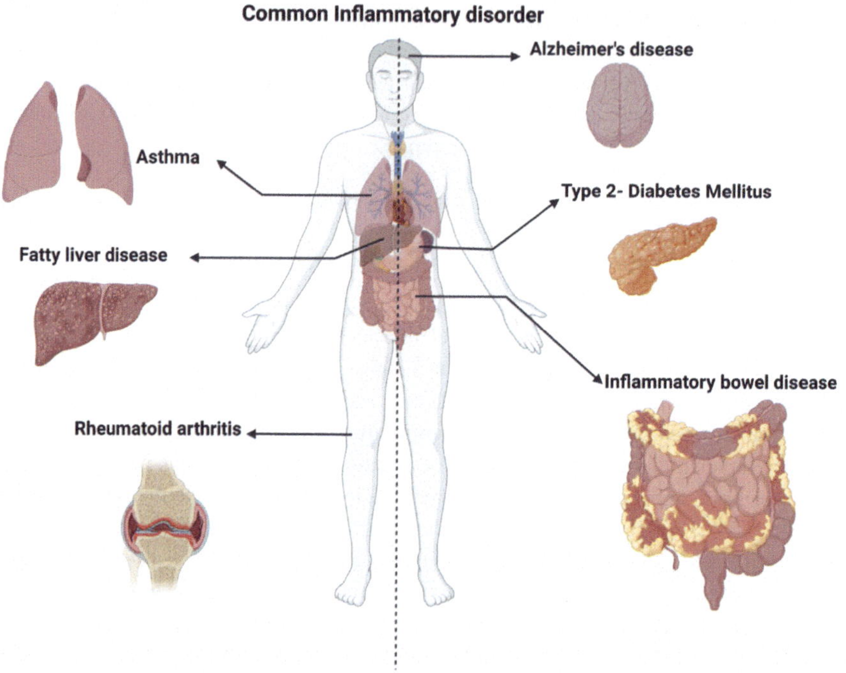

Fig. 2 Common inflammatory disorder in the human body

4 Application of Nanomaterials in the Diagnosis and Treatment of Common Inflammatory Disorders

4.1 Rheumatoid Arthritis

Rheumatoid arthritis is a complex multifactorial inflammatory disease that affects the synovial joints to a great extent [79]. It is mainly associated with progressive disability, systemic complications, high morbidity, and disease burden on society [80]. The Etiology of RA is still unknown. However, several studies have explained that RA results from the interaction of genetics, environment, and the immune system [81]. The first-line clinical diagnosis is mostly based on physical examination and observation of distinctive symptoms, such as pain, stiffness and edema of several joints in bilaterally symmetrical patterns [82]. Serological tests are used to identify numerous biomarkers in blood, such as antibodies (RF, ACPA) caused by immune response and/or acute phase reactants (C-reactive protein (CRP), erythrocyte sedimentation rate (ESR)) that are enhanced under inflammatory conditions [83]. Conventional radiography, ultrasonography (US), computed tomography (CT), and magnetic resonance imaging (MRI) are imaging diagnostic approaches for RA. However, there are significant drawbacks to conventional diagnosis, including (1) the inability to detect RA early in pure clinical diagnosis. (2) X-ray radiation is hazardous and insensitive. (3) CT is bright, costly, and incapable of detecting active inflammations such as synovitis and tenosynovitis. Conventional RA treatment, on the other hand, is mostly focused on (1) early standardized therapy with frequent monitoring and follow-up. (2) Inhibiting cytokine generation and activity. (3) Conventional anti-rheumatic medications (DMARDs), glucocorticoids (GCs), nonsteroidal anti-inflammatory medicines (NSAIDs), and biological agents [84–86]. Although these therapies are successful, their use is limited due to their harmful side effects, which include cardiac difficulties, ulcers, and gastrointestinal damage, as well as immunological suppression, which leads to the development of opportunistic infections [85]. As a result, there is an urgent need to suggest safer and more effective alternatives to RA therapy. Nowadays, rheumatoid arthritis (RA) diagnosis and therapy have shown promise for using nanomaterials. Quantum dots, superparamagnetic nanoparticles, carbon nanoallotropes, micelles, and liposomes are a few examples of nanomaterials that can be created to have certain features that facilitate precise imaging and diagnosis of RA [86–88]. These nanomaterials may be functionalized with targeted ligands that precisely bind to RA-related markers, enabling improved joint inflammation detection and imaging [88]. Additionally, nanomaterials such as colorimetric nanosensors, organic–inorganic hybrid nanomaterials including silver/gold (Ag@Au) core–shell NPs)-poly(aniline) hybrid nanostructures (CBCPHNs), surface-enhanced Raman scattering (SERS)-based multiplexed detection of auto-antibodies (RF IgM and anti-CCP) can be used in the development of biosensors for the detection of specific biomarkers associated with RA, enabling early and precise diagnosis [89, 90]. Nanomaterials can be employed as carriers for targeted medicine delivery to RA patients' swollen joints. It is conceivable to increase therapeutic efficacy and decrease adverse effects by encapsulating pharmaceuticals into nano-sized carriers such as liposomes, dendrimers, or polymeric nanoparticles [91]. These nanocarriers may be programmed to release medications at the site of inflammation, increasing local concentrations while reducing systemic exposure. NSAID-based metallic and polymeric NP conjugates (chitosan-dexamethasone NPs, or CS-Dex NPs), (hyaluronic-gold-tocilizumab NPs), human serum albumin NPs (arginine-glycine aspartic acid attached with gold nano half-shells conjugated with methotrexate (MTX)), solid lipid NPs (curcumin CUR-loaded SL [92]. These NPs could function better at sizes between 100 and 400 nm. In order to properly target macrophages, which are crucial to the inflammatory processes that take place in RA, NPs can also be coated with certain antibodies [93–95].

4.2 Asthma

Asthma is a long-term inflammatory lung disease characterized by encompasses reversible airway obstruction and bronchial hyper-responsiveness [96]. With just symptomatic relief as a therapy option, it affects more than 230 million people globally [97]. Despite improvements in diagnosis, increased knowledge of the pathophysiology of asthma, and the development of new treatment approaches such as—anti-IgE monoclonal antibodies, long-acting bronchodilators, phosphodiesterase inhibitors, and leukotriene inhibitors—asthma remains a serious public health issue with no signs of its prevalence decreasing [98–100]. Alternative therapy approaches are therefore necessary to reduce both inflammatory and remodeling processes in various asthma subtypes and improve lung function without causing negative side effects like immunosuppression. Furthermore, nanomaterials have great potential in the diagnosis and treatment of asthma. Nanomaterials can be used in order to provide sensitive diagnostic methods for asthma. For example, nanosensors can detect and measure specific biomarkers associated with asthma, such as exhaled breath markers or inflammatory cytokines [101]. Additionally, these nanosensors have the potential to provide inexpensive, quick, non-invasive methods for monitoring airway inflammation, assisting in determining the severity of the condition and improving therapy and diagnostic options [102].

Nanomaterials can also be effective as drug delivery vehicles specific to the lungs, enhancing the efficacy of asthma treatments. In addition, it demonstrated the potential for delayed liberation of drugs, increased drug accessibility and solubility, decreased drug adverse effects, and the capacity to navigate around physical limitations in targeted therapy and diagnosis [100]. Further, bronchodilators, anti-inflammatory drugs, or immunomodulators can be enclosed in nano-sized drug carriers, such as liposomes or nanoparticles, allowing for targeted and prolonged release in the airways. This method of targeted delivery improves drug concentration at the site of action, lowers systemic adverse effects, and improves patient compliance. Using a mouse model, researchers evaluated the influence of liposomal budesonide on asthma therapy; the finding demonstrated that this drug delivery method significantly decreased lung inflammation and the toxicity of inhaled steroid asthma medications [103].

Liposomes were created for packing of salbutamol sulfate (SBS) in an aerosol form and the aggregations displayed pro-longed anti-asthmatic outcomes than the unbound SBS [102].

Further, in vitro cell tests revealed that liposome-encapsulated curcumin inhibited LPS-induced airway inflammation [104]. Using cell and animal tests, [103] discovered that liposomal voids (UTS 001) might be utilized to treat chronic respiratory disorders via reducing eosinophils and decreasing epithelial pro-inflammatory cytokines. Several organizations have successfully changed liposomes to enhance their characteristics, such as cellular absorption, stability, and targeting, in addition to the basic liposomal nanocarriers. For instance, [105] revealed that in vitro cell studies and rat animal models may be used to develop starch-coated magnetic liposomes as an inhalable carrier to administer fasudil to treat pulmonary hypertension. Furthermore, prior research found that after intra-tracheal injection of mannosylated liposomes, cellular absorption by alveolar macrophages was greater than that of non-modified liposomes, both in vivo and in vitro [106]. In vitro experiments have shown that octa arginine-coated liposomes might promote cytosolic targeting, promote cellular absorption, and lower drug toxicity in airway cells [107]. Combining nanotechnology has quickened the evolution of therapeutics for respiratory illness based on the advantages of inhaled delivery. However, while preclinical research has revealed promising development possibilities, most applicable investigations are yet in the initial testing stages, and the clinical impacts must be confirmed [108].

Additional research and development are required to assess their safety, efficacy, and feasibility for clinical usage. Regulatory considerations and long-term effects should also be carefully assessed before implementing nano-based solutions for asthma management.

4.3 Inflammatory Bowel Diseases (IBD)

Inflammatory bowel disease (IBD) is a chronic relapsing gastrointestinal (GI) ailment with two basic types, ulcerative colitis (UC) and Crohn's disease (CD), both of which are caused by dysbiosis of the gut microbiota. To date, neither of these two kinds is currently permanently curable [109]. Nanomaterials have shown promise in the diagnosis and therapeutics of inflammatory bowel diseases (IBD). Nanomaterials can be used to create diagnostic tools for IBD. Fluorescent nanoparticles or quantum dots, for example, can be coupled with particular antibodies or ligands to target and visualize inflammatory areas of the gastrointestinal system. This customized imaging technique can help in IBD diagnosis, localization, and monitoring, allowing for more accurate therapy planning and evaluation.

Furthermore, nanotechnology and imaging technologies can be used together at the cellular or molecular level to provide early IBD diagnosis and disease severity monitoring [110, 111]. Wu et al. improved SPECT/CT and MRI imaging quality by labeling macrophages with indium (^{111}In)/iron oxide nanoparticles [112]. Due to colonoscopy's limitations in evaluating the whole intestine and the discomfort of patients with IBD, Basirat et al. looked into the accuracy of 99mTc(V)-dimercaptosuccinic acid (DMSA) and fecal calprotectin with ileocolonoscopy as a new approach for inflammatory localization [113]. Dextran-coated cerium oxide NPs produced strong CT contrast and helped detect colitis in the gut. To observe hypoxia associated with IBD, Zhou et al. developed an in-vivo hypoxia-activatable and cytoplasmic protein-powered fluorescence cascade amplifier (HCFA) [114]. Nanomaterials can be integrated into biosensors or diagnostic systems to identify biomarkers linked to IBD. Functionalized nanostructures, such as nanowires or nanotubes, can be engineered to identify and bind to disease-specific biomarkers in blood or stool samples. This allows for the detection of extremely sensitive and specific IBD biomarkers, which aids in early diagnosis, illness monitoring, and therapy response evaluation [115, 116]. To create a stool analysis method, Shepherd et al. combined a chemo-resistive metal oxide gas sensor with gas chromatography. An artificial neural network program was given data on resistance variation over time from several samples for IBD detection [117]. With an overall mean prediction accuracy of 76%, the computer was able to discriminate samples from IBS and IBD with a sensitivity and specificity of 76% and 88%, respectively. Prior to now, miRNAs and silver nanoparticles were used to diagnose inflammation via fluorescence quenching [118]. Due to biomatrix interference, these probes have low limits of detection (LODs). By creating a novel DNA/AgNC-cDNA nanosensor that operated by increasing fluorescence, Fang et al. were able to fix this issue [118]. To combat IBD infection, they employed miR-223 (an IBD biomarker). In comparison to the traditional fluorescence method, the newly designed probe demonstrated excellent specificity and sensitivity (10 times). Gholami et al. created a highly sensitive gold-coated copper oxide nanomaterial surface-enhanced Raman spectroscopy (SERS) quenching nanosensor for the detection of TNF- in blood [119]. Although the exact mechanism of action of IBD is still unknown, research suggests that oxidative stress, chronic illnesses, genetic susceptibility, and changes in the microbiome might all contribute to its development [120–122]. Traditional treatments for IBD include negative side effects, such as an increased risk of infections and several malignancies [123]. Nanomedicine is a form of therapy that delivers drugs to specific areas of the body using nanoparticles or nanocarriers [124, 125]. In the experimental study, Miroliaee and colleagues revealed that freshly synthesized selenium-NPs coupled with silymarin had ameliorative effects [126]. Laroui et al. transported KPV, an anti-inflammatory tripeptide, to the colon in a mouse model of UC using modified NPs. They discovered that this tripeptide may be administered to the inflamed area utilizing NP-based DDS at a far lower dosage than the tripeptide alone [127]. Theiss and colleagues effectively administered prohibitin, a protein that is widely expressed, to the colon using an encapsulated NP/hydrogel system, and they noted the encouraging ameliorating effects of NMs in drug administration tripeptide alone [128]. Ocansey et al. recommended employing exosome-like NPs in IBD models to deliver proteins, functional RNAs, and synthetic chemicals and medications to the colon [129]. Likewise, grape exosome-like nanoparticles elicited good defense from UC in a mouse model [130]. Additionally, NPs have been used to develop novel oral DDS, in which the pH, the presence of enzymes, or the presence of reactive oxygen species are the main factors influencing drug release in the GI tract [131–133]. As efficient drug delivery platforms for IBD treatment, RNAi-based NPs with a variety of therapeutic applications have also been proposed [134]. Excellent biocompatibility has also been shown for synthetic chondroitin sulfate-curcumin nanoparticles, which suggests that they might be a potential IBD treatment method [135]. Intriguingly, intestine organoids containing poly (lactide-co-glycolide) (PLGA) NPs have been proposed as a possible DDS for delivering traditional IBD drugs to inflamed regions [136].

5 Clinical Translation of Nanomaterials

The administration of nanoparticles via the parenteral route is the predominant method for treating inflammatory disease in clinical settings. This is supported by the fact that the majority of nanoparticles utilized in the clinic are intended for this purpose. The structural foundation of these entities is rooted in rudimentary nanomedicine platforms, surface charge-altered nanoparticles, specifically fundamental nanoparticles, and PEGylated nanoparticles [137]. Despite the promising experimental results, the progress of clinical applications of nanotechnology for non-cancer diseases has been hindered by various barriers in both the preclinical and clinical stages of development. Long-term toxicity, as well as affordability, is among the problems that need to be addressed. Complexity in manufacturing and characterization is another challenge [138].

The process of getting regulatory approval for nanomedicine is notoriously difficult. When compared to more traditional formulations, nanoscale material management often provides higher scientific and technical obstacles [139]. The category of nanomaterials comprises a diverse range of nanoparticles and nanostructures, posing a significant obstacle in the development of consistent regulatory procedures and methodologies for ensuring adherence to Good Manufacturing Practices (GMP), as well as for conducting safety and toxicology assessments and designing clinical trials. The implementation of these procedures is crucial in verifying the therapeutic effectiveness and safety of a drug before it is granted marketing approval for use in a broader patient population. To achieve successful clinical translation, it is imperative to adopt an interdisciplinary approach that facilitates the creation of innovative infrastructure, procedures, and tests for producing and characterizing nanomaterials [140]. The successful execution of this project will require the involvement of professionals with expertise in a number of fields such as pharmaceutics, engineering, biology, medicine, and toxicology, both from academic and industrial backgrounds. One potential strategy for expediting the progression of promising novel nanomaterials (NNMs) to clinical trials involves the creation or coordination of specialized laboratories and centers. These entities would possess the necessary expertise to (i) NNM platforms must be fully characterized, (ii) Preclinical research on NNMs must be carried out and submitted to regulatory bodies, (iii) laboratory preparation of nanomaterials must be scaled up in accordance with industry and regulations for early clinical studies, and (iv) clinical trials that carried out for NNM platforms must be planned [139].

6 Potential Risks and Limitations of Nanomaterials

Despite the approval of numerous drugs based on nanomaterials by the FDA, the efficacy of these treatments has not met anticipated levels. The numerical value lacks impressiveness due to the ongoing need for standardization of methods and protocols utilized in evaluating the efficacy, toxicity, and safety of clinical applications [141]. It may be difficult to translate nanomaterial-based treatments without established scientific methods for evaluating risk. A rigorous approach to nanoparticle production, material optimization, screening, and the discovery of new therapeutic-specific targets is necessary for the successful clinical translation of nanomaterials and ultimate FDA approval of such materials. Researchers have dedicated significant time and effort toward the establishment of protocols for the consistent synthesis of reproducible batches of nanoparticles. For many years, this has been a subject of research. It is still difficult to scale up nanomedicine platforms for industrial usage in a way that complies with Good Manufacturing Practice (GMP) and legal requirements for human use [142]. Implementing nanoparticles in therapeutic settings has been significantly hampered by the management of their biodistribution and targeted administration in vivo. The anticipated outcome of these studies is to offer a deeper understanding of the distribution of nanoparticles within organs, which could potentially facilitate expedited drug trials. The application of organ-on-a-chip platforms, which are in vitro models of human organs, is a promising emerging field of study for examining the pharmacokinetics and pharmacodynamics of nanoparticles in pre-clinical studies. This method has the potential to significantly improve nanoparticle-based medicinal toxicity studies [143].

7 Future Perspective

The field of nanotechnology has shown considerable potential in the diagnosis and treatment of inflammatory disorders. With their distinct physical and chemical characteristics at the nanoscale, nanomaterials provide an abundance of benefits. It promotes drug accumulation at the site of inflammation while reducing the systemic side effects. It ensures monitored release and enables sustained drug release at the site of inflammation, extending the therapeutic effect and minimizing the dose frequency. Furthermore, the nanoscale size of the carriers enhances their entry into tissues and cells, improving the anti-inflammatory effect. By conjugating imaging agents to nanoparticles, they can aggregate at the site of inflammation and offer improved contrast for the identification and monitoring of inflammatory disorders. It allows for the sensitive and precise identification of inflammatory biomarkers, such as cytokines or chemokines, in patients suffering from inflammatory disorders, such as asthma. These biomarkers may be captured and concentrated by functionalized nanoparticles, enhancing their detection signal by several sensing modalities such as electrochemical, optical, or magnetic techniques. This permits early diagnosis, disease progression tracking, and therapy response evaluation. However, structure activity relationship (SAR) researches are required to further advance this area of nanotechnology. Despite the numerous advantages that nanoparticle holds as drug carriers, there are still some drawbacks [144]. For instance, because of their tiny size, NPs may cause toxicity by widening the distribution of the drug throughout the body. Hence, increasing the drugs' potency, although beneficial, may also have an impact on the body's immunological response and result in toxicity. Therefore, nanomaterials-based treatment and diagnostics will undoubtedly play a significant role in inflammatory disorders in the future. Future research should, however, concentrate more on toxicity and on pinpointing crucial factors including dosage, mode of administration, physicochemical characteristics, and composition that may provoke toxicity. Better understanding will allow for more effective treatment and diagnosis of inflammatory diseases and reduce the disease burden.

8 Conclusion

In conclusion, the exhaustive utilization of nanomaterials in the early diagnosis and treatment of inflammatory disorders such as asthma, and inflammatory bowel disease, rheumatoid arthritis holds great promise for improving patient outcomes. The unique and notable properties of nanomaterials enable targeted drug delivery, controlled release of therapeutics, and sensitive detection of inflammatory biomarkers. These properties can be employed by nanotechnology-based methods to enhance the effectiveness and safety of therapies while minimizing systemic side effects particularly in inflammatory disorders. Additionally, the use of nanomaterials offers opportunities for early diagnosis, non-invasive imaging, and better monitoring of inflammatory disorders. However, questions on safety, efficacy, dose, and cost are still points of concern. Therefore, additional research is needed to evaluate safety concerns, optimize nanomaterial design, and validate their efficacy in clinical studies. To ensure the correct implementation of nanotechnology into clinical practice, regulatory and ethical concerns additionally need to be taken into account. With continued development and multidisciplinary collaborations, nanomaterials have the great potential to transform the diagnosis and treatment of inflammatory disorders, leading to improved patient care and decreasing the disease burden.

References

1. Bonifácio BV, Da Silva PB, dos Santos A, Ramos M, Maria Silveira Negri K, Maria Bauab T, Chorilli M (2014) Nanotechnology-based drug delivery systems and herbal medicines: a review. Int J Nanomed 9(1):1–15
2. Gatoo MA, Naseem S, Arfat MY, Mahmood Dar A, Qasim K, Zubair S (2014) Physicochemical properties of nanomaterials: implication in associated toxic manifestations. BioMed Res Int
3. Kumar R, Lal S (2014) Synthesis of organic nanoparticles and their applications in drug delivery and food nanotechnology: a review. J Nanomater Mol Nanotechnol 3:4 of 11,2
4. Madkour LH, Madkour LH (2019) Introduction to nanotechnology (NT) and nanomaterials (NMs). In: Nanoelectronic materials: fundamentals and applications, pp 1–47
5. Medzhitov R (2010) Inflammation 2010: new adventures of an old flame. Cell 140(6):771–776
6. Ferrero-Miliani L, Nielsen OH, Andersen PS, Girardin S (2007) Chronic inflammation: importance of NOD2 and NALP3 in interleukin-1β generation. Clin Exp Immunol 147(2):227–235
7. Nathan C, Ding A (2010) Nonresolving inflammation. Cell 140:871–882
8. Zhou Y, Hong Y, Huang H (2016) Triptolide attenuates inflammatory response in membranous glomerulo-nephritis rat via downregulation of NF-κB signaling pathway. Kidney Blood Press Res 41(6):901–910

9. Jo J, Tian C, Xu G, Sarazin J, Schiopu E, Gandikota G, Wang X (2018) Photoacoustic tomography for human musculoskeletal imaging and inflammatory arthritis detection. Photoacoustics 12:82–89
10. Glaudemans AW, Quintero AM, Signore A (2012) PET/MRI in infectious and inflammatory diseases: will it be a useful improvement? Eur J Nucl Med Mol Imaging 39:745–749
11. Cormode DP, Skajaa T, Fayad ZA, Mulder WJ (2009) Nanotechnology in medical imaging: probe design and applications. Arterioscler Thromb Vasc Biol 29(7):992–1000
12. Dardzinski BJ, Schmithorst VJ, Holland SK, Boivin GP, Imagawa T, Watanabe S, Hirsch R et al (2001) MR imaging of murine arthritis using ultrasmall superparamagnetic iron oxide particles☆. Magn Reson Imaging 19(9):1209–1216
13. Chen CL, Siow TY, Chou CH, Lin CH, Lin MH, Chen YC, Chang C et al (2017) Targeted superparamagnetic iron oxide nanoparticles for in vivo magnetic resonance imaging of T-cells in rheumatoid arthritis. Mol Imag Biol 19:233–244
14. Finkelstein AE, Walz DT, Batista V, Mizraji M, Roisman F, Misher A (1976). Auranofin. New oral gold compound for treatment of rheumatoid arthritis. Ann Rheum Dis 35(3):251–257
15. Ma J, Liu X, Wang R, Zhang J, Jiang P, Wang Y, Tu G (2020) Bimetallic core–shell nanostars with tunable surface plasmon resonance for surface-enhanced Raman scattering. ACS Appl Nano Mater 3(11):10885–10894
16. Zhao Y, Liu Y, Li X, Wang H, Zhang Y, Ma H, Wei Q (2018) Label-free ECL immunosensor for the early diagnosis of rheumatoid arthritis based on asymmetric heterogeneous polyaniline-gold nanomaterial. Sens Actuators, B Chem 257:354–361
17. Fournelle M, Bost W, Tarner IH, Lehmberg T, Weiß E, Lemor R, Dinser R (2012) Antitumor necrosis factor-α antibody-coupled gold nanorods as nanoprobes for molecular optoacoustic imaging in arthritis. Nanomed Nanotechnol Biol Med 8(3):346–354
18. Rieter WJ, Kim JS, Taylor KM, An H, Lin W, Tarrant T, Lin W (2007) Hybrid silica nanoparticles for multimodal imaging. Angew Chem Int Ed 46(20):3680–3682
19. Jayanthi VSA, Das AB, Saxena U (2017) Recent advances in biosensor development for the detection of cancer biomarkers. Biosens Bioelectron 91:15–23
20. Chinen AB, Guan CM, Ferrer JR, Barnaby SN, Merkel TJ, Mirkin CA (2015) Nanoparticle probes for the detection of cancer biomarkers, cells, and tissues by fluorescence. Chem Rev 115(19):10530–10574
21. Shao H, Chung J, Balaj L, Charest A, Bigner DD, Carter BS, Lee H et al (2012) Protein typing of circulating microvesicles allows real-time monitoring of glioblastoma therapy. Nat Med 18(12):1835–1840
22. Perrier A, Gligorov J, Lefèvre G, Boissan M (2018) The extracellular domain of Her2 in serum as a biomarker of breast cancer. Lab Invest 98(6):696–707
23. Triulzi T, Bianchi GV, Tagliabue E (2016) Predictive biomarkers in the treatment of HER2-positive breast cancer: an ongoing challenge. Future Oncol 12(11):1413–1428
24. Contreras-Naranjo JC, Wu HJ, Ugaz VM (2017) Microfluidics for exosome isolation and analysis: enabling liquid biopsy for personalized medicine. Lab Chip 17(21):3558–3577
25. Srivastava A, Moxley K, Ruskin R, Dhanasekaran DN, Zhao YD, Ramesh R (2018) A non-invasive liquid biopsy screening of urine-derived exosomes for miRNAs as biomarkers in endometrial cancer patients. AAPS J 20:1–11
26. Tu M, Chia D, Wei F, Wong D (2016) Liquid biopsy for detection of actionable oncogenic mutations in human cancers and electric field induced release and measurement liquid biopsy (eLB). Analyst 141(2):393–402
27. Pazos E, Garcia-Algar M, Penas C, Nazarenus M, Torruella A, Pazos-Perez N, Alvarez-Puebla RA et al (2016) Surface-enhanced Raman scattering surface selection rules for the proteomic liquid biopsy in real samples: efficient detection of the oncoprotein c-MYC. J Am Chem Soc 138(43):14206–14209
28. Niu S, Lv Z, Liu J, Bai W, Yang S, Chen A (2014) Colorimetric aptasensor using unmodified gold nanoparticles for homogeneous multiplex detection. PLoS ONE 9(10):e109263
29. Pei X, Zhang B, Tang J, Liu B, Lai W, Tang D (2013) Sandwich-type immunosensors and immunoassays exploiting nanostructure labels: a review. Anal Chim Acta 758:1–18

30. Pan LH, Kuo SH, Lin TY, Lin CW, Fang PY, Yang HW (2017) An electrochemical biosensor to simultaneously detect VEGF and PSA for early prostate cancer diagnosis based on graphene oxide/ssDNA/PLLA nanoparticles. Biosens Bioelectron 89:598–605
31. Posthuma-Trumpie GA, Korf J, van Amerongen A (2009) Lateral flow (immuno) assay: its strengths, weaknesses, opportunities and threats. a literature survey. Anal Bioanal Chem 393:569–582
32. Mohd Bakhori N, Yusof NA, Abdullah J, Wasoh H, Md Noor SS, Ahmad Raston NH, Mohammad F (2018) Immuno nanosensor for the ultrasensitive naked eye detection of tuberculosis. Sensors 18(6):1932
33. Suresh L, Brahman PK, Reddy KR, Bondili JS (2018) Development of an electrochemical immunosensor based on gold nanoparticles incorporated chitosan biopolymer nanocomposite film for the detection of prostate cancer using PSA as biomarker. Enzyme Microb Technol 112:43–51
34. Patolsky F, Zheng G, Hayden O, Lakadamyali M, Zhuang X, Lieber CM (2004) Electrical detection of single viruses. Proc Natl Acad Sci 101(39):14017–14022
35. Singh R, Mukherjee MD, Sumana G, Gupta RK, Sood S, Malhotra BD (2014) Biosensors for pathogen detection: a smart approach towards clinical diagnosis. Sens Actuators, B Chem 197:385–404
36. Bui MPN, Brockgreitens J, Abbas A (2018) Gold nanoplate-enhanced chemiluminescence and macromolecular shielding for rapid microbial diagnostics. Adv Healthc Mater 7(13):1701506
37. Khateb H, Klös G, Meyer RL, Sutherland DS (2020) Development of a label-free LSPR-apta sensor for Staphylococcus aureus detection. ACS Appl Bio Mater 3(5):3066–3077
38. Qiu G, Gai Z, Tao Y, Schmitt J, Kullak-Ublick GA, Wang J (2020) Dual-functional plasmonic photothermal biosensors for highly accurate severe acute respiratory syndrome coronavirus 2 detection. ACS Nano 14(5):5268–5277
39. Scheler O, Kindt JT, Qavi AJ, Kaplinski L, Glynn B, Barry T, Bailey RC et al (2012) Label-free, multiplexed detection of bacterial tmRNA using silicon photonic microring resonators. Biosens Bioelectron 36(1):56–61
40. Shin Y, Perera AP, Tang WY, Fu DL, Liu Q, Sheng JK, Park MK et al (2015) A rapid amplification/detection assay for analysis of Mycobacterium tuberculosis using an isothermal and silicon bio-photonic sensor complex. Biosens Bioelectron 68:390–396
41. Ozalp VC, Bayramoglu G, Erdem Z, Arica MY (2015) Pathogen detection in complex samples by quartz crystal microbalance sensor coupled to aptamer functionalized core–shell type magnetic separation. Anal Chim Acta 853:533–540
42. Vedarethinam V, Huang L, Xu W, Zhang R, Gurav DD, Sun X, Qian K et al (2019) Detection and inhibition of bacteria on a dual-functional silver platform. Small 15(3):1803051
43. Chu HW, Lai CS, Ko JY, Harroun SG, Chuang CI, Wang RY, Huang CC et al (2019) Nanoparticle-based LDI-MS immunoassay for the multiple diagnosis of viral infections. ACS Sensors 4(6):1543–1551
44. Qian WY, Sun DM, Zhu RR, Du XL, Liu H, Wang SL (2012) pH-sensitive strontium carbonate nanoparticles as new anticancer vehicles for controlled etoposide release. Int J Nanomed:5781–5792
45. Kingsley JD, Dou H, Morehead J, Rabinow B, Gendelman HE, Destache CJ (2006) Nanotechnology: a focus on nanoparticles as a drug delivery system. J Neuroimmune Pharmacol 1:340–350
46. Sun T, Zhang YS, Pang B, Hyun DC, Yang M, Xia Y (2014) Engineered nanoparticles for drug delivery in cancer therapy. Angew Chem Int Ed 53(46):12320–12364
47. How CW, Rasedee A, Manickam S, Rosli R (2013) Tamoxifen-loaded nanostructured lipid carrier as a drug delivery system: characterization, stability assessment and cytotoxicity. Colloids Surf, B 112:393–399
48. Malam Y, Loizidou M, Seifalian AM (2009) Liposomes and nanoparticles: nanosized vehicles for drug delivery in cancer. Trends Pharmacol Sci 30(11):592–599
49. Torchilin VP (2005) Recent advances with liposomes as pharmaceutical carriers. Nat Rev Drug Discov 4(2):145–160

50. Wang AZ, Langer R, Farokhzad OC (2012) Nanoparticle delivery of cancer drugs. Annu Rev Med 63:185–198
51. Bamrungsap S, Zhao Z, Chen T, Wang L, Li C, Fu T, Tan W (2012) Nanotechnology in therapeutics: a focus on nanoparticles as a drug delivery system. Nanomedicine 7(8):1253–1271
52. Wang X, Wang Y, Chen ZG, Shin DM (2009) Advances of cancer therapy by nanotechnology. Cancer Res Treat Off J Korean Cancer Assoc 41(1):1–11
53. Zhu Y, Liao L (2015) Applications of nanoparticles for anticancer drug delivery: a review. J Nanosci Nanotechnol 15(7):4753–4773
54. Wennerström H, Lindman B (1979) Micelles: physical chemistry of surfactant association. Phys Rep 52(1):1–86
55. Bianco A (2004) Carbon nanotubes for the delivery of therapeutic molecules. Expert Opin Drug Deliv 1(1):57–65
56. Ng CM, Loh HS, Muthoosamy K, Sridewi N, Manickam S (2016) Conjugation of insulin onto the sidewalls of single-walled carbon nanotubes through functionalization and diimide-activated amidation. Int J Nanomed 11:1607
57. Lu W, Huang Q, Ku G, Wen X, Zhou M, Guzatov D, Li C et al (2010) Photoacoustic imaging of living mouse brain vasculature using hollow gold nanospheres. Biomaterials 31(9):2617–2626
58. Von Maltzahn G, Park JH, Lin KY, Singh N, Schwöppe C, Mesters R, Bhatia SN et al (2011) Nanoparticles that communicate in vivo to amplify tumour targeting. Nat Mater 10(7):545–552
59. Newkome GR, Moorefield CN, Vögtle F, Vögtle F, Vögtle F, Chemist G (2001) Dendrimers and dendrons: concepts, syntheses, applications, vol 623. Wiley-vch, Weinheim
60. Mignani S, Tripathi RP, Chen L, Caminade AM, Shi X, Majoral JP (2018) New ways to treat tuberculosis using dendrimers as nanocarriers. Pharmaceutics 10(3):105
61. Lai PS, Lou PJ, Peng CL, Pai CL, Yen WN, Huang MY, Shieh MJ et al (2007) Doxorubicin delivery by polyamidoamine dendrimer conjugation and photochemical internalization for cancer therapy. J Control Release 122(1):39–46
62. Bhadra D, Bhadra S, Jain S, Jain NK (2003) A PEGylated dendritic nanoparticulate carrier of fluorouracil. Int J Pharm 257(1–2):111–124
63. Lee CC, Gillies ER, Fox ME, Guillaudeu SJ, Fréchet JM, Dy EE, Szoka FC (2006) A single dose of doxorubicin-functionalized bow-tie dendrimer cures mice bearing C-26 colon carcinomas. Proc Natl Acad Sci 103(45):16649–16654
64. Alivisatos AP (1996) Semiconductor clusters, nanocrystals, and quantum dots. Science 271(5251):933–937
65. Bruchez Jr M, Moronne M, Gin P, Weiss S, Alivisatos AP (1998) Semiconductor nanocrystals as fluorescent biological labels. Science 281(5385):2013–2016
66. Geißler D, Charbonnière LJ, Ziessel RF, Butlin NG, Löhmannsröben HG, Hildebrandt N (2010) Quantum dot biosensors for ultrasensitive multiplexed diagnostics. Angew Chem Int Ed 49(8):1396–1401
67. Zensi A, Begley D, Pontikis C, Legros C, Mihoreanu L, Wagner S, Kreuter J et al (2009) Albumin nanoparticles targeted with Apo E enter the CNS by transcytosis and are delivered to neurones. J Control Release 137(1):78–86
68. Moghimi SM, Hunter AC, Murray JC (2001) Long-circulating and target-specific nanoparticles: theory to practice. Pharmacol Rev 53(2):283–318
69. Perez C, Sanchez A, Putnam D, Ting D, Langer R, Alonso MJ (2001) Poly (lactic acid)-poly (ethylene glycol) nanoparticles as new carriers for the delivery of plasmid DNA. J Control Release 75(1–2):211–224
70. Langer R (1990) New methods of drug delivery. Science 249(4976):1527–1533
71. Langer R, Folkman J (1976) Polymers for the sustained release of proteins and other macromolecules. Nature 263(5580):797–800
72. Fong JF, Ng YH, Ng SM (2018) Carbon dots as a new class of light emitters for biomedical diagnostics and therapeutic applications. In: Fullerens, graphenes and nanotubes. William Andrew Publishing, pp 227–295

73. Wang P, Li A, Yu L, Chen Y, Xu D (2020) Energy conversion-based nanotherapy for rheumatoid arthritis treatment. Front Bioeng Biotechnol 8:652
74. Mahajan S, Raval N, Kalyane D, Anup N, Maheshwari R, Tambe V, Tekade RK et al (2020) NanoGold-core dendrimeric seeds for combined chemo-, photothermal-, and photodynamic therapy of cancer. J Drug Deliv Sci Technol 58:101814
75. Huang X, Jain PK, El-Sayed IH, El-Sayed MA (2008) Plasmonic photothermal therapy (PPTT) using gold nanoparticles. Lasers Med Sci 23:217–228
76. Jain PK, El-Sayed MA (2007) Universal scaling of plasmon coupling in metal nanostructures: extension from particle pairs to nanoshells. Nano Lett 7(9):2854–2858
77. Willets KA, Van Duyne RP (2007) Localized surface plasmon resonance spectroscopy and sensing. Annu Rev Phys Chem 58:267–297
78. Mayer KM, Hafner JH (2011) Localized surface plasmon resonance sensors. Chem Rev 111(6):3828–3857
79. McInnes IB, Schett G (2011) The pathogenesis of rheumatoid arthritis. N Engl J Med 365(23):2205–2219
80. Firestein GS (2003) Evolving concepts of rheumatoid arthritis. Nature 423(6937):356–361
81. Aletaha D, Smolen JS (2018) Diagnosis and management of rheumatoid arthritis: a review. JAMA 320(13):1360–1372
82. Majithia V, Geraci SA (2007) Rheumatoid arthritis: diagnosis and management. Am J Med 120(11):936–939
83. Dolati S, Sadreddini S, Rostamzadeh D, Ahmadi M, Jadidi-Niaragh F, Yousefi M (2016) Utilization of nanoparticle technology in rheumatoid arthritis treatment. Biomed Pharmacother 80:30–41
84. Umar S, Asif M, Sajad M, Ansari MM, Hussain U, Ahmad W, Khan HA et al (2012) Anti-inflammatory and antioxidant activity of Trachyspermum ammi seeds in collagen induced arthritis in rats. Int J Drug Dev Res 4(1):210–219
85. Mitragotri S, Yoo JW (2011) Designing micro-and nano-particles for treating rheumatoid arthritis. Arch Pharmacal Res 34:1887–1897
86. Kalangi SK, Swarnakar NK, Sathyavathi R, Narayana Rao D, Jain S, Reddanna P (2018) Synthesis, characterization, and biodistribution of quantum dot-celecoxib conjugate in mouse paw edema model. Oxid Med Cell Longev
87. Carneiro MFH, Machado ART, Antunes LM, Souza TE, Freitas VA, Oliveira LC, Barbosa F et al (2020) Gold-coated superparamagnetic iron oxide nanoparticles attenuate collagen-induced arthritis after magnetic targeting. Biol Trace Elem Res 194:502–513
88. Hosseinikhah SM, Barani M, Rahdar A, Madry H, Arshad R, Mohammadzadeh V, Cucchiarini M (2021) Nanomaterials for the diagnosis and treatment of inflammatory arthritis. Int J Mol Sci 22(6):3092
89. Veigas B, Matias A, Calmeiro T, Fortunato E, Fernandes AR, Baptista PV (2019) Antibody modified gold nanoparticles for fast colorimetric screening of rheumatoid arthritis. Analyst 144(11):3613–3619
90. Hwang EY, Lee JH, Lim DW (2020) Compartmentalized bimetal cluster-poly (aniline) hybrid nanostructures for multiplexed detection of autoantibodies in early diagnosis of rheumatoid arthritis. Sens Actuators, B Chem 321:128482
91. Zheng M, Jia H, Wang H, Liu L, He Z, Zhang Z, Gao F et al (2021) Application of nanomaterials in the treatment of rheumatoid arthritis. RSC Adv 11(13):7129–7137
92. Schiavi S, Ocampo-Pineda M, Barakovic M, Petit L, Descoteaux M, Thiran JP, Daducci A (2020) A new method for accurate in vivo mapping of human brain connections using microstructural and anatomical information. Sci Adv 6(31):eaba8245
93. Feng N, Guo F (2020) Nanoparticle-siRNA: a potential strategy for rheumatoid arthritis therapy? J Control Release 325:380–393
94. Wu J, Liang H, Li Y, Shi Y, Bottini M, Chen Y, Liu L (2020) Cationic block copolymer nanoparticles with tunable DNA affinity for treating rheumatoid arthritis. Adv Func Mater 30(27):2000391

95. Aden K, Rehman A, Waschina S, Pan WH, Walker A, Lucio M, Rosenstiel P et al (2019) Metabolic functions of gut microbes associate with efficacy of tumor necrosis factor antagonists in patients with inflammatory bowel diseases. Gastroenterology 157(5):1279–1292
96. Holgate ST, Wenzel S, Postma DS, Weiss ST, Renz H, Sly PD (2015) Asthma (Primer). Nat Rev Dis Primers 1(1)
97. GINA (2019) Global strategy for asthma management and prevention 2019. https://ginasthma.org/wp-content/uploads/2019/06/GINA-2019-main-reportJune-2019-wms.pdf
98. Anandan C, Nurmatov U, Van Schayck OCP, Sheikh A (2010) Is the prevalence of asthma declining? Systematic review of epidemiological studies. Allergy 65(2):152–167
99. Chung KF (2013) New treatments for severe treatment-resistant asthma: targeting the right patient. Lancet Respir Med 1(8):639–652
100. Papierniak ES, Lowenthal DT, Harman E (2013) Novel therapies in asthma: leukotriene antagonists, biologic agents, and beyond. Am J Ther 20(1):79–103
101. Horváth I, Hunt J, Barnes PJ (2005) Exhaled breath condensate: methodological recommendations and unresolved questions. Eur Respir J 26(3):523–548
102. Cheng WC, Chen CH (2019) Nanotechnology bring a new hope for asthmatics. Ann Transl Med 7(20)
103. Konduri KS, Nandedkar S, Rickaby DA, Düzgüneş N, Gangadharam PR (2005) The use of sterically stabilized liposomes to treat asthma. In: Methods in enzymology, vol 391. Academic Press, pp 413–427
104. Ng ZY, Wong JY, Panneerselvam J, Madheswaran T, Kumar P, Pillay V, Chellappan DK et al (2018) Assessing the potential of liposomes loaded with curcumin as a therapeutic intervention in asthma. Colloids Surf, B 172:51–59
105. Nahar K, Absar S, Patel B, Ahsan F (2014) Starch-coated magnetic liposomes as an inhalable carrier for accumulation of fasudil in the pulmonary vasculature. Int J Pharm 464(1–2):185–195
106. Wijagkanalan W, Kawakami S, Takenaga M, Igarashi R, Yamashita F, Hashida M (2008) Efficient targeting to alveolar macrophages by intratracheal administration of mannosylated liposomes in rats. J Control Release 125(2):121–130
107. Cryan SA, Devocelle M, Moran PJ, Hickey AJ, Kelly JG (2006) Increased intracellular targeting to airway cells using octaarginine-coated liposomes: in vitro assessment of their suitability for inhalation. Mol Pharm 3(2):104–112
108. Luo MX, Hua S, Shang QY (2021) Application of nanotechnology in drug delivery systems for respiratory diseases. Mol Med Rep 23(5):1–17
109. Fiocchi C (2012) Towards a 'cure' for IBD. Dig Dis 30(4):428–433
110. Liu Z, Liu J, Wang R, Du Y, Ren J, Qu X (2015) An efficient nano-based theranostic system for multi-modal imaging-guided photothermal sterilization in gastrointestinal tract. Biomaterials 56:206–218
111. Wu Y, Briley K, Tao X (2016) Nanoparticle-based imaging of inflammatory bowel disease. Wiley Interdisc Rev Nanomed Nanobiotechnol 8(2):300–315
112. Wu Y, Briley-Saebo K, Xie J, Zhang R, Wang Z, He C, Tao X et al (2014) Inflammatory bowel disease: MR-and SPECT/CT-based macrophage imaging for monitoring and evaluating disease activity in experimental mouse model—pilot study. Radiology 271(2):400–407
113. Basirat V, Azizi Z, Anbardan SJ, Asl MT, Farbod Y, Teimouri A, Daryani NE (2016) Accuracy of 99mTc (V)-dimercaptosuccinic acid scintigraphy and fecal calprotectin compared with colonoscopy in localizing active lesions in inflammatory bowel disease. Middle East J Dig Dis 8(3):212
114. Zhou Y, Yang S, Guo J, Dong H, Yin K, Huang WT, Yang R (2020) In vivo imaging of hypoxia associated with inflammatory bowel disease by a cytoplasmic protein-powered fluorescence cascade amplifier. Anal Chem 92(8):5787–5794
115. Arasaradnam RP, McFarlane M, Daulton E, Skinner J, O'Connell N, Wurie S, Covington J et al (2016) Non-invasive exhaled volatile organic biomarker analysis to detect inflammatory bowel disease (IBD). Dig Liver Dis 48(2):148–153

116. Monasta L, Pierobon C, Princivalle A, Martelossi S, Marcuzzi A, Pasini F, Perbellini L (2017) Inflammatory bowel disease and patterns of volatile organic compounds in the exhaled breath of children: a case-control study using Ion molecule reaction-mass spectrometry. PLoS ONE 12(8):e0184118
117. Shepherd SF, McGuire ND, de Lacy Costello BPJ, Ewen RJ, Jayasena DH, Vaughan K, Ratcliffe NM et al (2014) The use of a gas chromatograph coupled to a metal oxide sensor for rapid assessment of stool samples from irritable bowel syndrome and inflammatory bowel disease patients. J Breath Res 8(2):026001
118. Fang B, Guo P, Yang M, Ma Y, Yan X, Jia Z, Zhang M et al (2020) A novel fluorescent enhancing platform based on DNA-scaffolded silver nanoclusters for potential inflammatory bowel disease-associated microRNA detection. Talanta 218:121122
119. Gholami MD, Sonar P, Ayoko GA, Izake EL (2020) A highly sensitive SERS quenching nanosensor for the determination of tumor necrosis factor alpha in blood. Sens Actuators, B Chem 310:127867
120. Gersemann M, Wehkamp J, Stange EF (2012) Innate immune dysfunction in inflammatory bowel disease. J Intern Med 271(5):421–428
121. Lu Y, Li X, Liu S, Zhang Y, Zhang D (2018) Toll-like receptors and inflammatory bowel disease. Front Immunol 9:72
122. Neuman MG (2007) Immune dysfunction in inflammatory bowel disease. Transl Res 149(4):173–186
123. Bonovas S, Fiorino G, Allocca M, Lytras T, Nikolopoulos GK, Peyrin-Biroulet L, Danese S (2016) Biologic therapies and risk of infection and malignancy in patients with inflammatory bowel disease: a systematic review and network meta-analysis. Clin Gastroenterol Hepatol 14(10):1385–1397
124. Sargazi G, Afzali D, Mostafavi A, Shadman A, Rezaee B, Zarrintaj P, Mozafari M et al (2019) Chitosan/polyvinyl alcohol nanofibrous membranes: towards green super-adsorbents for toxic gases. Heliyon 5(4):e01527
125. Sargazi G, Khajeh Ebrahimi A, Afzali D, Badoei-dalfard A, Malekabadi S, Karami Z (2019) Fabrication of PVA/ZnO fibrous composite polymer as a novel sorbent for arsenic removal: design and a systematic study. Polym Bull 76:5661–5682
126. Miroliaee AE, Esmaily H, Vaziri-Bami A, Baeeri M, Shahverdi AR, Abdollahi M (2011) Amelioration of experimental colitis by a novel nanoselenium–silymarin mixture. Toxicol Mech Methods 21(3):200–208
127. Laroui H, Dalmasso G, Nguyen HTT, Yan Y, Sitaraman SV, Merlin D (2010) Drug-loaded nanoparticles targeted to the colon with polysaccharide hydrogel reduce colitis in a mouse model. Gastroenterology 138(3):843–853
128. Theiss AL, Laroui H, Obertone TS, Chowdhury I, Thompson WE, Merlin D, Sitaraman SV (2011) Nanoparticle-based therapeutic delivery of prohibitin to the colonic epithelial cells ameliorates acute murine colitis. Inflamm Bowel Dis 17(5):1163–1176
129. Ocansey DK, Zhang L, Wang Y, Yan Y, Qian H, Zhang X, Mao F et al (2020) Exosome-mediated effects and applications in inflammatory bowel disease. Biol Rev 95(5):1287–1307
130. Ju S, Mu J, Dokland T, Zhuang X, Wang Q, Jiang H, Zhang HG et al (2013) Grape exosome-like nanoparticles induce intestinal stem cells and protect mice from DSS-induced colitis. Mol Ther 21(7):1345–1357
131. Azandaryani AH, Kashanian S, Jamshidnejad-Tosaramandani T (2019) Recent insights into effective nanomaterials and biomacromolecules conjugation in advanced drug targeting. Curr Pharm Biotechnol 20(7):526–541
132. Teruel AH, Gonzalez-Alvarez I, Bermejo M, Merino V, Marcos MD, Sancenon F, Martinez-Mañez R et al (2020) New insights of oral colonic drug delivery systems for inflammatory bowel disease therapy. Int J Mol Sci 21(18):6502
133. Zhang M, Merlin D (2018) Nanoparticle-based oral drug delivery systems targeting the colon for treatment of ulcerative colitis. Inflamm Bowel Dis 24(7):1401–1415
134. Guo J, Jiang X, Gui S (2016) RNA interference-based nanosystems for inflammatory bowel disease therapy. Int J Nanomed 11:5287

135. Gou S, Huang Y, Wan Y, Ma Y, Zhou X, Tong X, Xiao B et al (2019) Multi-bioresponsive silk fibroin-based nanoparticles with on-demand cytoplasmic drug release capacity for CD44-targeted alleviation of ulcerative colitis. Biomaterials 212:39–54
136. Davoudi Z, Peroutka-Bigus N, Bellaire B, Wannemuehler M, Barrett TA, Narasimhan B, Wang Q (2020) Intestinal organoids containing PLGA nanoparticles for the treatment of inflammatory bowel. In: Intestinal organoids as new treatments for inflammatory bowel disease, 70
137. Sainz V, Conniot J, Matos AI, Peres C, Zupanŏiŏ E, Moura L, Gaspar RS et al (2015) Regulatory aspects on nanomedicines. Biochem Biophys Res Commun 468(3):504–510
138. Tinkle S, McNeil SE, Mühlebach S, Bawa R, Borchard G, Barenholz Y, Desai N et al (2014) Nanomedicines: addressing the scientific and regulatory gap. Ann NY Acad Sci 1313(1):35–56
139. Hafner A, Lovrić J, Lakoš GP, Pepić I (2014) Nanotherapeutics in the EU: an overview on current state and future directions. Int J Nanomed 9:1005
140. Kumar Teli M, Mutalik S, Rajanikant GK (2010) Nanotechnology and nanomedicine: going small means aiming big. Curr Pharm Des 16(16):1882–1892
141. Grainger DW, Bae YH (2015) The 17th international symposium on recent advances in drug delivery, 14–17 June 2015, Salt Lake City, UT, USA. Preface. J Control Release Off J Controlled Release Soc 220(Pt B):569–570
142. Hofmann-Amtenbrink M, Grainger DW, Hofmann H (2015) Nanoparticles in medicine: Current challenges facing inorganic nanoparticle toxicity assessments and standardizations. Nanomed Nanotechnol Biol Med 11(7):1689–1694
143. Bhise NS, Ribas J, Manoharan V, Zhang YS, Polini A, Massa S, Khademhosseini A et al (2014) Organ-on-a-chip platforms for studying drug delivery systems. J Control Release 190:82–93
144. Chabib L, Ikawati Z, Martien R, Ismail H, Wahyudi MDP, Arimurni DA, Hidayat A et al (2018) Rheumatoid arthritis and the challenge of using nanoparticles for its treatment. In: MATEC web of conferences, vol 154. EDP Sciences, p 04005

Nanotechnology-Based Therapeutic Approaches for the Management of Osteoarthritis

Mrinmoy Ghosh, Md. Meraj Ansari, and Young-Ok Son

Abstract Osteoarthritis (OA) is very common among the elderly, affecting 33% of people aged 60 to 70 and 43.7% of those over 80. This high prevalence emphasizes the importance of appropriate treatment and management measures for an aging population. Both non-pharmacological and pharmaceutical treatments are employed for OA non-pharmacological treatments, such as physiotherapy and exercise, are intended to improve joint function and alleviate symptoms whereas nonsteroidal anti-inflammatory medications, opioids, and glucocorticoids are all pharmacological treatments that help with pain and inflammation management. Medications that help control inflammation and lessen pain include glucocorticoids, opioids, and nonsteroidal anti-inflammatory drugs (NSAIDs). NSAIDs can reduce pain by 30–40%, but prolonged usage might cause serious negative effects. Joint replacement surgery is still the only choice for severe instances when other therapies fail to offer relief. Hence due to the limitations of the current available therapies, there is an urgent need for therapeutic techniques that can successfully reduce the progression of OA and avoid irreversible cartilage deterioration. Nanotechnology has tremendous potential due to its excellent biocompatibility and degradability, as well as the ability to provide specific therapeutic molecules. Drugs can be delivered to targeted locations using nanoparticle-based drug delivery systems, increasing therapeutic benefits

M. Ghosh · Md. M. Ansari · Y.-O. Son (✉)
Department of Animal Biotechnology, Faculty of Biotechnology, College of Applied Life Sciences, Jeju National University, Jeju Special Self-Governing Province, Jeju, South Korea
e-mail: sounagi@jejunu.ac.kr

M. Ghosh
Department of Biotechnology, School of Bio, Chemical and Processing Engineering (SBCE), Kalasalingam Academy of Research and Educational, Krishnankoil, India

Md. M. Ansari · Y.-O. Son
Interdisciplinary Graduate Program in Advanced Convergence Technology and Science, Jeju National University, Jeju Special Self-Governing Province, Jeju, South Korea

Y.-O. Son
Bio-Health Materials Core-Facility Center, Jeju National University, Jeju, South Korea

Practical Translational Research Center, Jeju National University, Jeju, South Korea

Md. M. Ansari et al. (eds.), *Emergence of Sustainable Biomaterials in Tackling Inflammatory Diseases*, Smart Nanomaterials Technology,
https://doi.org/10.1007/978-981-96-2112-5_10

and reducing unintended side effects. Thus, incorporating nanotechnology into OA therapy is a promising area that could lead to more effective and precise treatment.

Keywords Cartilage degradation · Chondrocytes · Drug delivery · Inflammation · Nanocarriers · Nanotechnology · Osteoarthritis

1 Introduction

Arthritis is joint joint-related condition that commonly causes joint inflammation and deterioration. By 2030, the World Health Organization projects that a large percentage of adults over 60 will have osteoarthritis (OA). As a result, OA will have a substantial effect on morbidity, mortality, and healthcare expenses for people of all income levels. In people with OA, physiotherapy and exercise can help reduce pain by 10–20% and improve function by 20–25%. However, it is critical to understand the anabolic and catabolic factors that influence OA progression and other cartilage disorders, as well as the treatment's therapeutic potential in managing these conditions by both mitigating and enhancing cartilage breakdown and synthesis [1, 2].

The OA progression is complex and multifunctional, involving the entire joint structure. The aging process has a significant impact on joint cell senescence. Age-relation accumulation of senescent cells, preventing cellular proliferation and regeneration [3]. These factors contribute to the degradation of the cartilage, meniscus, synovial membrane, subchondral bone, ligaments, joint capsule, and periarticular musculature, resulting in pain, stiffness, and joint dysfunction [3, 4]. Several inflammatory mediators, including cytokines, ROS, reactive nitrogen species, and matrix metalloproteinase (MMP), influence the progression of OA. A recent study on salt therapy discovered that it has a concentration-dependent effect on chondrocytes stimulated by pro-inflammatory cytokines. It specifically increased anabolic factors like aggrecan, SOX9, and COL2A1. Aggrecan is an essential component of the cartilage extracellular matrix, aiding in its structural integrity and function. The transcription factor SOX9 regulates the growth of chondrocytes and the synthesis of cartilage, while COL2A1 encodes the alpha-1 chain of type II collagen, which is a vital structural element of cartilage. Catabolic proteins such as MMP3, MMP13, ADAMTS4, and ADAMTS5 were all dramatically downregulated in expression at the same time by the therapy. MMP3 and MMP13 are enzymes that dissolve extracellular matrix components, causing cartilage degradation. The aggrecanases ADAMTS4 and ADAMTS5 are highly effective at breaking down aggrecan, resulting in cartilage matrix breakdown [5–7]. Pharmacological treatments, such as nonsteroidal anti-inflammatory medicines, can assist in alleviating pain; but, if used for an extended period, these drugs may have substantial side effects. A significant risk of osteoporosis, a 2–4% rise in the risk of gastrointestinal problems, and a 10–50% increase in coronary heart disease are linked to long-term NSAID use.

Numerous nanomedicines have been developed to successfully treat OA as drug delivery devices based on nanotechnology advances. These nanomedicines enable therapeutic substances to enter deeply and stay there for a long time while it administered directly to the site of action [8]. Nanotechnologies, with a small footprint, high surface area-to-volume ratio, and controllable surface qualities, can avoid many of the drawbacks associated with traditional therapies. Drug release can be precisely controlled using systems based on nanotechnology by targeting specific body locations and maintaining appropriate drug concentrations, increasing therapeutic efficacy while reducing unwanted effects. Thus, treating osteoarthritis and other joint disorders might be significantly advanced, patient outcomes and quality of life could be improved, and nanotechnology could be further explored and integrated into medicine [9, 10].

2 Pathogenesis of the Development of OA

OA is a chronic disease and it has a complicated pathophysiological mechanism of initiation and progression. Researchers have extensively studied in preclinical and clinical settings on the mechanism of disease initiation and progression but still the exact mechanism underlying its initiation and progression is elusive. After extensive studies, researchers revealed that the structural alteration in the synovial joint is a well-established factor for OA [11]. Age-related senescent cell accumulation, joint tissue alteration, signaling pathways, and some protein molecules play significant roles in OA pathogenesis.

2.1 Interactions of Tissue in OA

As we know, age-related factor plays a significant role in OA's pathogenesis. The subchondral bone sclerosis is the characteristic feature of OA. Researchers demonstrated that subchondral bone sclerosis is one of the major causes of age-related OA and that aberrant bone remodeling caused by dysregulation of osteoblasts and osteoclasts plays an important role in initiation and progression of OA [12–14]. The early phases of OA were characterized by lower bone density, increased subchondral bone porosity and remodeling, and abnormal matrix organization in bone mineralization. These findings were attributed to subchondral pores and vascular invasion as the cause of bone-cartilage crosstalk [15, 16]. Based on findings and facts, researchers conclude that these factors play an important role in the early initiation of OA. In the advanced stage or late stage of OA, researchers revealed that remarkable alteration in subchondral bone morphology due to loss of bone structure and subchondral bone densification that intensify sclerosis [17]. With time, researchers revealed that the conventional theory of subchondral bone sclerosis in OA pathogenesis has a strong correlation with age-related joint tissue deterioration. Additionally, it has been demonstrated that OA is associated with early bone loss because of fast bone remodeling followed by delayed turnover, which causes the subchondral plate to densify and lose all of its cartilage [18]. However, it is unknown how subchondral bone interacts with articular cartilage and joint problems to cause the initiation and progression of OA. Further research is necessary to clarify this. Another tissue-related factor is synovitis in which inflammation of the synovium membrane occurs. Researchers revealed that before the onset of articular cartilage deterioration, joint tissue-related inflammation has remarkable changes in the form of synovitis in the very early phase of OA [19]. Researchers have revealed the involvement of synovium in the development and course of OA for the past 20 years. Histological alterations in the early phases of arthritis include synovial lining expansion and hyperplasia, increased angiogenesis, modest levels of synovial inflammation, and synovial fibrosis [20]. Synovitis is inflammation of the synovial membrane that causes joint pain and swelling. T cells, macrophages, and synoviocytes that resemble fibroblasts are some of the stages that lead to the influx of inflammatory cells to synovial tissue. At the last stage of OA, synovitis could show an increased amount of macrophage infiltration [21]. These histopathological findings are used to calculate synovitis scores, which are one of the OA assessment tools [22]. Low-grade synovial inflammation can be found in more than half of OA patients in both the early and late phases of OA [23]. As a result, the OA research community has focused the most attention on synovial inflammation. Macrophages in the synovium tissue play a significant role in joint inflammation by releasing chemicals that cause fibroblasts to create enzymes that degrade the cartilage matrix. They can also defend the synovium from inflammatory attacks by decreasing immune cell infiltration, lowering inflammation, and digesting neutrophils. Proinflammatory cytokines, including IL-1 and TNF-α, play a crucial role in synovitis and influence other inflammatory cytokines and enzymes [21, 24] (Fig. 1). However, the interactions of different cell types in the synovial joint, as well as synovial characteristics at different stages of OA, need to be thoroughly studied.

2.2 Changes in Articular Cartilage in OA

A particularly specialized connective tissue found in synovial joints is known as articular cartilage [25]. It is composed of a dense extracellular matrix, hyaline cartilage, and chondrocytes [26, 27]. These chondrocytes are involved in the production of enzymes and proteoglycan, collagen, and respond to mechanical and chemical stimuli. The breakdown of articular cartilage is a feature of the vast majority of OA cases, particularly knee OA. Chondrocytes in degradative articular cartilage exhibit greater active cell death than those in normal articular cartilage. The earliest recorded form of cell death is apoptosis, which is a carefully regulated planned cell death [28]. Whereas the TUNEL assay detects low levels of apoptosis in normal articular cartilage, the number of apoptotic cells increases considerably in the early and late phases of OA articular chondrocytes. Several studies have indicated that death receptors (e.g., FAS, TNFR) [29], cytokines (e.g., TNF-α, IL-1β) [30–32], aberrant mechanical stress, and physicochemical variables, such as ultraviolet and ionizing radiation, can trigger chondrocyte apoptosis [33]. Furthermore, some signaling mechanisms

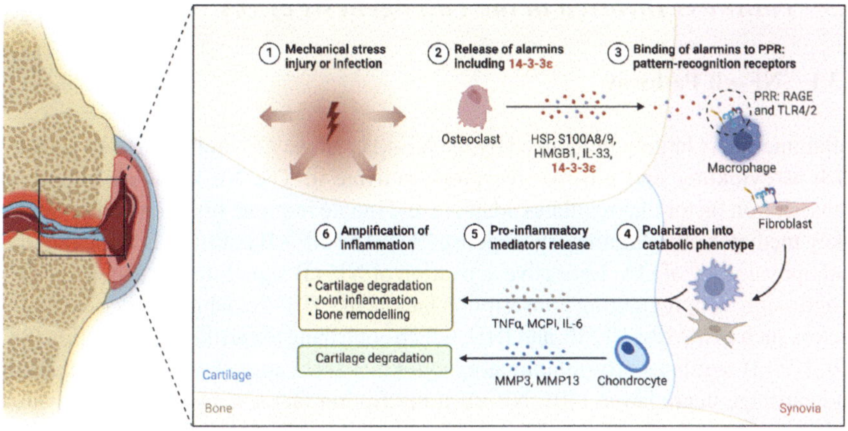

Fig. 1 The figure illustrates the pathway involved in the pathogenesis of OA. Numerous interrelated processes, including mechanical, metabolic, and genetic variables, are involved in the development of OA. Mechanical stress has a substantial impact on OA development. Obesity, joint injury, or repetitive activity can all cause excessive mechanical stress, resulting in cartilage degradation. Chondrocytes, or cartilage cells, respond to mechanical stress by generating matrix-degrading enzymes such as MMPs and ADAMTS. Increased activity of these enzymes causes the extracellular matrix to break down, resulting in cartilage deterioration. Inflammation is another important mechanism in OA. Joint damage or cartilage debris are common causes of synovial inflammation. Pro-inflammatory cytokines, including IL-1β and TNF-α, are released during inflammation. Image created with Biorender.com

have been identified as activators of apoptosis of articular chondrocytes in OA are mitochondrial-mediated caspase-dependent pathways, the death receptor route, and the endoplasmic reticulum stress-induced unfolded protein response. The aberrant buildup of unfolded or misfolded proteins can cause endoplasmic reticulum stress, which is involved in the pathophysiology of OA. The unfolded protein response is a signaling system that is activated by the accumulation of unfolded or misfolded proteins in the ER. This reaction helps to restore homeostasis by upregulating chaperone proteins, among other things [34, 35]. However, it is unclear whether OA is caused by or results from apoptosis.

The morphology of chondrocytes is a round shape situated in a deeper layer, which is divided and differentiated by chondrocytes from the superficial layer [36]. OA is revealed by the enhanced production of ECM-degrading enzymes and the removal of articular cartilage, similar to endochondral ossification, and altered articular chondrocyte hypertrophic differentiation is involved in the onset and progression of OA. Thus, factors that accelerate this process are probably important hallmarks for inducing OA. In recent decades, more and more data suggest that metabolism plays an important role in maintaining the energy balance of articular chondrocytes and that aberrations in chondrocyte metabolism might lead to the initiation and progression of OA. Persistent inflammation in the synovium membrane, angiogenesis, hypoxia, and oxidative stress are constantly contributing to the pathogenesis of OA.

2.3 Pathways Involved in the Pathogenesis of OA

2.3.1 NF-κB Pathway

Inflammation is largely influenced by the NF-κB pathway. Proinflammatory signals, such as cytokines and toll-like receptors, activate it. The NF-κB is an important transcription factor that regulates adaptive and innate immune function and serves as a key mediator of inflammatory response [37]. The NF-κB pathway is critical in the pathophysiology of OA. Excessive activation of NF-κB signaling can accelerate OA progression by upregulating transcription factors like HIF-2α, which induce catabolic factors such as MMPs, VEGF, and IHH, which contribute to cartilage degradation [38, 39]. NF-κB regulates pro-inflammatory cytokines that cause chondrocyte cell death and cartilage degradation [40]. NF-κB transcription factor is abnormally activated in OA and contributes to the disease, according to studies [41–43]. NF-κB plays a role in several OA-related events, including chondrocyte catabolism, survival, and synovial inflammation. Potential targets for treating OA include NF-κB, its upstream regulators, co-factors, and downstream effectors [44].

2.3.2 MTOR Pathway

A nutrient-sensing signaling cascade that controls cell growth and metabolism in response to the dietary environment is the target of rapamycin (TOR). The drug, rapamycin, was identified prior to the TOR pathway. Researchers demonstrated the significant value of the mTOR pathway in chronic inflammation and inflammatory diseases. In a study, researchers demonstrated that this pathway plays a remarkable function in the recruitments and activation of immune cells contributing to inflammatory diseases like OA [45]. Researchers revealed that the mTOR pathway is involved in the progression of OA in a variety of ways. It controls chondrocyte proliferation, apoptosis, autophagy, inflammation, and ECM homeostasis. Furthermore, researchers conclude that dysregulation of these cellular mechanisms remarkably contributes to the onset and progression of disease. Researchers have explored that mechanical stress degrades articular cartilage by activating mTORC1 signaling. In a study, researchers demonstrated that mechanical loading is a critical factor in the development and progression of OA [46]. Furthermore, researchers revealed that these factors modulate chondrocyte apoptosis and autophagy in the temporomandibular joint in vitro and in vivo [47]. Now it is clear from extensive studies that mTOR contributes to synovial inflammation [48]. We already know that Rapamycin is a mTORC1 inhibitor and researchers demonstrated that it lowers ADAMTS5 and IL-1β levels in synovitis in OA mice models. It is also revealed by scientists that micro RNAs (miRNAs) as an epigenetic mechanism that regulates gene expression and play a remarkable role in OA. In a study, researchers revealed that the miR-7, miR-27a, and miR-218-5p suppress the PI3K/Akt/mTOR signaling pathway, promoting autophagy and preventing apoptosis in chondrocytes [49, 50]. In another study, researchers revealed that exosomes produced from MSCs in the sub-patellar fat pad reduced DMM-induced OA in mice by increasing autophagy and suppressing mTOR via miR-100-5p [51]. Targeting the mTOR pathway is being investigated as a potential treatment for OA [45].

2.3.3 HIF Pathway

Hypoxia-inducible factors (HIFs) are essential modulators of the body's adaptation to low oxygen concentrations, or hypoxia, a characteristic frequently observed at inflammatory locations. They regulate the expression of genes related to a number of different activities, including glycolysis, angiogenesis, and the prevention of apoptosis. By influencing neutrophil function and encouraging the resolution of inflammation through a variety of pathways, HIFs also have an impact on the control of immune cells during inflammation. Furthermore, it has been discovered that HIF1α and HIF2α overexpress in inflammatory bowel illness, suggesting their important involvement in inflammatory diseases. The extensive studies by researchers revealed that HIFs play a remarkable role in OA pathogenesis. HIF-1α promotes chondrocyte development and survival, required for healthy cartilage. We already know that HIF-1α promotes the development of chondrocytes and their survival for healthy cartilage. Whereas, HIF-2α promotes OA because it is involved in the regulation of genes for chondrocyte hypertrophy, differentiation, and degradation of extracellular matrix. HIF is a transcription factor that is critical in the cellular response to low oxygen, orchestrating a metabolic transition that enables cells to survive in this environment. In immunity, diseased and inflamed tissues are frequently hypoxic, and HIF aids immune cells in adapting. During immunity and inflammation, HIF-α stabilizes and regulates metabolism as well as immune gene expression under normal conditions [52]. Researchers revealed that from in vivo study, in hypoxia conditions HIF-1α was a remarkable survival factor [53]. In another study, researchers demonstrated that in promoting chondrocytes survival in hypoxia conditions, HIF-1α suppresses mitochondrial respiration in the embryonic development plate [54]. As we know articular cartilage is a vascular tissue, it receives a limited supply of oxygen [55, 56]. As a result, researchers revealed that under the physiological condition, the articular chondrocytes are hypoxic in both mice and humans, accumulating HIF-1α and HIF-2α proteins were found in articular cartilage [57, 58]. Early in post-traumatic OA mouse models, researchers revealed less expression of HIF-1α-positive articular chondrocytes [59]. Furthermore, researchers revealed that HIF-1α and its target genes are expressed in the articular cartilage of normal human knee joints and are markedly elevated in the cartilage and synovial fluid of OA patients' joints [60]. Unlike HIF-1α, HIF-2α has been identified as a risk factor that directly stimulates the expression of catabolic genes such as MMP1, MMP3, MMP13, ADAMTS4, and NOS2 as well as the OA risk factor Runx2. This process leads to the breakdown of cartilage [58].

2.4 *Molecules Involved in the Pathogenesis of OA*

2.4.1 Matrix Metalloproteinases

MMPs, commonly referred to as matrixins, are a class of zinc-dependent proteins that are believed to be essential for the degradation of extracellular matrix. Major components cleaved by MMPs include collagen, elastin, gelatin, and casein [61]. MMPs regulate the bioavailability and activity of cytokines, chemokines, and growth factors, which in turn regulate inflammation, as well as the integrity of physical tissue barriers. Moreover, MMPs control inflammation in the tumor microenvironment and aid in the immune evasion of tumor cells [62]. Researchers revealed that MMPs are remarkable molecular markers of OA pathophysiology in humans. Collagenase is a class of multimeric proteases that includes MMPs 1, 8, and 13. Collagenase is an enzyme that breaks down collagen and aids in the catabolic turnover of articular cartilage. Researchers demonstrated that MMP13 has a remarkable connection with ECM degradation of articular cartilage in OA. As we know that collagen Type II is most prevalent in articular cartilage is also demonstrated by researchers as a substrate for MMP13. The DMM- induced OA is well studied animal model. Researchers revealed that MMP13 deleted mice have less articular cartilage degradation as compared with wild type [63]. Furthermore, researchers demonstrated that remarkably elevated level of MMP13 in the synovium and articular cartilage at early phase of disease in the articular cartilage of DMM-induced OA [64], MMP2, MMP3, and MMP14 are highly elevated at one-day, one-week, and six-week time points [65]. During the early stages of OA, inhibiting MMP13 or other MMPs may be a useful therapeutic approach to stop cartilage degradation and the OA formation process.

2.4.2 ADAMTS and ADAMs

The proteins known as ADAMTS mediate the chondrocytes' ECM breakdown, which results in the distraction of cartilage integrity. These enzymes are known to play important roles in OA and are therefore viable targets for the development of OA therapies. It is revealed by scientists that these enzymes contribute to OA pathogenesis and are viable targets for the development of new drugs against OA therapies. There are 19 secreted zinc metalloproteinases protein members in the ADAMTS family. The primary proteases involved in the breakdown of aggrecan, the main building block of cartilage extracellular matrix, are ADAMTS4/5, which are expressed in the tissues of OA cartilage in both humans and animals. The aforementioned enzymes may have functions in cartilage matrix anabolism, such as the procollagen N-proteinases ADAMTS-2, ADAMTS-3, and ADAMTS-14; they may also be involved in chondrocyte differentiation and proliferation, such as ADAM9, ADAM10, and ADAM12; additionally, they may be involved in cartilage catabolism, such as the enzymes that degrade cartilage oligomeric protein, which is ADAMTS-7 and ADAMTS-12 [66]. Even in the early stages of osteoarthritis, articular chondrocytes had much higher expression of ADAMTS4/5. An OA model in mice created surgically showed that ADAMTS5 deletion prevented cartilage degradation in joints [67, 68].

2.4.3 Lubricin/Proteoglycan-4 (PRG4)

Proteoglycan 4 (PRG4) or lubricin is first discovered in synovial fluid. Researchers revealed that it serves as a structural protein of the extracellular matrix of joints that has a remarkable contribution to adhesion-dependent synovial growth regulation. Furthermore, it involves bulk protein deposition onto the cartilage and more importantly protects cartilage from wear and tear [69]. On the basis of researchers finding, it is concluded that the expression and function of PRG4 have a strong correlation with inflammatory joint complications such as OA [70]. Furthermore, extensive study researchers revealed that embryonic joint progenitors express Prg4 as a remarkable factor that protects joints. The more intensified studies revealed that the PRG4-positive articular chondrocytes found on the surface of the joint cartilage in the adult mice are precursors of mature articular cartilage present at deeper layers [36, 71]. In this direction, more intensified studies are further required to explore whether the case of enhanced expression of PRG4 in the synovial joint might promote the regeneration of articular cartilage or inhibit the OA onset. Nowadays, we can consider the PRG4 as an important marker for OA. Recently, in a clinical study researchers revealed that lubricin a possible biomarker in human synovial fluid for the purpose of diagnosis and OA progression in patients (NCT02664870).

3 Nanoparticles in the Treatment of OA

Intra-articular and systemic injections used to treat OA are challenged by the pharmacological agent's disorganized diffusion within the body. This has led to a significant focus of current research on the targeted delivery of medications using trustworthy OA targeting techniques. Nanomaterials are ideal carriers for enhancing cargo stability and achieving targeted therapies due to their size advantage. NP carriers are classified into three types: lipid-based, polymeric, and inorganic NPs (Table 1). By improving drug localization and retention in the affected place, targeted biomaterials can significantly enhance the overall treatment outcomes for OA.

By utilizing nano polymers such as chitosan, sodium alginate, and hyaluronic acid (HA), along with synthetic polymers like poly (ethylene glycol) (PEG), poly(ε-caprolactone) (PCL), and poly (lactic-co-glycolic acid) (PLGA), therapeutic efficacy is enhanced through site-specific targeting and side effects are minimized. Additional kinds of polymeric NPs, such as dendrimers, nanospheres, and nanocapsules, are often employed as carriers. A wide range of drug delivery applications can benefit from the improved stability, targeted dispersion, and precisely controlled release of therapeutic components provided by these diverse polymeric nanoparticle formulations and morphologies.

Table 1 Application of nanotechnology in osteoarthritis

Aspect	Details	Example	Advantages	Limitation
Therapeutic approaches	Drug delivery: nanocarriers such as liposomes, micelles, dendrimers, and polymeric NPs can transfer medications directly to the damaged joints, increasing efficacy while lowering side effects	Liposomal dexamethasone	Improved targeting, reduced systemic toxicity	Potential for immunogenicity, stability issues during storage
		PLGA nanoparticles delivering NSAIDs	Sustained release, reduced dosing frequency	Potential toxicity, complex manufacturing process
		Chitosan nanoparticles with hyaluronic acid	Enhanced joint lubrication, reduced pain	Limited long-term stability, potential for allergic reactions
		Dendrimer-based methotrexate	High drug loading capacity, reduced side effects	High cost of production, potential toxicity
Targeted therapy	Targeting inflammation: nanoparticles can be specific to target inflammatory tissues, enabling targeted treatment while reducing systemic exposure	Gold nanoparticles (AuNPs)	Precise targeting, anti-inflammatory properties	Potential toxicity, high cost of production
		Cerium oxide nanoparticles	Antioxidant properties, reduce oxidative stress	Long-term safety unknown, potential accumulation in the body
		Silica nanoparticles	Targeted drug delivery, biocompatibility	Potential toxicity, challenges in large-scale production
		PLGA-PEG nanoparticles with TNF-alpha inhibitors	Enhanced targeting of inflammatory cytokines	Stability issues, potential for immune response
Regenerative medicine	Scaffold-based therapies: nanofibers and nanoporous scaffolds can help develop new cartilage tissue by providing a condition suitable to cell proliferation and differentiation	Nanofibrous scaffolds	Promotes tissue regeneration, mimics natural extracellular matrix	Potential immune response, complex fabrication methods
		Hydrogel nanocomposites	Enhanced mechanical properties, support cell growth	Degradation rate control, potential for infection

(continued)

Table 1 (continued)

Aspect	Details	Example	Advantages	Limitation
		Carbon nanotubes	High strength, promotes cell adhesion	Biocompatibility issues, potential toxicity
		Collagen nanofibers	Biocompatible, promotes chondrocyte attachment and growth	Risk of immune rejection, high production costs
Types of nanomaterials used	Metallic nanoparticles: gold and silver nanoparticles have anti-inflammatory and antibacterial characteristics, which could be effective in treating joint infections and inflammation	Silver nanoparticles (AgNPs)	Anti-inflammatory, antimicrobial	Potential toxicity, environmental concerns
		Gold nanoparticles (AuNPs)	Anti-inflammatory, biocompatible	High cost, potential long-term toxicity
	Polymeric nanoparticles: biodegradable polymers such as PLGA (polylactic-co-glycolic acid) are employed for long-term medication release and precise distribution	PLGA nanoparticles	Biodegradable, controlled drug release	Complex manufacturing, potential immune response
		Chitosan nanoparticles	Biocompatible, enhances drug stability	Limited long-term stability, potential for allergic reactions
		Albumin nanoparticles	Natural, non-toxic, and biodegradable	Stability issues, potential for rapid clearance from the body
	Lipid-based nanoparticles: liposomes and solid lipid nanoparticles improve drug solubility and stability, which increases bioavailability	Solid lipid nanoparticles (SLNs)	Enhanced drug solubility, stable drug formulation	Stability issues, potential for rapid clearance from the body
		Liposomes	Biocompatible, versatile drug delivery system	Stability issues, potential for rapid clearance from the body
		Nanoemulsions	Enhanced bioavailability, improved drug delivery	Stability issues, potential for phase separation

(continued)

Table 1 (continued)

Aspect	Details	Example	Advantages	Limitation
Imaging and diagnosis	Enhanced imaging: nanoparticles can be employed as contrast agents in imaging techniques such as MRI and CT scans to aid in the detection and monitoring of osteoarthritis	Iron oxide nanoparticles	Enhanced MRI contrast, better diagnostic accuracy	Potential toxicity, accumulation in organs
		Gold nanoparticles (AuNPs)	CT imaging contrast, high biocompatibility	High cost, potential long-term toxicity
		Quantum dots	Fluorescent imaging, high sensitivity	Potential toxicity, environmental concerns
		Silica-coated iron oxide nanoparticles	Dual-modality imaging, improved resolution	Complex synthesis, potential toxicity

An additional OA therapy option is the synthetic polymer poly (lactic-co-glycolic acid) (PLGA). An anthraquinone medication called diacerein, which can halt the course of osteoarthritis OA, was loaded into almost spherical diacerein/PLGA NPs. In both in vitro and in vivo OA rat models, these NPs showed stronger anti-inflammatory advantages over free diacerein, significantly reducing pro-inflammatory variables such as IL-1, ADAMTS-5, and MMP-13 [72, 73]. In addition, Bedingfield et al. developed PLGA microtiter plates with MMP-13 siRNA NPs, which changed the way PLGA was used. It has been shown that this nano microcarrier technology releases siRNA NPs with high bioactivity and effectiveness continuously for five weeks [74]. These findings show that PLGA-based medicines have the potential to improve drug delivery by providing enhanced stability, targeted treatment, and enduring therapeutic effects for OA.

In a study, Zhao et al. revealed how to alter chitosan such that it functions as a near-infrared light-responsive nanocarrier by using molybdenum disulfide. This technique has been demonstrated to lower TNF-α and IL-1β levels in animal models of OA by prolonging the time that dexamethasone remains in the joint due to the effects of NIR light. Maudens et al. additionally created heat-responsive NPs that prevented OA mice models from producing VEGF, IL-1β, and TNF-α. When the HA (Hyaluronic acid)-pNiPAM system was used instead of traditional hyaluronic acid formulations, it was successful in prolonging the residence time of dexamethasone and intra-articular injectable materials [75, 76].

The characteristics of hyaluronic acid NPs (HA-NPs) for treating OA in normal mice were determined in another investigation, which also showed the NPs' penetration and distribution. Time-dependent alterations in the surface charge and particle size of HA-NPs were evaluated in Dulbecco's Modified Eagle Medium and phosphate-buffered saline. Comprehending the stability and behavior of HA-NPs in various contexts is contingent upon these investigations [77] (Fig. 2). These comprehensive investigations demonstrate the superior therapeutic agent delivery capabilities of HA-NPs for osteoarthritis, guaranteeing improved penetration, retention, and focused activity within the afflicted joint locations [78].

Liposome-based nanocarriers provide adaptability and improved capabilities for delivering a diverse range of biologically active substances. Liposomes have a significant benefit in that they can encapsulate a wide range of cargo types, with drug loading percentages ranging from 0.5% to 20%. This flexibility enables effective therapeutic encapsulation while retaining nanoparticle stability and integrity. A lipid NP-siRNA system has been shown to increase type II collagen levels while suppressing pro-inflammatory proteins like MMP-13, which slows cartilage deterioration [79]. Bhalekar et al. created lipid NPs containing chloroquine and examined them in OA models in rats in a different study. Mice treated with chloroquine-lipid NPs showed significantly less bone damage and cartilage loss than those treated with free chloroquine, according to radiological data [80]. These findings demonstrate how lipid nanoparticles (NPs) can target and lessen the underlying causes of inflammation and cartilage deterioration, improving the administration and effectiveness of therapeutic medications for osteoarthritis and providing improved outcomes. In response to IL-1β activation, soybean phosphatidylcholine liposomes containing curcumin a water-soluble compound with therapeutic promise for OA boosted the OPG/RANKL ratio. This assisted in reducing the expression of pro-inflammatory genes in osteoblasts upon stimulation and inhibiting osteoclastogenesis [81].

For targeted therapy, modified mesoporous silica NPs, metals, and metal oxides are also employed as inorganic nanoparticles. Modified mesoporous silica nanoparticles (NPs) functionalized with polyacrylic acid (PAA) were created by researchers to improve the therapeutic administration of andrographolide, a medication known for its antioxidant and anti-inflammatory properties [82, 83]. These modifications provided the NPs pH-responsive qualities, which are essential for managing drug release in various inflammatory settings. The objective of the functionalized mesoporous silica NPs is to reduce andrographolide's unwanted release in low-inflammatory conditions. By ensuring that the medication is predominantly given where it is needed, this specific release mechanism reduces the possibility of side effects and improves the effectiveness of treatment. In vitro tests showed that the modified NPs outperformed both free andrographolide and andrographolide loaded in non-functionalized mesoporous silica NPs in terms of chondrocyte cell survival. This shows that the PAA-functionalized NPs were less cytotoxic and better at controlling the drug's release, both of which improved cell conditions. The benefits of this approach were subsequently validated by in vivo studies using surgically-induced rat models of OA [82]. In comparison to conventional treatment modalities, the modified mesoporous silica NPs demonstrated superior therapeutic effects, as demonstrated by enhanced cartilage health and decreased inflammation. This implies that the pH-responsive drug delivery technique not only increases the stability and efficacy of andrographolide but also significantly slows the progression of osteoarthritis by delivering particular anti-inflammatory and antioxidant properties.

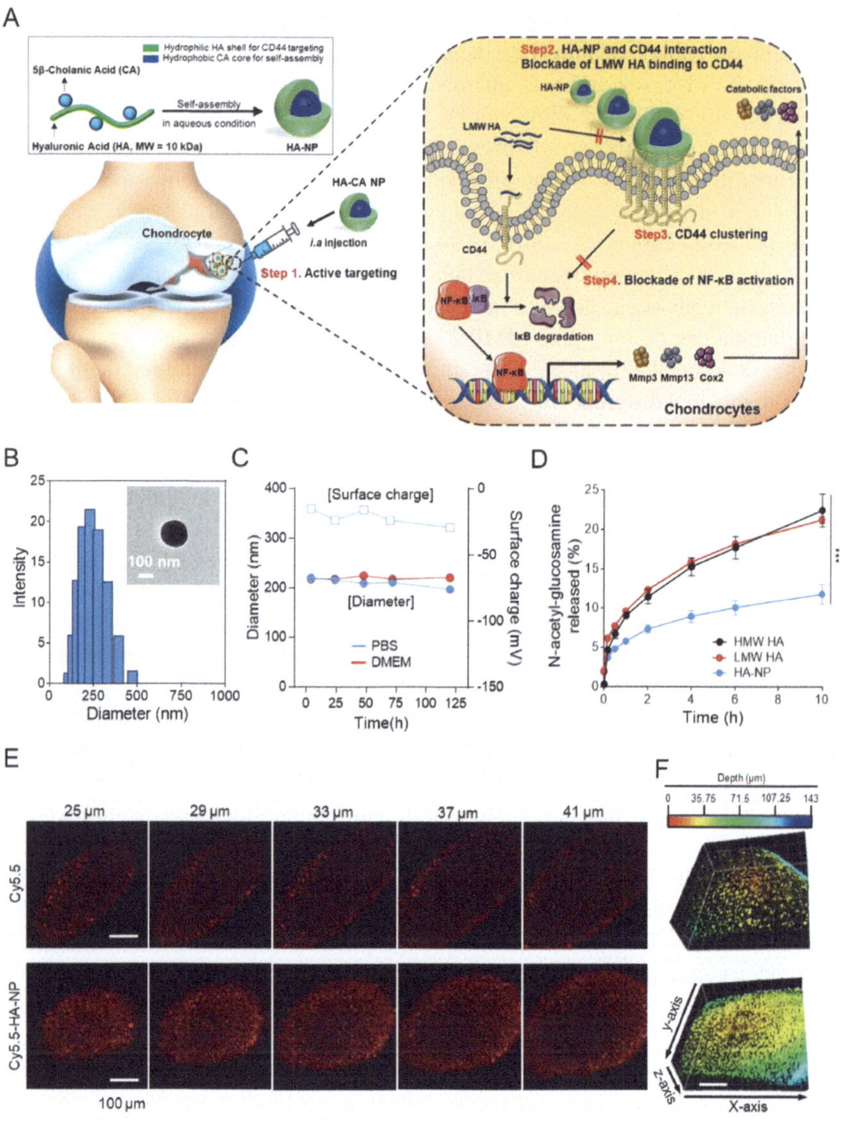

Fig. 2 A schematic image depicts the development and application of HA-NPs, showing their structural components and mode of action in treating OA-affected areas. Transmission electron microscopy images show the shape and homogeneity of HA-NPs, with a scale bar of 100 nm to emphasize their nanoscale size. A size distribution graph depicts the consistent size range of these particles. The amount of N-acetylglucosamine produced was determined after treating with 1 mg/ml free HAs (10 kDa low molecular weight and 2000 kDa high molecular weight) or HA-NP with 100 IU/ml hyaluronidase-II (HYAL-II). The data, shown as mean ± SEM (n = 4), demonstrated substantial differences (***$P < 0.001$), suggesting the efficiency of HA-NPs in preserving stability and therapeutic activity [77]. Copyright 2021, with permission from Elsevier

Micelles are composed of amphiphilic polymers that self-assemble in aqueous liquids. Their sizes range from 20 to 200 nm. Hydrophobic pharmaceuticals are trapped in the core of the micelle, however, hydrophilic drugs can also be connected with it. Hydrophobic tiny molecules like indomethacin, triamcinolone, and kartogenin are commonly used in micellar drug delivery systems for osteoarthritis (OA) and can be administered either locally or systemically. The indomethacin loading efficiency of indomethacin-loaded poly(N-isopropylacrylamide)-polyphosphazene micelles functionalized with ethyl 4-aminobenzoate is higher than that of nonfunctionalized micelles [84]. Saadat et al. introduced a polymeric micelle system based on hyaluronic acid and 1,2-distearoylphosphatidyl-ethanolamine for the sustained delivery of triamcinolone. Triamcinolone is a well-known corticosteroid that is frequently used to treat OA and rheumatoid arthritis. It works by inhibiting the synthesis of COX1 and COX2, two enzymes that are involved in inflammation [85].

In addition to NPs, hydrogels, and scaffolds are excellent at delivering bioactive chemicals. Medicinal chemicals can be encapsulated and released with the aid of hydrogels, which are three-dimensional networks of cross-linked polymers that have a high water absorption capacity. Therapeutic compounds can be released in a controlled manner thanks to the ability of these biomaterials to be engineered with certain pore sizes and degradation rates. A novel drug-delivery hydrogel incorporating liposomes loaded with celecoxib with dynamic covalent bond-based HA was described by Huang et al. Tests conducted both in vivo and in vitro showed that this hydrogel functioned as a dependable drug delivery mechanism for the sustained release of celecoxib. This new hydrogel demonstrated the ability to maintain the anabolic-catabolic balance, enhance cartilage deterioration, and delay the onset of OA [86].

4 Drug Delivery System for OA Therapy

There are commonly four routes employed for the treatment of OA such as oral administration, intravenous injection, intra-articular injection and topical or transdermal application. The oral route of drug administration is most preferred and easily acceptable by patients because this route can easily bypass the fear of the patient from injection.

Intravenous injections face challenges in effectively distributing drugs to tissues like articular cartilage. As a result, it's possible that the therapeutic drugs won't reach the target location in high enough concentrations, which would decrease their effectiveness and necessitate bigger doses that raise the possibility of systemic side effects. When it comes to lowering systemic exposure and raising local drug concentration, intra-articular delivery works better than intravenous administration. This technique reduces the danger of systemic adverse effects and minimizes exposure to other organs by injecting directly into the joint. One major problem, though, is that the medication is quickly eliminated into the lymphatic system due to the continuous interchange of synovial fluid within the joint. Similar to intravenous injection, bigger molecules are evacuated by the lymphatics and smaller molecules by the synovial capillaries. This rapid clearance can reduce the drug's effectiveness, necessitating frequent injections or higher doses [78, 83, 87–89].

Oral administration is another systemic route that is often the most convenient for patients due to its ease of use. However, the environment of the gastrointestinal system, first-pass metabolism in the liver, and the possibility of gastrointestinal side effects can all affect the bioavailability of medications taken orally. In a study, researchers revealed that nanocarrier of yeast cell wall loaded with miR365 antagomir can easily translocated into knee joints via macrophage phagocytosis in intestine-associated lymphoid tissues and have better therapeutic effects [90, 91].

Transdermal delivery is an appealing alternative to injections, offering painless administration and improved patient compliance. This method allows for prolonged drug release and, unlike oral administration, bypasses the gastrointestinal tract, reducing the risk of metabolic loss and gastrointestinal disturbances. Transdermal delivery also eliminates concerns about cardiovascular complications associated with some oral medications. But still. Transdermal medication delivery is hampered by the skin's natural barrier qualities [87, 92].

Targeted delivery systems, such as NPs and functionalized nanocarriers, are engineered to recognize and bind to specific markers in the OA-affected tissues. This targeted approach not only maximizes the therapeutic effects of the drug but also reduces its systemic distribution, thereby lowering the risk of side effects (Table 2).

5 Current Status of Nano-based Therapy for OA Treatment

Several drugs are commonly used in the treatment of OA, each with specific functions and potential adverse effects. Acetaminophen, ibuprofen, and naproxen sodium are NSAID drugs that are frequently employed to reduce inflammation and pain by inhibiting the production of inflammatory hormones. While generally well-tolerated, excessive use can lead to liver toxicity, cardiovascular problems, bleeding issues, and damage to the liver and kidneys (Table 3).

Researchers formulated Liposomal Dexamethasone for the delivery of dexamethasone targeting synovial macrophage of inflamed joint tissue. According to a study, the lipophilic prodrug dexamethasone palmitate (DXP) significantly enhanced drug loading and burst release following intravenous administration when it was encapsulated in poly (DL-lactide-co-glycolide)-polyethylene glycol (PLGA-PEG) NPs. These DXP-loaded NPs (DXP-NPs) had a drug loading of 7.5% and a size of around 150 nm, with low cytotoxicity and a moderate hemolytic profile, indicating their suitability for application. DXP-NPs successfully suppressed macrophage inflammatory cytokine release in response to lipopolysaccharide (LPS), indicating their potential for targeted anti-inflammatory treatment in disorders such as osteoarthritis. This technique may modulate inflammation in OA, improving dexamethasone's therapeutic efficacy while decreasing systemic side effects [103].

Table 2 Targeted biomaterials for osteoarthritis treatment

Materials	Targeted strategies	Targeted cells	Particle size	Drugs	Biological functions	Advantages	Limitations
Liposomal dexamethasone	Targeted delivery to inflamed joints	Synovial macrophages	100–200 nm	Dexamethasone	Anti-inflammatory, reduces joint inflammation	Improved targeting, reduced systemic toxicity	Potential for immunogenicity, stability issues during storage
PLGA nanoparticles	Sustained release, surface modification for targeting	Chondrocytes, synoviocytes	100–500 nm	NSAIDs, methotrexate	Prolonged drug release, anti-inflammatory, cartilage protection	Biodegradable, controlled drug release	Potential toxicity, complex manufacturing process
Chitosan nanoparticles	Conjugation with hyaluronic acid for joint targeting	Chondrocytes	50–300 nm	Hyaluronic acid, curcumin	Enhanced joint lubrication, anti-inflammatory, promotes cartilage repair	Biocompatible, enhances drug stability	Limited long-term stability, potential for allergic reactions
Gold nanoparticles (AuNPs)	Surface modification with targeting ligands	Synovial macrophages, chondrocytes	10–100 nm	TNF-alpha inhibitors, IL-1 inhibitors	Targeted anti-inflammatory, reduces cytokine levels	Precise targeting, anti-inflammatory properties	Potential toxicity, high cost of production
Cerium oxide nanoparticles	Reactive oxygen species scavenging	Synoviocytes, chondrocytes	5–50 nm	None (intrinsic antioxidant properties)	Reduces oxidative stress, protects cartilage from degradation	Antioxidant properties, reduces oxidative stress	Long-term safety unknown, potential accumulation in the body

(continued)

Table 2 (continued)

Materials	Targeted strategies	Targeted cells	Particle size	Drugs	Biological functions	Advantages	Limitations
Silica nanoparticles	Functionalization for targeting inflamed tissues	Synovial macrophages, chondrocytes	50–200 nm	NSAIDs, Dexamethasone	Targeted drug delivery, reduces inflammation, improves drug bioavailability	Targeted drug delivery, biocompatibility	Potential toxicity, challenges in large-scale production
PLGA-PEG nanoparticles	PEGylation for enhanced circulation time and targeting	Synovial macrophages	100–200 nm	TNF-alpha inhibitors	Enhanced targeting of inflammatory cytokines, prolonged circulation	Enhanced targeting, prolonged circulation time	Stability issues, potential for immune response
Nanofibrous scaffolds	Scaffold-based cell delivery	Chondrocytes	Scaffold structure	Growth factors, stem cells	Supports tissue regeneration, promotes cell attachment and proliferation	Promotes tissue regeneration, mimics natural extracellular matrix	Potential immune response, complex fabrication methods
Hydrogel nanocomposites	Injectable gels with nanoparticle integration	Chondrocytes	Hydrogel structure	NSAIDs, Growth factors	Enhanced mechanical properties, sustained drug release, supports cell growth	Enhanced mechanical properties, supports cell growth	Degradation rate control, potential for infection
Carbon nanotubes	High surface area for drug loading	Chondrocytes, synoviocytes	1–100 nm	Anti-inflammatory drugs	High drug loading capacity, promotes cell adhesion	High strength, promotes cell adhesion	Biocompatibility issues, potential toxicity

(continued)

Table 2 (continued)

Materials	Targeted strategies	Targeted cells	Particle size	Drugs	Biological functions	Advantages	Limitations
Collagen nanofibers	Mimics extracellular matrix	Chondrocytes	Scaffold structure	Growth factors	Biocompatible, promotes chondrocyte attachment and growth	Biocompatible, promotes chondrocyte attachment and growth	Risk of immune rejection, high production costs
Iron oxide nanoparticles	Magnetic targeting for enhanced localization	Synovial macrophages, chondrocytes	10–100 nm	Dexamethasone, methotrexate	Improved MRI contrast, site-specific drug delivery, anti-inflammatory	Improved MRI contrast, site-specific drug delivery	Potential toxicity, accumulation in organs
Silver nanoparticles (AgNPs)	Antimicrobial targeting for infected joints	Synovial macrophages	10–50 nm	Antibiotics	Antimicrobial, reduces joint infection	Anti-inflammatory, antimicrobial	Potential toxicity, environmental concerns
Quantum dots	Fluorescent imaging for diagnostic targeting	Chondrocytes, synoviocytes	2–10 nm	Imaging agents	High-sensitivity imaging, tracks disease progression	High-sensitivity imaging, tracks disease progression	Potential toxicity, environmental concerns
Silica-coated iron oxide NPs	Dual-modality imaging and drug delivery	Synovial macrophages, chondrocytes	50–200 nm	NSAIDs, Imaging agents	Dual imaging capability, targeted drug delivery, enhances diagnostic accuracy	Dual imaging capability, targeted drug delivery	Complex synthesis, potential toxicity

(continued)

Table 2 (continued)

Materials	Targeted strategies	Targeted cells	Particle size	Drugs	Biological functions	Advantages	Limitations
Dendrimer-based nanoparticles	Multivalent drug loading for targeted delivery	Synovial macrophages	5–10 nm	Methotrexate, NSAIDs	Enhanced drug encapsulation capacity, site-targeted delivery, reduced side effects	High drug loading capacity, targeted delivery	High cost, potential toxicity
Albumin nanoparticles	Natural targeting due to endogenous protein nature	Chondrocytes, synoviocytes	100–200 nm	Curcumin, dexamethasone	Biocompatible, enhances drug solubility and stability, natural targeting properties	Natural, non-toxic, and biodegradable	Stability issues, potential for rapid clearance from the body
Solid lipid nanoparticles	Enhanced drug solubility and stability	Chondrocytes, synoviocytes	50–300 nm	NSAIDs, dexamethasone	Stable drug formulation, sustained release	Enhanced drug solubility, stable drug formulation	Stability issues, potential for rapid clearance from the body
Liposomes	Versatile drug delivery system	Synoviocytes	50–200 nm	Dexamethasone, curcumin	Biocompatible, versatile drug delivery system	Biocompatible, versatile drug delivery system	Stability issues, potential for rapid clearance from the body
Nanoemulsions	Improved drug delivery and bioavailability	Synovial macrophages, chondrocytes	100–500 nm	NSAIDs, curcumin	Enhanced bioavailability, improved drug delivery	Enhanced bioavailability, improved drug delivery	Stability issues, potential for phase separation

(continued)

Table 2 (continued)

Materials	Targeted strategies	Targeted cells	Particle size	Drugs	Biological functions	Advantages	Limitations
Curcumin nanoparticles	Enhanced solubility and stability	Chondrocytes	50–200 nm	Curcumin	Increased solubility, enhanced bioavailability	Increased solubility, enhanced bioavailability	Stability issues, potential for rapid clearance from the body
Resveratrol nanoparticles	Improved stability and therapeutic effect	Chondrocytes, synoviocytes	50–200 nm	Resveratrol	Improved stability, enhanced therapeutic effect	Improved stability, enhanced therapeutic effect	Stability issues, potential for rapid clearance from the body
Hydrogel-based nanoparticles	Sustained drug release, minimized dosing frequency	Chondrocytes	Hydrogel structure	NSAIDs, growth factors	Sustained release, supports cell growth	Sustained release, supports cell growth	Degradation rate control, potential for infection
Gene-editing nanoparticles	Personalized treatment options, gene-specific therapy	Synoviocytes	20–150 nm	CRISPR/Cas9 components	Personalized medicine, targeted gene editing	Personalized treatment options, potential for gene-specific therapy	Ethical concerns, potential off-target effects
RNA interference nanoparticles	Specific gene silencing, targeted treatment	Synoviocytes	20–150 nm	siRNA, miRNA	Specific gene silencing, targeted treatment	Specific gene silencing, targeted treatment	Stability issues, potential immune response

Table 3 List of the certain drugs commonly used for OA treatment

Drugs	Function	Adverse effect	References
Acetaminophen (non-opioid)	Analgesic, antipyretic, relief from mild to moderate pain	GI complications, hepatotoxicity	[93, 94]
Ibuprofen (NSAID)	Anti-inflammatory, relief from pain	Stomach upset, cardiovascular problems, hepatotoxicity and nephrotoxicity	[95, 96]
Naproxen sodium (NSAID)	Anti-inflammatory, relief from pain	Stomach upset, hepatotoxicity and nephrotoxicity	[96–98]
Duloxetine (cymbalta)	Antidepressant, relief from chronic pain	Nausea, fatigue, dizziness, increased blood pressure, sweating, and decreased appetite	[99, 100]
Methotrexate (MTX)	Relief from pain	GI complications, hepatotoxicity	[101, 102]

Liposomal dexamethasone (Dex-Lips) has been shown to inhibit the inflammatory response and alleviate pain symptoms in OA by affecting the polarization of macrophages. Dex-Lips were also found to reduce inflammation in both the cartilage and serum of the treated mice. This broad anti-inflammatory effect underscores the potential of Dex-Lips as a therapeutic approach for managing OA by targeting multiple aspects of the inflammatory cascade [104]. However, the usage of liposomal dexamethasone can lead to problems. The primary concern is immunogenicity, in which the immune system responds to the liposomal formulation, thereby limiting its effectiveness and generating serious immune system responses. Furthermore, stability difficulties during storage can be challenging, as liposomes may vary in size, aggregation, or drug leakage with time, affecting their efficacy or safety.

Peptides provide novel opportunities for changing certain intracellular targets with superior efficacy, consistency, and minimal side effects compared to small-molecule treatments. Therapeutic peptides and gene therapy formulations based on RNA have been studied for their potential to control intracellular targets. In mice models of post-traumatic OA, Duan et al. showed that intra-articular delivery of nanoformulated peptide-siRNA significantly reduced the periostin synthesis [105]. Moreover, researchers revealed that MMP-13 expression, subchondral bone sclerosis, bone volume/total volume, volumetric bone mineral density, and heterotopic ossification were all significantly reduced in treated animals. In order to maintain siRNA-loaded NPs (siNPs) in the joint for a lot longer than when they were supplied gratis, the researchers also made siRNA-loaded siNPs that were enclosed in PLGA-based microplates with particular forms. They inhibited MMP13, enabling post-traumatic OA to be treated for an extended period of time [106, 107]. Bedingfield et al. suppressed MMP-13 expression while preserving the integrity of the cartilage and joint structure by locally infusing NPs modified with antibodies that target type II collagen and small interfering RNA that targets MMP-13 [108].

In order to increase the effective functional duration of gene carriers and enable prompt disease treatment through microenvironmental reactions, gene therapy for OA faces a number of challenges. One of the challenges lies in the ability of gene carriers to move through the complex biological environment of the body, avoiding immune-mediated rejection and nuclease breakdown. The development of delivery mechanisms that can protect the genetic material until it reaches the intended cells is necessary to achieve this. One more important factor is stability in vivo. For therapeutic genes or nucleic acids to have the intended impact, they must stay stable and functional for a considerable amount of time after being delivered to the target site. Researchers are investigating a variety of techniques for addressing these difficulties. A possible strategy is to create gene carriers that respond to the joint's microenvironmental conditions. Hydrogels and NPs, for example, can respond to changes in pH, temperature, or enzymatic activity to offer immediate release of therapeutic genes in response to specific microenvironmental stimuli associated with OA.

By directly delivering therapeutic genes to the afflicted joint, gene-hydrogel microspheres present a potential treatment option for osteoarthritis. G5-AHP and miR-140 are combined in one method to create gene-hydrogel microspheres, which are then injected straight into the joint's articular region to transport miR-140 to the desired location and slow the progression of osteoarthritis (Fig. 3). Chondrocytes internalized it by endocytosing it. Once inside the cells, the polyplexes release miR-140, which has therapeutic effects. miR-140 has been demonstrated to regulate genes involved in cartilage homeostasis and inflammation, reducing cartilage degradation and increasing joint function[109].

6 Limitations and Prospects

The field of nanotherapeutics for OA treatment is still emerging. Growth factors, nucleic acids, and OA medication that modifies the disease can be delivered specifically with the help of NPs composed of polymers, lipids, and other components. These methods offer delayed, sustained, or triggered drug release, increased joint retention, and reduced off-target side effects. Nanostructured scaffolds imitate cartilage, enhance biochemical constituents, and improve biomechanical properties. However, many nanocarriers face limitations, such as the need for effective drug loading and the rarity of sustained-release systems in clinical use [106, 110]. Nanotechnology is also expensive and unknown to the majority of OA patients, which limits its therapeutic application. Researchers must continue to address these challenges. There are still several obstacles in the way of using nanotechnology to treat OA. These include developing more effective nanodrug delivery methods, switching from intra-articular injections to oral and topical medications, and treating OA ultra-early by addressing the illness at the genetic level. Additionally, researchers revealed that to hasten bone healing and increase the longevity of joint replacement prostheses, implant materials require improvement. To enhance nanoparticle efficacy and functionality, polymer materials are frequently used for surface modification [111, 112].

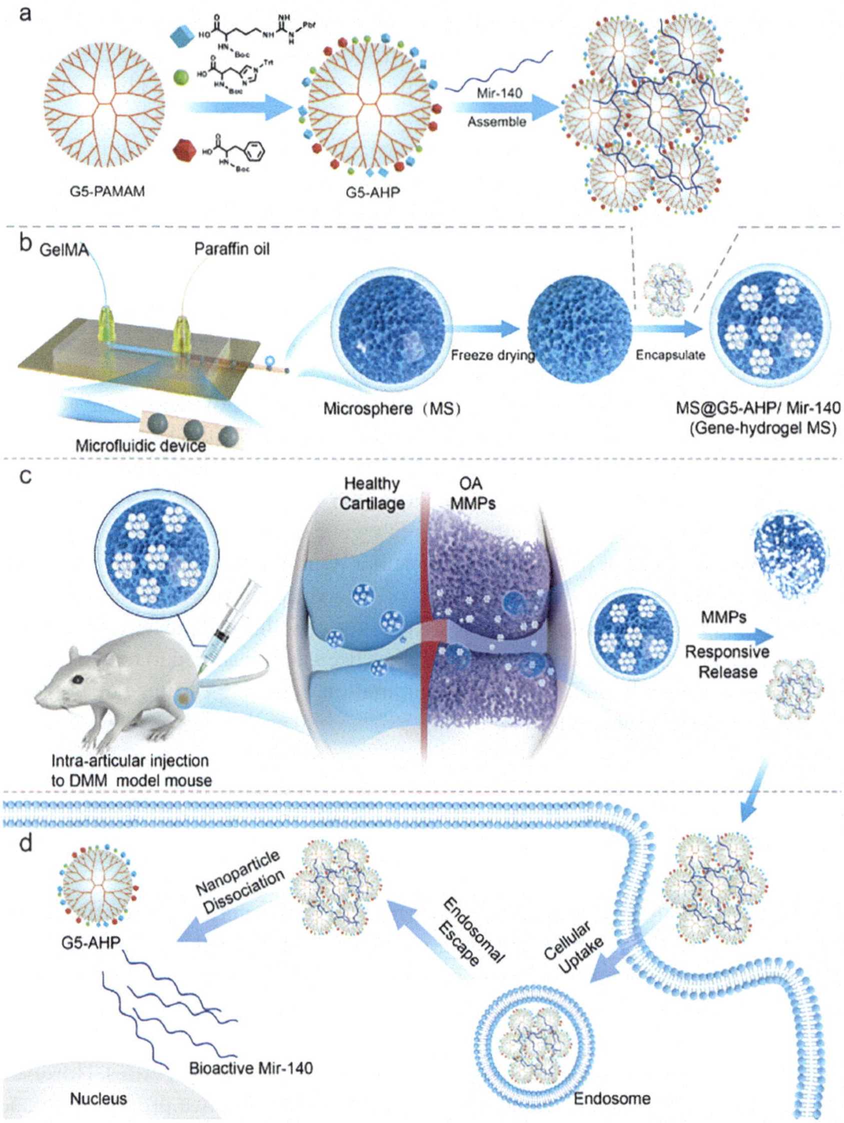

Fig. 3 The representative micrograph depicted novel strategies for the treatment of OA with the application of gene-hydrogel microspheres. **a** synthesis of G5-AHP and G5-AHP/miR-140. **b** Gene-hydrogel MSs. **c** Injection with MS@G5-AHP/miR-140 into the articular space to alleviate the progression of OA. (D) Endocytosis of G5-AHP/miR-140 polyplexes and the release of miR-140. Reproduced with permission [109]. Copyright 2022, Nature Publishing Group

Together, engineers, scientists, and medical professionals are improving patient care by using nanotechnology to reveal new chemical properties and activities.

Conflict of Interest The Authors do not have any conflict of interest.

References

1. Nuki G (1999) Z Rheumatol 58:142–147
2. Pacifici M, Golden EB, Oshima O, Shapiro IM, Leboy PS, Adams SL (1990) Ann N Y Acad Sci 599:45–57
3. Ansari MM, Ghosh M, Lee D-S, Son Y-O (2024) Ageing Res Rev 96:102275
4. Motta F, Barone E, Sica A, Selmi C (2023) Clin Rev Allergy Immunol 64:222–238
5. Mueller MB, Tuan RS (2011) PM&R 3:S3–S11
6. Kim H, Kang D, Cho Y, Kim J-H (2015) Mol Cells 38:677–684
7. Kim M, Kim MH, Kim J, Kang K, Lee J, Ghosh M, Son Y-O (2024) Appl Biol Chem 67:18
8. Holyoak DT, Tian YF, van der Meulen MC, Singh A (2016) Ann Biomed Eng 44:2062–2075
9. Wang Y, Liu L, Le Z, Tay A (2022) Adv NanoBiomed Res 2:2200085
10. Wang Y, Tay A (2023) ACS Nano 17:9850–9869
11. Tong L, Yu H, Huang X, Shen J, Xiao G, Chen L, Wang H, Xing L, Chen D (2022) Bone Res 10:60
12. Hu W, Chen Y, Dou C, Dong S (2021) Ann Rheum Dis 80:413–422
13. Han X, Cui J, Xie K, Jiang X, He Z, Du J, Chu L, Qu X, Ai S, Sun Q, Wang L, Wu H, Zhang W, Yu Z, Yan M (2020) Arthritis Res Ther 22:203
14. Zhang L, Kirkwood CL, Sohn J, Lau A, Bayers-Thering M, Bali SK, Rachala S, Marzo JM, Anders MJ, Beier F, Kirkwood KL (2021) Arthritis Res Ther 23:287
15. Xilin C, Yan G, Juan LU, Luxue Q, Tingyao HU, Xin Z, Xinyue W, Anran Z, Yuxin Z, Honggang Z, Changqing G (2023) J Tradit Chin Med 43:734–743
16. Hügle T, Geurts J (2017) Rheumatology (Oxford) 56:1461–1471
17. Fazzalari NL, Parkinson IH (1997) J Bone Miner Res 12:632–640
18. Burr DB, Gallant MA (2012) Nat Rev Rheumatol 8:665–673
19. Mathiessen A, Conaghan PG (2017) Arthritis Res Ther 19:18
20. Maglaviceanu A, Wu B, Kapoor M (2021) Wound Repair Regen 29:642–649
21. Thomson A, Hilkens CMU (2021) Front Immunol 12:678757
22. Krenn V, Perino G, Rüther W, Krenn VT, Huber M, Hügle T, Najm A, Müller S, Boettner F, Pessler F, Waldstein W, Kriegsmann J, Casadonte R, Häupl T, Wienert S, Krukemeyer MG, Sesselmann S, Sunitsch S, Tikhilov R, Morawietz L (2017) Pathol Res Pract 213:874–881
23. Bonnet CS, Walsh DA (2005) Rheumatology (Oxford) 44:7–16
24. Bondeson J, Wainwright SD, Lauder S, Amos N, Hughes CE (2006) Arthritis Res Ther 8:R187
25. Sophia Fox AJ, Bedi A, Rodeo SA (2009) Sports Health 1:461–468
26. Eschweiler J, Horn N, Rath B, Betsch M, Baroncini A, Tingart M, Migliorini F (2021) Life (Basel) 11
27. Żylińska B, Sobczyńska-Rak A, Lisiecka U, Stodolak-Zych E, Jarosz Ł, Szponder T (2021) In Vivo 35:1355–1363
28. Hwang HS, Kim HA (2015) Int J Mol Sci 16:26035–26054
29. Janczi T, Böhm BB, Fehrl Y, DeGiacomo P, Kinne RW, Burkhardt H (2019) Arthritis Rheumatol 71:63–72
30. Del Carlo Jr M, Loeser RF (2002) Arthritis Rheum 46:394–403
31. Wang G, Li Y, Yang G, Yang T, He L, Wang Y (2021) Hum Genomics 15:24
32. Zhang J, Hao X, Chi R, Qi J, Xu T (2021) J Cell Physiol 236:7504–7515

33. Sun Y, Leng P, Guo P, Gao H, Liu Y, Li C, Li Z, Zhang H (2021) Mol Med 27:96
34. Huang J, Zhou Q, Ren Q, Luo L, Ji G, Zheng T (2022) Front Endocrinol (Lausanne) 13:1094394
35. Hughes A, Oxford AE, Tawara K, Jorcyk CL, Oxford JT (2017) Int J Mol Sci 18
36. Kozhemyakina E, Zhang M, Ionescu A, Ayturk UM, Ono N, Kobayashi A, Kronenberg H, Warman ML, Lassar AB (2015) Arthritis Rheumatol 67:1261–1273
37. Liu T, Zhang L, Joo D, Sun S-C (2017) Signal Transduct Target Ther 2:17023
38. Saito T, Tanaka S (2017) Arthritis Res Ther 19:94
39. Rigoglou S, Papavassiliou AG (2013) Int J Biochem Cell Biol 45:2580–2584
40. Ye Y, Zhou J (2023) Front Endocrinol (Lausanne) 14:1117489
41. Marcu KB, Otero M, Olivotto E, Borzi RM, Goldring MB (2010) Curr Drug Targets 11:599–613
42. Nam J, Aguda BD, Rath B, Agarwal S (2009) PLoS ONE 4:e5262
43. Gu R, Liu N, Luo S, Huang W, Zha Z, Yang J (2016) Medicine (Baltimore) 95:e4315
44. Roman-Blas JA, Jimenez SA (2006) Osteoarthritis Cartilage 14:839–848
45. Pal B, Endisha H, Zhang Y, Kapoor M (2015) Drugs R D 15:27–36
46. Bijlsma JW, Berenbaum F, Lafeber FP (2011) Lancet 377:2115–2126
47. Yang H, Wen Y, Zhang M, Liu Q, Zhang H, Zhang J, Lu L, Ye T, Bai X, Xiao G, Wang M (2020) Autophagy 16:271–288
48. Karonitsch T, Kandasamy RK, Kartnig F, Herdy B, Dalwigk K, Niederreiter B, Holinka J, Sevelda F, Windhager R, Bilban M, Weichhart T, Säemann M, Pap T, Steiner G, Smolen JS, Kiener HP, Superti-Furga G (2018) Cell Rep 23:2157–2167
49. Jiang H, Dai J, Zhang C, Sun H, Tang X (2021) J Bioenerg Biomembr 53:97–107
50. Zhou X, Li J, Zhou Y, Yang Z, Yang H, Li D, Zhang J, Zhang Y, Xu N, Huang Y, Jiang L (2020) Aging (Albany NY) 12:20163–20183
51. Wu J, Kuang L, Chen C, Yang J, Zeng WN, Li T, Chen H, Huang S, Fu Z, Li J, Liu R, Ni Z, Chen L, Yang L (2019) Biomaterials 206:87–100
52. McGettrick AF, O'Neill LAJ (2020) Cell Metab 32:524–536
53. Schipani E, Ryan HE, Didrickson S, Kobayashi T, Knight M, Johnson RS (2001) Genes Dev 15:2865–2876
54. Yao Q, Khan MP, Merceron C, LaGory EL, Tata Z, Mangiavini L, Hu J, Vemulapalli K, Chandel NS, Giaccia AJ, Schipani E (2019) Dev Cell 49:748-763.e747
55. Pfander D, Gelse K (2007) Curr Opin Rheumatol 19:457–462
56. Ummarino D (2016) Nat Rev Rheumatol 12:315–315
57. Yang F, Huang R, Ma H, Zhao X, Wang G (2020) Med Sci Monit 26:e921155
58. Yang S, Kim J, Ryu JH, Oh H, Chun CH, Kim BJ, Min BH, Chun JS (2010) Nat Med 16:687–693
59. Bouaziz W, Sigaux J, Modrowski D, Devignes CS, Funck-Brentano T, Richette P, Ea HK, Provot S, Cohen-Solal M, Haÿ E (2016) Proc Natl Acad Sci U S A 113:5453–5458
60. Qing L, Lei P, Liu H, Xie J, Wang L, Wen T, Hu Y (2017) Exp Ther Med 13:63–68
61. Zitka O, Kukacka J, Krizkova S, Huska D, Adam V, Masarik M, Prusa R, Kizek R (2010) Curr Med Chem 17:3751–3768
62. Nissinen L, Kähäri VM (2014) Biochim Biophys Acta 1840:2571–2580
63. Little CB, Barai A, Burkhardt D, Smith SM, Fosang AJ, Werb Z, Shah M, Thompson EW (2009) Arthritis Rheum 60:3723–3733
64. Salazar-Noratto GE, De Nijs N, Stevens HY, Gibson G, Guldberg RE (2019) Osteoarthritis Cartilage 27:294–303
65. Chang JC, Sebastian A, Murugesh DK, Hatsell S, Economides AN, Christiansen BA, Loots GG (2017) J Orthop Res 35:474–485
66. Yang CY, Chanalaris A, Troeberg L (2017) Osteoarthritis Cartilage 25:1000–1009
67. Glasson SS, Askew R, Sheppard B, Carito B, Blanchet T, Ma HL, Flannery CR, Peluso D, Kanki K, Yang Z, Majumdar MK, Morris EA (2005) Nature 434:644–648
68. Stanton H, Rogerson FM, East CJ, Golub SB, Lawlor KE, Meeker CT, Little CB, Last K, Farmer PJ, Campbell IK, Fourie AM, Fosang AJ (2005) Nature 434:648–652

69. Das N, Schmidt TA, Krawetz RJ, Dufour A (2019) BioEssays 41:e1800166
70. Das N, de Almeida LGN, Derakhshani A, Young D, Mehdinejadiani K, Salo P, Rezansoff A, Jay GD, Sommerhoff CP, Schmidt TA, Krawetz R, Dufour A (2023) Nat Commun 14:1910
71. Lefebvre V, Bhattaram P (2015) Arthritis Rheumatol 67:1151–1154
72. Pavelka K, Bruyère O, Cooper C, Kanis JA, Leeb BF, Maheu E, Martel-Pelletier J, Monfort J, Pelletier JP, Rizzoli R, Reginster JY (2016) Drugs Aging 33:75–85
73. Jung JH, Kim SE, Kim H-J, Park K, Song GG, Choi SJ (2020) Int J Pharm 581:119249
74. Bedingfield SK, Colazo JM, Di Francesco M, Yu F, Liu DD, Di Francesco V, Himmel LE, Gupta MK, Cho H, Hasty KA, Decuzzi P, Duvall CL (2021) ACS Nano 15:14475–14491
75. Maudens P, Meyer S, Seemayer CA, Jordan O, Allémann E (2018) Nanoscale 10:1845–1854
76. Zhao Y, Wei C, Chen X, Liu J, Yu Q, Liu Y, Liu J (2019) ACS Appl Mater Interfaces 11:11587–11601
77. Kang LJ, Yoon J, Rho JG, Han HS, Lee S, Oh YS, Kim H, Kim E, Kim SJ, Lim YT, Park JH, Song WK, Yang S, Kim W (2021) Biomaterials 275:120967
78. Liu L, Tang H, Wang Y (2023) Int J Nanomed 18:4969–4983
79. Wei G, Wang Y, Yang G, Wang Y, Ju R (2021) Theranostics 11:6370–6392
80. Bhalekar MR, Upadhaya PG, Madgulkar AR (2016) Eur J Pharm Sci 84:1–8
81. Yeh CC, Su YH, Lin YJ, Chen PJ, Shi CS, Chen CN, Chang HI (2015) Drug Des Dev Ther 9:2285–2300
82. He M, Qin Z, Liang X, He X, Zhu B, Lu Z, Wei Q, Zheng L (2021) Regen Biomater 8:rbab020
83. Li J, Zhang H, Han Y, Hu Y, Geng Z, Su J (2023) Theranostics 13:931–954
84. Zhang JX, Yan MQ, Li XH, Qiu LY, Li XD, Li XJ, Jin Y, Zhu KJ (2007) Pharm Res 24:1944–1953
85. Saadat E, Shakor N, Gholami M, Dorkoosh FA (2015) Int J Pharm 489:218–225
86. Lei Y, Wang X, Liao J, Shen J, Li Y, Cai Z, Hu N, Luo X, Cui W, Huang W (2022) Bioact Mater 16:472–484
87. Fenton OS, Olafson KN, Pillai PS, Mitchell MJ, Langer R (2018) Adv Mater 30:1705328
88. Yu C, Li X, Hou Y, Meng X, Wang D, Liu J, Sun F, Li Y (2019) Molecules 24
89. Alam MM, Han HS, Sung S, Kang JH, Sa KH, Al Faruque H, Hong J, Nam EJ, Kim IS, Park JH, Kang YM (2017) J Control Release 252:62–72
90. Zhang L, Peng H, Zhang W, Li Y, Liu L, Leng T (2020) Theranostics 10:8479–8493
91. Ensign LM, Cone R, Hanes J (2012) Adv Drug Deliv Rev 64:557–570
92. Jeong WY, Kwon M, Choi HE, Kim KS (2021) Biomater Res 25:24
93. Towheed TE, Maxwell L, Judd MG, Catton M, Hochberg MC, Wells G (2006) Cochrane Database Syst Rev 2006:Cd004257
94. Conaghan PG, Arden N, Avouac B, Migliore A, Rizzoli R (2019) Drugs Aging 36:7–14
95. Adatia A, Rainsford KD, Kean WF (2012) J Pharm Pharmacol 64:626–636
96. Magni A, Agostoni P, Bonezzi C, Massazza G, Menè P, Savarino V, Fornasari D (2021) Pain Ther 10:783–808
97. Golden HE, Moskowitz RW, Minic M (2004) Am J Ther 11:85–94
98. Hsueh MF, Bolognesi MP, Wellman SS, Kraus VB (2020) Osteoarthritis Cartilage 28:639–645
99. Osani MC, Bannuru RR (2019) Korean J Intern Med 34:966–973
100. Ma X, Zhou S, Sun W, Sun J, Li G, Wang L, Guo Y (2023) BMC Musculoskelet Disord 24:394
101. Kingsbury SR, Tharmanathan P, Arden NK, Batley M, Birrell F, Cocks K, Doherty M, Edwards CJ, Garrood T, Grainger AJ, Green M, Hewitt C, Hughes R, Moots R, O'Neill TW, Roddy E, Scott DL, Watt FE, Torgerson DJ, Conaghan PG (2015) Trials 16:77
102. Cronstein BN, Aune TM (2020) Nat Rev Rheumatol 16:145–154
103. Simón-Vázquez R, Tsapis N, Lorscheider M, Rodríguez A, Calleja P, Mousnier L, de Miguel Villegas E, González-Fernández Á, Fattal E (2022) Drug Deliv Transl Res 12:1270–1284
104. Teng H, Chen S, Fan K, Wang Q, Xu B, Chen D, Zhao F, Wang T (2023) Mol Pharm 20:3843–3853
105. Duan X, Cai L, Pham CT, Abu-Amer Y, Pan H, Brophy RH, Wickline SA, Rai MF (2021) Arthritis Rheumatol 73:2249–2260

106. Guo X, Lou J, Wang F, Fan D, Qin Z (2022) Front Pharmacol 13:924387
107. Bedingfield SK, Colazo JM, Di Francesco M, Yu F, Liu DD, Di Francesco V, Himmel LE, Gupta MK, Cho H, Hasty KA (2021) ACS Nano 15:14475–14491
108. Bedingfield SK, Colazo JM, Yu F, Liu DD, Jackson MA, Himmel LE, Cho H, Crofford LJ, Hasty KA, Duvall CL (2021) Nat Biomed Eng 5:1069–1083
109. Li B, Wang F, Hu F, Ding T, Huang P, Xu X, Liang J, Li C, Zhou Q, Lu M, Deng L, Guo L, Cui W (2022) NPG Asia Mater 14:1
110. Jiang T, Kan H-M, Rajpura K, Carbone EJ, Li Y, Lo KW-H (2018) J Nanosci Nanotechnol 18:2310–2317
111. Xiao L, Cui J, Sun Z, Liu Y, Zheng J, Dong Y (2022) Front Pharmacol 13:920824
112. Lawson TB, Mäkelä JTA, Klein T, Snyder BD, Grinstaff MW (2021) J Orthop Res 39:473–484

Nanomaterial in the Management of Diabetes

Shubhrat Maheshwari, Aditya Singh, Amita Verma, Juber Akhtar, Purba Mandal, Jagat Pal Yadav, and Bhupendra G. Prajapati

Abstract Diabetes mellitus is a chronic metabolic disorder characterized by persistent hyperglycemia, stemming from defects in insulin secretion, insulin action, or both. The disorder is categorized into two primary types: Type 1 diabetes (T1D), an autoimmune condition resulting in the destruction of insulin-producing pancreatic beta cells, and Type 2 diabetes (T2D), characterized by insulin resistance and impaired insulin secretion. Chronic hyperglycemia is associated with severe complications, including cardiovascular and neurological disorders, while hypoglycemia can result in life-threatening conditions such as unconsciousness or death. Traditional diabetes management strategies, particularly for T1D and progressive T2D, often involve frequent insulin injections and regular blood glucose monitoring. These methods are invasive and pose challenges related to patient compliance and precise glycemic control. Nanotechnology has emerged as a promising frontier in diabetes management, offering innovative approaches to drug delivery and glucose monitoring. Nanomaterials, including silica nanoparticles, solid lipid nanoparticles (SLNs), quantum dots (QDs), polymeric nanoparticles, and carbon-based materials like graphene oxide (GO) and carbon nanotubes (CNTs), provide enhanced drug bioavailability, stability, targeted delivery, and improved patient compliance. Moreover, nanotechnology-enabled biosensors and continuous glucose monitoring systems utilizing nanosensors and quantum dots enhance the sensitivity and specificity of glucose detection, enabling real-time monitoring and early diagnosis of

S. Maheshwari (✉) · A. Verma · J. P. Yadav
Bioorganic and Medicinal Chemistry Research Laboratory, Department of Pharmaceutical Sciences, Sam Higginbottom University of Agriculture, Technology and Sciences, Prayagraj, Uttar Pradesh, India
e-mail: shubhrat.maheshwari@gmail.com

A. Singh · J. Akhtar · P. Mandal
Department of Pharmacy, Faculty of Pharmacy, Integral University, Lucknow, Uttar Pradesh, India

B. G. Prajapati (✉)
Shree S. K. Patel College of Pharmaceutical Education and Research, Ganpat University, Gujarat, India
e-mail: bhupen27@gmail.com

Faculty of Pharmacy, Silpakorn University, Nakhon Pathom 73000, Thailand

Md. M. Ansari et al. (eds.), *Emergence of Sustainable Biomaterials in Tackling Inflammatory Diseases*, Smart Nanomaterials Technology,
https://doi.org/10.1007/978-981-96-2112-5_11

diabetes-related complications. The application of nanoparticles in conjunction with phytochemicals also shows potential for managing MS and diabetes, highlighting the multifaceted benefits. This chapter delves into the role of nanomaterials in diabetes management, exploring their potential to revolutionize treatment protocols and improve outcomes for diabetic patients.

Keywords Type 1 diabetes · Drug delivery · Nanotechnology · Diabetes management · Nanomaterials

1 Introduction

Diabetes mellitus represents a metabolic disorder characterized by chronic hyperglycemia. This condition hinders the body's capacity to regulate blood glucose homeostasis, resulting in heightened blood glucose levels (BGLs) and severe health complications [1]. Type 1 diabetes (T1D) emerges from the autoimmune destruction of pancreatic beta cells responsible for insulin production, while Type 2 diabetes (T2D) involves insulin resistance or a deficit in the cellular response to insulin. Both forms of diabetes can lead to notable deviations in BGLs, presenting as hyperglycemia or hypoglycemia, each carrying substantial health risks [2]. Chronic hyperglycemia is linked to cardiovascular and neurological complications, whereas hypoglycemia may lead to energy deficiencies, unconsciousness, or fatality. As diabetes prevalence escalates globally, impacting millions worldwide and is anticipated to affect 592 million individuals by 2035, the necessity for enhanced management strategies intensifies. Current conventional treatments for diabetes, especially for T1D and progressive T2D [3], necessitate frequent subcutaneous insulin injections and periodic blood glucose monitoring, both of which are invasive and may lead to noncompliance and imprecise glycemic control. With the advent of globalization and modernization [4], there has been a significant transformation in the lifestyles of individuals, characterized by increased working hours and decreased leisure time. The ubiquitous use of electronic devices has also played a pivotal role in streamlining lifestyles, leading to the emergence of various lifestyle-related ailments, with metabolic syndrome (MS) standing out as a prevalent condition [5]. MS encompasses a cluster of disorders such as glucose intolerance [6]. The global prevalence of MS is estimated to affect approximately 30% of the population, thus posing a significant public health challenge. Its etiology is multifactorial involving metabolic, genetic, and environmental determinants [7]. Individuals with MS face a heightened risk of myocardial infarction [8].

Modification of lifestyle habits and the use of pharmaceutical interventions targeting blood pressure, glucose levels, and triglycerides represent the conventional approaches to managing MS [9]. The utilization of various nanomaterials, precise targeting with sustained release, enhanced drug availability, improved solubility, stability, intracellular delivery, and reduced toxicity [10]. Given the adverse effects associated with medications used in MS treatment, the application of nanoparticles is recommended [11]. Furthermore, nanoparticles combined with phytochemicals demonstrate superior efficacy in MS management. Pathophysiology refers to the functional alterations accompanying a particular syndrome or disease. Investigating pathophysiology aims to elucidate the modified biological processes in an organism induced by specific diseases or syndromes [12, 13]. Hyperglycemia triggers increased insulin secretion, leading to hyperinsulinemia. Prolonged hyperinsulinemia ultimately disrupts compensatory mechanisms, culminating in diminished insulin production [14]. Therefore, a comprehensive understanding of the pathophysiology of diseases is essential for effective treatment. In 2016, the WHO reported that the number of adults living with diabetes reached 422 million (approximately 8.5% of the adult population), a fourfold increase compared to 1980 [15].

The rise can be primarily ascribed to type 2 diabetes and its associated risk factors, notably overweight and obesity. Annually, approximately 1.5 million fatalities are directly linked to diabetes, positioning it as the ninth principal cause of death on a global scale. In 2014, data from Portugal unveiled that 9.2% of the populace was afflicted with diabetes, contributing to 5% of total deaths across all age groups. The escalating incidence rate and the profound health hazards related to diabetes have propelled investigations towards formulating novel therapeutic methodologies to combat the growing prevalence of this ailment and to enhance the manifestations, standard of living, and life expectancy of diabetic individuals [16].

Novel therapeutic tactics have been suggested for the oral administration of antidiabetic medications or for attaining a more regulated hypoglycemic impact. Various nanomaterials with diverse compositions, such as lipid and polymer nanoparticles, metal–organic frameworks, and silica nanoparticles [17, 18]. Nonetheless, they are prone to disintegration in biological fluids, constraining the oral bioavailability of the drug and possibly inducing hypoglycemia if there is a sudden release in the bloodstream [19]. Diabetic therapy, albeit to a lesser extent, leads to numerous uncertainties concerning stability, intestinal permeability, and safety. Conversely, silica nanoparticles exhibit exceptional physicochemical stability, favorable biocompatibility [20], and substantial loading capacity (especially with porous silica) offering customizable particle dimensions, surface area, and pore volume, along with a readily modifiable surface. Consequently, several investigations showcase the promise of silica nanoparticles as an alternative for diabetes treatment. Among inorganic coating materials, amorphous silica (SiO2) is predominantly utilized for bio-applications. The silica coating enables a platform for direct functionalization with easily accessible organofunctional silane coupling agents for subsequent linkage with drugs [21]. Through MSNs, numerous compounds with antidiabetic properties have been examined, with a specific emphasis on insulin. Despite the well-recognized antidiabetic properties of various polyphenols, nanosilica-loaded polyphenols for diabetes treatment have not been reported thus far [22]. Nevertheless, these systems have been assessed in the treatment of various other ailments, encompassing diverse forms of cancer like breast cancer, prostate cancer, melanoma, and cardiovascular disorders. The standard pharmacological management of type 2 Diabetes includes metformin

Table 1 Present anti-diabetic medication (not including insulin therapy) for type 2 diabetes

Drug class	Treatment	Drugs	Action
Sulfonylureas	First/second line	Chlorpropamide, glycyclamide, metahexamide, tolazamide, tolbutamide glibornuride, gliclazide, glipizide, gliquidone, glisoxepide, glimepiride	Glucose-independent insulin [23]
Biguanide	First line	Metformin	Increase hepatic insulin [25]
Thiazolidinediones	Second/third line	Pioglitazone, rosiglitazone, lobeglitazone	Activate PPARs—decrease insulin resistance [26]
SGLT2	First/second/ third line	canagliflozin, dapagliflozin, empagliflozin	Increase glucose excretion [27]
Incretin mimetics/ analogs	Second line	Exetatide, liraglutide	GLP-1 receptors activate [28]
DDP4 inhibitors	Second/third line	Sitagliptin, vildagliptin, linagliptin, algogliptin	Stimulate insulin release [29]

as the first-line treatment, sulfonylureas as a backup if the patient tolerates the medication well or has a normal weight, but these are typically used as second-line treatments, and thiazolidinediones as a third-line therapy that can also be used as second-line [23]. These can be used as monotherapy (in the event that the patient finds metformin intolerable) or as second- and third-line medications. Table 1 describes the features of these medications [24].

2 Types of Nanomaterials Used in Diabetes Management

Innovative approaches to measuring insulin and blood glucose levels for diabetes management are enabled by the field of nanotechnology. Moreover, the utilization of biosensors incorporating nanotechnology has the potential to enhance insulin detection, facilitating early disease identification related to insulin (Fig. 1) [30]. Several examples are presented for illustration. Continuous Glucose Monitoring (CGM) through nanosensors involves wearable devices equipped with nanoscale sensors to deliver immediate glucose data, eliminating the need for frequent finger-stick tests in effectively managing diabetes. Utilizing Quantum Dot Nanoparticles for Blood Glucose detection offers increased sensitivity and specificity in glucose monitoring, utilizing glucose-sensitive quantum dots that emit fluorescent signals [31, 32]. By encapsulating medications within nanocarriers, which are nano-sized delivery systems, the effectiveness, bioavailability, and targeted delivery of the medications are improved [33]. Nanocarriers present a promising approach for managing diabetes mellitus by offering various advantages for medication delivery in type 2 diabetes [34]. The utilization of nanocarriers for drug delivery provides the following advantages: Medications delivered via nanocarriers exhibit improved absorption and protection from degradation in the gastrointestinal tract, thereby enhancing the bioavailability of the drugs [35]. Nanocarriers designed to target specific tissues or cells, such as insulin-resistant tissues or pancreatic beta cells, can enhance drug delivery to the desired sites of action [36]. Solid lipid nanoparticles (SLNs) have emerged as a leading option for applying nanotechnology in diabetes treatment, offering benefits over traditional drug delivery methods. Composed of biocompatible lipids, SLNs provide advantages such as improved drug release, enhanced stability of encapsulated pharmaceuticals, and reduced toxicity compared to conventional approaches [37]. To enhance insulin bioavailability, Lipid-based nanocapsules have shown promise as a therapeutic delivery system for various applications, including diabetes management. Typically, these nanocapsules encapsulate therapeutic agents like insulin or other antidiabetic medications within a lipid bilayer. One study successfully created such nanocapsules for the oral administration of peptides. Polymeric nanoparticles, also known as NPs, consist of solid particles made from biodegradable polymers [38]. Graphene oxide (GO) is a versatile carbon-based nanomaterial renowned for its large surface area, excellent biocompatibility, and functionalization potential. In diabetes management, GO is primarily utilized for glucose sensing and monitoring. Its high electrical conductivity and sensitivity make it an ideal material for developing non-enzymatic glucose sensors. These sensors leverage GO's ability to interact with glucose molecules, leading to measurable changes in electrical signals [39]. Additionally, GO has been explored for drug delivery, where its functional groups enable the conjugation of anti-diabetic drugs, ensuring controlled and targeted release, which is crucial for maintaining optimal glucose levels. Carbon nanotubes (CNTs) exhibit unique electrical, mechanical, and thermal properties, making them suitable for various biomedical applications [40]. In diabetes management, CNTs are used in the development of highly sensitive glucose sensors. Their high surface-to-volume ratio and excellent electron transfer capabilities enhance the performance of glucose biosensors, allowing for rapid and accurate glucose level detection. Moreover, CNTs have been employed in the delivery of insulin and other therapeutic agents. Functionalized CNTs can efficiently load and transport these drugs, ensuring their stability and bioavailability, which is essential for effective diabetes treatment [41].

Liposomes are spherical vesicles composed of lipid bilayers, known for their ability to encapsulate both hydrophilic and hydrophobic drugs. In diabetes management, liposomes are particularly useful for insulin delivery [42]. They can protect insulin from enzymatic degradation in the gastrointestinal tract, making oral insulin delivery feasible. Additionally, liposomes facilitate the sustained release of insulin, maintaining blood glucose levels within the desired range. Their biocompatibility and ability to fuse with cell membranes further enhance the delivery efficiency, reducing the frequency of insulin administration and improving patient compliance [43].

Solid lipid nanoparticles (SLNs) are composed of solid lipids, stabilized by surfactants. These nanoparticles offer a promising alternative for drug delivery in diabetes management due to their ability to enhance the bioavailability of poorly soluble

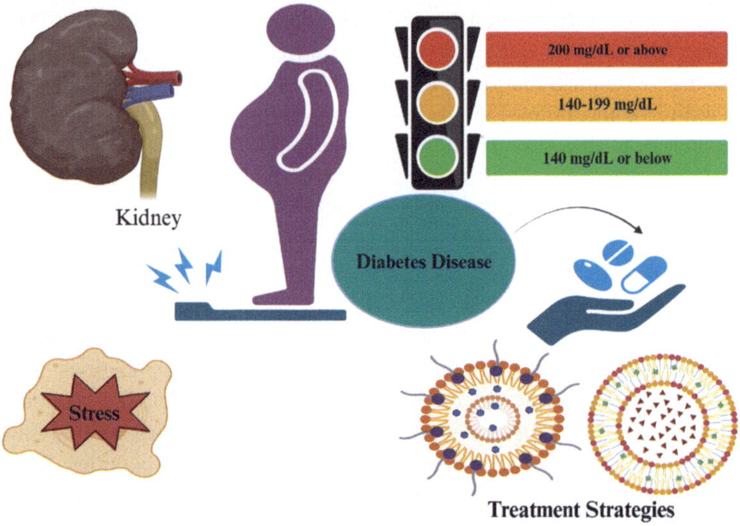

Fig. 1 Nanomaterial used in the management of diabetes

drugs. SLNs can encapsulate anti-diabetic drugs, ensuring their controlled release and targeted delivery. This targeted approach minimizes side effects and maximizes therapeutic efficacy [44]. Furthermore, SLNs' biocompatibility and biodegradability make them suitable for repeated administration, providing a sustained therapeutic effect and better management of diabetes [45].

Quantum dots (QDs) are semiconductor nanocrystals with unique optical properties, such as size-tunable fluorescence and high photostability. In diabetes management, QDs are employed in the development of glucose sensors. Their fluorescence properties enable the creation of highly sensitive and selective sensors for glucose detection [46]. QDs can be functionalized with glucose oxidase, an enzyme that catalyzes the oxidation of glucose, leading to fluorescence changes proportional to glucose concentration. This allows for real-time monitoring of blood glucose levels, offering a non-invasive and efficient approach to diabetes management [47].

Poly(lactic-co-glycolic acid) (PLGA) is a biodegradable and biocompatible polymer widely used in drug delivery systems. In diabetes management, PLGA nanoparticles are used to encapsulate insulin and other anti-diabetic drugs, providing controlled and sustained release. This encapsulation protects the drugs from degradation and ensures their prolonged therapeutic effect. PLGA nanoparticles can also be engineered to target specific tissues, such as pancreatic cells, enhancing the treatment's efficacy [48]. The use of PLGA nanoparticles reduces the frequency of drug administration, improving patient adherence to treatment regimens [49].

Chitosan, a natural polymer derived from chitin, is known for its biocompatibility, biodegradability, and mucoadhesive properties. In diabetes management, chitosan nanoparticles are used for oral delivery of insulin and other drugs. These nanoparticles protect insulin from enzymatic degradation and enhance its absorption through the intestinal mucosa. The mucoadhesive property of chitosan ensures prolonged retention at the absorption site, improving drug bioavailability. Additionally, chitosan nanoparticles can be functionalized to achieve targeted delivery, reducing systemic side effects and enhancing therapeutic outcomes [50].

Gold nanoparticles (AuNPs) are renowned for their biocompatibility and unique optical properties. In diabetes management, AuNPs are used in glucose sensors due to their excellent conductivity and ability to enhance the performance of electrochemical sensors. AuNPs can be functionalized with glucose oxidase or other sensing molecules, providing highly sensitive and specific glucose detection [51]. Moreover, AuNPs are explored for insulin delivery, where their ability to form stable conjugates with insulin ensures controlled and targeted release, improving glucose regulation [52].

Silver Nanoparticles Silver nanoparticles (AgNPs) exhibit potent antimicrobial properties and are used in wound-healing applications for diabetic patients. Chronic wounds and infections are common complications of diabetes, and AgNPs' antimicrobial action helps prevent and treat these infections. AgNPs can be incorporated into wound dressings, providing a sustained release of silver ions that combat microbial growth. Additionally, AgNPs' anti-inflammatory properties promote faster wound healing, improving the quality of life for diabetic patients [53].

Iron oxide nanoparticles (IONPs) are known for their magnetic properties and biocompatibility. In diabetes management, IONPs are explored for their potential in magnetic resonance imaging (MRI) to monitor pancreatic islet transplantation. IONPs can label transplanted islets, allowing non-invasive tracking and assessment of their viability and function [54]. This imaging capability provides valuable insights into the success of islet transplantation, facilitating early intervention in case of complications. Moreover, IONPs are being investigated for their ability to deliver drugs and therapeutic agents, leveraging their magnetic properties for targeted delivery [55] (Table 2).

3 Mechanisms of Action of Nanomaterials in Diabetes

Nanomaterials have revolutionized drug delivery in diabetes management by significantly enhancing drug bioavailability and absorption. Conventional drug delivery systems often suffer from poor solubility, stability, and rapid degradation, leading to suboptimal therapeutic outcomes. Nanomaterials, due to their minor size and enlarged surface area, improve the solubility of hydrophobic drugs, protect drugs from degradation, and facilitate their absorption through biological membranes. Nanoparticles, for instance, can be engineered to encapsulate insulin or other hypoglycemic agents, protecting them [63]. This leads to a more efficient systemic circulation of the drug. Nanomaterials such as liposomes can also be surface-modified with ligands or polymers to enhance mucosal adhesion and retention, further increasing bioavailability. This ability to improve drug absorption and bioavailability makes nanomaterials a promising tool for diabetes therapies.

Table 2 Various categories of nano-formulations designed for the management of type II diabetes

Formulation type	Mechanism/ evaluation	Objective	Outcome	References
Nanocapsule	Targeted lipid	Enhanced oral peptide delivery	Increased endogenous GLP-1 levels	[56]
Micelles	Box–Behnken response curve	Quercetin-loaded micelles	Pharmacokinetic	[57]
Gold NPs	Evaluation in in vivo study	Anti-diabetic	Reduce blood sugar levels	[58]
SLNs	In vitro study	Oral insulin administration	Bioavailability	[59]
Nanocapsule	In vivo study	Oral delivery	Bioavailability	[60]
Silver NPs	Anti-diabetic efficacy	Anti-diabetic	Bioavailability	[61]
Micelles	Living radical polymerization	Creation of polysaccharide-based micelle	Bioavailability	[62]

One of the most significant advantages of nanomaterials in diabetes treatment is their capacity for site targeting. Traditional drug delivery systems release drugs in a non-specific manner, often leading to systemic side effects and fluctuating blood drug levels [64]. Nanomaterials can be engineered to release their payload in a controlled manner, responding to specific stimuli such as pH, temperature, or glucose levels. For example, glucose-responsive nanoparticles can release insulin in response to elevated blood glucose levels, mimicking the physiological release of insulin and providing tight glycemic control. This controlled release not only reduces the frequency of drug administration but also minimizes the risk of hypoglycemia, a common complication in diabetes management. By ensuring that the drug is released at the right place and time, nanomaterials offer a sophisticated solution to the challenges of diabetes therapy [65].

Nanomaterials also play a crucial role in overcoming biological barriers that limit the efficacy of traditional drug delivery systems. The gastrointestinal tract, blood–brain barrier, and cellular membranes are some of the significant barriers that drugs must traverse to reach their target sites. Nanomaterials, due to their nanoscale size and surface functionalization capabilities, can penetrate these barriers more efficiently. For instance, nanocarriers can cross the blood–brain barrier, which is particularly relevant for the treatment of diabetes-associated neurodegenerative complications. Similarly, in oral drug delivery, nanomaterials can enhance the permeability of the intestinal epithelium, facilitating the systemic absorption of therapeutic agents that are otherwise poorly absorbed. Furthermore, surface modification of nanomaterials with specific ligands or antibodies enables receptor-mediated endocytosis, allowing drugs to enter cells through specific pathways, bypassing the usual resistance mechanisms. This ability to navigate through and bypass biological barriers significantly enhances the therapeutic potential of nanomaterials in diabetes management [47].

Nanomaterials not only serve as carriers for drug delivery but also actively interact with cellular pathways, potentially modulating biological processes to enhance therapeutic outcomes. These interactions can lead to the activation or inhibition of specific signaling pathways, influencing cellular behavior in ways that are beneficial for diabetes management. For instance, nanomaterials can be designed to interact with insulin receptors on the surface of cells, enhancing insulin sensitivity and improving glucose uptake. Similarly, nanoparticles can be engineered to modulate inflammatory pathways, reducing chronic inflammation that is often associated with diabetes [52]. Nanomaterials can influence gene expression by delivering RNA or DNA molecules to specific cells, offering a potential approach for gene therapy in diabetes. These interactions with cellular pathways open up new avenues for the development of therapies that not only manage the symptoms of diabetes but also address the underlying molecular mechanisms of the disease. By modulating cellular pathways, nanomaterials offer a promising strategy for achieving more comprehensive and long-lasting control over diabetes [66].

4 Preclinical and Clinical Studies

Animal models help researchers understand the efficacy, safety, and mechanisms of action of nanotechnology-based therapies before they are tested in humans. Commonly used animal models in diabetes research include mice and rats are frequently used due to their genetic similarity to humans and their ability to develop diabetes either spontaneously (in the case of genetic models like the NOD mouse) or through chemical induction (using streptozotocin or alloxan). Pigs and non-human primates are occasionally used due to their closer physiological resemblance to humans, especially in terms of metabolism and insulin regulation. In these models [67], nanomaterials are used to deliver insulin, oral hypoglycemics, or gene therapy agents. Studies typically assess glucose levels, insulin sensitivity, and histopathological changes in tissues to determine the therapeutic potential of these nanomaterials. Animal models also provide insights into the biodistribution, pharmacokinetics, and potential toxicity of the nanomaterials, guiding the design of safer and more effective therapies. Preclinical studies in nanotechnology for diabetes management have yielded promising results, demonstrating the potential of nanomaterials to improve the delivery and efficacy of diabetes therapies. These have been used to encapsulate insulin or other drugs, providing sustained and controlled release [68]. Preclinical studies have shown improved glucose regulation and reduced dosing frequency compared to conventional therapies. Liposomes and solid lipid nanoparticles have been studied for their ability to deliver insulin and other therapeutic agents orally, overcoming the challenge of gastrointestinal degradation. These studies have reported enhanced bioavailability and better glycemic control in diabetic animal models. These are designed to release insulin in response to high glucose levels, providing a dynamic and physiologically relevant approach to diabetes management. Preclinical trials have demonstrated their ability to mimic the natural insulin response, reducing the risk of hypoglycemia [69]. Preclinical studies have explored the use of nanoparticles for the delivery of genetic material aimed at modulating gene expression or correcting genetic defects associated with diabetes. These studies have shown potential in reversing or mitigating the progression of diabetes in animal models. While most research on nanomaterials for diabetes treatment is still in the preclinical stage, there are a growing number of clinical trials that are testing the safety and efficacy of these advanced therapies in humans. Clinical trials are assessing the use of nanoparticle formulations to improve the pharmacokinetics of insulin, aiming for more stable blood glucose levels and reduced injection frequency [70]. Some studies are also exploring non-invasive routes of administration, such as inhaled or oral insulin. Trials are being conducted on nanocarriers that can protect oral hypoglycemic drugs from degradation in the gastrointestinal tract and enhance their absorption, offering a more convenient alternative to injections. Clinical studies are evaluating the effectiveness of glucose-responsive nanoparticles in delivering insulin or other therapies in a controlled manner, responding dynamically to blood glucose levels. In addition to efficacy, clinical trials focus on the safety and tolerability of nanomaterials, assessing potential side effects, immunogenicity, and long-term impact on health [71].

5 Challenges and Future Perspectives

The development and application of nanotechnology in diabetes management face several technical and manufacturing challenges that must be addressed before these technologies can be widely adopted. One of the primary challenges is the complexity of synthesizing nanomaterials with consistent size, shape, and functional properties. Manufacturing nanomaterials at a commercial scale often requires specialized equipment and stringent quality control measures to prevent contamination and ensure uniformity, which can drive up costs and limit accessibility. Another technical challenge lies in the stability of nanomaterials during storage and administration. Nanoparticles and other nanocarriers are often prone to aggregation, degradation, or premature release of their payloads, which can compromise their efficacy and safety. Developing stable formulations that can withstand physiological conditions without losing functionality is critical. The interaction of nanomaterials with biological systems is complex and not fully understood which can lead to unpredictable behavior such as off-target effects, toxicity, or immunogenicity. Biocompatibility and toxicity are also significant concerns. Nanomaterials designed for diabetes therapy must be biocompatible, meaning they should not induce adverse immune responses or toxicity in the body [72]. Long-term studies on the safety and biodistribution of these materials are essential, as some nanomaterials may accumulate in organs such as the liver, spleen, or kidneys, potentially causing harm. The lack of standardized protocols for evaluating the safety of nanomaterials further complicates their development and regulatory approval [73].

6 Regulatory and Ethical Considerations

Regulatory challenges are closely tied to the technical hurdles of nanotechnology in diabetes treatment. The novel nature of nanomaterials means that existing regulatory frameworks, which are often designed for conventional drugs and medical devices, may not be entirely applicable. Regulatory agencies like the FDA and EMA require comprehensive data on the safety, efficacy, and manufacturing processes of nanotechnology-based products, but the lack of standardized testing methods and criteria specific to nanomaterials poses significant challenges for developers. Establishing clear guidelines for the evaluation of nanomedicines is crucial to facilitate their regulatory approval and ensure their safe use in clinical settings. Ethical considerations also play a critical role in the development and deployment of nanotechnology in diabetes management [74]. The use of nanomaterials raises questions about patient consent, especially when the long-term effects of these materials are not fully understood. Patients need to be informed about the potential risks and benefits of nanotechnology-based therapies, and their autonomy in making treatment decisions must be respected. Additionally, there is a concern about the equitable distribution of these advanced therapies. Nanotechnology has the potential to revolutionize diabetes care, but if access is limited to those who can afford it, this could exacerbate existing health disparities. Ensuring that nanotechnology-based treatments are accessible and affordable to all segments of the population is a key ethical challenge that needs to be addressed. Looking ahead, the future of nanotechnology in diabetes management holds great promise, with several exciting avenues for exploration and development. One potential direction is the further refinement of glucose-responsive nanoparticles that can deliver insulin or other hypoglycemic agents in response to real-time blood glucose levels [75]. Smart delivery mimics the natural regulatory mechanisms of the pancreas, providing more precise control over blood glucose levels and reducing the risk of hypoglycemia. Another promising area is the development of non-invasive delivery systems, such as transdermal patches or inhalable nanoparticles, which could offer a more convenient alternative to traditional insulin injections. These systems not only improve patient compliance but also aim to enhance the bioavailability and absorption of therapeutic agents. Research into the use of nanotechnology for islet cell transplantation and pancreatic tissue engineering is also gaining traction. Nanoscaffolds and nanomaterials can provide structural support and create a conducive environment for the growth and function of transplanted cells, potentially offering a cure for type 1 diabetes. Gene therapy is another frontier where nanotechnology could make a significant impact. Nanocarriers can be used to deliver genetic material that can correct or modulate the expression of genes involved in the pathogenesis of diabetes. This approach has the potential to address the underlying causes of the disease rather than just managing its symptoms. Furthermore, the integration of nanotechnology with emerging fields in personalized nanomedicines. AI-driven algorithms could analyze individual patient data to optimize the design of nanocarriers, tailoring therapies to the specific needs of each patient. The integration of nanotechnology with personalized medicine represents a transformative approach to diabetes management. Personalized medicine aims to tailor treatment strategies to individual patients based on their genetic makeup, lifestyle, and other personal factors. Nanotechnology can enhance this approach by enabling the design of highly specific and targeted therapies [46, 76–78]. For example, nanoparticle-based drug delivery systems can be engineered to carry multiple therapeutic agents, each targeting different aspects of a patient's disease profile. These systems can be personalized based on the patient's genetic information, ensuring that the therapy is optimized for their specific needs. Additionally, nanotechnology can be used to develop personalized diagnostic tools, such as nanosensors that monitor biomarkers in real time. The convergence of nanotechnology and personalized medicine also opens up possibilities for more effective disease prevention. By identifying individuals at high risk for diabetes and developing personalized nanotherapeutics or lifestyle interventions, it may be possible to prevent the onset of the disease altogether. Moreover, as our understanding of the molecular mechanisms underlying diabetes improves, nanotechnology could enable the development of therapies that not only treat the symptoms but also address the root causes of the disease. However, the implementation of personalized nanomedicine in clinical practice is not without challenges. It requires sophisticated infrastructure, advanced diagnostic tools, and a deep understanding of the interplay between nanomaterials and biological systems. Additionally, the cost of developing and implementing personalized nanomedicine can be high, raising concerns about its accessibility and affordability for all patients [79].

7 Conclusion

Nanotechnology offers a promising frontier in the management of diabetes, providing innovative solutions that address many of the limitations associated with conventional therapies. The application of nanomaterials in drug delivery systems enhances drug bioavailability, ensures targeted and controlled release, and overcomes biological barriers, making treatments more effective and patient-friendly. Preclinical studies and emerging clinical trials underscore the potential of these technologies to revolutionize diabetes care, offering new ways to monitor and treat the disease. However, the path to fully realizing the potential of nanotechnology in diabetes management is not without challenges. Technical and manufacturing hurdles, such as the scalability of nanomaterial production and ensuring their stability, must be addressed to transition from laboratory research to widespread clinical use. Regulatory and ethical considerations also present significant obstacles, requiring clear guidelines and a commitment to equity in access to these advanced therapies. The future of diabetes treatment lies in the integration of nanotechnology with personalized medicine, tailoring therapies to individual patient profiles for more precise and effective care. This approach promises not only to manage symptoms more efficiently but also to tackle the underlying causes of diabetes, potentially transforming the prognosis of this chronic condition. In the contemporary materialistic society of today, individuals are exhibiting reduced levels of physical activity, leading to the onset of various ailments like diabetes, atherosclerosis, and heart attacks, among others, thereby underscoring the crucial need for seeking appropriate treatments. Research indicates that naturally derived nanoparticles exhibit considerable potential as therapeutic agents, offering a promising outlook in the battle against these diseases. The distinctive dimensions and adjustable characteristics of nanomaterials have captured the interest of a wide array of stakeholders, including researchers and medical professionals. Furthermore, biologically synthesized nanomaterials are found to possess lower levels of toxicity and higher levels of compatibility with living organisms in comparison to their chemically produced counterparts. In essence, it is evident that nanotechnology assumes a significant role in facilitating drug delivery for the management of various metabolic disorders. The utilization of nanomaterials can prove advantageous in minimizing drug dosages or mitigating adverse effects in the treatment of a multitude of conditions, such as metabolic disorders and cancer. Nonetheless, the realm of nanoformation presents numerous challenges, encompassing aspects like particle size, production costs, reproducibility, stability, and regulatory frameworks, which continue to lack clarity. Addressing the issue of poor oral medication bioavailability necessitates a thorough exploration of the potentials and constraints associated with nanoparticle formulations, indicating a pressing requirement for further investigative endeavors in this domain. This comprehensive review paper serves to shed light on diverse nanomaterials and their efficacious contributions to combatting metabolic disorders. Moreover, the review paper distinguishes itself by providing an in-depth analysis of the associated constraints and dietary considerations, thereby paving the way for extensive research initiatives and exploration in this field.

References

1. Yun SJ, Jeong IK, Cha JH, Lee J, Cho HC, Choi SH, Chun S, Jeon HJ, Kang HC, Kim SS, Ko SH (2022) Current status of low-density lipoprotein cholesterol target achievement in patients with type 2 diabetes mellitus in Korea compared with recent guidelines. Diabetes Metab J 46(3):464
2. Ninčević V, Omanović Kolarić T, Roguljić H, Kizivat T, Smolić M, Bilić ĆI (2019) Renal benefits of SGLT 2 inhibitors and GLP-1 receptor agonists: evidence supporting a paradigm shift in the medical management of type 2 diabetes. Int J Mol Sci 20(23):5831
3. Natali A, Nesti L, Tricò D, Ferrannini E (2021) Effects of GLP-1 receptor agonists and SGLT-2 inhibitors on cardiac structure and function: a narrative review of clinical evidence. Cardiovasc Diabetol 20(1):196

4. Puglisi S, Rossini A, Poli R, Dughera F, Pia A, Terzolo M, Reimondo G (2021) Effects of SGLT2 inhibitors and GLP-1 receptor agonists on renin-angiotensin-aldosterone system. Front Endocrinol 21(12):738848
5. Lee MM, Petrie MC, McMurray JJ, Sattar N (2020) How do SGLT2 (sodium-glucose cotransporter 2) inhibitors and GLP-1 (glucagon-like peptide-1) receptor agonists reduce cardiovascular outcomes? Completed and ongoing mechanistic trials. Arterioscler Thromb Vasc Biol 40(3):506–522
6. Górriz JL, Soler MJ, Navarro-González JF, García-Carro C, Puchades MJ, D'Marco L, Martínez Castelao A, Fernández-Fernández B, Ortiz A, Górriz-Zambrano C, Navarro-Pérez J (2020) GLP-1 receptor agonists and diabetic kidney disease: a call of attention to nephrologists. J Clin Med 9(4):947
7. Cherney DZ, Udell JA, Drucker DJ (2021) Cardiorenal mechanisms of action of glucagon-like-peptide-1 receptor agonists and sodium-glucose cotransporter 2 inhibitors. Med 2(11):1203–1230
8. Bertoccini L, Baroni MG (2021) GLP-1 receptor agonists and SGLT2 inhibitors for the treatment of type 2 diabetes: new insights and opportunities for cardiovascular protection. Diabetes Res Clin Pract 4:193–212
9. Chang SS, Jalal K, Charest AF (2020) Renin-angiotensin-aldosterone system (RAAS) blockade does not affect kidney progression in patients with CKD without diabetes and without proteinuria: PO0442. J Am Soc Nephrol 31(10S):184
10. Ferrario CM, Mullick AE (2017) Renin angiotensin aldosterone inhibition in the treatment of cardiovascular disease. Pharmacol Res 1(125):57–71
11. Pugliese NR, Masi S, Taddei S (2020) The renin-angiotensin-aldosterone system: a crossroad from arterial hypertension to heart failure. Heart Fail Rev 25(1):31–42
12. Balakumar P, Jagadeesh G (2015) Drugs targeting RAAS in the treatment of hypertension and other cardiovascular diseases. Pathophysiol Pharmacother Cardiovasc Dis:751–806
13. Chan GC, Tang SC (2016) Diabetic nephropathy: landmark clinical trials and tribulations. Nephrol Dial Transplant 31(3):359–368
14. Taraji H (2021) A study of risk factors linked to depression among the elderly population of iran: a systematic review across 20 years. Doctoral dissertation, Morgan State University
15. Fatima A, Rasool S, Devi S, Talha M, Waqar F, Nasir M, Khan MR, Jaffari SM, Haider A, Shah SU, Sapna FN (2023) Exploring the cardiovascular benefits of sodium-glucose cotransporter-2 (SGLT2) inhibitors: expanding horizons beyond diabetes management. Cureus 15(9)
16. Sena CM, Bento CF, Pereira P, Seiça R (2010) Diabetes mellitus: new challenges and innovative therapies. EPMA J 1:138–163
17. Verspohl EJ (2009) Novel therapeutics for type 2 diabetes: incretin hormone mimetics (glucagon-like peptide-1 receptor agonists) and dipeptidyl peptidase-4 inhibitors. Pharmacol Ther 124(1):113–138
18. Khursheed R, Singh SK, Wadhwa S, Kapoor B, Gulati M, Kumar R, Ramanunny AK, Awasthi A, Dua K (2019) Treatment strategies against diabetes: Success so far and challenges ahead. Eur J Pharmacol 5(862):172625
19. Wagman AS, Nuss JM (2001) Current therapies and emerging targets for the treatment of diabetes. Curr Pharm Des 7(6):417–450
20. Dahlén AD, Dashi G, Maslov I, Attwood MM, Jonsson J, Trukhan V, Schiöth HB (2022) Trends in antidiabetic drug discovery: FDA approved drugs, new drugs in clinical trials and global sales. Front Pharmacol 19(12):807548
21. Chao EC, Henry RR (2010) SGLT2 inhibition—a novel strategy for diabetes treatment. Nat Rev Drug Discov 9(7):551–559
22. Ansari MA, Chauhan W, Shoaib S, Alyahya SA, Ali M, Ashraf H, Alomary MN, Al-Suhaimi EA (2023) Emerging therapeutic options in the management of diabetes: recent trends, challenges and future directions. Int J Obes 47(12):1179–1199
23. Tang J, Su T, Huang K, Dinh PU, Wang Z, Vandergriff A, Hensley MT, Cores J, Allen T, Li T, Sproul E (2018) Targeted repair of heart injury by stem cells fused with platelet nanovesicles. Nat Biomed Eng 2(1):17–26

24. Mandal D, Sarmah JK, Gupta J (2023) NanoRevolution: pioneering applications of nanotechnology in type II diabetes care. Eng Proc 56(1):56
25. Soccio RE, Chen ER, Lazar MA (2014) Thiazolidinediones and the promise of insulin sensitization in type 2 diabetes. Cell Metab 20(4):573–591
26. Hansen KB, Vilsbøll T, Knop FK (2010) Incretin mimetics: a novel therapeutic option for patients with type 2 diabetes–a review. Diabetes Metab Syndr Obes Targets Ther 17:155–163
27. Dicker D (2011) DPP-4 inhibitors: impact on glycemic control and cardiovascular risk factors. Diabetes Care 34(Suppl 2):S276
28. Singh M, Kumar A (2018) Risks associated with SGLT2 inhibitors: an overview. Curr Drug Saf 13(2):84–91
29. Bhatia S (2016) Natural polymer drug delivery systems. Springer International Publishing, Basel, Switzerland
30. Wicki A, Witzigmann D, Balasubramanian V, Huwyler J (2015) Nanomedicine in cancer therapy: challenges, opportunities, and clinical applications. J Control Release 28(200):138–157
31. Le LV, Chendke GS, Gamsey S, Wisniewski N, Desai TA (2020) Near-infrared optical nanosensors for continuous detection of glucose. J Diabetes Sci Technol 14(2):204–211
32. Chang CC, Chen CP, Wu TH, Yang CH, Lin CW, Chen CY (2019) Gold nanoparticle-based colorimetric strategies for chemical and biological sensing applications. Nanomaterials 9(6):861
33. Ghasemi Y, Peymani P, Afifi S (2009) Quantum dot: magic nanoparticle for imaging, detection and targeting. Acta Biomed 80(2):156–165
34. Badugu R, Reece EA, Lakowicz JR (2018) Glucose-sensitive silicone hydrogel contact lens toward tear glucose monitoring. J Biomed Opt 23(5):057005
35. Xu Y, De Keersmaecker H, Braeckmans K, De Smedt S, Cani PD, Préat V, Beloqui A (2020) Targeted nanoparticles towards increased L cell stimulation as a strategy to improve oral peptide delivery in incretin-based diabetes treatment. Biomaterials 1(255):120209
36. Singh J, Mittal P, Vasant Bonde G, Ajmal G, Mishra B (2018) Design, optimization, characterization and in-vivo evaluation of quercetin enveloped Soluplus®/P407 micelles in diabetes treatment. Artif Cells Nanomed Biotechnol 46(sup3):S546–S555
37. Oladipo IC, Lateef A, Azeez MA, Asafa TB, Yekeen TA, Ogunsona SB, Irshad HM, Abbas SH (2020)Antidiabetic properties of phytosynthesized gold nanoparticles (AuNPs) from Datura stramonium seed. In: IOP conference series: materials science and engineering (vol 805(1). IOP Publishing, p 012035
38. Ansari MJ, Anwer MK, Jamil S, Al-Shdefat R, Ali BE, Ahmad MM, Ansari MN (2016) Enhanced oral bioavailability of insulin-loaded solid lipid nanoparticles: pharmacokinetic bioavailability of insulin-loaded solid lipid nanoparticles in diabetic rats. Drug Deliv 23(6):1972–1979
39. Xu Y, Van Hul M, Suriano F, Préat V, Cani PD, Beloqui A (2020) Novel strategy for oral peptide delivery in incretin-based diabetes treatment. Gut 69(5):911–919
40. Sengottaiyan A, Aravinthan A, Sudhakar C, Selvam K, Srinivasan P, Govarthanan M, Manoharan K, Selvankumar T (2016) Synthesis and characterization of solanum nigrum-mediated silver nanoparticles and its protective effect on alloxan-induced diabetic rats. J Nanostruct Chem 6:41–48
41. Wen N, Lü S, Xu X, Ning P, Wang Z, Zhang Z, Gao C, Liu Y, Liu M (2019) A polysaccharide-based micelle-hydrogel synergistic therapy system for diabetes and vascular diabetes complications treatment. Mater Sci Eng C 1(100):94–103
42. Mishra M, Kumar H, Singh RK, Tripathi K (2008) Diabetes and nanomaterials. Dig J Nanomater Biostruct 3(3):109–113
43. Jiang H, Xia C, Lin J, Garalleh HA, Alalawi A, Pugazhendhi A (2023) Carbon nanomaterials: a growing tool for the diagnosis and treatment of diabetes mellitus. Environ Res 15(221):115250
44. Bai Q, Han K, Dong K, Zheng C, Zhang Y, Long Q, Lu T (2020) Potential applications of nanomaterials and technology for diabetic wound healing. Int J Nanomed 3:9717–9743

45. Shoaib A, Darraj A, Khan ME, Azmi L, Alalwan A, Alamri O, Tabish M, Khan AU (2023) A nanotechnology-based approach to biosensor application in current diabetes management practices. Nanomaterials 13(5):867
46. Verma J, Dahiya S (2024) Nanomaterials for diabetes: diagnosis, detection and delivery. Nanotechnology 35(39):392001
47. Liu Y, Zeng S, Ji W, Yao H, Lin L, Cui H, Santos HA, Pan G (2022) Emerging theranostic nanomaterials in diabetes and its complications. Adv Sci 9(3):2102466
48. Mohsen AM (2019) Nanotechnology advanced strategies for the management of diabetes mellitus. Curr Drug Targets 20(10):995–1007
49. Chauhan PS, Yadav D, Tayal S, Jin JO (2020) Therapeutic advancements in the management of diabetes mellitus with special reference to nanotechnology. Curr Pharm Des 26(38):4909–4916
50. Jeevanandam J, Danquah KM, Debnath S, Meka VS, Chan YS (2015) Opportunities for nano-formulations in type 2 diabetes mellitus treatments. Curr Pharm Biotechnol 16(10):853–870
51. Si S, Pal A, Mohanta J, Satapathy SS (2017) Gold nanostructure materials in diabetes management. J Phys D Appl Phys 50(13):134003
52. Priyam A, Singh PP, Gehlout S (2018) Role of endocrine-disrupting engineered nanomaterials in the pathogenesis of type 2 diabetes mellitus. Front Endocrinol 26(9):704
53. Kesharwani P, Gorain B, Low SY, Tan SA, Ling EC, Lim YK, Chin CM, Lee PY, Lee CM, Ooi CH, Choudhury H (2018) Nanotechnology based approaches for anti-diabetic drugs delivery. Diabetes Res Clin Pract 1(136):52–77
54. Rashid R, Naqash A, Bader GN, Sheikh FA (2020) Nanotechnology and diabetes management: recent advances and future perspectives. Appl Nanotechnol Biomed Sci:99–117
55. Pathak K, Saikia R, Sarma H, Pathak MP, Das RJ, Gogoi U, Ahmad MZ, Das A, Wahab BA (2023) Nanotheranostics: application of nanosensors in diabetes management. J Diabetes Metab Disord 22(1):119–133
56. Wang T, Wang N, Song H, Xi X, Wang J, Hao A, Li T (2011) Preparation of an anhydrous reverse micelle delivery system to enhance oral bioavailability and anti-diabetic efficacy of berberine. Eur J Pharm Sci 44(1–2):127–135
57. Agnish S, Sharma AD, Kaur I (2022) Nanoemulsions (O/W) containing cymbopogon pendulus essential oil: development, characterization, stability study, and evaluation of in vitro anti-bacterial, anti-inflammatory, anti-diabetic activities. Bionanoscience 12(2):540–554
58. Maity S, Mukhopadhyay P, Kundu PP, Chakraborti AS (2017) Alginate coated chitosan core-shell nanoparticles for efficient oral delivery of naringenin in diabetic animals—an in vitro and in vivo approach. Carbohyd Polym 15(170):124–132
59. Akhtar N, Mohammed SA, Khan RA, Yusuf M, Singh V, Mohammed HA, Al-Omar MS, Abdellatif AA, Naz M, Khadri H (2020) Self-Generating nano-emulsification techniques for alternatively-routed, bioavailability enhanced delivery, especially for anti-cancers, anti-diabetics, and miscellaneous drugs of natural, and synthetic origins. J Drug Deliv Sci Technol 1(58):101808
60. Roy M, Pal R, Chakraborti AS (2017) Pelargonidin-PLGA nanoparticles: fabrication, characterization, and their effect on streptozotocin induced diabetic rats
61. Agrawal M, Saraf S, Saraf S, Dubey SK, Puri A, Patel RJ, Ravichandiran V, Murty US, Alexander A (2020) Recent strategies and advances in the fabrication of nano lipid carriers and their application towards brain targeting. J Control Release 10(321):372–415
62. Shamekhi F, Tamjid E, Khajeh K (2018) Development of chitosan coated calcium-alginate nanocapsules for oral delivery of liraglutide to diabetic patients. Int J Biol Macromol 1(120):460–467
63. Martínez-Esquivias F, Guzmán-Flores JM, Párez-Larios A, Rico JL, Becerra-Ruiz JS (2021) A review of the effects of gold, silver, selenium, and zinc nanoparticles on diabetes mellitus in murine models. Mini Rev Med Chem 21(14):1798–1812
64. Mariadoss AV, Sivakumar AS, Lee CH, Kim SJ (2022) Diabetes mellitus and diabetic foot ulcer: etiology, biochemical and molecular based treatment strategies via gene and nanotherapy. Biomed Pharmacother 1(151):113134

65. Alomari G, Hamdan S, Al-Trad B (2021) Gold nanoparticles as a promising treatment for diabetes and its complications: current and future potentials. Braz J Pharm Sci 26(57):e19040
66. Uma Suganya KS, Govindaraju K, Veena Vani C, Premanathan M, Ganesh Kumar VK (2019) In vitro biological evaluation of anti-diabetic activity of organic–inorganic hybrid gold nanoparticles. IET Nanobiotechnol 13(2):226–229
67. Abdel-Halim AH, Fyiad AA, Aboulthana WM, El-Sammad NM, Youssef AM, Ali MM (2020) Assessment of the anti-diabetic effect of Bauhinia variegata gold nano-extract against streptozotocin induced diabetes mellitus in rats. J Appl Pharm Sci 10(05):077–091
68. Al-Radadi NS (2022) Biogenic proficient synthesis of (Au-NPs) via aqueous extract of red dragon pulp and seed oil: characterization, antioxidant, cytotoxic properties, anti-diabetic anti-inflammatory, anti-alzheimer and their anti-proliferative potential against cancer cell lines. Saudi J Biol Sci 29(4):2836–2855
69. Seetharaman PK, Ramalingam P, Chandrika M, Rajan R, Chelliah J (2023) Antidiabetic potential of gymnemic acid mediated gold nanoparticles (Gym@ AuNPs) on streptozotocin-induced diabetic rats—an implication on in vivo approach. Int J Pharm 5(636):122843
70. Tan Y, Zhang Z, Zheng C, Wintergerst KA, Keller BB, Cai L (2020) Mechanisms of diabetic cardiomyopathy and potential therapeutic strategies: preclinical and clinical evidence. Nat Rev Cardiol 17(9):585–607
71. Engel SS, Williams-Herman DE, Golm GT, Clay RJ, Machotka SV, Kaufman KD, Goldstein BJ (2010) Sitagliptin: review of preclinical and clinical data regarding incidence of pancreatitis. Int J Clin Pract 64(7):984–990
72. Calcutt NA, Cooper ME, Kern TS, Schmidt AM (2009) Therapies for hyperglycaemia-induced diabetic complications: from animal models to clinical trials. Nat Rev Drug Discov 8(5):417–430
73. Kiencke S, Handschin R, von Dahlen R, Muser J, Brunner-LaRocca HP, Schumann J, Felix B, Berneis K, Rickenbacher P (2010) Pre-clinical diabetic cardiomyopathy: prevalence, screening, and outcome. Eur J Heart Fail 12(9):951–957
74. D'Souza R (2017) Ethical Issues associated with diagnosing and managing diabetes. Glob Bioeth Enquiry 2017:52–56
75. Quintal A, Messier V, Rabasa-Lhoret R, Racine E (2020) A qualitative study of the views of individuals with type 1 diabetes on the ethical considerations raised by the artificial pancreas. Narrat Inq Bioeth 10(3):237–261
76. Baker HB, McQuilling JP, King NM (2016) Ethical considerations in tissue engineering research: case studies in translation. Methods 15(99):135–144
77. Mohi-ud-din R, Mir RH, Pottoo FH, Sawhney G, Masoodi MH, Bhat ZA (2020) Nanophytomedicine ethical issues, regulatory aspects, and challenges. Nanophytomed Concept Clin:173–192
78. Caturano A, Nilo R, Nilo D, Russo V, Santonastaso E, Galiero R, Rinaldi L, Monda M, Sardu C, Marfella R, Sasso FC (2024) Advances in nanomedicine for precision insulin delivery. Pharmaceuticals 17(7):945
79. Addissouky T, Ali M, El Sayed IE, Wang Y (2023) Revolutionary innovations in diabetes research: from biomarkers to genomic medicine. Iran J Diabetes Obes

Emergence of Nanomaterials in the Management of Multiple Sclerosis

Shubhrat Maheshwari, Amita Verma, Aditya Singh, Rufaida Wasim, Mohammad Shariq, Juber Akhtar, and Bhupendra G. Prajapati

Abstract The emergence of nanomaterials has marked a transformative shift in the management of multiple sclerosis (MS), a chronic and progressive neurological disorder characterized by immune-mediated damage to myelin in the central nervous system. This chapter explores the significant advancements that nanotechnology brings to the field of MS treatment, emphasizing how nanomaterials are addressing the complex challenges associated with the disease. The introduction provides an overview of MS and current treatment challenges, setting the stage for the discussion on how nanomaterials offer innovative solutions. The chapter delves into the pathophysiology of MS, highlighting the role of immune system dysregulation, blood–brain barrier (BBB) dysfunction, and neuroinflammation. It then examines various types of nanomaterials, including liposomes, dendrimers, nanoparticles, carbon nanotubes, and polymeric nanoparticles, detailing their unique properties and therapeutic applications. The focus is on how these nanomaterials enhance drug solubility and stability, enable targeted drug delivery, facilitate BBB crossing, and provide controlled and sustained release of therapeutic agents. Further, the chapter discusses nanomaterial-based therapeutic approaches tailored for MS,

S. Maheshwari · A. Verma (✉)
Bioorganic and Medicinal Chemistry Research Laboratory, Department of Pharmaceutical Sciences, Sam Higginbottom University of Agriculture, Technology and Sciences, Prayagraj, Uttar Pradesh, India
e-mail: amita.verma@shiats.edu.in

A. Verma
Department of Allied Sciences (Chemistry), Graphic Era (Deemed to be University), Dehradun, India

A. Singh · R. Wasim · M. Shariq · J. Akhtar
Department of Pharmacy, Faculty of Pharmacy, Integral University, Lucknow, Uttar Pradesh, India

B. G. Prajapati (✉)
Shree S. K. Patel College of Pharmaceutical Education and Research, Ganpat University, Gujarat, India
e-mail: bhupen27@gmail.com

Faculty of Pharmacy, Silpakorn University, Nakhon Pathom, Thailand

Md. M. Ansari et al. (eds.), *Emergence of Sustainable Biomaterials in Tackling Inflammatory Diseases*, Smart Nanomaterials Technology,
https://doi.org/10.1007/978-981-96-2112-5_12

such as anti-inflammatory therapies, immunomodulatory strategies, neuroprotection and remyelination, and gene therapy delivery systems.

Keywords Multiple sclerosis · Immunomodulatory strategies · Neuroprotection · Neurological disorder · Liposomes · Dendrimers

1 Introduction

Multiple sclerosis (MS) is a chronic, inflammatory, and neurodegenerative disease characterized by the demyelination of nerve fibers in the central nervous system (CNS). It is an autoimmune disorder wherein the immune system erroneously targets the myelin sheath, the protective covering of nerve fibers, leading to disrupted communication between the brain and the rest of the body [1]. The etiology of MS is multifactorial, involving genetic predisposition, environmental factors, and possibly viral infections [2]. Common symptoms include fatigue, motor and sensory disturbances, visual impairments, and cognitive deficits, significantly impacting patients' quality of life. The current therapeutic landscape for MS includes disease-modifying therapies (DMTs) such as interferon-beta, glatiramer acetate, and monoclonal antibodies like natalizumab and ocrelizumab [3]. These treatments primarily aim to reduce the frequency and severity of relapses, delay disease progression, and manage symptoms. Despite advancements in DMTs, significant challenges persist [4]. The efficacy of these treatments varies among patients, with many experiencing suboptimal responses or adverse effects. Moreover, the blood–brain barrier (BBB) poses a formidable obstacle to effective drug delivery, limiting the CNS bioavailability of therapeutic agents [5]. Long-term use of immunosuppressive therapies also raises concerns about increased risks of infections and malignancies [6–9]. The disease mechanisms underlying MS involve a series of pathological processes that lead to progressive neurological dysfunction [10]. At its core, MS is driven by an autoimmune response where the immune system becomes aberrantly activated against central nervous system (CNS) components, particularly myelin, the protective sheath around nerve fibers. This autoimmune attack initiates an inflammatory cascade, ultimately causing demyelination, axonal damage, and neuronal loss. The progression of MS can be broadly categorized into relapsing–remitting and progressive forms [11]. Primary progressive MS (PPMS) involves a steady progression of disability from the onset without distinct relapses while progressive-relapsing MS (PRMS) is marked by progressive disease with superimposed relapses. The pathological hallmarks of MS include multifocal areas of demyelination (plaques) [12], axonal injury, gliosis (scar tissue formation), and inflammation. Immune system dysregulation plays a central role in the pathogenesis of MS. The disease is driven by the infiltration of autoreactive T cells into the CNS, which are activated in the peripheral immune system by unknown triggers, potentially involving molecular mimicry or bystander activation. These T cells cross the blood–brain barrier (BBB) and recognize myelin antigens, leading to the activation of microglia and macrophages, the resident immune cells of the CNS. This activation results in the release of pro-inflammatory cytokines, chemokines, and other cytotoxic mediators that perpetuate the inflammatory response [13]. B cells also contribute to MS pathology by producing autoantibodies against myelin components and forming ectopic lymphoid follicles within the CNS, which sustain chronic inflammation. Moreover, dysregulation of regulatory T cells (Tregs), which normally suppress immune responses, exacerbates the autoimmune attack in MS. The dysfunction of the BBB is a critical factor in the pathophysiology of MS. The BBB is a highly selective permeability barrier that protects the CNS from potentially harmful substances in the bloodstream while allowing the selective passage of essential molecules. In MS, the integrity of the BBB is compromised, primarily due to the action of pro-inflammatory cytokines and matrix metalloproteinases (MMPs) released by activated immune cells [14]. These molecules degrade the tight junction proteins that maintain BBB integrity, leading to increased permeability and facilitating the entry of immune cells and inflammatory mediators into the CNS [15]. The resultant breach of the BBB amplifies the inflammatory milieu within the CNS, promoting further demyelination and neurodegeneration. Oxidative stress and mitochondrial dysfunction are also implicated in the pathogenesis of MS. The chronic inflammation in MS generates reactive oxygen species (ROS), which cause damage to myelin, axonal structures, and cellular components. Mitochondrial dysfunction, resulting from oxidative damage, contributes to neuronal injury and loss, exacerbating disease progression [16]. Nanomaterials are engineered materials with dimensions on the nanoscale [17]. Their small size and high surface area-to-volume ratio make them particularly advantageous for biomedical applications, including drug delivery, diagnostics, and therapeutic interventions. In the context of MS, nanomaterials offer promising potential for overcoming existing treatment limitations [18–21], providing new avenues for more effective and targeted therapies [22].

2 Types of Nanomaterials Application in MS

Various types of nanomaterials are being explored for their application in MS management, each with unique characteristics and advantages [23]. The core lipid composition of solid lipid nanoparticles (SLNs) commonly exhibits variations in the lipid matrix composition and structure, leading to distinct morphological characteristics [24]. Monostearin is a commonly employed solid lipid in nanostructured lipid carrier (NLC) formulations, while liquid lipids such as olive oil, sesame oil, almond oil, peanut oil, and soybean oil are also utilized [25, 26].

Gadhave et al. recently developed a dihydroorotate dehydrogenase inhibitor with optimization of the formulation conducted through the Box–Behnken statistical design. These lipid nanocarriers were evaluated based on parameters such as particle size, entrapment efficiency, in vitro and ex vivo permeation studies, as well as pharmacological and toxicological assessments [27]. The optimized formulation demonstrated a particle size of 99.82 nm, a surface charge, and an entrapment efficiency of 83.39%. The incorporation of mucoadhesive and gelling agents enhanced the formulation, and the results indicated superior permeation flux of the mucoadhesive nanosystem compared to the plain formulation. The intranasal delivery of these nanostructures promoted rapid remyelination of damaged neurons in a rat model of cuprizone-induced injury, without notable alterations in hepatic biomarkers or sub-acute toxicity, signifying its efficacy and safety for treating brain disorders [28].

Another investigation involved the formulation of methylthioadenosine-loaded SLNs for oral administration targeting the brain to manage MS [29–32]. Comparing passive and active targeting using short peptide fragments of ApoE or β-amyloid, researchers prepared PEG-DSPE-2000 lipid for stealthiness, and peptide-conjugated sterically stabilized liposomes by linking ApoE or β-amyloid to dioleoyl (DO)-succinate for BBB transport. Physicochemical characterization was conducted, and EAE mice models were employed to assess therapeutic efficacy and mechanism of action in acute and adoptive transfer EAE [33]. Results indicated no superiority of active targeting for liposomes carrying methylprednisolone, with both methylprednisolone and TFM formulations ameliorating EAE clinical signs and pathology. The authors suggested that these nano-drugs, meeting disease-modifying drug criteria, should be further developed and evaluated as potential therapeutic agents for inflammatory diseases. The prevalence of autoimmunity is increasing, with no cure for autoimmune diseases stemming from self-tolerance loss. Liposomes, beyond drug delivery, have been utilized for immunotherapy due to their capacity to induce immunological tolerance through apoptosis-mediated mechanisms involved in maintaining self-tolerance, particularly efficient apoptotic cell clearance [34].

Dendrimers are highly branched, tree-like macromolecules with a well-defined, three-dimensional architecture. Their unique structure provides numerous functionalizable surface groups, allowing for precise drug loading and release control. Dendrimers can be designed to carry multiple therapeutic agents simultaneously, offering a multifaceted approach to MS treatment by combining anti-inflammatory, neuroprotective, and immunomodulatory effects [35]. The ability to modify dendrimer surfaces with targeting ligands and imaging agents further enhances their potential for targeted drug delivery and diagnostic applications in MS.

Nanoparticles, including gold, silver, and quantum dots, exhibit unique optical, electronic, and catalytic properties due to their nanoscale size. Gold nanoparticles, for example, can be easily functionalized with various biomolecules, making them suitable for targeted drug delivery and diagnostic imaging in MS [36]. Silver nanoparticles possess antimicrobial properties, which could be beneficial in preventing infections associated with immunosuppressive MS therapies. Quantum dots, semiconductor nanoparticles, offer superior fluorescent properties for imaging applications, enabling the tracking of disease progression and therapeutic efficacy.

Carbon nanotubes are cylindrical nanostructures composed of rolled-up graphene sheets. Their high aspect ratio, mechanical strength, and electrical conductivity make them attractive for various biomedical applications, including drug delivery and neural tissue engineering. Carbon nanotubes can be functionalized to improve their biocompatibility and targeted delivery to the CNS [37], and their ability to penetrate cell membranes enhances their potential for delivering therapeutic agents directly to affected neural tissues in MS.

Polymeric nanoparticles are colloidal particles made from biodegradable polymers. These nanoparticles can encapsulate a wide range of drugs, providing controlled and sustained release profiles. Polymeric nanoparticles can be engineered to enhance their stability, bioavailability, and targeting capabilities, making them suitable for delivering therapeutic agents across the BBB in MS [38]. These nanoparticles can also be designed to respond to specific triggers, such as pH or temperature changes, for on-demand drug release.

The synthesis and characterization of nanomaterials are crucial steps in ensuring their efficacy and safety for biomedical applications. Various methods are employed to synthesize nanomaterials, including chemical vapor deposition, sol–gel processes, and self-assembly techniques. Each method offers specific advantages in terms of control over the size, shape, and surface properties of the nanomaterials. Characterization techniques, such as electron microscopy, dynamic light scattering, and spectroscopy, are used to analyze the properties of nanomaterials. These techniques provide valuable insights into the size distribution, surface charge, chemical composition, and functionalization efficiency of nanomaterials, which are essential for optimizing their performance in MS treatment [39] (Fig. 1).

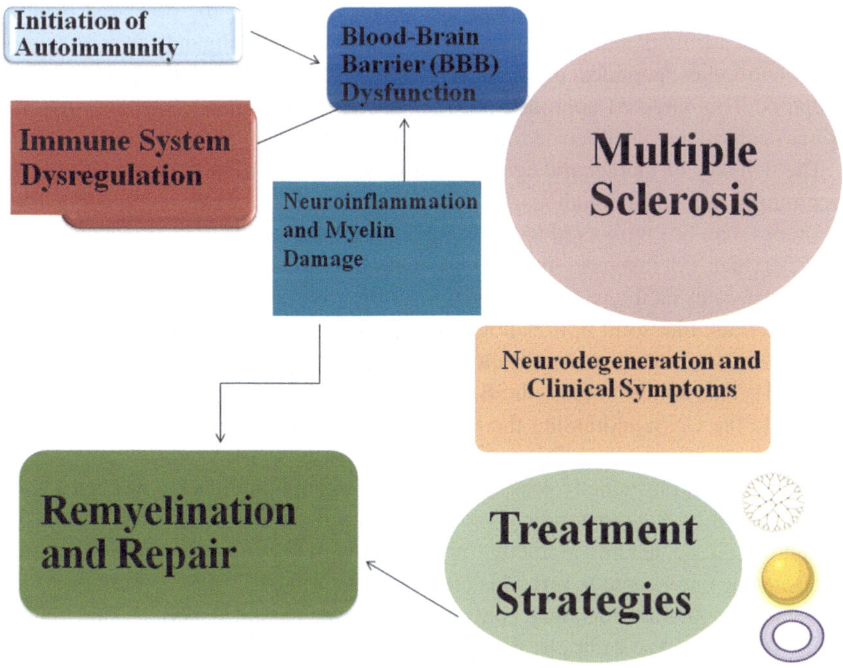

Fig. 1 The pathophysiology of MS involves a complex interplay of immune system dysregulation, blood–brain barrier (BBB) dysfunction, and neuroinflammatory processes. Understanding the MS pathway is crucial for developing effective therapeutic strategies

3 Nanomaterials in Drug Delivery for MS

Nanomaterials have revolutionized drug delivery by enhancing targeting and controlled release. Their unique physicochemical properties allow for innovative solutions that significantly improve the efficacy and safety of therapeutic agents [40]. Many therapeutic agents, especially those with poor aqueous solubility, face challenges in achieving optimal bioavailability [41]. Nanomaterials can encapsulate these poorly soluble drugs within their core or on their surface, effectively increasing their solubility. For example, liposomes, composed of phospholipid bilayers, can solubilize hydrophobic drugs within their lipid layers, while dendrimers can provide a stable environment for both hydrophilic and hydrophobic drugs within their branched structure [42]. Similarly, polymeric nanoparticles can encapsulate drugs within a biodegradable polymer matrix, protecting them from degradation and enhancing their stability in biological environments. This improvement in solubility and stability not only enhances the bioavailability of therapeutic agents but also extends their shelf life and efficacy.

Targeted drug delivery is another critical advantage of nanomaterials. Conventional drug delivery systems often suffer from non-specific distribution, leading to systemic side effects and suboptimal therapeutic outcomes [43]. For instance, in the context of multiple sclerosis (MS), nanomaterials can be designed to target activated immune cells or inflamed CNS tissues, improving the precision of drug delivery. Dendrimers and liposomes can be conjugated with ligands such as monoclonal antibodies, peptides, or small molecules that selectively bind to cell surface receptors. This targeted approach not only increases therapeutic efficacy but also reduces potential side effects associated with systemic drug administration.

The delivery of therapeutic agents. Nanomaterials offer innovative solutions for overcoming this barrier. Nanoparticles, such as liposomes, dendrimers, and carbon nanotubes, can be engineered to cross the BBB by exploiting their size and surface characteristics. For instance, surface modification with ligands that interact with BBB transporters can facilitate active transport across the barrier. Additionally, certain nanomaterials, like solid lipid nanoparticles, can be designed to cross the BBB via endocytosis, wherein the nanoparticles are internalized by endothelial cells lining the BBB [44]. This ability to cross the BBB enhances the delivery of therapeutic agents directly to the CNS, addressing the limitations of traditional drug delivery systems in treating neurological conditions like MS.

Controlled and sustained release of therapeutic agents is another significant advantage provided by nanomaterials. Traditional drug delivery systems often result in rapid drug release, leading to fluctuating drug levels and potential toxicity. Nanomaterials can be engineered to provide controlled and sustained release profiles, which help maintain therapeutic drug concentrations over extended periods and reduce dosing frequency. For example, polymeric nanoparticles can be designed to release their payload in a controlled manner in response to specific physiological triggers, such as pH changes or enzymatic activity [45]. Similarly, liposomes can be engineered to release their contents gradually through passive diffusion or triggered by environmental factors. This controlled release not only improves patient compliance but also enhances therapeutic efficacy and reduces the risk of side effects [46].

4 Nanomaterial-Based Therapeutic Approaches for MS

These approaches harness the unique properties of nanomaterials to enhance therapeutic efficacy, minimize side effects, and provide targeted interventions for key aspects of MS. This exploration covers anti-inflammatory therapies, immunomodulatory strategies, neuroprotection and remyelination, and gene therapy delivery systems [47], highlighting how nanomaterials can revolutionize MS treatment. Anti-inflammatory therapies are fundamental in managing MS due to their characteristic chronic inflammation and autoimmune attack on myelin. Nanomaterials can significantly improve the delivery and effectiveness of anti-inflammatory agents. For instance, liposomes are widely used to encapsulate corticosteroids or NSAIDs, allowing for targeted delivery to sites of inflammation within the CNS [48]. The liposomal bilayer protects these drugs from degradation and enhances their bioavailability at the site of disease activity, reducing systemic exposure and potential side effects. Similarly, polymeric nanoparticles can be engineered to release anti-inflammatory drugs in a controlled manner, ensuring sustained drug levels and targeted action at inflammatory sites. Functionalized dendrimers offer another approach by carrying anti-inflammatory agents with high precision, allowing for the modulation of specific immune responses involved in MS [49]. By delivering small molecules or RNA-based therapies that inhibit key inflammatory mediators, these nanomaterials can effectively dampen the autoimmune response and mitigate tissue damage. Immunomodulatory strategies focus on altering the aberrant immune responses that drive MS pathology. Nanomaterials enable targeted delivery of immunomodulatory agents or vaccines to modulate the immune system. For example, dendrimers can be functionalized with peptides or antigens that promote tolerance or induce regulatory T cells, thereby suppressing autoreactive immune responses. This approach can help retrain the immune system to recognize self-antigens as benign, reducing autoimmune activity. Polymeric nanoparticles can deliver vaccines or immunomodulatory agents that specifically target immune cells involved in MS, such as T cells or B cells [50]. These nanoparticles can also be used to deliver checkpoint inhibitors or small molecules that modulate immune cell signaling, providing a means to fine-tune the immune response and alter disease progression. Neuroprotection and remyelination are crucial therapeutic goals in MS, aimed at preserving neuronal function and promoting the repair of damaged myelin. Nanomaterials offer novel strategies for supporting neuroprotection and facilitating remyelination. Liposomes and polymeric nanoparticles can be designed to deliver neuroprotective agents such as antioxidants, anti-apoptotic factors, or neurotrophic factors directly to neurons and glial cells. For example, nanoparticles can be loaded with compounds like BDNF or NGF [51], which support neuronal survival and function, thereby mitigating neurodegeneration. Furthermore, nanomaterials can facilitate remyelination by delivering agents or gene therapies that promote myelin repair. Dendrimers and nanoparticles can carry plasmids or viral vectors encoding myelin-associated proteins or growth factors that stimulate oligodendrocyte differentiation and myelin regeneration. By promoting remyelination, these nanomaterials help restore neuronal function and improve clinical outcomes for MS patients. Gene therapy delivery systems represent a cutting-edge approach to treating MS by directly modifying genetic material to address disease-related defects or introduce therapeutic genes [52]. Nanomaterials play a critical role in the delivery of gene therapy constructs, ensuring efficient and targeted gene transfer. Liposomes, polymeric nanoparticles, and dendrimers can encapsulate genetic materials such as plasmid DNA, RNAi molecules, or CRISPR/Cas9 systems, facilitating their delivery across cellular membranes and the BBB. For example, lipid-based nanoparticles can deliver RNAi molecules that specifically target and silence genes involved in MS pathology, such as those encoding inflammatory cytokines or immune modulators. Similarly, dendrimers can be designed to deliver gene editing tools or therapeutic genes that correct genetic defects or modulate disease pathways. The ability to deliver gene therapies with precision and efficiency holds significant promise for personalized and targeted treatment strategies in MS, potentially correcting the underlying causes of the disease and providing long-lasting therapeutic benefits [53].

The therapeutic objectives of utilizing disease-modifying agents in individuals diagnosed with multiple sclerosis (MS) include reducing the duration of acute exacerbations, minimizing their frequency, and providing symptomatic relief. Currently, there are no curative treatments for MS that have been endorsed by the Food and Drug Administration (FDA) [54]. Symptomatic interventions are geared towards preserving functionality and improving quality of life [55]. Despite the chronic nature of MS, FDA-approved therapeutic agents have shown efficacy in managing disease activity and slowing progression in patients with relapsing forms of MS, including those with secondary progressive MS who continue to experience relapses, as outlined in Table 1. Initially introduced for MS treatment, these agents have evolved over time. A multitude of novel drug therapies have received FDA approval post-2010, thereby providing more options for individuals with MS [56].

5 Nanotechnology-Based Diagnosis in MS

MS represents a multifaceted challenge in the realm of neurology, characterized by the progressive deterioration of the central nervous system's myelin sheath. This myelin degradation disrupts neural communication and leads to a spectrum of neurological symptoms that vary in severity and presentation. The identification and management of MS is complex due to the lack of definitive biomarkers beyond the oligoclonal IgG bands found in cerebrospinal fluid. Although these bands assist in diagnosis, they are detected through an invasive lumbar puncture, and their correlation with disease activity and therapeutic response remains inadequately understood. The reliance on (MRI) for monitoring, diagnosing, and managing MS, while

Table 1 FDA-approved drugs for MS management

Brand name	Year of approval	Drug	Mechanism of action	Dosage and route of administration
Avonex	1996	Interferon β-1a	inhibits T cell activation	30 μg, IM, once a week [57]
Novantrone	2000	Mitoxantrone	Suppresses T cells, B-cells, and macrophages	12 mg/m^2 once in every 3 months [58]
Betaseron, Extavia	1993	Interferon β-1b	inhibits T cell activation	250 μg, SC, [59]
Copaxone	1997	Glatiramer acetate	Activates T-lymphocyte suppressor cells	20 mg SC [60]
Gilenya	2010	Fingolimod	Prevent lymphocyte entry to CNS	0.5 mg, oral, daily. [61]
Lemtrada	2014	Alemtuzumab	Depletes B-cells, T-cells	IV infusion [62]
Aubagio	2012	Teriflunomide	Prrimidine synthesis inhibitor halts proliferation of T-lymphocytes and B-lymphocytes	7 mg or 14 mg oral, daily [63]
Plegridy	2014	Peg interferon β-1a	Inhibits T cell activation	125 μg, SQ [64]

instrumental, is hindered by its high cost and limited accessibility. Consequently, the search for more effective, non-invasive diagnostic and therapeutic modalities is crucial [65]. Nanotechnology, with its ability to operate at the molecular level, presents a transformative approach to MS management by offering innovative solutions across diagnosis, treatment, and disease monitoring. Nanomaterials, owing to their unique physicochemical properties and biocompatibility, have emerged as powerful tools in addressing the limitations of traditional MS diagnostics and therapies. The application of nanoparticles and nanoplatforms can potentially overcome many of the constraints associated with current methodologies, thereby advancing our understanding and management of MS [66]. One of the primary challenges in MS is the early and accurate detection of the disease. Traditional diagnostic methods, such as MRI and lumbar puncture, are essential but have limitations in sensitivity and specificity. Nanotechnology offers new possibilities for enhancing disease detection through the development of advanced nanomaterials designed to target specific biomarkers or pathological features of MS. For example, nanoparticles can be engineered to bind selectively to myelin proteins or other molecules associated with MS pathology. These nanoparticles can be utilized in various imaging techniques, including MRI and optical imaging, to provide enhanced contrast and resolution [67]. By improving the sensitivity and specificity of imaging, these nanomaterials can facilitate earlier detection of MS and more precise monitoring of disease progression. The therapeutic landscape for MS is equally challenging, as current treatments primarily focus on managing symptoms and slowing disease progression rather than addressing the underlying causes. Nanotechnology provides a novel approach to therapy through targeted drug delivery systems. Nanoparticles can be designed to encapsulate therapeutic agents and deliver them directly to the sites of inflammation or demyelination in the CNS. This targeted delivery minimizes systemic exposure and enhances the efficacy of the drugs while reducing potential side effects. Additionally, nanomaterials can be engineered to cross the (BBB), a major obstacle in CNS drug delivery. This ability to penetrate the BBB allows for more effective treatment of MS lesions and inflammation, potentially leading to improved therapeutic outcomes. Beyond drug delivery, nanotechnology offers innovative therapeutic strategies such as immunomodulation and neuroprotection. Nanoparticles can be functionalized with anti-inflammatory agents or molecules that modulate immune responses, targeting the aberrant immune activity characteristic of MS. For instance, nanoparticles loaded with anti-inflammatory drugs can reduce neuroinflammation by delivering these drugs specifically to the inflamed regions of the brain. Moreover, nanomaterials with antioxidant properties can help mitigate oxidative stress, another key contributor to neuronal damage in MS. By addressing both inflammation and oxidative stress, nanotechnology holds promise for more comprehensive therapeutic approaches [68]. Gene therapy is another exciting area where nanotechnology can make a significant impact. Nanoparticles can serve as delivery vehicles for gene-editing tools, such as CRISPR/Cas9, enabling precise modifications of genes involved in MS pathogenesis. This approach offers the potential for targeted genetic interventions that could correct underlying genetic predispositions or modulate gene expression to slow disease progression. The development of effective nanotechnology-based gene delivery systems is still in its early stages, but it represents a promising frontier for personalized medicine in MS [69]. The integration of nanotechnology into neuroimaging also addresses several limitations associated with existing techniques. In vivo neuroimaging methods, such as MRI and PET scans, are crucial for observing biochemical, structural, and functional changes in the brain. However, these techniques often suffer from issues related to sensitivity, resolution, and artifact interference. Nanoparticles can enhance imaging modalities by providing improved contrast and enabling more detailed visualization of brain structures and functions. For example, nanoparticles can be conjugated with imaging agents that enhance the signal intensity of MRI or optical imaging, allowing for better visualization of pathological changes associated with MS. Moreover, advances in nanotechnology have led to the development of multifunctional imaging agents that combine diagnostic and therapeutic capabilities. These nanoplatforms can simultaneously provide high-resolution imaging and deliver therapeutic agents, offering a more integrated approach to MS management. Such multifunctional platforms could facilitate real-time monitoring of treatment efficacy and disease progression, leading to more personalized and adaptive therapeutic strategies [70].

Despite the promising potential of nanotechnology in MS diagnosis and treatment, several challenges need to be addressed. Ensuring the safety and biocompatibility of nanomaterials is critical, as their unique properties may lead to unforeseen toxicological effects. Rigorous preclinical and clinical testing is required to assess the long-term safety of nanoparticles and their potential interactions with biological systems. Additionally, the scalability and cost-effectiveness of nanotechnology-based solutions need to be considered to ensure their accessibility and feasibility for widespread clinical use.

6 Emerging Concept of Neuroprotection Through Nanotechnology

The concept of neuroprotection through nanotechnology represents an exciting frontier in the field of neuroscience and therapeutic innovation. Neuroprotection, fundamentally aimed at preserving neuronal function and preventing neuronal damage, is crucial in a range of neurodegenerative diseases and acute neurological injuries. Nanotechnology, with its ability to manipulate materials at the nanometer scale, offers transformative potential in delivering neuroprotective agents with unprecedented precision and efficacy. This integration of nanotechnology into neuroprotection is driven by the necessity to address the limitations of traditional therapeutic approaches and to harness the unique properties of nanomaterials to target complex neurological conditions more effectively. At the core of nanotechnology's contribution to neuroprotection is its capacity to enhance drug delivery and optimize therapeutic interventions. Nanoparticles, due to their size and surface properties, can be engineered to cross the (BBB), a major challenge in treating neurological disorders. The BBB, a selective permeability barrier, restricts the entry of many therapeutic agents into (CNS). Nanoparticles, including liposomes, dendrimers, and solid lipid nanoparticles, can be designed to circumvent this barrier, either by passive diffusion through disrupted BBB regions or by active targeting using ligands that bind to specific receptors on endothelial cells. This ability to deliver drugs directly to the brain opens new avenues for targeting neurodegenerative diseases, such as Alzheimer's disease, Parkinson's disease, and multiple sclerosis, with enhanced precision and reduced systemic side effects [71] to improving drug delivery, nanotechnology enables the development of targeted and controlled release systems. By encapsulating neuroprotective agents within nanoparticles, researchers can achieve sustained and localized release of therapeutics. This controlled release is crucial for managing chronic neurodegenerative conditions, where consistent therapeutic levels are necessary over extended periods. Furthermore, nanoparticles can be engineered to release their payload in response to specific stimuli, such as changes in pH, temperature, or the presence of certain enzymes, which are often associated with pathological conditions in the CNS. This responsive release mechanism ensures that the therapeutic agents are delivered precisely when and where they are needed, thereby maximizing their efficacy while minimizing potential side effects. The anti-inflammatory and antioxidant properties of nanomaterials also play a significant role in neuroprotection. Neuroinflammation and oxidative stress are common pathological features in many neurodegenerative diseases, contributing to neuronal damage and disease progression. Nanoparticles can be designed to possess inherent antioxidant properties or to deliver antioxidants and anti-inflammatory agents directly to the sites of inflammation and oxidative damage. For example, metal nanoparticles, such as gold and silver nanoparticles, have been shown to exhibit strong antioxidant activity, while polymeric nanoparticles can be loaded with anti-inflammatory drugs to mitigate neuroinflammation. By targeting these underlying pathological processes, nanotechnology-based therapies can help preserve neuronal function and slow disease progression [72]. Moreover, nanotechnology facilitates the development of advanced imaging and diagnostic tools that enhance the early detection and monitoring of neurological diseases. Nanoparticles can be functionalized with imaging agents, such as fluorescent dyes or magnetic nanoparticles, to enable high-resolution imaging of brain structures and processes. This capability is particularly valuable for detecting early changes in the brain that precede clinical symptoms, allowing for timely intervention. Additionally, nanotechnology can contribute to the development of diagnostic assays that can detect biomarkers associated with neurodegenerative diseases with high sensitivity and specificity. Early and accurate diagnosis is critical for implementing effective neuroprotective strategies and improving patient outcomes. The potential for nanotechnology to enable gene therapy and personalized medicine in the realm of neuroprotection is another groundbreaking advancement. Nanoparticles can serve as delivery vehicles for gene therapies, including gene editing technologies like CRISPR/Cas9. By delivering genetic material or gene-editing tools directly to specific neurons or brain regions, nanotechnology can help correct genetic mutations or modulate gene expression to combat neurodegenerative diseases. Personalized medicine approaches, facilitated by nanotechnology, can tailor treatments based on individual genetic profiles, ensuring that therapies are optimized for each patient's unique needs. Despite the promising potential of nanotechnology in neuroprotection, several challenges remain. The long-term safety and biocompatibility of nanomaterials need thorough evaluation, as the unique properties of nanoparticles can sometimes lead to unexpected toxicological effects. Ensuring that nanomaterials do not accumulate in the brain or other organs and that they are safely metabolized and excreted is crucial for their successful translation into clinical practice. Additionally, regulatory frameworks for nanotechnology-based therapies need to evolve to address the specific challenges associated with these advanced materials. Comprehensive preclinical and clinical studies are necessary to establish the safety, efficacy, and optimal use of nanotechnology in neuroprotection. The utilization and implementation of nanotechnological approaches could potentially address and even eradicate the challenges associated with neurotherapeutics. CNS injuries often result in an elevated presence of reactive oxygen species. Restoring nerve defects consistently poses a significant challenge for medical practitioners, and inadequate recovery from nerve cell injuries can result in various adverse consequences, such as sensory or motor function impairment, numbness, profound permanent disability, and the potential development of chronic pain. Nanotechnology-based drug delivery systems offer a promising approach to mitigate these issues [73] (Tables 2 and 3)

Table 2 Illustration of the utilization of nanoparticles in the treatment of MS

Site of applications	Formulations	Outcomes	References
Therapeutics	Oxide nanoparticles	Inhibition of MMP-9	[74]
Therapeutics	Liposomes	EAE abrogation	[75]
Therapeutics	Suspension	Antioxidant activity	[76]
Therapeutics	Microparticles	Sustained drug release	[77]
Therapeutics	Polyamidoamine dendrimers	Cerebral palsy	[78]
Surgery	Quantum dots	Imaging	[79]
Diagnosis	Carbon nanotubes	Epithelial cell chirality	[80]
Diagnosis	USPIO	Detects inflammation	[81]
Diagnosis	Nanoparticles	Molecular imaging	[82]
Diagnosis	Nanotubes	Multiple sclerosis	[83]

Table 3 An overview of the most recent patents concerning treatments for multiple sclerosis

Patent no	Short patent title	References
WO2011065867A3	MS	[84]
CA2430897C	Inflammatory disorders	[85]
EP2282719A4	MS	[86]
EP2836229A1	MS	[87]
WO2005117978A2	MS	[88]
US20080063702	Liposomal formulations for treatment of neurodegenerative conditions	[89]
US 7955618 B2	Inflammatory disorders	[90]
WO 2015042294 A1	Nanoparticle compositions	[91]
WO2005117978A2	MS	[92]

7 Conclusion

MS remains a challenging disease with significant unmet medical needs, characterized by a complex interplay of immune system dysregulation and neuroinflammatory processes. The autoimmune attack against myelin, coupled with subsequent damage to the central nervous system (CNS), leads to the characteristic axonal injury and neurodegeneration observed in MS. Understanding these intricate mechanisms is crucial for developing strategies to preserve blood–brain barrier (BBB) integrity and protect neural tissues, ultimately improving patient outcomes.

By overcoming traditional limitations in drug delivery, nanomaterials have the potential to significantly enhance therapeutic outcomes, optimize drug efficacy, and reduce side effects. Nanomaterial-based therapeutic approaches provide innovative solutions for managing MS by addressing key aspects of the disease. Anti-inflammatory therapies delivered through nanomaterials can target specific sites of inflammation, reducing systemic exposure and enhancing treatment efficacy. Immunomodulatory strategies utilizing nanomaterials can modulate aberrant immune responses and promote immune tolerance. Neuroprotective and remyelination strategies facilitated by nanomaterials offer the potential to support neuronal function and repair damaged myelin. Additionally, gene therapy delivery systems enabled by nanomaterials hold promise for precise and targeted genetic modifications, offering long-term therapeutic benefits. Ongoing research and development are crucial for evaluating the effectiveness of nanotechnology applications. As the field of nanotechnology advances, it holds the promise of revolutionizing MS management by providing more effective, targeted, and personalized therapeutic options. Ultimately, these advancements offer hope against this debilitating disease, potentially improving clinical outcomes for patients.

References

1. Singh AV, Khare M, Gade WN, Zamboni P (2012) Theranostic implications of nanotechnology in multiple sclerosis: a future perspective. Autoimmun Dis 2012(1):160830
2. Nuzzo D, Picone P (2021) Multiple sclerosis: focus on extracellular and artificial vesicles, nanoparticles as potential therapeutic approaches. Int J Mol Sci 22(16):8866
3. Zhang Q, Dai X, Zhang H, Zeng Y, Luo K, Li W (2021) Recent advances in development of nanomedicines for multiple sclerosis diagnosis. Biomed Mater 16(2):024101
4. Ghosh S, Bhatti GK, Sharma PK, Kandimalla R, Mastana SS, Bhatti JS (2024) Potential of nano-engineered stem cells in the treatment of multiple sclerosis: a comprehensive review. Cell Mol Neurobiol 44(1):6
5. Subhash S, Chaurawal N, Raza K (2024) Promises of lipid-based nanocarriers for delivery of dimethyl fumarate to multiple sclerosis brain. In: Neuroprotection: method and protocols. Springer US, New York, NY, pp 457–475
6. Tüzün E (2017) Nanotechnology-based management of neurological autoimmune diseases. Nanotechnol Methods Neurol Dis Brain Tumors 1:279–290
7. Saxena H, Kumar A, Solanki P, Gowtham Bhandari K (2024) Rise of nanotechnology for neurological disorders management. In: Theranostic applications of nanotechnology in neurological disorders. Springer Nature, Singapore, pp 31–42
8. Zhang B, Guo Y, Lu Y, Ma D, Wang X, Zhang L (2024) Bibliometric and visualization analysis of the application of inorganic nanomaterials to autoimmune diseases. Biomater Sci
9. Nally FK, De Santi C, McCoy CE (2019) Nanomodulation of macrophages in multiple sclerosis. Cells 8(6):543
10. Tarannum S, Jain K (2024) Drug delivery strategies in multiple sclerosis, Huntington's disease and other neurodegenerative diseases. In: Drug delivery strategies in neurological disorders: challenges and opportunities. Springer Nature, Singapore, pp 375–403
11. Yetkin MF (2024) Pathophysiology of multiple sclerosis. Gazi Med J 2:35
12. Maghzi AH, Borazanci A, McGee J, Alexander JS, Gonzalez-Toledo E, Minagar A (2011) Multiple sclerosis: pathophysiology, clinical features, diagnosis, and management. In: Neuroinflammation. Elsevier, pp 1–23
13. Salvetti M, Landsman D, Schwarz-Lam P, Comi G, Thompson AJ, Fox RJ (2015) Progressive MS: from pathophysiology to drug discovery. Mult Scler J 21(11):1376–1384
14. Racke MK, Frohman EM, Frohman T (2022) Pain in multiple sclerosis: understanding pathophysiology, diagnosis, and management through clinical vignettes. Front Neurol 13(12):799698
15. Mado H, Adamczyk-Sowa M, Sowa P (2023) Role of microglial cells in the pathophysiology of MS: synergistic or antagonistic? Int J Mol Sci 24(3):1861

16. Guerrero JM, Aguirre FS, Mota ML, Carrillo A (2021) Advances for the development of in vitro immunosensors for multiple sclerosis diagnosis. BioChip J 15(3):205–215
17. Ashraf H, Solla P, Sechi LA (2022) Current advancement of immunomodulatory drugs as potential pharmacotherapies for autoimmunity based neurological diseases. Pharmaceuticals 15(9):1077
18. Pearson RM, Podojil JR, Shea LD, King NJ, Miller SD, Getts DR (2019) Overcoming challenges in treating autoimmuntity: development of tolerogenic immune-modifying nanoparticles. Nanomed Nanotechnol Biol Med 18:282–291
19. Mishra A, Ahsan R, Islam A, Tiwari RK, Dash PP (2024) Hybrid nanocarriers for neurological disorders: diagnostic & therapeutic approach. Recent Patents Nanotechnol 18(2):164–178
20. He Z, Lai J, Wang H, He Y, Zhang C, Gao L, Huang H, Zheng L, Hao J, Gao X, Gao F (2024) The therapeutic effect of gold clusters in an EAE model of multiple sclerosis by regulating T cell differentiation. Nano Today 1(54):102128
21. Shamsher E, Khan RS, Davis BM, Dine K, Luong V, Cordeiro MF, Shindler KS (2024) Intranasal resveratrol nanoparticles enhance neuroprotection in a model of multiple sclerosis. Int J Mol Sci 25(7):4047
22. Krsek A, Baticic L (2024) Nanotechnology-driven therapeutic innovations in neurodegenerative disorders: a focus on Alzheimer's and Parkinson's disease. Future Pharmacol 4(2):352–379
23. Rana P, Kaur G, Chirayimmel AJ, Rajput A, Rath SK, Dwibedi V (2024) Nanocarriers and nano drug delivery therapy in neuro diseases. In: Nanomaterials for drug delivery and neurological diseases management. Springer Nature, Singapore, pp 219–251
24. Rath SK, Dwibedi V, Husen A, Akhtar N. Nanomaterials for drug delivery and neurological diseases management
25. Arab FL, Yousefi F, Jaafari MR, Rajabian A, Dana H, Tabasi N, Behboodifar S, Faridzadeh A, Mahmoudi M (2024) Evaluation of the immune-modulatory, anti-oxidant, proliferative, and anti-apoptotic effects of nano-silymarin on mesenchymal stem cells isolated from multiple sclerosis patients' adipose tissue sources. J Funct Foods 1(113):105958
26. Cano A, Sanchez-Lopez E, Ettcheto M, Lopez-Machado A, Espina M, Souto EB, Galindo R, Camins A, García ML, Turowski P (2020) Current advances in the development of novel polymeric nanoparticles for the treatment of neurodegenerative diseases. Nanomedicine 15(12):1239–1261
27. Gadhave DG, Kokare CR (2019) Nanostructured lipid carriers engineered for intranasal delivery of teriflunomide in multiple sclerosis: optimization and in vivo studies. Drug Dev Ind Pharm 45(5):839–851
28. George KS, Jacob A, Anagha KK, Nair AR (2023) Ameliorative efficacy of bionanoparticles against neuronal dysfunctions. In: Green sustainable process for chemical and environmental engineering and science. Elsevier, pp 277–290
29. Mahmoudi M, Rastin M, Arababadi MK, Anaeigoudari A, Nosratabadi R (2022) Enhancing the efficacy of hypericum perforatum in the treatment of an experimental model of multiple sclerosis using gold nanoparticles: an in vivo study. Avicenna J Phytomed 12(3):325
30. Chountoulesi M, Demetzos C (2020) Promising nanotechnology approaches in treatment of autoimmune diseases of central nervous system. Brain Sci 10(6):338
31. Shah V, Panchal B, Gona C, Shah M, Prajapati M (2024) A comprehensive study on applications of nanomaterials in petroleum upstream and downstream industry. Environ Sci Pollut Res 31(10):14406–14423
32. Mwema A, Muccioli GG, Des RA (2023) Innovative drug delivery strategies to the CNS for the treatment of multiple sclerosis. J Control Release 1(364):435–457
33. Rashid AB, Haque M, Islam SM, Labib KR (2024) Nanotechnology-enhanced fiber-reinforced polymer composites: recent advancements on processing techniques and applications. Heliyon
34. Gul MM, Ahmad KS, Thomas AG, Tighezza AM (2024) NiZr2S4 bimetallic sulphide thin films: synthesis and multifunctional applications in nanotechnology. J Appl Electrochem 54(7):1667–1681

35. Vijayaram S, Razafindralambo H, Sun YZ, Vasantharaj S, Ghafarifarsani H, Hoseinifar SH, Raeeszadeh M (2024) Applications of green synthesized metal nanoparticles—a review. Biol Trace Elem Res 202(1):360–386
36. Liu F, Liu X, Abdiryim T, Gu H, Astruc D (2024) Heterometallic macromolecules: synthesis, properties and multiple nanomaterial applications. Coord Chem Rev 1(500):215544
37. Zhou S, Cai H, He X, Tang Z, Lu S (2024) Enzyme-mimetic antioxidant nanomaterials for ROS scavenging: design, classification, and biological applications. Coord Chem Rev 1(500):215536
38. Javed Z, Rasool RT, Alhummiany H, Majeed A, Gulbadan S, Ashraf GA, Mana Al-Anazy M, Irfan M, Akhtar MN, Arshad M, Khan MA (2024) Structural, morphological, dielectric, and spectral properties of Sr-Mg-Ho X-type magnetic nano materials suitable for microwave absorption application. Vacuum 222:112965
39. Cen T, Torrent L, Testino A, Ludwig C (2024) Rotating disk diluter hyphenated with single particle ICP-MS as an online dilution and sampling platform for metallic nanoparticles characterization in ambient aerosol. J Aerosol Sci 1(175):106283
40. Teleanu DM, Negut I, Grumezescu V, Grumezescu AM, Teleanu RI (2019) Nanomaterials for drug delivery to the central nervous system. Nanomaterials 9(3):371
41. Kumar A, Chaudhary JS, Dubey A, Pachorkar SS (2024) Nanotechnology in ankylosing spondylitis: advancements in drug delivery and targeted therapy. Int J Drug Deliv Technol 14(2):1162–1173
42. Karami MH, Abdouss M, Maleki B (2024) The state of the art metal nanoparticles in drug delivery systems: a comprehensive review. Nanomed J 11(3)
43. Takei EK (2024) The use of nanoparticles in the diagnosis and treatment of multiple sclerosis: a scoping review. STEM Fellowship J 23:1–5
44. Lv Y, Li W, Liao W, Jiang H, Liu Y, Cao J, Lu W, Feng Y (2024) Nano-drug delivery systems based on natural products. Int J Nanomed 31:541–569
45. Kang W, Xu Z, Lu H, Liu S, Li J, Ding C, Lu Y (2024) Advances in biomimetic nanomaterial delivery systems: harnessing nature's inspiration for targeted drug delivery. J Mater Chem B
46. Mou X, Ali Z, Li S, He N (2015) Applications of magnetic nanoparticles in targeted drug delivery system. J Nanosci Nanotechnol 15(1):54–62
47. Mahmood S, Mandal UK, Chatterjee B, Taher M (2017) Advanced characterizations of nanoparticles for drug delivery: investigating their properties through the techniques used in their evaluations. Nanotechnol Rev 6(4):355–372
48. Forooque F, Mughees MM, Wasi M, Khan MS (2024) Green sustainable nanoparticles as a drug delivery system—an updated review. Sustain Nanomater Synth Environ Appl 21:171–201
49. Udhayakumari D (2024) A review of nanotechnology-enabled fluorescent chemosensors for environmental toxic ion detection. J Fluoresc 1:1–9
50. Alwael H, Alsulami AN, Abduljabbar TN, Oubaha M, El-Shahawi MS (2024) Innovative Sol-gel functionalized polyurethane foam for sustainable water purification and analytical advances. Front Chem 8(12):1324426
51. Vatika K, Kumbhar S, Yasmeen M, Akare S, Sangode VK (2024) Nanoparticle of Azadirachta Indica (neem) mitigating lead acetate induced reproductiive toxicity. Biochem Cell Arch 24(1)
52. Vikas, Mehata AK, Singh C, Malik AK, Setia A, Muthu MS (2024) Alginate-based hydrogels as drug carriers. In: Biomaterial-based hydrogels: therapeutics carrier and tissue regeneration. Springer Nature, Singapore, pp 39–65
53. Li S, Hou Z, Ye T, Song X, Hu X, Chen J (2024) Saponin components in Polygala tenuifolia as potential candidate drugs for treating dementia. Front Pharmacol 10(15):1431894
54. Frischer JM, Bramow S, Dal-Bianco A, Lucchinetti CF, Rauschka H, Schmidbauer M, Laursen H, Sorensen PS, Lassmann H (2009) The relation between inflammation and neurodegeneration in multiple sclerosis brains. Brain 132(5):1175–1189
55. Singh A (2010) Multiple sclerosis takes venous route: CCSVI and liberation therapy. Indian J Med Sci 64(7):337
56. Kasper LH, Shoemaker J (2010) Multiple sclerosis immunology: the healthy immune system vs the MS immune system. Neurology 74(1_supplement_1):S2–S8

57. Nave KA, Trapp BD (2008) Axon-glial signaling and the glial support of axon function. Annu Rev Neurosci 31(1):535–561
58. O'Connor KC, Appel H, Bregoli L, Call ME, Catz I, Chan JA, Moore NH, Warren KG, Wong SJ, Hafler DA, Wucherpfennig KW (2005) Antibodies from inflamed central nervous system tissue recognize myelin oligodendrocyte glycoprotein. J Immunol 175(3):1974–1982
59. Bielekova B, Goodwin B, Richert N, Cortese I, Kondo T, Afshar G, Gran B, Eaton J, Antel J, Frank JA, McFarland HF (2000) Encephalitogenic potential of the myelin basic protein peptide (amino acids 83–99) in multiple sclerosis: results of a phase II clinical trial with an altered peptide ligand. Nat Med 6(10):1167–1175
60. Spagnuolo G, Piavis A, Williams T (2017) Involvement of immune regulation in multiple sclerosis. Immunol Immunogenet Insights 3(9):1178634517734175
61. Gurav N, Mhatre S (2019) Nanoscience in multiple sclerosis. Bombay Technol 31:16–24
62. Jansen MI, Thomas Broome S, Castorina A (2022) Exploring the pro-phagocytic and anti-inflammatory functions of PACAP and VIP in microglia: implications for multiple sclerosis. Int J Mol Sci 23(9):4788
63. Ojha S, Kumar B (2018) A review on nanotechnology based innovations in diagnosis and treatment of multiple sclerosis. J Cell Immunother 4(2):56–64
64. McManus RM, Mills KH, Lynch MA (2015) T cells—protective or pathogenic in Alzheimer's disease? J Neuroimmune Pharmacol 10:547–560
65. Elliott R, Li F, Dragomir I, Chua MM, Gregory BD, Weiss SR (2013) Analysis of the host transcriptome from demyelinating spinal cord of murine coronavirus-infected mice. PLoS ONE 8(9):e75346
66. Calvillo-Robledo A, Ramírez-Farías C, Valdez-Urias F, Huerta-Carreón EP, Quintanar-Stephano A (2023) Arginine vasopressin hormone receptor antagonists in experimental autoimmune encephalomyelitis rodent models: a new approach for human multiple sclerosis treatment. Front Neurosci 14(17):1138627
67. Nair SÁ, Dileep A, Rajanikant GK (2012) Nanotechnology based diagnostic and therapeutic strategies for neuroscience with special emphasis on ischemic stroke. Curr Med Chem 19(5):744–756
68. Wen MM, El-Salamouni NS, El-Refaie WM, Hazzah HA, Ali MM, Tosi G, Farid RM, Blanco-Prieto MJ, Billa N, Hanafy AS (2017) Nanotechnology-based drug delivery systems for Alzheimer's disease management: technical, industrial, and clinical challenges. J Control Release 10(245):95–107
69. Rahman M, Beg S, Verma A, Kazmi I, Kumar Patel D, Anwar F, A Al Abbasi F, Kumar V (2017) Therapeutic applications of liposomal based drug delivery and drug targeting for immune linked inflammatory maladies: a contemporary view point. Curr Drug Targets 18(13):1558–1571
70. Rahiman N, Mohammadi M, Alavizadeh SH, Arabi L, Badiee A, Jaafari MR (2022) Recent advancements in nanoparticle-mediated approaches for restoration of multiple sclerosis. J Control Release 1(343):620–644
71. Fonseca-Santos B, Gremião MP, Chorilli M (2015) Nanotechnology-based drug delivery systems for the treatment of Alzheimer's disease. Int J Nanomed 4:4981–5003
72. Jan Z, Mollazadeh S, Abnous K, Taghdisi SM, Danesh A, Ramezani M, Alibolandi M (2022) Targeted delivery platforms for the treatment of multiple sclerosis. Mol Pharm 19(7):1952–1976
73. Soursou G, Alexiou A, Md Ashraf G, Ali Siyal A, Mushtaq G, A Kamal M (2015) Applications of nanotechnology in diagnostics and therapeutics of Alzheimer's and Parkinson's disease. Curr Drug Metab 16(8):705–712
74. Dolati S, Babaloo Z, Jadidi-Niaragh F, Ayromlou H, Sadreddini S, Yousefi M (2017) Multiple sclerosis: therapeutic applications of advancing drug delivery systems. Biomed Pharmacother 1(86):343–353
75. Al-Jamal WT, Kostarelos K (2007) Liposome–nanoparticle hybrids for multimodal diagnostic and therapeutic applications. Nanomedicine 2(1):85–98
76. Meng Y, Niu X, Li G (2023) Liposome nanoparticles as a novel drug delivery system for therapeutic and diagnostic applications. Curr Drug Deliv 20(1):41–56

77. Silindir-Gunay M, Ozer AY (2018) Liposomes and micelles as nanocarriers for diagnostic and imaging purposes. In: Design of nanostructures for theranostics applications. William Andrew Publishing, pp 305–340
78. Saeedi M, Eslamifar M, Khezri K, Dizaj SM (2019) Applications of nanotechnology in drug delivery to the central nervous system. Biomed Pharmacother 1(111):666–675
79. Shaw TK, Paul P (2022) Recent approaches and success of liposome-based nano drug carriers for the treatment of brain tumor. Curr Drug Deliv 19(8):815–829
80. Gu MJ, Li KF, Zhang LX, Wang H, Liu LS, Zheng ZZ, Han NY, Yang ZJ, Fan TY (2015) In vitro study of novel gadolinium-loaded liposomes guided by GBI-10 aptamer for promising tumor targeting and tumor diagnosis by magnetic resonance imaging. Int J Nanomed 17:5187–5204
81. Zeng H, Li J, Hou K, Wu Y, Chen H, Ning Z (2022) Melanoma and nanotechnology-based treatment. Front Oncol 9(12):858185
82. He R, Li L, Zhang T, Ding X, Xing Y, Zhu S, Gu Z, Hu H (2023) Recent advances of nanotechnology application in autoimmune diseases–a bibliometric analysis. Nano Today 1(48):101694
83. Barani M, Mukhtar M, Rahdar A, Sargazi S, Pandey S, Kang M (2021) Recent advances in nanotechnology-based diagnosis and treatments of human osteosarcoma. Biosensors 11(2):55
84. Barenholz Y, Ovadia H, Kizelsztein P, inventors; Hadasit Medical Research Services, Development Co, Yissum Research Development Co of Hebrew University of Jerusalem, Assignee (2011) Liposomal formulations comprising an amphipathic weak base like tempamine for treatment of neurodegenerative conditions. United States patent application US 12/903,266
85. Singh AV, Subhashree L, Milani P, Gemmati D, Zamboni P (2010) Interplay of iron metallobiology, metalloproteinases, and FXIII, and role of their gene variants in venous leg ulcer. Int J Low Extrem Wounds 9(4):166–179
86. Singh AV, Mehta KK, Worley K, Dordick JS, Kane RS, Wan LQ (2014) Carbon nanotube-induced loss of multicellular chirality on micropatterned substrate is mediated by oxidative stress. ACS Nano 8(3):2196–2205
87. Millward JM, Schnorr J, Taupitz M, Wagner S, Wuerfel JT, Infante-Duarte C (2013) Iron oxide magnetic nanoparticles highlight early involvement of the choroid plexus in central nervous system inflammation
88. Raymond MJ, Ray P, Kaur G, Singh AV, Wan LQ (2016) Cellular and nuclear alignment analysis for determining epithelial cell chirality. Ann Biomed Eng 44:1475–1486
89. Luchetti A, Milani D, Ruffini F, Galli R, Falini A, Quattrini A, Scotti G, Comi G, Martino G, Furlan R, Politi LS (2012) Monoclonal antibodies conjugated with superparamagnetic iron oxide particles allow magnetic resonance imaging detection of lymphocytes in the mouse brain. Mol Imaging 11(2):7290–2011
90. Richards JM, Shaw CA, Lang NN, Williams MC, Semple SI, MacGillivray TJ, Gray C, Crawford JH, Alam SR, Atkinson AP, Forrest EK (2012) In vivo mononuclear cell tracking using superparamagnetic particles of iron oxide: feasibility and safety in humans. Circ Cardiovas Imaging 5(4):509–517
91. Weissleder R, Elizondo G, Wittenberg J, Rabito CA, Bengele HH, Josephson L (1990) Ultrasmall superparamagnetic iron oxide: characterization of a new class of contrast agents for MR imaging. Radiology 175(2):489–493
92. Gabivov A, Belogurov A, Ponomarenko N, Smirnov I, Bacon A, Gregoriadis G, Inventors (2013) Liposomes containing oligopeptide fragments of myelin basic protein, a pharmaceutical composition and a method for treatment of multiple sclerosis. United States patent application US 13/444,788

Nanocarriers for the Management of Rheumatoid Arthritis

Md. Meraj Ansari, Nadeem Akhtar, Zamurrad Nayab, and Nemat Ali

Abstract Rheumatoid arthritis (RA) is a chronic autoimmune disease. It is characterized by severe joint destruction due to inflammatory synovitis. There are several factors that have a strong link in the initiation and progression of RA. Albeit the exact cause of disease initiation is still elusive. It is currently understood that RA is a multifactorial disease in which the total risk of disease susceptibility, persistence, and severity are determined by intricate interactions between the host and particular inflammatory mediators in order to lessen symptoms. Doctors are generally using the drugs such as disease-modifying anti-rheumatic drugs (DMARDs), non-steroidal anti-inflammatory drugs (NSAIDs), glucocorticoids (GCs), and biological agents in the treatment of RA patients. These traditional drugs have some disadvantages including low patient compliance, the failure to halt the disease's development, and safety concerns. Therefore, to overcome these issues, it is imperative to create innovative drug delivery vehicles that may improve safety evaluation and deliver therapeutics drugs to inflammatory synovial regions efficiently. From this perspective, nanotechnology shows great promise in developing new carriers for the management of RA and maybe a future alternative to conventional treatment approaches for RA. Many flexible nanocarriers with distinct physicochemical properties, and individualized drug release patterns were developed in order to increase the efficacy of drugs in

Md. M. Ansari (✉)
Department of Animal Biotechnology, Faculty of Biotechnology, College of Applied Life Sciences, Jeju National University, Jeju Special Self-Governing Province, Jeju-si, South Korea
e-mail: rajmeraj385@gmail.com

Interdisciplinary Graduate Program in Advanced Convergence Technology and Science, Jeju National University, Jeju Special Self-Governing Province, Jeju-si, South Korea

N. Akhtar
Department of Animal Biosciences, University of Guelph, Guelph, Canada

Z. Nayab
Department of Biotechnology, Science College Patna, Patna University, Patna, Bihar, India

N. Ali
Department of Pharmacology and Toxicology, College of Pharmacy, King Saud University, Riyadh, Saudi Arabia

Md. M. Ansari et al. (eds.), *Emergence of Sustainable Biomaterials in Tackling Inflammatory Diseases*, Smart Nanomaterials Technology,
https://doi.org/10.1007/978-981-96-2112-5_13

the treatment of RA. In this chapter, we have covered in-depth the nanocarriers that are currently being used to treat RA in a preclinical setting and its future perspective.

Keywords Nanocarrier · Autoimmune disease · Rheumatoid arthritis · Chronic inflammatory disease · Targeted delivery

1 Introduction

Rheumatoid arthritis (RA) is a chronic inflammatory autoimmune disease of joint tissue [1, 2]. In the case of RA, pathologists observed persistent inflammation of joints, infiltration of neutrophils and macrophages at the site of inflammation, infiltration of cytokines, synovial membrane hyperplasia, damage of articular cartilage and bone. The pannus formation due to hyperplasia of the synovial membrane is another characteristic of RA. Initially, RA starts from mild to moderate inflammation affecting a few joints. However, if it is not treated early then RA worsens and affects more joints. The exact cause of the onset of RA is still elusive, but researchers revealed that factors such as infection, genetic and the environment as hallmarks in the development of RA [3, 4]. According to [5], about 18 million people around the globe were living with RA [5]. It is found that females are more affected than males with RA. According to [5], approximately 70% of people having RA are female, and 55% those patients his/her age are more than 55 years [5]. It is found that 13 million RA patients have severity levels from moderate to severe. These patients can benefit from natural rehabilitation [6].

In the current treatment approaches of RA, relief from chronic pain, reduction in joint inflammation and disease severity are prime concerns of pharmaceutical therapy. The European League Against Rheumatism has suggested many drugs for the treatment of RA patients [7]. Furthermore, the guidelines suggest that start treatment of RA after confirmation with therapeutics drugs. The conventional drugs that are being used in the treatment of RA patients are synthetic drugs and biological agents. The synthetic drugs are DMARDs, NSAIDs and GCs. These synthetic drugs have revealed strong side effects in clinical settings while biological agents are very expensive. If you are an RA patient, you have to take medicine for a long time. During the course of RA treatment, these medicines have adverse effects on our vital organs. Researchers revealed that the use of DMARDs, NSAIDs and GCs has adverse effects on our body's gastrointestinal system, liver, kidney and heart. Researchers revealed the severer adverse effect of taking long-term glucocorticoids in the form of gastric ulcers, buffalo hump, and bone loss [8]. Patient compliance and safety are significantly reduced by these off-target consequences. Researchers have looked closely at the parenteral route of drug delivery as a means of overcoming these restrictions. In order to cure RA, it is imperative to overcome the limits of these therapeutic treatments [9]. There are several steroidal and non-steroidal drugs are being used in clinical settings for the treatment of RA are summarized in Table 1 with their classification, advantages and disadvantages on the human body during the treatment of RA.

The emerging field of nanotechnology is nanomedicine as a future alternative to conventional approaches to RA treatment. The aim of nanomedicine is to deliver drugs at the site of inflammation with a controlled release pattern and very less off-target. For drug delivery purposes, nanomedicine is using smart drug carriers. These carriers are based on natural biopolymers or FDA-approved biocompatible materials. In a study, researchers demonstrated certolizumab pegol enclosed in PEGylation-encapsulated nanocarriers increases the half-life period to fourteen days and shows promising results in patients with RA [10]. However, GCs can be used as a stopgap measure until the clearly therapeutic effects of conventional synthetic DMARDs are observed [11]. As we know in RA inflammation, infiltration of neutrophils and macrophages at the site of inflammation occurs and these cells contribute to disease pathology. Researchers hypothesized that targeting these cells with targeted drug delivery may be effective in the treatment of RA. Researchers observed that many nanoparticles can deliver therapeutics drugs at the site of inflammation in joints by modulating macrophages [12]. By using the nanomedicine technique, the drugs become more accessible and soluble, avoid overdosing, and patient adherence is enhanced. Nowadays, varieties of nanocarriers are explored in the treatment of

Table 1 List of clinical drugs of RA with classification, advantage and disadvantage

Classification	Drugs	Advantage	Disadvantage
NSAIDs)	Naproxen Indomethacin Ibuprofen Oxaprozin Diclofenac sodium	Significantly helps in the treatment of chronic pain and reduced joint inflammation	Stomach complication, gastrointestinal bleeding, and kidney problem
DMARDs	Sulfasalazine Methotrexate Hydroxychloroquine Leflunomide Azathioprine	Decrease pain and inflammation	Adverse effects on liver and kidney
Glucocorticoids (GCs)	Prednisolone Betamethasone Dexamethasone	Great potential to inhibit inflammation	Have some adverse effects in the human body like high blood pressure, immunosuppression and osteoporosis
Biological agents	Adalimumab Anakinra Rituximab Tocilizumab Etanercept Infliximab	Highly specific in target, reduce inflammation	Highly expensive and economically weak patients cannot prefer, low responsiveness, and the major drawback is increased risk of infection

inflammatory diseases in preclinical studies. These nanocarriers are customized as desired and we may call them smart carriers. The specificity and targetability are the main motives of these carriers in drug delivery systems. In the targeted delivery of therapeutics drugs with the help of nanocarriers offer precise drug administration to cells or tissue display molecular markers unique to that tissue, enabling them to differentiated from healthy ones [13]. In the nanocarrier-based targeted delivery of therapeutics drugs, medication released by the delivery system proceeds to pass through several epithelial and biological barriers to reach the site of inflammation. An efficient drug delivery plan guarantees that the medicine is bioavailable at the intended site [14]. The lowest effective drug concentration and harmful drug concentration is based on the concentration of drugs at the site of inflammation [15]. Nowadays, drug delivery systems are very advanced, but still, many drugs have undesirable side effects due to off-target of organs and tissues [16]. For targeted drug delivery in the treatment of RA, there is a huge demand for innovative drug carriers such as liposomes, dendrimers, niosomes, solid lipid NPs [17]. For the treatment of tumor cells, researchers extensively working on tumor-targeted approach in which they use the enhanced permeability and retention effect on the target tumor to enhance drug accumulation at the site of tumor, similarly therapeutic approach is required on urgent based on delivery of drugs on the basis of the affected region, microenvironment of synovial hyperplasia, angiogenesis marks and others signal requested by inflammatory cells. This mechanism may be called the inflammatory cell-mediated sequestration effect, which follows extravasation through leaky vasculature [18]. Ki-67 is a strong angiogenesis marker and researchers revealed that in the case of RA, the enhanced expression of Ki-67 in joint inflammatory tissue. This gives a clear indication of new blood vessels at the site of inflammation. Researchers revealed enhanced permeability of blood at the joint barrier due to arthritis inflammation [19]. There are many routes of drug delivery in the human body, which we have explained in detail in the next section of this chapter. Oral drug delivery is the choice of drug administration route and is most accepted by patients. On the basis of the finding, researchers revealed that some complications in oral drug delivery due to a lack of drug specificity such as gastrointestinal bleeding, hepatotoxicity and nephrotoxicity [20].

Researchers are extensively working on the direct delivery of drugs into the synovial joint tissue to avoid unnecessary complications in human body. In a study, researchers demonstrated the effectiveness of intravenous betamethasone sodium phosphate loaded PLGA NPs. They reported this nanocarrier is effective in the targeted and delayed release of loaded cargo [21]. In another study, researchers reported against adjuvant-induced arthritis in rats, PLGA NPs containing triptolide have strong ant-inflammatory efficacy [22]. To avoid rapid drug clearance from the oral route, researchers are extensively working to explore other routes of drug delivery. Several other studies have reported that intra-articular injection of corticosteroids can ameliorate joint inflammation with less adverse effect. The nanotechnology-based drug delivery system has become more advanced in targeted drug delivery. Researchers have developed smart micelles as nanocarriers for intra-articular drug delivery for the treatment of RA in the preclinical setting. In this regard, we have discussed in the next section of this chapter. Recently researchers have developed smart injectable hydrogel for the treatment of collagen-induced rheumatoid arthritis in rats. They revealed that reduced joint inflammation, and decrease expression of inflammatory cytokines in joint tissues and other mediators [23]. We hypothesized that these nanocarriers will be definitely a future ray of hope in the management of RA.

2 Pathophysiology of RA

RA is well well-known immunological disorder disease. It is a chronic autoimmune disease that arises more frequently in females than males, predominantly in aged people [24]. The precise cause of RA is still elusive, despite the fact that it is one of the most prevalent forms of inflammatory arthritis and has long been investigated in great detail as a model for autoimmune disorders. There are several factors that are involved in the onset of RA in which RA is more influenced by genetic and environmental factors. The pathophysiology of RA is a very complex cascade of immunological response. In response to inflammation, infiltration of neutrophils and macrophages occurs at the site of inflammation. In the immunological response and inflammatory cascades, rapid cellular proliferation and activation of cells that contribute to synovial hyperplasia and pannus formation occurs. Professional antigen-presenting cells (APC) present antigens to T cells that cause activation of autoreactive T cells. Apart from antigen presentation, B cells secret autoantibodies such as anti-citrullinated protein antibodies (ACPA) and rheumatoid factor (RF) in blood which can attack self tissue. Furthermore, during these inflammatory cascades, others cell also produces a number of inflammatory cytokines (TNF-α, IL-6, etc.) and invasive proteases that lead to the destruction of cartilage and bone (Fig. 1). Furthermore, one characteristic of RA is the gradual degeneration of joints coated with synovium. The principal agents responsible for the degradation are the fibroblast-like synoviocytes. In a recent study, researchers revealed that intracellular scaffold proteins (14-3-3n are widely distributed; increasing the development of integradosomes by upregulating Snail expression [25]. Researchers revealed that 14-3-3n a new regulator of synovial cells invadosome and matrix breakdown. The extracellular traps by neutrophils are called NETosis. NETosis plays a very remarkable role in the pathophysiology of RA. Recently researchers revealed that citrullinated autoantigens produced in RA mainly due to NETosis. In a study, researchers demonstrated that peptide GPI-induced arthritis has enhanced NETosis [26]. Furthermore, extensive studies by researchers revealed that ant-IL-6 receptor antibody may be an RA ameliorative approach due to reduction in citrullinated protein in plasma, NETosis in joint tissue, and infiltration of neutrophils. Furthermore, researchers reviewed the molecular pathophysiology of systemic osteoporosis in the case of RA [27]. The inflammatory cytokines such as TNF-α, IL-6 and IL-17 have remarkable roles in localized osteoporosis in RA patients. In RA, senescent T cells generate more RNAKL. Moreover, osteoporosis and anti-citrullinated protein antibodies (ACPAs)

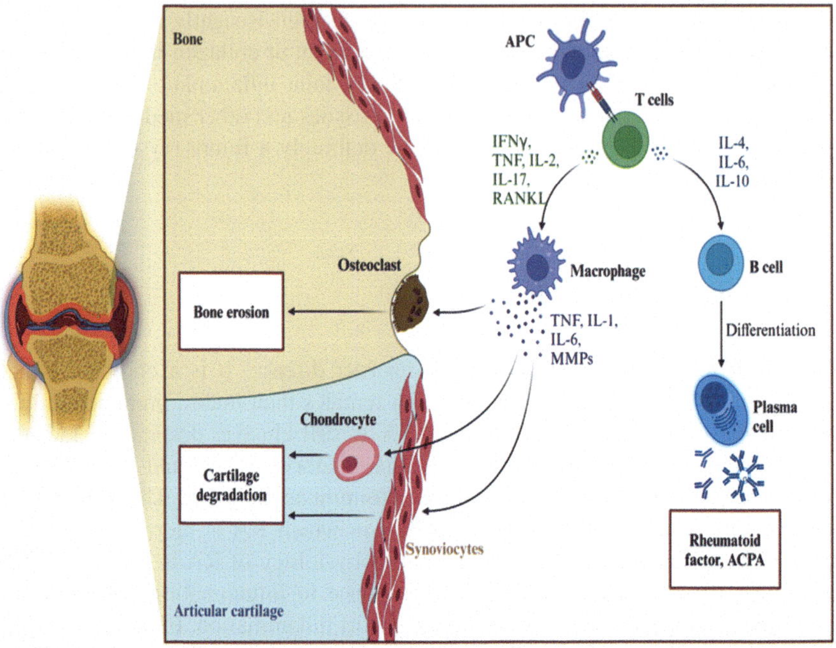

Fig. 1 **Pathophysiology of RA**. During the pathophysiology analysis, enhanced cellular proliferation and activation of cells contribute to synovial hyperplasia and pannus formation. Professional antigen-presenting cells (APC) present antigens to T cells that cause activations of autoreactive T cells. Apart from antigen presentation, B cells secret autoantibodies such as anti-citrullinated protein antibodies (ACPA) and rheumatoid factor (RF) in blood which can attack self tissue. Furthermore, during these inflammatory cascades, others cell also produces a number of inflammatory cytokines (TNF-α, IL-6, etc.) and invasive proteases that lead to the destruction of cartilage and bone. Image created with Biorender.com

are linked. Omata Y et al.'s summary of the available data on type 2 immunity's regulatory function in bone metabolism and arthritis focused on the remarkable role of type 2 innate lymphoid cells and type 2 cytokines [28].

3 Nanotechnology: An Overview

Nowadays, the nanotechnology platform is gaining more attention to tackle inflammatory diseases. It is a bright option to bypass the adverse effects of traditional drugs. The nanotechnologically derived nanoscale particles, engineered particles or nanoparticles size in the range of 1–100 nm with the help of a variety of techniques, particles' morphology, particles' surface charge, particles' size modifications are studied in the scientific field of nanotechnology [29, 30]. Researchers revealed a dramatic change in the potential of therapeutic drugs in the form of nanoscale due to sustained and controlled release of drugs. Extensive studies have revealed the growing applications of nanoparticles in biological science and nanomedicine. The physiochemical qualities of bulk materials are very different from nanomaterials. After the formulation of bulk materials into nanomaterials, its range of application in biological sciences increased. The small size of nanoparticles, high surface-to-volume ratio and ability to interface with cellular or molecular processes are some specific features that make it more suitable for a wide range of biomedical applications. The nanoscale materials primary benefit in the medical industry is that they can make medications that are insoluble in water more soluble by increasing surface area and surface contact, which in turn speeds up the pace of dissolution. The aim of nanoscale materials is to specific drug delivery, controlled drug release, short drug half-life in blood circulation and minimum adverse effects [31]. By the extensive studies, it is revealed that nanomaterials can serve as drug nanocarriers in targeted drug delivery systems. The nanomaterials may be of biological origin and FDA-approved biomaterials. There are many types of polymeric nanoparticles that are employed as nanocarriers to tackle inflammatory disease. The polymeric nanoparticles have less antigenic properties. Nanotechnologically derived nanomaterials or polymers in the form of nanocarriers such as liposomes, micelles, nanosomes, and solid lipid nanoparticles are employed in tackling inflammatory disease and RA.

Researchers revealed that targeted drug delivery can bypass the adverse effects of traditional drugs such as low drug bioavailability, off-target effect and high-dose-related toxicity. Researchers revealed that active and passive targeting as well as encapsulation of lipophilic, hydrophilic or combination of drugs delivery with the help of liposome as nanocarrier in the treatment of RA [32].

4 Drug Delivery in RA

Nanotechnology application in biomedical science has increased recent years. It has advantages over traditional drugs such as site specific, sustained release with high efficiency. There are several routes for drug delivery for RA such as intravenous, oral, transdermal, and intra-articular. In this section, we have discussed these routes of drug delivery in detail (Fig. 2).

4.1 Intravenous Drug Delivery

The drug administration through injection into tissue and circulatory system is known as the parenteral route of drug administration. The drug delivery through injection into the blood circulation is known as intravenous drug delivery. This mode of drug administration is not very common but this route offers more efficacy. This mode of drug delivery may be recommended by doctors for many reasons. They can easily bypass the first-pass effect of the digestive process and have a stronger

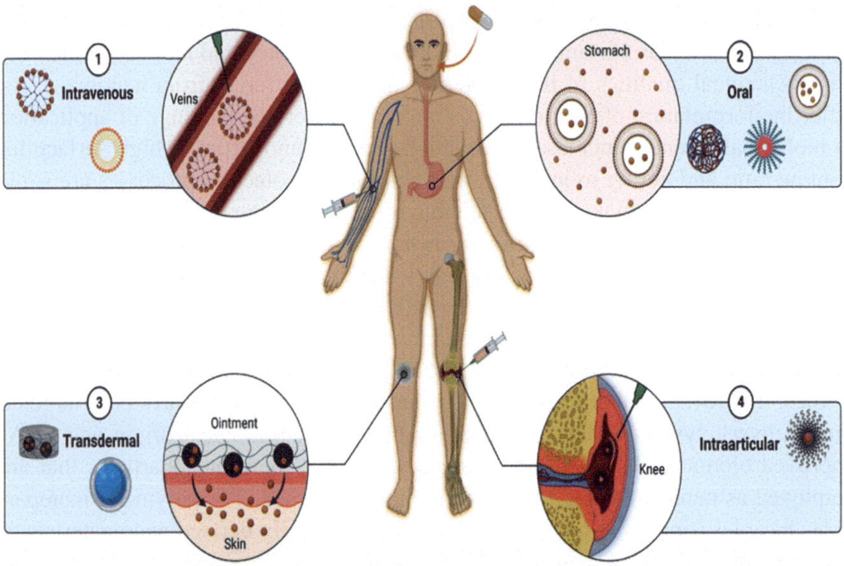

Fig. 2 Schematic presentation of routes of drug delivery. Image created with Biorender.com

effect. The intravenous route of drug administration is the most popular parenteral route. While injection of venous injection drug, first-pass hepatic processing can be circumvented. For intravenous drug administration, generally doctors use peripheral veins due to easy access to the blood circulatory system. Nowadays, these routes of drug administration are being used in drug delivery and for the enhancement of drug efficacy. In a recent study, researchers demonstrated that aminocellulose grafted polycaprolactone-coated gelatin NPs for dual drug delivery through an intravenous route. They delivered glycyrrhizic acid and budesonide in the RA model of rats. They demonstrated enhanced anti-arthritic efficacy in pre-clinical findings [33]. In a recent study, researchers demonstrated the strong anti-inflammatory effect of glucocorticoid dexamethasone phosphate encapsulated in large non-PEGylated liposomes in rat models of antigen-induced arthritis. It was found that the non-PEGylated liposomal medication administered intravenously (i.v.) totally reduced joint oedema [34, 35]. Furthermore, another study by Rauchhaus et al. extensively studies the effectiveness of liposomal dexamethasone phosphate and free dexamethasone for the treatment of mice models of collagen-induced RA [34, 35]. Furthermore, researchers revealed that intravenous injection of methotrexate-loaded liposome effectively blocks the release of PGE2 and IL-1β from macrophages in the collagen-induced RA model. Long-circulating polyethylene glycol liposomes did not seem to offer significant therapeutic potential in the management of arthritis in vivo [36].

4.2 Oral Drug Delivery

Oral route of drug administration is the most common route and drug delivery through the oral route is called oral drug delivery. This route of drug administration is the most common and preferred choice by patients. There is no requirement for injection from this route, which reduces pain and discomfort, and it also gets rid of contaminations. The severe physiological physicochemical conditions of the body's GI tract pose a major obstacle to the focused design and bioavailability of the oral route [37]. Researchers revealed that drug administration through this route has some limitations such as first-pass effects, fast metabolism, inadequate absorption, and adverse effects. The application of smart nanocarriers for drug delivery through oral routes has surfaced as a viable substitute for these obstacles in recent years. Nowadays, researchers manipulated a number of advanced nanocarriers for oral route drug delivery. These carriers are robust and stimuli-responsive for targeted drug delivery. The restrictions of standard administration routes can be avoided by using smart nanocarriers. Overcoming resistance or intolerance to traditional antirheumatic drugs is one of the main benefits of smart nanocarrier. Researchers revealed that nanocarriers can enhance the therapeutic potential of drugs by better drug stability, controlled released and enhanced solubility. Additionally, they can shield the medicine that has been encapsulated, extending its duration of circulation and promoting steady release at the intended location [38]. In a recent study, researchers revealed the enhanced anti-arthritic efficacy of oral delivery of zinc gluconate-loaded chitosan nanoparticles in collagen-induced arthritis [39].

4.3 Transdermal Drug Delivery

In the transdermal route of drug administration, the drug is absorbed by the epidermis or outer layer of skin and into the blood circulation. From this route, there is no involvement of mucous membranes. Transdermal drug delivery systems are a viable alternative for the treatment of RA since they can circumvent first-pass effect, GI manifestation and increase patient compliance. Naturally, transdermal drug delivery also confronts particular difficulties due to drug administration being partially restricted by the skin's physiological barrier. Researchers have extensively studied on transdermal facilitation principle to get around present obstacles and maximize drug delivery efficiency. From the latest study, it has been revealed that several generations of transdermal drug delivery remarkably enhanced delivery efficiency and controlled release [40]. It has been reported that in transdermal drug administration, a drug can reach the action site through the bloodstream via the skin barrier at a relatively low concentration [41]. Researchers are extensively working on gel nanoemulsion techniques for topical drug delivery. Gel Nanoemulsion can be created by water phase, oil phase, surfactant, and co-surfactant in appropriate combination. Researchers optimized that the solubility and stability of loaded cargo can be increased by the surface charge, reduced droplet size, and superior kinetic stability against flocculation [42]. Researchers revealed that some dark side of nanoemulsion in topical applications such as water phase evaporation, low viscosity, and excessive fluidity. The preparation of nanoemulsion gel involved mixing nanoemulsions with a gel matrix in order to increase the gel's transdermal cargo delivery efficacy [43].

4.4 Intra-articular Drug Delivery

The poor accumulation and/or retention of conventional medicines in articular cartilage caused by systemic or intra-articular administration leads to dose-limiting toxicities and decreased efficacy. As a result, it is imperative to create better medication delivery strategies, especially through the nanotechnology platform. Researchers revealed that intra-articular drug delivery prevents fast cellular and enzymatic clearance and decreases drug efflux from the articular joint. Furthermore, researchers revealed that intra-articular drug administration improves drug retention in the articular cartilage region, has fewer side effects, and boosts therapeutic potential. Researchers revealed that due to sustained drug release, the frequency of intra-articular drug administration is decreased. In a study, researchers revealed that intra-articular injection of sivelestat-loaded nanomicelle ameliorates collagen-induced arthritis [44]. In a study, researchers revealed that liposomal formulation had a longer retention period [45]. Researchers are extensively working on multilamellar and unilamellar vesicles for drug delivery objectives. In a study, researchers revealed that the multilamellar vesicles prevented the cellular infiltration linked to arthritis [46].

5 Nanocarriers

Nanocarriers are vehicles generally derived from natural polymeric materials and biomaterials. There are several types of nanocarriers are employed for drug delivery. The nanocarriers in the field of pharmaceutical research have made steady progress in improving drug stability, bioavailability, and specific targeting. Pharmaceutical nanocarriers are extremely versatile in the delivery of cargo from submicron to nanometre-sized. These comprise liposomes, polymeric, lipidic, hydrogel, micelles, and inorganic nanoparticles, among many more (Fig. 3). It is theoretically possible to affix ligands to the surface of nanocarriers to improve their absorption and targetability [47–49].

The nanocarriers are vehicles for drugs and drugs are loaded inside the nanocarrier's layers or distributed across the nanocarrier matrix. Researchers revealed that it is very simple to manipulate nanocarrier size, surface charge, surface characteristics, surface modification, controlled uptakes and targeting for benefits for standard pharmacological therapies [50]. Currently, many drug delivery methods are available for therapy such as parenteral [51], nasal [52], topical [53, 54], and oral [55, 56]. Due

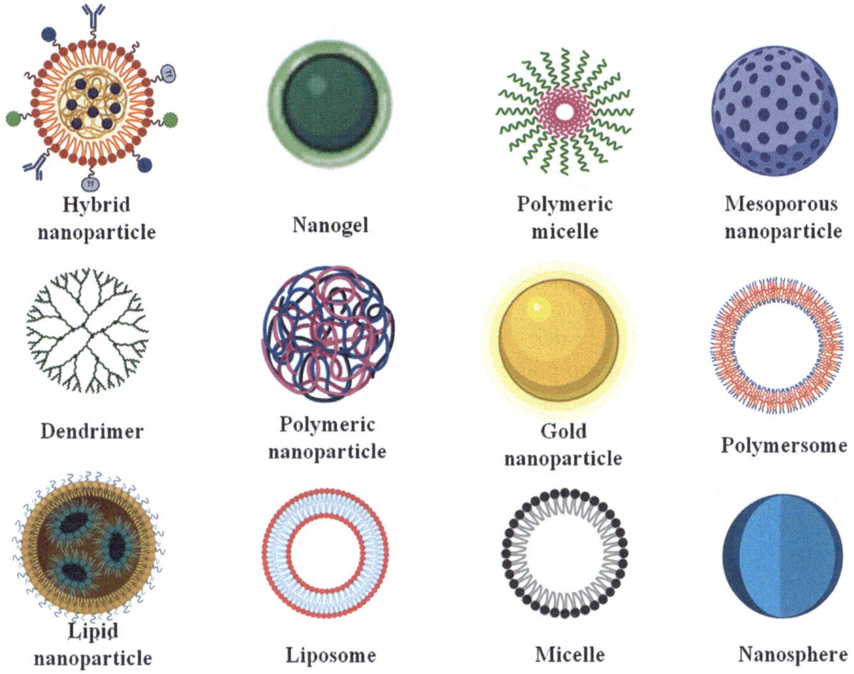

Fig. 3 Different types of nanocarriers for the treatment of RA. Image created with Biorender.com

to the benefits mentioned above, there is a growing need to create nanocarriers that have a wide variety of characteristics and can target several diseases.

5.1 Nanoparticles

Nowadays, the application of nanoparticles in drug delivery systems to enhance the effectiveness of both novel and pre-existing drugs is growing. They provide tailored distribution, which might lessen unwanted effects and boost therapeutic impact. Many nanocarriers have been created for medicinal applications recently, including tissue engineering and disease screening. The development of nanoparticles that may either actively or passively target inflamed joints has been the focus of recent research on nanoparticle-based medication delivery systems for RA. With the ability to deliver drugs targeted to the site of inflammation, these nanocarriers may lessen systemic adverse effects and enable the re-evaluation of drugs that were previously considered as very dangerous for systemic administration. The main goal of researchers is to enhance drugs bioavailability and therapeutic efficacy [57]. From research, it is revealed that M1 macrophages are abundant in RA joints and have a role in the long-term progression of inflammation and damage. It is revealed that as anti-inflammation potential by their apoptosis and polarization toward M2 phenotypes. Three different types of biomimetic NPs produced from distinct cells were developed in a study for dual-targeting to M1 macrophages and inflammatory sites. These NPs were designed to target drug delivery to both inflammatory joints and M1 macrophages, with the goal of inducing M1 macrophage death. It was found that NPs had high uptakes in M1 macrophage and huge accumulation in the inflammatory site. It is revealed that methotrexate-loaded NPs significantly reduced the pro-inflammatory cytokines expression and macrophage viability. In a recent study, researchers revealed that NPs effectively the macrophage quantity, release of pro-inflammatory cytokines, and cartilage damage in an adjuvant-induced arthritic mouse model [58]. Researchers after extensive studies revealed that apoptosis of M1 macrophages and the scavenging of ROS helps in the polarisation of M2 macrophages that may be helpful in RA treatment [59]. In a study, researchers created perfect decoys of biological agents that target neutrophils. They demonstrated that decoys of biological agents have the ability to target cartilage matrix, decrease synovial inflammation, neutralize proinflammatory cytokines, and offer robust chondroprotection against joint injury [60].

5.2 *Polymer Lipid Hybrid*

The features of liposomes and polymeric nanoparticles are combined in polymer-lipid hybrid nanoparticles, which are sophisticated nanoconstructs with a polymeric core and a lipidic coating. Researchers have extensively studied on the therapeutic potential of hybrid nanocarriers in the management of RA. They can be made to respond to stimuli, target certain inflammatory regions, and provide medication in a regulated way. Recently researchers revealed that nanoparticles coated by peptide and folate have promising features in the treatment of RA. To further target active macrophages in synovial joints, magnetic fibrin nanoparticles coupled with folic acid have been employed as MRI contrast agents. Additionally, developments in nano-targeted drug delivery systems have been made that use lipid-polymer hybrid nanoparticles loaded with siRNAs targeting TNF to reduce inflammation in animal models of experimental arthritis. In a study, the researcher presents the creation of hybrid nanoparticles (NPs) that promote tolerance in B and T cells and subsequently the suppression of RA [61].

In a study, researchers constructed a hybrid NPs gel and evaluated ex vivo and in vivo the therapeutic potential in RA [62]. One of the primary causes of autoimmune illnesses, which exacerbate the symptoms of RA, is an abnormally high quantity of cell-free DNA (cfDNA). The use of cationic polymeric nanoparticles for cfDNA scavenging has recently come to light as a potentially effective therapeutic approach for the management of RA. However, there is a chance that cationic polymeric nanoparticles will dissociate and become hazardous when injected intravenously into the bloodstream. We used a series of silica particles grafted with poly

(2-(dimethylamino) ethyl methacrylate) (PDMA) (SiNP@PDMA) brush to selectively scavenge cfDNA in an inflamed joint cavity. These particles have adjustable PDMA content and particle size. The outcomes showed that SiNP100-PDMA200 performed better than its counterparts, which is consistent with their more advantageous distribution in inflammatory articular cavities. All things considered, this research demonstrated the structure–property link of cfDNA scavengers, paving the way for the creation of more secure and efficient cfDNA scavenging systems [63].

5.3 Micelles

A micelle is a collection of molecules of surfactant scattered across a liquid colloid. A typical micelle's shape is typically bilayer, ellipsoid, cylinder, global, and generally spherical. Long-term blood circulation is made possible by the hydrophilic shells and supramolecular structure of micelles. Drugs are released continuously from the hydrophobic core of polymer micelles, but the release pattern can be adjusted by varying the length of the hydrophobic block and the kind of monomer. One kind of nanoparticulate drug delivery technology that has attracted a lot of interest recently is polymeric micelles, or PMs. Their reduced toxicity, improved circulation half-life, biocompatibility, and capacity to solubilize hydrophobic medicines are well-known attributes. Researchers are investigating the use of polymeric micelles in sophisticated rheumatoid arthritis medication delivery systems. They may increase the effectiveness of treatment and lessen negative effects by extending the half-life of medications and enabling targeted distribution into the joints. In order to enable more targeted medication delivery, recent research has concentrated on creating micelles that react to the enhanced levels of ROS in inflammatory joints. RA aetiology and development are significantly influenced by activated macrophages, which produce pro-inflammatory substances. Dextamethasone (DXM) is useful in treating RA's uncontrollable pain and inflammatory development. Furthermore, after intravenous injection, the potent anti-inflammatory action and efficient accumulation of DXM@DS-cad-DXM micelles were noted in the inflammatory joints of rats [64]. When administered intravenously, the PPT micelles exhibit superior therapeutic activity and preferentially aggregate in arthritic joints in AIA [65].

5.4 Dendrimers

Dendrimers are well-employed nanocarriers in drug delivery. They are generally radially symmetrical nanosized molecules with monodisperse structures made up of tree-like branches. Artificial tree-like structures can be utilized to distribute medications throughout the body. Their great degree of branching enables them to transport several medicinal molecules and release them at specific locations. By extensive studies by researchers revealed that dendrimers are good in drug loading capacity and have capability to conjugate with targeting moieties which makes them suitable candidates for drug delivery in RA. In a study, to decrease inflammation, miR-23b is delivered via a fluorinated polyamidoamine dendrimer (FP), which both triggers and inhibits the inflammatory response in macrophages. In experimental RA models, FP/miR-23b nanoparticles are injected intravenously. The nanoparticles exhibit therapeutic efficacy with decreased bone and cartilage attrition, decreased inflammatory response, suppressed synoviocyte infiltration, and restored mobility. Additionally, the nanoparticles build up in the inflammatory joint and are non-specifically absorbed by synoviocytes, which causes the synovium to express miR-23b to return and encourage anti-inflammatory and anti-proliferative responses [66]. In a recent study, the researcher develops a unique multifunctional dendrimer with a nanogold core for pulsatile photothermal, photochemotherapeutic, and photodynamic therapy of RA. The results of this study indicate that multifunctional targeted NPs that have been produced may be useful as treatments for the better management of RA [67]. In other studies, a multifunctional nanoplatform based on generation 5 (G5) poly (amidoamine) dendrimer-entrapped gold nanoparticles (Au DENPs) is presented for effective RA therapy. The multifunctional nanoplatform that was developed may be used in combination therapy for RA that combines anti-inflammatory and antioxidant agents [68].

5.5 *Liposomes*

The liposomes are spherical lipid vesicles. They are formed by emulsification of natural or synthetic lipids in an aqueous solution. The lipid vesicles contain one or more lipid bilayers. The liposomes have some silent features like compatibility, stability, ease of synthesis, high drug-loading efficiency that make them suitable for the preparation of liposome nanoemulsions. The liposome nanoemulsion has wide applications in nanomedicine. Recently, liposomes have been employed as potential drug delivery carriers due to their very small size and ability to carry hydrophilic as well as hydrophobic drug molecules.

In the field of medication delivery, liposomes are thought to be promising drug vesicles. Compared to conventional medication delivery methods, they provide a number of benefits, including targeted distribution to particular body locations, controlled medication release, enabling long-lasting therapeutic benefits, medication defense against deterioration and clearance, Because of their biocompatibility, there are fewer harmful side effects. Numerous uses are being investigated for them, such as therapeutics and diagnostics (theranostics). Using anti-IL-23 Abs, a stable suspension of monodispersed liposomes measuring about 130 nm is effectively biofunctionalized. The ability to collect and neutralize IL-23 is verified by the use of activated macrophages. As a result, the created formulation might be a useful RA treatment plan [69]. In a recent study, researchers demonstrated the anti-inflammatory potential in RA of dimeric artesunate phospholipid conjugate liposomes [70]. In a study, a liposomal system was created by researchers for the targeted distribution of tofacitinib citrate to joint tissue and they revealed enhanced efficacy [71]. In a recent study, researchers revealed that liposomes have the ability to extend the triptolide in body circulation with anti-inflammatory and cartilage protective properties and offer promising efficacy against RA [72]. In another study, researchers created and tested a triple therapy that effectively treats RA both in vivo and in vitro. In addition to protecting cartilage, folate-modified liposomes + laser therapy decreased clinical arthritis scores and serum cytokine levels [73].

5.6 Hydrogel

Hydrogel as a drug delivery is gaining popularity nowadays. The water-loving polymeric network that can absorb large amount of water is known as hydrogel. They can serve as controlled drug delivery by expanding and contracting. The hydrogel is hydrophilic in nature and has good compatibility and porosity in aquatic conditions that make it a biocompatible drug delivery nanocarrier [74]. Researchers revealed the enhanced therapeutic efficacy of hydrogel-based nanocarrier's application to tackle inflammatory diseases. The application of hydrogel nanocarriers as drug delivery is gradually expanding in the management of disease. Researchers revealed that some novel characteristics of hydrogel carriers such as malleable and controlled degradation may be significant for encapsulated drugs for controlled release [75]. In a recent study, Kumar et al. formulated smart injectable hydrogel for the management of RA. They revealed that the budesonide-loaded enzyme responsive-hydrogel improved joint physiology and decreased inflammatory markers as compared to control rats in collagen-induced rheumatoid arthritis [23]. In another study, in order to reduce the inflammatory response in RA, a DNase-functionalized hydrogel has been created for the therapy of the disease by continuously breaking down the unique therapeutic target of neutrophil extracellular traps [76]. The Nrf-2 pathway is the mechanism via which the beneficial phytochemical phenethyl isothiocyanate (PEITC) demonstrates its chemopreventive, antioxidant, and anti-inflammatory properties. Furthermore, researchers revealed that the low bioavailability of PEITC hamper their applications in clinics due to less water solubility, instability and short half-life. The PEITC-loaded chitosan smart hydrogel performed satisfactorily as a thermosensitive injectable in studies involving healthy rats. It also showed adequate pharmacokinetics (drug release that lasts) and biosafety (nontoxicity to major organs), as well as injectability (easy administration). When PEITC-hydrogel (PH50) was injected into the knee joint of the adjuvant-induced arthritis rat, RA-related symptoms were dramatically reduced. Additionally, there was a noticeable decrease in inflammation-specific biomarkers and bone erosion [77].

6 Functionalized Nanoparticles

Functionalized nanoparticles are designed to have particular surface characteristics that improve their functionality in a range of applications. For the achievement of desirable interaction and stability, researchers customized the nanoparticles to desired size, functionality, and surface morphology. Due to their special qualities, which include low toxicity, chemical stability, surface activity, and biocompatibility, these nanoparticles can be used in a variety of industrial, medicinal, and environmental contexts. Functionalized nanoparticles exhibit improved mechanical properties compared to virgin nanoparticles, including increased strength, stiffness, hardness, elasticity, and stiffness. Researchers are extensively exploring the applicability of functionalized nanoparticles to tackle inflammatory diseases including RA. Because of their optical and magnetic characteristics, they have a wide range of applications in the diagnosis and treatment of inflammatory diseases. Researchers are extensively working to investigate the hyaluronate-functionalized hydroxyapatite NPs as a potential feature drug delivery system for methotrexate and teriflunomide delivery to ameliorate inflammation against RA. It's still unclear exactly how these nanoparticles work to target RA. Furthermore, researchers are extensively exploring poly nanosystems functionalized by RGD peptide for the delivery of STAT1 siRNA to the joint tissue of the RA animal model. The stability of siRNA after encapsulation and the characteristics of the nanoparticles were evaluated. Compared to control mice, arthritic mice showed an increase in lung transport of nanoparticles with RGD functionalization. In the animals treated with STAT1siRNA, the disease improved, but in the other groups, it worsened. In the paws of the treated mice, researchers revealed that the enhanced levels of Mrc-1 and IL-10 mRNA whereas STAT1 mRNA levels decreased. Researchers observed that these functionalized nanocarriers loaded STAT1-targeted siRNAs in animal models of arthritis are effective, probably due to a selective suppression of dendritic and macrophage activation [78].

7 Advantage of Nanocarrier

Nanocarriers are vehicles that may carry drugs for the treatment of inflammatory diseases. We cannot ignore the importance of nanocarriers in the advancement of drug delivery. Nowadays, there are several nanocarriers employed in drug delivery in targeted delivery of drugs at the site of inflammation in RA. In this section, we have discussed in detail the merits of nanocarriers that make them a viable candidate for smart drug delivery in tackling inflammatory diseases.

7.1 Pharmacokinetics

Pharmacokinetics is an important feature in drug delivery systems. It is the interaction of drugs with the human body. In this regard, we can better understand the response of the body against given drugs. The physiological characteristics of drugs like solubility in an aqueous solution are regulatory factors for their pharmacokinetics. Better pharmacokinetic drugs may have better efficacy. The nanoformulated drugs generally have better pharmacokinetics. The nanoformulated drug or drug-loaded nanocarrier can make medications more stable and soluble, especially if they are insoluble.

7.2 Targeted Delivery

The main purpose of nanocarriers is to deliver cargo to specific sites or targeted delivery. From targeted delivery, we can limit from undesirable delivery of drugs and their adverse effects. It is a sophisticated approach. Nowadays, the targeted delivery of drug has become more advanced through the use of smart nanocarriers. These smart nanocarriers are very specific in drug delivery. They work in response to stimuli.

7.3 Controlled Release

Nanomedicine become more advanced in drug delivery recently. Due to advancements in drug delivery carriers, precise drug release and dosage management are made possible by them. These advantages reduce the possibility of negative side effects while boosting the effectiveness of both new and current medications. Sustained release makes it possible for medications to be released gradually and under strict supervision, which can enhance patient compliance and treatment results.

7.4 Enhanced Efficacy

The main goal of nanocarrier is to deliver drugs very specifically. By ensuring that medications reach the site of inflammation in higher quantities, they can increase the therapeutic efficacy of those medications.

7.5 Decreased Drug Resistance

Nowadays, drug resistance is a global concern. Drug resistance can limit the efficacy of drugs. Recently, there has been a lot of research on the application of nanocarrier-mediated drug delivery as a potential strategy to combat drug resistance. As a result, these strategies can lessen the payload's buildup off-target, which can lower cytotoxicity and improve targeting. By more efficiently delivering a combination of medicines, nanocarriers can aid in the fight against drug resistance.

8 Disadvantage of Nanocarriers

As we know that nanocarriers have a significant role in targeted drug delivery to tackle inflammatory diseases like RA. However, we cannot ignore the issues related to nanocarriers. While using nanocarriers for drug delivery has many benefits, there are certain drawbacks as well. In this section, we have discussed some possible drawbacks one by one.

8.1 Complexity

Polymers and biomaterials are generally used in the creation of nanocarriers. These polymers are very complex and require proper guidelines during nanocreation. There are several factors involved in the creation of desire carriers. The optimization process of nanocarriers is very complex. The creation and design of nanocarriers can be expensive and time-consuming.

8.2 Stability Concerns

For drug delivery purposes, the nanocarrier should be stable. Stability is a major concern in drug delivery system. The good stability of a nanocarrier is the specificity determining factor in the drug delivery system. Certain nanocarriers may experience stability challenges in the bloodstream or during storage. Some nanocarriers are not stable in long-term storage.

8.3 Safety Concerns

Nanoscale particles are very small in size. Due to its very small size, these particles may cross the blood–brain barrier. Once they cross the barrier, they might have some hazardous impact on the human neuronal system and sometimes dangerous for the human brain.

8.4 Regulatory Obstacles

Nanocarriers have toxic effects on humans and animals. Before preclinical and clinical testing, all toxic aspects of nanocarrier must be considered. Nanocarriers must pass strict regulations in order to be approved.

8.5 Low Drug Loading Capacity

Drug loading capacity is the prime concern in efficient drug delivery. Low drug loading is a big obstacle in drug delivery. This may result in the delivery of fewer medications to the intended location.

8.6 Crystallization During Storage

This may have an impact on the medicine's potency and stability. This may result in adverse effects and lower drug efficacy. These problems emphasize the necessity of meticulous nanocarrier design and testing to guarantee their efficacy and safety for rheumatoid arthritis patients.

9 Conclusion and Future Directions

Nanocarriers are a promising technique for drug delivery systems. The purpose of nanocarriers in drug delivery is specific drug delivery to the target, enhanced cellular uptakes, controlled drug release and minimum adverse effects. These unique features of nanocarriers make them potential vehicles in targeted drug delivery in inflammatory diseases and RA. Drugs for RA are being delivered using nanocarriers, which may lessen systemic side effects. Prodrug-based nanomedicines and focused therapy approaches are examples of these advancements. The use of nanocarriers in drug delivery appears to have a bright future. Customizing drug distribution to meet each patient's unique needs in order to improve results is known as personalized medicine. Increasing the accuracy with which drugs are delivered to particular tissues or cells may be a future ray of hope in tackling inflammation. Combining various medicinal substances into a single nanocarrier to produce a synergistic effect is known as combination therapy. To increase the uptake and efficacy of nanocarrier drug delivery systems, issues including scalability, legal barriers, and characterization techniques are also being addressed. From the future point of view, researchers should focus on stimuli-based drug delivery by personalization of nanocarriers. Researchers should also focus on the fate of nanocarriers on the human body before their application in nanomedicine for drug delivery. A strong regulatory guideline should be followed in the synthesis of nanocarriers.

References

1. Radu AF, Bungau SG (2021) Management of rheumatoid arthritis: an overview. Cells 10(11):2857. https://doi.org/10.3390/cells10112857
2. Hyndman IJ (2017) Rheumatoid arthritis: past, present and future approaches to treating the disease. Int J Rheum Dis 20(4):417–419. https://doi.org/10.1111/1756-185X.12823
3. Dedmon LE (2020) The genetics of rheumatoid arthritis. Rheumatology (Oxford) 59(10):2661–2670. https://doi.org/10.1093/rheumatology/keaa232
4. Karami J, Aslani S, Jamshidi A, Garshasbi M, Mahmoudi M (2019) Genetic implications in the pathogenesis of rheumatoid arthritis; an updated review. Gene 702:8–16. https://doi.org/10.1016/j.gene.2019.03.033
5. GBD (2019) Global burden of 369 diseases and injuries in 204 countries and territories, 1990–2019: a systematic analysis for the Global Burden of Disease Study 2019. https://vizhub.healthdata.org/gbd-results/
6. Cieza A, Causey K, Kamenov K, Hanson SW, Chatterji S, Vos T (2021) Global estimates of the need for rehabilitation based on the global burden of disease study 2019: a systematic analysis for the global burden of disease study 2019. Lancet (London, England) 396(10267):2006–2017. https://doi.org/10.1016/S0140-6736(20)32340-0
7. Bullock J, Rizvi SAA, Saleh AM, Ahmed SS, Do DP, Ansari RA, Ahmed J (2018) Rheumatoid arthritis: a brief overview of the treatment. Med Princ Pract Int J Kuwait Univ, Health Science Centre 27(6):501–507. https://doi.org/10.1159/000493390
8. Drini M (2017) Peptic ulcer disease and non-steroidal anti-inflammatory drugs. Aust Prescr 40(3):91–93. https://doi.org/10.18773/austprescr.2017.037
9. Buttgereit F (2020) Views on glucocorticoid therapy in rheumatology: the age of convergence. Nat Rev Rheumatol 16(4):239–246. https://doi.org/10.1038/s41584-020-0370-z
10. Stevenson M, Archer R, Tosh J, Simpson E, Everson-Hock E, Stevens J, Hernandez-Alava M, Paisley S, Dickinson K, Scott D, Young A, Wailoo A (2016) Adalimumab, etanercept, infliximab, certolizumab pegol, golimumab, tocilizumab and abatacept for the treatment of rheumatoid arthritis not previously treated with disease-modifying antirheumatic drugs and after the failure of conventional disease-modifying antirheumatic drugs only: systematic review and economic evaluation. Health Technol Assess (Winchester, England) 20(35):1–610. https://doi.org/10.3310/hta20350
11. Chatzidionysiou K, Emamikia S, Nam J, Ramiro S, Smolen J, van der Heijde D, Dougados M, Bijlsma J, Burmester G, Scholte M, van Vollenhoven R, Landewé R (2017) Efficacy of glucocorticoids, conventional and targeted synthetic disease-modifying antirheumatic drugs: a systematic literature review informing the 2016 update of the EULAR recommendations for

the management of rheumatoid arthritis. Ann Rheum Dis 76(6):1102–1107. https://doi.org/10.1136/annrheumdis-2016-210711
12. Xiao S, Tang Y, Lv Z, Lin Y, Chen L (2019) Nanomedicine—advantages for their use in rheumatoid arthritis theranostics. J Controlled Release Off J Controlled Release Soc 316:302–316. https://doi.org/10.1016/j.jconrel.2019.11.008
13. Li S, Su J, Cai W, Liu JX (2021) Nanomaterials manipulate macrophages for rheumatoid arthritis treatment. Front Pharmacol 12:699245. https://doi.org/10.3389/fphar.2021.699245
14. Singh AP, Biswas A, Shukla A, Maiti P (2019) Targeted therapy in chronic diseases using nanomaterial-based drug delivery vehicles. Signal Transduct Target Ther 4:33. https://doi.org/10.1038/s41392-019-0068-3
15. Perrie Y, Rades TT (2012) FASTtrack pharmaceutics: drug delivery and targeting. Pharmaceutical Press
16. Imming P, Sinning C, Meyer A (2006) Drugs, their targets and the nature and number of drug targets. Nat Rev Drug Discov 5(10):821–834. https://doi.org/10.1038/nrd2132
17. Pirmardvand Chegini S, Varshosaz J, Taymouri S (2018) Recent approaches for targeted drug delivery in rheumatoid arthritis diagnosis and treatment. Artif Cells Nanomed Biotechnol 46(sup2):502–514. https://doi.org/10.1080/21691401.2018.1460373
18. Yang M, Feng X, Ding J, Chang F, Chen X (2017) Nanotherapeutics relieve rheumatoid arthritis. J Controlled Release Off J Controlled Release Soc 252:108–124. https://doi.org/10.1016/j.jconrel.2017.02.032
19. Shaji JE, Lal MA (2013) Nanocarriers for targeting in inflammation. Asian J Pharm Clin Res 6(Suppl 3):3–12
20. Goldstein JL, Cryer B (2015) Gastrointestinal injury associated with NSAID use: a case study and review of risk factors and preventative strategies. Drug Healthc Patient Saf 7:31–41. https://doi.org/10.2147/DHPS.S71976
21. Higaki M, Ishihara T, Izumo N, Takatsu M, Mizushima Y (2005) Treatment of experimental arthritis with poly(D, L-lactic/glycolic acid) nanoparticles encapsulating betamethasone sodium phosphate. Ann Rheum Dis 64(8):1132–1136. https://doi.org/10.1136/ard.2004.030759
22. Liu M, Dong J, Yang Y, Yang X, Xu H (2005) Anti-inflammatory effects of triptolide loaded poly(D, L-lactic acid) nanoparticles on adjuvant-induced arthritis in rats. J Ethnopharmacol 97(2):219–225. https://doi.org/10.1016/j.jep.2004.10.031
23. Kumar A, Ali A, Kanika, Vyawahare A, Ahmad A, Mishra RK, Ansari MM, Nadeem A, Siddiqui N, Raza SS, Khan R (2023) Highly biocompatible smart injectable hydrogel for the management of rheumatoid arthritis. ACS Biomater Sci Eng 9(9):5312–5321. https://doi.org/10.1021/acsbiomaterials.3c00514
24. Guo Q, Wang Y, Xu D, Nossent J, Pavlos NJ, Xu J (2018) Rheumatoid arthritis: pathological mechanisms and modern pharmacologic therapies. Bone Res 6:15. https://doi.org/10.1038/s41413-018-0016-9
25. Kadiri M, Charbonneau M, Lalanne C, Harper K, Balg F, Marotta A, Dubois CM (2021) 14–3-3η promotes invadosome formation via the FOXO3-snail axis in rheumatoid arthritis fibroblast-like synoviocytes. Int J Mol Sci 23(1):123. https://doi.org/10.3390/ijms23010123
26. Ohyama A, Osada A, Kawaguchi H, Kurata I, Nishiyama T, Iwai T, Ishigami A, Kondo Y, Tsuboi H, Sumida T, Matsumoto I (2021) Specific increase in joint neutrophil extracellular traps and its relation to interleukin 6 in autoimmune arthritis. Int J Mol Sci 22(14):7633. https://doi.org/10.3390/ijms22147633
27. Pietschmann P, Butylina M, Kerschan-Schindl K, Sipos W (2022) Mechanisms of systemic osteoporosis in rheumatoid arthritis. Int J Mol Sci 23(15):8740. https://doi.org/10.3390/ijms23158740
28. Omata Y, Frech M, Saito T, Schett G, Zaiss MM, Tanaka S (2022) Inflammatory arthritis and bone metabolism regulated by type 2 innate and adaptive immunity. Int J Mol Sci 23(3):1104. https://doi.org/10.3390/ijms23031104
29. Garde-Cerdán T, Souza-da Costa B, Rubio-Bretón P, Pérez-Álvarez EP (2021) Nanotechnology: recent advances in viticulture and enology. J Sci Food Agric 101(15):6156–6166. https://doi.org/10.1002/jsfa.11406

30. Li S, Xu S, Liang X, Xue Y, Mei J, Ma Y, Liu Y, Liu Y (2021) Nanotechnology: breaking the current treatment limits of lung cancer. Adv Healthc Mater 10(12):e2100078. https://doi.org/10.1002/adhm.202100078
31. Chen S, Li R, Li X, Xie J (2018) Electrospinning: An enabling nanotechnology platform for drug delivery and regenerative medicine. Adv Drug Deliv Rev 132:188–213. https://doi.org/10.1016/j.addr.2018.05.001
32. Rahman M, Beg S, Anwar F, Kumar V, Ubale R, Addo RT, Ali R, Akhter S (2017) Liposome-based nanomedicine therapeutics for rheumatoid arthritis. Crit Rev Ther Drug Carrier Syst 34(4):283–316. https://doi.org/10.1615/CritRevTherDrugCarrierSyst.2017016067
33. Ansari MM, Ahmad A, Kumar A, Alam P, Khan TH, Jayamurugan G, Raza SS, Khan R (2021) Aminocellulose-grafted-polycaprolactone coated gelatin nanoparticles alleviate inflammation in rheumatoid arthritis: a combinational therapeutic approach. Carbohyd Polym 258:117600. https://doi.org/10.1016/j.carbpol.2020.117600
34. Rauchhaus U, Schwaiger FW, Panzner S (2009a) Separating therapeutic efficacy from glucocorticoid side-effects in rodent arthritis using novel, liposomal delivery of dexamethasone phosphate: long-term suppression of arthritis facilitates interval treatment. Arthritis Res Ther 11(6):R190. https://doi.org/10.1186/ar2889
35. Rauchhaus U, Kinne RW, Pohlers D, Wiegand S, Wölfert A, Gajda M, Bräuer R, Panzner S (2009b) Targeted delivery of liposomal dexamethasone phosphate to the spleen provides a persistent therapeutic effect in rat antigen-induced arthritis. Ann Rheum Dis 68(12):1933–1934. https://doi.org/10.1136/ard.2009.108985
36. Williams A, Goodfellow R, Topley N, Amos N, Williams B (2000) The suppression of rat collagen-induced arthritis and inhibition of macrophage derived mediator release by liposomal methotrexate formulations. Inflamm Res Off J Eur Histamine Res Soc 49(4):155–161. https://doi.org/10.1007/s000110050575
37. Lou J, Duan H, Qin Q, Teng Z, Gan F, Zhou X, Zhou X (2023) Advances in oral drug delivery systems: challenges and opportunities. Pharmaceutics 15(2):484. https://doi.org/10.3390/pharmaceutics15020484
38. Garhwal A, Kendya P, Soni S, Kori S, Soni V, Kashaw SK (2024) Drug delivery system approaches for rheumatoid arthritis treatment: a review. Mini Rev Med Chem 24(7):704–720. https://doi.org/10.2174/1389557523666230913105803
39. Ansari MM, Ahmad A, Mishra RK, Raza SS, Khan R (2019) Zinc gluconate-loaded chitosan nanoparticles reduce severity of collagen-induced arthritis in wistar rats. ACS Biomater Sci Eng 5(7):3380–3397. https://doi.org/10.1021/acsbiomaterials.9b00427
40. Xu Y, Zhao M, Cao J, Fang T, Zhang J, Zhen Y, Wu F, Yu X, Liu Y, Li J, Wang D (2023) Applications and recent advances in transdermal drug delivery systems for the treatment of rheumatoid arthritis. Acta Pharm Sin B 13(11):4417–4441. https://doi.org/10.1016/j.apsb.2023.05.025
41. Balata GF, Faisal MM, El-ghamry HA, Sabry SA (2020) Preparation and characterization of ivabradine HCl transfersomes for enhanced transdermal delivery. J Drug Deliv Sci Technol 60:101921
42. Nalamachu S, Gudin J (2020) Characteristics of analgesic patch formulations. J Pain Res 13:2343–2354. https://doi.org/10.2147/JPR.S270169
43. Nozawa T, Nishimura K, Ohara A, Hara R, Ito S (2019) Primary varicella infection in children with systemic juvenile idiopathic arthritis under tocilizumab therapy. Mod Rheumatol 29(3):558–562. https://doi.org/10.1080/14397595.2016.1254314
44. Vyawahare A, Ansari MM, Kumar A, Ahmad A, Mishra RK, Jori C, Nadeem A, Siddiqui N, Raza SS, Khan R (2023) Enzyme targeted delivery of sivelestat loaded nanomicelle inhibits arthritic severity in experimental arthritis. Life Sci 334:122206. https://doi.org/10.1016/j.lfs.2023.122206
45. López-García F, Vázquez-Autón JM, Gil F, Latoore R, Moreno F, Villalaín J, Gómez-Fernández JC (1993) Intra-articular therapy of experimental arthritis with a derivative of triamcinolone acetonide incorporated in liposomes. J Pharm Pharmacol 45(6):576–578. https://doi.org/10.1111/j.2042-7158.1993.tb05603.x

46. Williams AS, Camilleri JP, Goodfellow RM, Williams BD (1996) A single intra-articular injection of liposomally conjugated methotrexate suppresses joint inflammation in rat antigen-induced arthritis. Br J Rheumatol 35(8):719–724. https://doi.org/10.1093/rheumatology/35.8.719
47. Pontes JF, Grenha A (2020) Multifunctional nanocarriers for lung drug delivery. Nanomaterials (Basel, Switzerland) 10(2):183. https://doi.org/10.3390/nano10020183
48. Zeb A, Rana I, Choi HI, Lee CH, Baek SW, Lim CW, Khan N, Arif ST, Sahar NU, Alvi AM, Shah FA, Din FU, Bae ON, Park JS, Kim JK (2020) Potential and applications of nanocarriers for efficient delivery of biopharmaceuticals. Pharmaceutics 12(12):1184. https://doi.org/10.3390/pharmaceutics12121184
49. Nawaz M, Sliman Y, Ercan I, Lima-Tenório MK, Tenório-Neto ET, Kaewsaneha C, Elaissari A (2019) Stimuli responsive polymeric nanocarriers for drug delivery applications. In: Magnetic and ph-responsive magnetic nanocarriers. Elsevier, Amsterdam, The Netherlands, pp 37–85
50. Majumder J, Taratula O, Minko T (2019) Nanocarrier-based systems for targeted and site specific therapeutic delivery. Adv Drug Deliv Rev 144:57–77. https://doi.org/10.1016/j.addr.2019.07.010
51. Kolluru LP, Atre P, Rizvi SAA (2021) Characterization and applications of colloidal systems as versatile drug delivery carriers for parenteral formulations. Pharmaceuticals (Basel, Switzerland) 14(2):108. https://doi.org/10.3390/ph14020108
52. Vachhani S, Kleinstreuer C (2021) Comparison of micron- and nano-particle transport in the human nasal cavity with a focus on the olfactory region. Comput Biol Med 128:104103. https://doi.org/10.1016/j.compbiomed.2020.104103
53. Tolentino S, Pereira MN, Cunha-Filho M, Gratieri T, Gelfuso GM (2021) Targeted clindamycin delivery to pilosebaceous units by chitosan or hyaluronic acid nanoparticles for improved topical treatment of acne vulgaris. Carbohyd Polym 253:117295. https://doi.org/10.1016/j.carbpol.2020.117295
54. Paiva-Santos AC, Herdade AM, Guerra C, Peixoto D, Pereira-Silva M, Zeinali M, Mascarenhas-Melo F, Paranhos A, Veiga F (2021) Plant-mediated green synthesis of metal-based nanoparticles for dermopharmaceutical and cosmetic applications. Int J Pharm 597:120311. https://doi.org/10.1016/j.ijpharm.2021.120311
55. Reboredo C, González-Navarro CJ, Martínez-Oharriz C, Martínez-López AL, Irache JM (2021) Preparation and evaluation of PEG-coated zein nanoparticles for oral drug delivery purposes. Int J Pharm 597:120287. https://doi.org/10.1016/j.ijpharm.2021.120287
56. Yao W, Xu Z, Sun J, Luo J, Wei Y, Zou J (2021) Deoxycholic acid-functionalised nanoparticles for oral delivery of rhein. Eur J Pharm Sci Off J Eur Fed Pharm Sci 159:105713. https://doi.org/10.1016/j.ejps.2021.105713
57. Wang Q, Qin X, Fang J, Sun X (2021) Nanomedicines for the treatment of rheumatoid arthritis: state of art and potential therapeutic strategies. Acta Pharm Sin B 11(5):1158–1174. https://doi.org/10.1016/j.apsb.2021.03.013
58. Hao X, Gai W, Zhang Y, Zhou W, Feng Y (2024) Biomimetic and inflammation-targeted nanoparticles for rheumatoid arthritis therapy via M1-macrophages apoptosis and polarization. Mater Des 239:112806. https://doi.org/10.1016/j.matdes.2024.112806
59. Yang Y, Guo L, Wang Z, Liu P, Liu X, Ding J, Zhou W (2021) Targeted silver nanoparticles for rheumatoid arthritis therapy via macrophage apoptosis and re-polarization. Biomaterials 264:120390. https://doi.org/10.1016/j.biomaterials.2020.120390
60. Zhang Q, Dehaini D, Zhang Y, Zhou J, Chen X, Zhang L, Fang RH, Gao W, Zhang L (2018) Neutrophil membrane-coated nanoparticles inhibit synovial inflammation and alleviate joint damage in inflammatory arthritis. Nat Nanotechnol 13(12):1182–1190. https://doi.org/10.1038/s41565-018-0254-4
61. Brzezicka KA, Arlian BM, Wang S, Olmer M, Lotz M, Paulson JC (2022) Suppression of autoimmune rheumatoid arthritis with hybrid nanoparticles that induce B and T cell tolerance to self-antigen. ACS Nano 16(12):20206–20221. https://doi.org/10.1021/acsnano.2c05643
62. Aman RM, Zaghloul RA, Elsaed WM, Hashim IIA (2023) In vitro-in vivo assessments of apocynin-hybrid nanoparticle-based gel as an effective nanophytomedicine for treatment of

rheumatoid arthritis. Drug Deliv Transl Res 13(11):2903–2929. https://doi.org/10.1007/s13346-023-01360-5

63. Liu X, Chen S, Liu L, Chen Y (2023) Cationic brush hybrid nanoparticles scavenge cell-free DNA to enhance rheumatoid arthritis treatment. Acta Biomater 170:215–227. https://doi.org/10.1016/j.actbio.2023.08.032
64. Han J, Na R, Zhao N, Yuan X, Fu L, Jing J, Qian A, Ye W (2023a) Macrophage-targeted dextran sulfate-dexamethasone conjugate micelles for effective treatment of rheumatoid arthritis. Molecules (Basel, Switzerland) 28(2):591. https://doi.org/10.3390/molecules28020591
65. He L, Qin X, Fan D, Feng C, Wang Q, Fang J (2021) Dual-stimuli responsive polymeric micelles for the effective treatment of rheumatoid arthritis. ACS Appl Mater Interfaces 13(18):21076–21086. https://doi.org/10.1021/acsami.1c04953
66. Han H, Xing J, Chen W, Jia J, Li Q (2023b) Fluorinated polyamidoamine dendrimer-mediated miR-23b delivery for the treatment of experimental rheumatoid arthritis in rats. Nat Commun 14(1):944. https://doi.org/10.1038/s41467-023-36625-7
67. Pandey PK, Maheshwari R, Raval N, Gondaliya P, Kalia K, Tekade RK (2019) Nanogold-core multifunctional dendrimer for pulsatile chemo-, photothermal- and photodynamic-therapy of rheumatoid arthritis. J Colloid Interface Sci 544:61–77. https://doi.org/10.1016/j.jcis.2019.02.073
68. Li J, Chen L, Xu X, Fan Y, Xue X, Shen M, Shi X (2020) Targeted combination of antioxidative and anti-inflammatory therapy of rheumatoid arthritis using multifunctional dendrimer-entrapped gold nanoparticles as a platform. Small 16(49):e2005661. https://doi.org/10.1002/smll.202005661
69. Lima AC, Campos CF, Cunha C, Carvalho A, Reis RL, Ferreira H, Neves NM (2021) Biofunctionalized liposomes to monitor rheumatoid arthritis regression stimulated by interleukin-23 neutralization. Adv Healthc Mater 10(2):e2001570. https://doi.org/10.1002/adhm.202001570
70. Zhang Y, He W, Du Y, Du Y, Zhao C, Zhang Y, Zhang H, Yin L, Li X (2020) Dimeric artesunate phospholipid-conjugated liposomes as promising anti-inflammatory therapy for rheumatoid arthritis. Int J Pharm 579:119178. https://doi.org/10.1016/j.ijpharm.2020.119178
71. Shen Q, Shu H, Xu X, Shu G, Du Y, Ying X (2020) Tofacitinib citrate-based liposomes for effective treatment of rheumatoid arthritis. Pharmazie 75(4):131–135. https://doi.org/10.1691/ph.2020.9154
72. Guo RB, Zhang XY, Yan DK, Yu YJ, Wang YJ, Geng HX, Wu YN, Liu Y, Kong L, Li XT (2022) Folate-modified triptolide liposomes target activated macrophages for safe rheumatoid arthritis therapy. Biomater Sci 10(2):499–513. https://doi.org/10.1039/d1bm01520f
73. Xue L, Wang D, Zhang X, Xu S, Zhang N (2020) Targeted and triple therapy-based liposomes for enhanced treatment of rheumatoid arthritis. Int J Pharm 586:119642. https://doi.org/10.1016/j.ijpharm.2020.119642
74. Narayanaswamy R, Torchilin VP (2019) Hydrogels and their applications in targeted drug delivery. Molecules (Basel, Switzerland) 24(3):603. https://doi.org/10.3390/molecules24030603
75. Li J, Mooney DJ (2016) Designing hydrogels for controlled drug delivery. Nat Rev Mater 1(12):16071. https://doi.org/10.1038/natrevmats.2016.71
76. Wang N, Ma J, Song W, Zhao C (2023) An injectable hydrogel to disrupt neutrophil extracellular traps for treating rheumatoid arthritis. Drug Deliv 30(1):2173332. https://doi.org/10.1080/10717544.2023.2173332
77. Haloi P, Chawla S, Konkimalla VB (2023) Thermosensitive smart hydrogel of PEITC ameliorates the therapeutic efficacy in rheumatoid arthritis. Eur J Pharm Sci Off J Eur Fed Pharm Sci 181:106367. https://doi.org/10.1016/j.ejps.2022.106367
78. Scheinman RI, Trivedi R, Vermillion S, Kompella UB (2011) Functionalized STAT1 siRNA nanoparticles regress rheumatoid arthritis in a mouse model. Nanomedicine (Lond) 6(10):1669–1682. https://doi.org/10.2217/nnm.11.90

Nano-Based Targeted Drug Delivery in Ulcerative Colitis

Md. Meraj Ansari, Nadeem Akhtar, Zamurrad Nayab, Nemat Ali, Md. Nazrul Islam, Abdullah F. AlAsmari, and Rehan Khan

Abstract Ulcerative colitis (UC) is a chronic inflammatory bowel disease. In treatment with traditional chemotherapy, the drug lost its effectiveness due to the first bypass effect and acidic environments of the gastrointestinal tract (GI). It will be best to administer the oral therapeutic drug directly to the colonic tissue or at the site of inflammation to enhance the efficiency and reduce systemic toxicity of the drug. In this regard, nanotechnology plays a significant role in terms of targeted drug delivery systems. Targeted drug delivery for the treatment of UC is challenging due to physiological and biochemical barriers. Researchers are trying to tackle all obstacles in drug delivery with the latest advancements in nanotechnology. In this chapter, we have explained the pathophysiology of UC, and physiological and biochemical barriers in detail. Furthermore, we have compared a traditional treatment system with a targeted drug delivery system. In this regard, we have explained some latest advanced drug delivery systems to tackle UC such as smart drug delivery systems, pH-responsive

Md. M. Ansari (✉)
Department of Animal Biotechnology, Faculty of Biotechnology, College of Applied Life Sciences, Jeju National University, Jeju Special Self-Governing Province, Jeju-Si, South Korea
e-mail: rajmeraj385@gmail.com

Interdisciplinary Graduate Program in Advanced Convergence Technology and Science, Jeju National University, Jeju Special Self-Governing Province, Jeju-Si, South Korea

N. Akhtar
Department of Animal Biosciences, University of Guelph, Guelph, Canada

Z. Nayab
Department of Biotechnology, Science College Patna, Patna University, Patna, Bihar, India

N. Ali · A. F. AlAsmari
Department of Pharmacology and Toxicology, College of Pharmacy, King Saud University, Riyadh, Saudi Arabia

Md. N. Islam
Chemistry Department, Jawahar Lal Nehru College, Chakradharpur, Jharkhand, India

R. Khan
Chemical Biology Unit, Institute of Nano Science and Technology (INST), Sahibzada Ajit Singh Nagar, Knowledge City, Punjab, India

Md. M. Ansari et al. (eds.), *Emergence of Sustainable Biomaterials in Tackling Inflammatory Diseases*, Smart Nanomaterials Technology,
https://doi.org/10.1007/978-981-96-2112-5_14

drug delivery systems, microbial-based drug delivery systems, hydrogel-based drug delivery systems, ROS-responsive drug delivery system and osmotically regulated drug delivery system. In the preclinical setting, these advanced drug delivery systems have proven their potential in the management of UC. On the basis of facts and findings, we can say that nano-based targeted drug delivery may be a ray of hope for UC patients.

Keywords Chronic inflammation · Ulcerative colitis · Nanoparticles · Targeted drug delivery

1 Introduction

A large number of people around the globe are affected by chronic inflammatory GI tract disorders known as UC and Crohn's disease which are collectively called inflammatory bowel disease (IBD) [1]. It is estimated that more than a million US people and about 2.5 million people in Europe are suffering from IBD [2, 3]. These inflammatory diseases have a great impact on people around the globe and leading cause of morbidity [4], economic burden from medical costs, and lost productivity [5, 6]. On the basis of the previous analysis of facts regarding these inflammatory diseases, researcher hypothesized that genetically predisposed hosts and commensal bacteria induce an inappropriate immune response that results in IBD. Risk factors related to nutrition, environment, and heredity may contribute to the abnormal immune response, while the exact aetiology is unknown. The colonic mucosa is the only area where UC causes inflammation, unlike Crohn's disease. The afflicted area of the colon varies. While some people have more proximal disease, others have ulcerative proctitis, which is rectum-specific inflammation. UC throughout the entire colon is referred to as pancolitis [7]. The genetic factor is associated with UC onset. Genome-wide association studies (GWAS) revealed almost 200 genetic susceptibility loci for the onset of IBD [8, 9]. Furthermore, the sex factor is associated with Crohn's disease and this disease is more prominent in females as compared to males while in the case of UC, both sexes are equally affected. In the US, the prevalence of UC varies little by race. UC is partly caused by genetic causes. If the other monozygotic twin has UC, the relative risk of one monozygotic twin developing the condition is 95 times higher, and having an afflicted sibling boosts the disease's risk by 4.6 times [10]. The incidence of ulcerative colitis is also influenced by dietary variables. The risk is increased by a diet heavy in meat, fat, and refined sugar, the risk is decreased by a diet high in vegetables [11]. According to the hygiene hypothesis, children in Western industrialized nations with excessive hygienic practices are shielded from natural exposure to bacterial and helminthic antigens, which alters the immune system's reactivity. Clinical suspicion of the diagnosis is supported by endoscopic biopsies. After diagnosis of UC, 5-aminosalicylic acid (5-ASA) is a choice of drug for the treatment. In the case, 5-ASA is unable to treat successfully, in that case, corticosteroid may be given to the patient for the treatment of UC. You can add infliximab to induce and maintain remission. The traditional drugs will give you relief from inflammation but they have systematic toxicity as well. Serious problems with current UC therapies include rapid clearance, significant first-pass metabolism, inadequate or limited medicine absorption, and numerous other problems that lower bioavailability [12, 13]. The major issue for traditional drugs is the GI physiological and biochemical barrier. To tackle this issue, nanotechnology developed a targeted drug delivery system in which main goal will be to deliver therapeutic drugs to colon tissue. In this regard, researchers developed several types of drug delivery systems such as pH-responsive drug delivery system, ROS-responsible drug delivery system, microbiota-targeted drug delivery system, and osmolarity-regulated intelligent drug delivery system. These drug delivery systems are explained in the next section in detail. After the utilization of these novel drug delivery systems for the treatment of UC, the therapeutic efficacy has been enhanced in the preclinical setting. This is due to the ability to overcome intrinsic limits associated with conventional pharmaceuticals [14]. In this chapter, we have explained the pathophysiology of UC, the physiological and biochemical barriers of traditional drug delivery systems and the latest smart drug delivery system for targeted drug delivery against UC.

2 Pathophysiology of UC

The mucosal immune system is the first line of defense against pathogens but does not affect the commensal microbiota under healthy conditions. The mucus layer also plays a significant role in immunological effects by binding with pathogens. The secretory mucin protects colonic tissue. In the case of UC, it is observed that very thin layer of mucin in colonic tissue as well as a depleted mucin layer. In a study, researchers demonstrated that reduced synthesis of human-predominant secretory colonic mucin (MUC2) in inflammatory UC [15]. As we know the epithelial barrier of colonic tissue is a protective barrier, in the case of damage to the epithelial barrier causes increased permeability and loss of control on tight junction. Increased uptake of luminal antigens is made possible by this barrier failure, albeit it's unclear if this dysfunction develops as a result of ongoing inflammation or comes before UC. The intestinal epithelium tissue is a protective barrier that produces an antibacterial peptide called defensin that protects from invader pathogens. Defensin is also a pathological marker of UC. From studies and clinical data, it was observed that elevated expression levels of human beta-defensin in the colonic sample of UC patients. To date, it is not clear that infectious cytokines, microbes, or both are responsible for the enhanced synthesis of defensin [16, 17]. In the pathophysiology of UC, non-pathogenic intestinal microorganisms play a significant role, in a study on genetically modified animal models that develop chronic intestinal inflammation after colonisation with commensal gut bacteria but become disease-free in the absence of bacteria [18]. Researches on humans further substantiate the significance of intestinal microbiota, not only in the pathophysiology of the illness but also possibly in the degree of intestinal inflammation and the clinical picture [19, 20].

In the case of the mucus layer and tight junction being disrupted the intestinal epithelium becomes more permeable which enhances the absorption of luminal antigens. The macrophage and dendritic cells are the principal antigen-presenting cells for T cells. When macrophages and dendritic cells recognise non-pathogenic bacteria through molecular pattern recognition receptors (TLR), they change from a tolerogenic to an active state. Proinflammatory genes are stimulated to be transcriptionally active when NF-kB pathways are activated, which increases the production of cytokines like tumor necrosis factor, interleukins.

Following the antigen presentation by macrophage and dendritic cells, naive CD4 T-cells are presented that facilitate their differentiation into Th2 effector cells, which are identified by the production of IL-4.

As we know the primary function of interleukin is cell-to-cell communication. The NK cell is the prime source of IL-13 and it has a strong correlation with a breakdown of the epithelial barrier. While $\alpha E\beta 7$ binding to its ligand E-cadherin mediates lymphocyte retention in mucosal tissues, interactions between $\alpha 4\beta 7$ and its ligand, mucosal addressin cellular adhesion molecule-1, enhances gut-selective leukocyte homing. In inflammatory bowel illness, therapies aimed at gut-selective trafficking have demonstrated effectiveness (Fig. 1). Circulating leucocytes are drawn in as a result of the over-expression of pro-inflammatory chemokines such as CXCL1, CXCL3, and CXCL8, which feed the cycle of inflammation.

3 Traditional Therapy Versus Nano-Targeted Drug Delivery

5-ASA is a choice of drug for the treatment of chronic UC in humans. In the event that 5-ASA therapy is unsuccessful, corticosteroids may be administered. In Table 1, we have discussed several drugs that are being used in the clinic with detailed features and side effects. You can add infliximab to induce and maintain remission. Serious problems with current UC treatments include rapid clearance, significant first-pass metabolism, limited or inadequate pharmaceutical absorption, and a plethora of other problems that lower bioavailability [12, 13]. On the basis of preclinical studies findings, researchers revealed that nanotechnology-based therapy has proven to enhance therapeutic potential. This is because they have the capacity to surpass intrinsic limitations connected to traditional medications. Furthermore, when it comes to UC therapy, conventional treatments like corticosteroids, immunosuppressive drugs, and antibiotics typically just provide symptomatic relief from inflammation but in long-term use have severe side effect [21, 22].

Some disadvantages of conventional treatment methods are low solubility, lower drug concentrations that enter the bloodstream and cause poor bioavailability, substantial cytochrome P450 (CYP) first-pass liver enzymatic metabolism, drug degradation in harsh gastric acidic environments, and more [23, 24]. Sulfasalazine is a choice drug that is used in the treatment of UC, a sulfa component linked to 5-ASA;

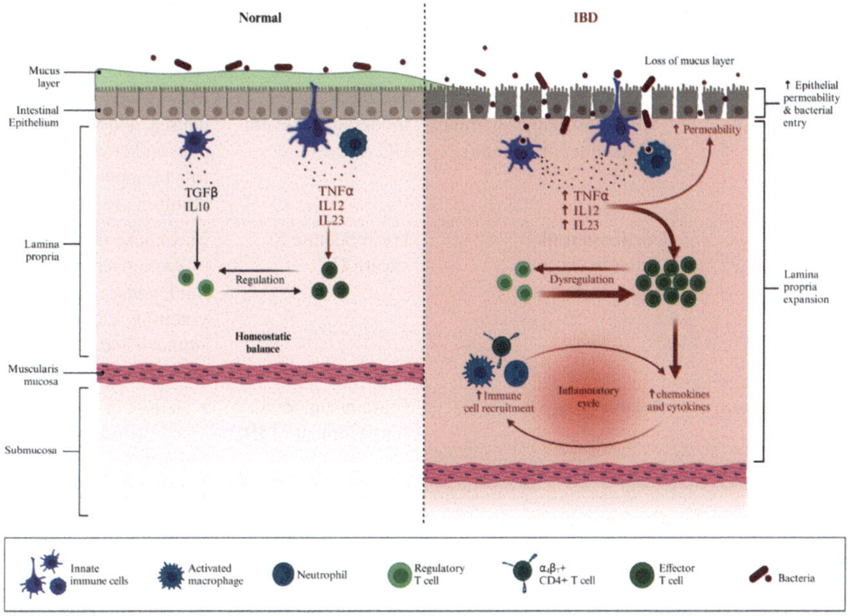

Fig. 1 Pathophysiology of ulcerative colitis or IBD. When mucus film covering the epithelial layer and tight connections are broken, the intestinal epithelium becomes more permeable, increasing the absorption of luminal antigens. Molecular pattern recognition receptors (TLRs) detect non-pathogenic germs, causing macrophages and dendritic cells to go from a tolerogenic to an active state. Image created with Biorender.com

nonetheless, it comes with a number of drawbacks [25]. The therapeutic efficiency of 5-ASA in clinical applications may be limited despite its exploration for treating ulcerative colitis in free form [26, 27]. This is because of several factors, such as poor absorption from the colon [28], extensive metabolism [29], allergy and intolerance [30], and serious unwanted side effects that require patients to discontinue their medication [31]. As a result, more recent approaches to drug targeting have concentrated on strategies to maximise contact between the drug and inflammatory colonic tissue [32, 33] & enhance drug protection against gastric acid and colon delivery selectively and locally [34–36]. In Fig. 2, we represented the different types of nanoparticles that play significant roles in stimuli-based drug released targeting the colon. In the next section of this chapter, we have explained smart drug delivery in detail targeting colitis.

Table 1 List of clinically used medicines for the treatment of IBD or UC

Categories	Drugs	Treatment	Adverse effects
Anti-inflammatory medication	5-ASA (Sulfasalazinemesalamine, balsalazide, olsalazine)	For the treatment of moderate to severe UC	May cause headache, nausea, abdominal pain and cramping, loss of appetite, vomiting, rash or fever
	Corticosteroid (Prednisone and Budesonide)	For moderate to severe UC	May cause retention of sodium salt and fluid, nausea, vomiting, ulcers in the stomach and duodenum
Immune system suppressors	Azathioprine	Immunosuppressant for treatment of IBD	• Nausea or feeling sick, vomiting, diarrhoea, loss of appetite
	Cyclosporine	Reserve for that patient who does not respond to other medicine	Side effects are tremors, restlessness, stomach upset, nausea, cramps, diarrhea, headache
Small molecules	Tofacitinib Upadacitinib Ozanimod	Use in IBD treatment	Increased risk of shingles infection and blood clot
Biological agents	Infliximab Adalimumab Golimumab	Use in the treatment of UC. These are inflammatory cytokine inhibitors	Chest pain, fever, chills, itching, hives, rash, dizziness, fainting, lightheadedness, headache, joint pain, difficulty with swallowing, trouble breathing, or swelling of the face, tongue, and throat

4 Nanoparticles

Nanoparticles (NPs) or nanoscale particles or engineered particles are a wide class of material substances with dimensions between 1 and 100 nm. They can have a broad variety of size-dependent features and be made up of one or more species of atoms [37]. In terms of energy state, nanoparticles are between tiny molecules and bulk materials due to their size range. The natural resource of water is the best example of the tiniest molecules with a diameter of 0.3 nm that helps us better grasp the nanoscale. On the other hand, most mammalian cells have a diameter of 104 nm, but most viruses have a diameter of 20 to 400 nm [38]. In Fig. 3, we represented

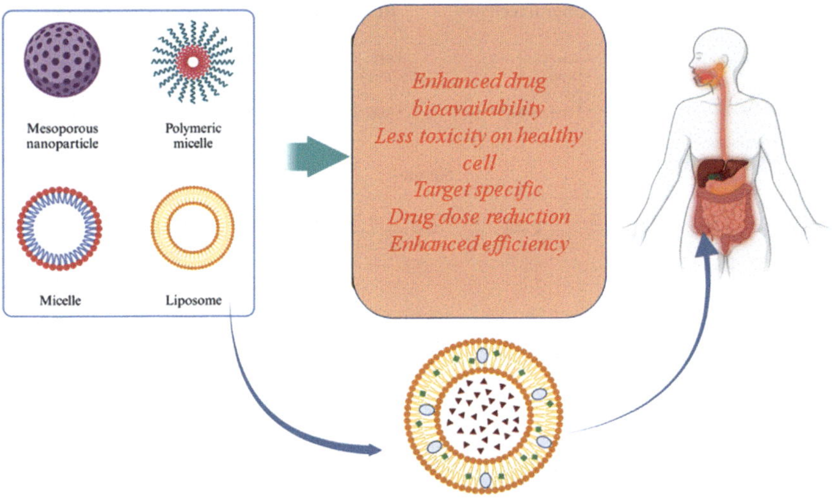

Fig. 2 Schematic representation of smart drug delivery system targeting colon. Image created with Biorender.com

different types of nanoparticles with main features. NPs size is mostly determined by the synthesis method employed. There are mainly two major approaches for making nanoscale particles or NPs. The first method in which the breaking of larger or bulk materials into fine particles is known as the top-down method. The second method which involves assembling gas or liquid atoms into ultrafine NPs through molecular condensation or atomic transformations is known as the bottom-up method [39]. After characterization of NPs, we can observe that great variation between the NPs and bulk materials. NPs have a high surface-to-volume ratio that makes for potential candidates for drug delivery. Furthermore, it has a high surface energy, and unique thermal, electrical, magnetic, mechanical and optical properties that make them more viable for drug delivery. Nowadays, nanomaterials and nanocarriers are more viable candidates in biological research in the form of nanomedicine due to their unique characteristics [40–42]. Day by day, nanotechnology applications in biological sciences in the form of nanomedicine becoming advanced. Nowadays, researches are working on making NPs more viable by surface modification with desired function groups for specific targetability in drug delivery systems. The specific targetability will be definitely a ray of hope for advanced drug delivery systems for critical disease treatment.

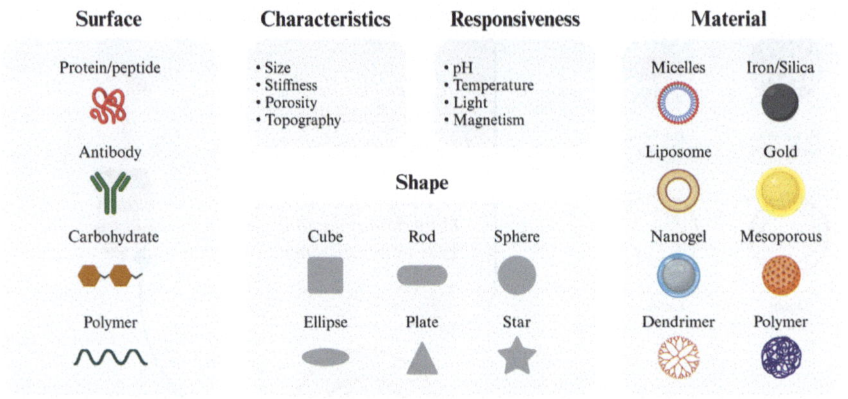

Fig. 3 Schematic representation of different types of nanoparticles with main features. Image created with Biorender.com

5 Targeted Drug Delivery to the Colon: General Barriers/ Challenges

5.1 Epithelium

The stomach is unsuitable for drug absorption due to its harsh acidic environment and enzyme content [43]. The intestinal epithelium, on the other hand, is very absorbent due to its many cell and structural types. The predominant epithelial cells that make up intestinal villi are enterocytes and goblet cells. Enterocytes allow nutrients from food to be absorbed during digestion while regulating the movement of macromolecules and infections [44]. Furthermore, M cells are made up of a few goblet cells and enterocytes, and they are found within the Peyer's patch epithelium. Claudins, occludins, and junction adhesion molecules make up the tight junctions that bind these cells together tightly and provide a robust barrier that prevents chemicals and pathogens from passing through [45]. Transport is impeded by the intercellular gaps' minuscule surface area and the epithelial cells' tightly packed connections via the paracellular pathway under normal conditions [46]. IBD, particularly ulcerative colitis, may alter intestinal permeability, making it easier for nanoparticles to pass past the intestinal epithelial barrier [47]. NPs are transported across cells by a process called transcytosis, in which the particles are absorbed by the cell.

5.2 Enzymes and pH

GI tract pH variation and enzymatic activity mediated biochemical reaction are the major barriers in oral drug delivery in the form of NPs. A wide range of variation in pH is present in the GI tract. It is found that a very acidic environment inside the stomach (pH 1.0–3.0), alkaline inside the duodenum and ileum (pH 6.6–7.5), and neutral inside the colon (pH 7–8) [48]. These variations in pH inside the GI tract is a major barrier to drug targetability because of pH-dependent reactions like oxidation, deamination or hydrolysis hamper drug efficiency [49]. Various liver enzymes are also involved in the detoxification and metabolism of xenobiotics and drug molecules significantly affect drug delivery and targetability of drugs. For instance, several medications that are based on peptides and proteins, like erythromycin, penicillin, and omeprazole, can be quickly broken down by the stomach's high acidity. The actions of digestive tract enzymes on drugs also reduce drug viability. The microbiota present inside the GI tract has a significant impact on the drug delivery system. Over 400 distinct kinds of both aerobic and anaerobic microbes can be found in the colon [50]. Numerous hydrolytic and reductive metabolising enzymes found in these bacteria are in charge of breaking down poly-, tri-, and di-saccharides [51]. Consequently, polysaccharides that target the colon, like guar gum, pectin, and chitosan, are frequently used as medication delivery vehicles.

5.3 *Mucus*

All of the body's mucosal surfaces are lined with viscoelastic mucus, which may act as a barrier to mucosal drug delivery. Mucosal barrier is a potent barrier to drug delivery. Mucus is generally made up of water and mucin, high molecular weight glycoproteins. Mucus creates a steric barrier as a result, and because of its hydrophilic overall structure and hydrophobic domains, it also offers an interaction barrier that restricts the free passage of constituents both through and within the mucus. The first physical barrier that oral drugs must overcome is the mucous layer [52]. Mucus is just like hydrogel containing complex molecules like protein, carbohydrates, lipids, inorganic ions, and dead particulate materials of bacteria [53]. Mucus's dense network of negatively charged, heavily glycosylated mucin fibres is the basis for its barrier qualities. Furthermore, hydrophobic domains that are consistently distributed across mucins bind hydrophobic particles very avidly [54]. Many small and big therapeutic molecules can be delivered orally without being directly blocked by mucus,nonetheless, encapsulating these medications into NPs carriers may enhance their oral bioavailability [55]. The whole and sole drug delivery depends upon the drug carrier capacity, how drug interacts with mucus and its ability to be absorbed by epithelial cells.

5.4 *P-glycoprotein*

P- glycoprotein (P-gp) is critical for active transport and drug delivery of substances with structure diversity that causes poor intestinal permeability and decreased bioavailability through oral administration of drugs. The P-gp is present at the apical surface of cells and plays a significant role in ATP-mediated drug efflux bumps [56]. The P-gp functions as a hydrophobic hoover cleaner. Its hydrophobic hoover cleaner properties help to remove hydrophobic substances or drug molecules from the lipid bilayer membrane. This hydrophobic hoover feature is a major barrier in the drug delivery system. On the basis of preclinical and clinical findings, there are various hydrophobic substance including steroid hormones, anticancer drugs, immunosuppressants, beta-blockers, calcium channel blockers, and cardiac glycosides. The small intestine of the human GI tract has significant expression of P-glycoprotein that play a critical role in controlling the oral bioavailability of candidate drug. The most popular anticancer drugs such as vincristine and paclitaxel are not optimally absorbed by oral administration due to P-glycoprotein [57].

5.5 *Microbiota*

The commensal microorganism also plays a significant barrier in drug delivery. The commensal microorganism has a great impact on GI tract physiological processes such as regulation of metabolism, inflammation, immunity, and local barrier homeostasis [58, 59]. The anaerobic colon contains the majority of the large intestine microbiome, where it ferments different substrates (such as di- and tri-polysaccharides and mucopolysaccharides) that were not broken down in the small intestine to meet its energy requirements. Commensal anaerobic bacteria play a role in the synthesis of hydrolytic enzymes in order to use this roughage as a carbon supply. Numerous reductive and hydrolytic enzymes are produced by the anaerobic bacteria in order to use utilising this roughage as a carbon supply. They can recognise a variety of substances and can respond to them after entry into the colon by producing suitable digestive enzymes [60]. Consequently, prodrugs and biodegradable polymers that are particularly broken down by the microbiota's enzymes can be incorporated into drug delivery systems. The intestinal Commensal microorganism can actively transform plant-based polysaccharides, biodegradable polymers, algae into simple saccharides. These saccharides are then used in drug-delivery systems based on NPs [61, 62]. Redox potential, a measure of total metabolic and microbiological activity, is notable, and can be altered by the diverse microbiota [63]. Thus, a highly precise strategy for colon-targeted medication administration could be employed, one that is induced by changes in the redox potential due to the microbiota.

5.6 *Pharmaceutical Factors*

Pharmaceutical factor is also significant role in drug delivery and must be taking into account during the drug design for targeted drug delivery systems. Peptides are examples of weakly absorbed materials that are more commonly absorbed in the colon due to the colon's extended retention period. Colons have a longer half-life for pharmaceuticals that are poorly absorbed, which results in higher absorption [64]. This is because pharmacological factors influence how long drugs remain inside the colonic tissue. Therefore, during drug designing for targeting the colon, pharmaceutical factors must be considered for effective targeted delivery of drugs. The type of drug and its planned application can be used as carrier screening parameters. The drug's active functional groups, the medication's chemistry, stability parameters, and some other factors that are at play include the drug's partition coefficient.

5.7 Comorbid GI Conditions

When developing a nanosystem or receiving treatment, conditions including diarrhoea, constipation, irritable bowel syndrome, and others should be considered since they can impact the colonic drug delivery system[65]. Irritable bowel syndrome thickens the mucosa, which reduces its surface area and, as a result, lowers the absorption of lipophilic medications in the setting of disorders as barriers. Furthermore, diarrhoea results in poorer absorption and a shorter medication retention period [66]. Therefore, while choosing a nanosystem for a particular patient, these are factors that need to be considered.

6 Nanocarriers for Targeted Drug Delivery

Nanocarriers are like vehicles to carry drugs for delivery purposes. Nowadays, nanomedicine is expanding with the use of nanocarriers for targeted delivery of drugs. There are various natural and synthetic polymers are being used for drug delivery purposes. Gelatin and chitosan are the most commonly used natural polymers for drug delivery. These polymers are very suitable for drug delivery due to their hydrophilic nature, biocompatible to the cell, biodegradable and almost no significant antigenic potentiality. During drug design, it is a very critical part of pharmaceuticals products to be categorised as safe materials by US FDA [67, 68]. Euragit-S100 is an anionic copolymer and generally being used in colon and gastrointestinal-related disease targeting. It has specific features like enhancing the less soluble drug dissolution at a pH value above 7.0. This copolymer is being used in pharmaceutical industries for enteric coating over drug or nanocarriers to bypass stomach acidity drug breakdown for targeted and specific drug delivery to inflamed colonic tissue [69, 70]. Nanotechnology-based platforms have been shown to be significantly helpful in enhancing the efficacy of candidate drugs for the management of inflammation by UC and other intestinal-related diseases. This is due to the ability to overcome intrinsic limits associated with conventional pharmaceuticals [14]. In a study by Ahmad et al. the formulated 5-ASA loaded gelatin NPs and coated with copolymer Eudragit-s 100 to bypass gastric acid surrounding for oral delivery of drug aimed at targeting inflamed colonic tissue. They reported that after delivery of the drug at the site of inflammation, significantly reduced colonic inflammation and significantly reduced inflammatory mediators against the UC animal model [12, 13].

7 Traditional Nanocarriers for the Treatment of UC

When it comes to conventional nanocarrier systems for UC, the use of nanogel can effectively shield the therapeutic drugs or nanocarrier from the acidic pH of the GI tract, allowing the payloads to be released once the nanocarrier successfully enters the inflamed colonic tissue [71]. To prevent surgery, a hydrogel based on an extracellular matrix was created and directed towards UC. The outcomes demonstrated a significant alteration in the macrophage phenotype that caused the colonic mucosa's barriers to be quickly replaced. The produced nanosystem adhered to the colonic tissue, speeding up the healing process [72]. Sun et al. in another study demonstrated that reducing the expression of pro-inflammatory cytokines is significantly helpful in the management of UC [73]. To overcome from the complication of oral delivery in UC treatment, researchers developed a hydrid delivery system. For making a hybrid delivery system, researchers prepared sodium alginate hydrogel and encapsulated it with a siRNA- loaded metal–organic framework. They demonstrated targeted delivery of drugs to colonic tissue and effective treatment of colonic inflammation without deterioration in the highly acidic surroundings of the stomach [74]. According to Singh et al. liposomes are nanoscaffolds that can be used to encapsulate drug candidates that are both hydrophobic and water soluble [75]. They also offer the option of surface modification, which aids in directing the medicine into the inflammatory colon. In a study, researchers demonstrated that solid-lipid nanoparticles have great potential in extended drug release patterns and enhanced protection as the matrix in managing inflammation [22]. Additionally, hydrophobic drugs that target GI tract illness problems are delivered via a self-micro emulsifying system [76]. Naturally occurring curcumin has also proven its potentiality against inflammatory diseases. In a study, researchers demonstrated that curcumin-loaded folate-modified self-emulsified liposomes have great potential against UC [77]. In a study, researchers demonstrated that liposome-loaded co-delivery of nitric oxide and oxymatrine suppressed inflammation in UC by inhibiting macrophage infiltration at the site of inflammation and pro-inflammatory cytokines release [78]. In several studies in preclinical findings, polymeric NPs have demonstrated their capability against inflammatory disease, but there are still some significant obstacles to overcome, including toxicity and scalability issues with production. This issue can be removed by using polymers derived from natural sources [79]. In a study, researchers demonstrated that NPs derived from natural sources such as ginger and *lyceum barbarum* have great anti-inflammatory potential through siRNA delivery [80, 81]. Results from an animal model of UC caused by acetic acid demonstrated that the drug-loaded nanocarrier system was efficacious at larger dosages than at lower doses, demonstrating the effectiveness of the previously described nanosystem

[82]. Similar to this, azathioprine-loaded Phragmites rhizoma polysaccharide selenium nanoparticles were created to target ulcerative colitis lesions. According to Cui et al. in-vivo data demonstrated a considerable down-regulation of cytokines at blood serum and colonic levels, which led to a reduction in UC clinical symptoms and lesions healing [83]. However, present traditional nanocarriers are unable to work in response to external and internal stimuli. Therefore, traditional nanocarriers are unable to release drugs at the right concentration at in site of inflammation. As a result, traditional nanocarriers lack intelligence and need to be functionalized or altered in order to acquire intelligence. The intelligent nanosystems are supposed to react to both internal and external stimuli by avoiding immune system clearance, concentrating in specific regions, and releasing the loaded cargo at the target spot at the required concentration. Furthermore, they must be able to co-deliver medications along with other materials, such as genetic components, imaging agents, etc. There are steps to turn traditional nanocarriers for drug delivery into smart ones, depending on the kinds and applications of nanocarriers [84].

8 Targeting UC with a Smart Nanosystem

Colitis nanotherapeutics are gaining rapid regard and are being used to overcome a number of limitations associated with traditional drug delivery methods, including medication targeting and non-specific distribution with low oral bioavailability. Novel tactics for targeting approaches that may converse new expectations for individuals with UC have been made possible by advances in medical science. The intended use determines the best way to transform conventional nanocarriers into intelligent ones. PEGylation is one method to prevent this kind of nanocarrier clearance; it was described previously. The process of PEGylation involves covalently attaching polymer chains of polyethylene glycol to another molecule, usually a medication or other bioactive component. This polymer has FDA approval and is extremely soluble in water, non-immunogenic, non-antigenic, and non-toxic. PEG-drug conjugates have a number of benefits, including a longer half-life in the body, a lower rate of breakdown by metabolic enzymes, and a decreased or eliminated immunogenicity of the proteins. PEGylation, however, also reduces the amount of medication that cells absorb. Furthermore, specialised ligands should be added to nanocarriers to enable them to discriminate between healthy and sick cells. By attaching functional groups or chemical bonds to nanocarriers, one can increase their sensitivity to both internal and external stimuli. The smart targeted drug delivery system can also take advantage of the co-delivery of drugs and other substances for synergistic and theranostics effects. The application of smart polymer-based drug delivery systems has proven potential for a wide range of use in drug administration in targeting inflammation of colonic tissue. Nowadays, the pH-sensitive anionic and cationic polymers are choices of interest and researchers extensively studied them. There are so many anionic pH-sensitive smart polymers available such as chitosan, poly ethylenemie and N, N-dimethylaminoethyl methacrylamide. The application of an intelligent delivery system for targeted delivery of drugs to the colon has been revealed in several preclinical studies in inflammatory diseases. Such intelligent nanosystems have demonstrated effectiveness in treating a number of GI tract conditions. On the basis of several clinical findings data against inflammatory disease, smart drug delivery system have proven their potential to deliver to specific targets with enhanced bioavailability of drugs.

9 ROS-Responsive Nanodrug Delivery

Reactive oxygen species (ROS) and oxidative stress are closely related. Excess ROS is responsible for oxidative stress and subsequently inflammation. It means that there is inflammation, there is enhanced ROS. If we create ROS-responsive drug delivery system that can easily target the inflammatory tissues. Researchers hypothesised that ROS-responsive drug delivery system may be feature feature-leading drug delivery systems in the management of inflammation and inflammatory diseases. Furthermore, for the management of UC, these types of smart drug delivery systems may be effective in clinical studies that can alter tissue redox potential [85]. The main cause of inflammation is oxidative stress in which researchers demonstrated that imbalance between ROS and the natural antioxidant system. Oxidative stress is the leading cause of activation of various transcription factors that have significant contributions in the activation of inflammatory pathways. Oxidative stress is a unique feature of the inflammatory process in which infiltrated inflammatory cells such as macrophages and neutrophils produce enhanced levels of ROS [86]. There has been a connection between UC and excessive ROS production. It was observed that ROS concentration in mucosal tissue of UC patients is 10–100-fold higher than normal, and this relation is strongly correlated with disease pathology [87]. IBD therapy has great promise for redox-responsive nano-delivery devices, which capitalise on the pathological circumstances associated with the disease. Thioketal NPs (TKNs) taken orally may have been utilised to direct TNF-αsiRNA distribution to intestinal inflammatory areas. Another example of a ROS-responsive polymer is 1, 4-phenyleneacetone dimethylene thioketal (PPADT), this polymer can degrade into TKNs in the response of ROS [88]. A low-molecular-weight TEMPOL was produced using nitroxide radical-containing NPs, which are made of the amphiphilic block copolymer methoxy-poly(ethylene glycol)-b-poly (4-[2,2,6,6-tetramethyl-piperidine-1-oxyl]oxymethylstyrene, or MeO-PEG-b- PMOT [89]. When these particles were given orally to animals suffering from colitis, their particular accumulation occurred in their colons. Researchers have been reported that ROS-responsive system a promising approach for the treatment of inflammatory diseases such as UC [90, 91]. For more validation of this drug delivery system extensive studies are required in preclinical and clinical settings.

10 Hydrogel-Based Drug Delivery

The application of nanotechnology in nanomedicine for drug delivery is due to nano-based drug delivery systems offering targeted delivery, minimum target and sustained release of drugs. The goal of nanotechnology can be fulfilled by creating a controlled drug delivery system by limiting drug dosage as well as frequency of delivery sustaining the drug concentration for therapeutic purposes for the extended time period [92]. In this regard, hydrogel may be a future suitable delivery system for the management of inflammation and inflammatory diseases such as IBD [93]. The hydrogel system is a cross-linked polymer network in water.

Hydrogels have usually between 70 and 99% water, great biocompatibility and can easily encapsulate hydrophilic medicines due to their physical resemblance to biological tissues. Furthermore, there is little chance that drugs may denaturate and aggregate when exposed to organic solvents because the majority of the hydrogel is generated in aqueous solutions [94]. Researchers demonstrated that a hydrogel that cross-linked chitosan and alginate that employed Ca^{2+} and $SO4^{2-}$ ions was sensitive to the colon's environment [95]. It is observed that the myeloperoxidase (MPO) enzymatic activity was decreased in DSS-induced colitis by using improved NPs in a hydrogel system and histopathological alteration also ameliorated by the treatment with this hydrogel system. In short, we can say that this hydrogel system ameliorates colitis symptoms. In a study, IBD was also treated with CD98 siRNA-containing hydrogel-embedded NPs [96]. It was observed that in the DSS-induced mice model of colitis, reduced expression of CD98 in colonic inflammatory tissue by oral treatment with CD98 siRNA/polyethyleneimine (PEI) NPs encapsulated in this hydrogel. A study validates the excellent potentiality of hydrogel in inflammatory disease [97].

11 Intelligent Drug Delivery Driven by Microbiology

The human gut microbiota is the reservoir of several types of microorganisms. The Commensal microorganism is involved in releasing several types of enzymes that help in the metabolism of carbohydrates and other reductive substances [98]. In the smart drug delivery system derived by microbiology, in the presence of colonic response, the delivery system releases the loaded drug after the breakdown of the system by the gut microflora and releases the drug inside the GI at a particular place. Due to this concept, the smart polymer not has broken at the early intestine level due to fewer microfloras. This case involved the use of metronidazole-loaded azo aromatase-based tablets to treat UC. The intended outcome was attained by means of polymer degradation at the colon level by floral enzymes. According to Sharif et al. the microbial flora's nitroreductase, arabinosidase, galactosidase, dehydroxylase, and xylosidase enzymes are among those engaged in these processes [99]. Researchers have reported the effects of administering a tri-peptide-loaded hydrogel against DSS-induced colitis in mice. In a research finding, researchers demonstrated that tripeptide had a lower delivery concentration compared to free drugs, but still, it had similar therapeutic efficacy and little bit side effects [100]. After being exposed to the microbial flora, a guar gum-based system loaded with 5-fluorouracil was administered to the colon. This system demonstrated an effective release of the medication, suggesting the possibility for enzyme responsiveness and the ability to use guar gum in the treatment of colitis [101]. Additionally, researchers reported that from in vitro experimentation, chitosan-modified lipid NPs revealed better drug release with colon-targeting effects loaded with additional esterase-responsive properties. When compared to free dexamethasone, the in vivo distribution results demonstrated substantial drug retention in the colitis-bearing mouse model. Furthermore, researchers revealed that this type of smart enzyme-responsive drug delivery system has great potential against colonic inflammation and colitis [102]. These types of smart drug delivery systems have significant potential in amelioration of inflammation by enhancing the healing process in the case of chronic UC [103]. For the treatment of ulcers, oral gene therapy by the delivery of nucleic acid medication has potent anti-inflammatory potential and a protective approach for UC. Alginate-based polymeric nanocapsules loaded with miR-320 were produced and directed towards the colon. Furthermore, an in vivo imaging result demonstrated that said type of smart drug delivery system reached the site of colonic inflammation after two hours and released miR-320 at the site of inflammation in colonic mucosa after enzyme broken down alginates [104]. The result of these investigations conducted in vitro showed that the medication discharged displayed features of an enzyme response, more precisely, α-amylase responsive release. Meanwhile, in vivo results revealed that the drug was therapeutically effective, reshaped the microbial flora, rapidly absorbed macrophages, and strongly biodistributed at the colonic level, all of which led to the healing of UC [105].

12 Osmotically Regulated Drug Delivery

Extensive research is still under process in the advancement in the field of nanomedicine for targeted delivery and tackling of inflammatory disease in effective ways. Researchers are working on enhancing drug absorption from the intestine and inflammatory site-directed drug delivery. Osmotically controlled drug delivery is a smart oral drug delivery controlled by osmotic pressure. This delivery system can deliver a higher concentration of drug at the site of inflammation with less off-target. Another scaffold that effectively addresses colonic illnesses such as UC is the osmotically regulated intelligent drug delivery system. The osmotic unit consists of five to six push–pull units, each covered in a stiff gelatin capsule [106]. The osmotically controlled system has a drug layer and an osmotic push layer that collectively form a bilayer system that is covered by a semipermeable membrane with an orifice in the drug layer. When the drug administered orally inside the body, the osmotic push compartment swells after absorption of water and form gels. Through an aperture, drugs are released outside [107]. When it comes to ulcerative colitis treatment, the delivery mechanism is made to prevent drug distribution in the small intestine a few hours after stomach emptying [108]. Additionally, after delivering the drug cargo within the colitis site for nearly twenty-four hours, the delivery method achieves a steady drug release rate when it enters the colon [109]. Calcium phosphate (CaP) is a naturally pH-responsive material and is being used in smart drug delivery materials. Researchers demonstrated that siRNA-loaded CaP/PLGA NPs are effective in the form of a smart drug delivery system [110]. These types of smart delivery systems may be a future ray of hope in the tackling of inflammation and inflammatory disease in an effective manner.

13 pH-Responsive Drug Delivery

As we know the GI tract has variations in pH. The delivery of drugs inside the body in response to pH is called pH-responsive drug delivery system. This type of smart drug delivery system has been developed by researchers recently. They offer delivery of optimum drug concentration at the site of inflammation. Such responsive smart nanocarriers, which limit the unwanted distribution of targeting pharmaceuticals into healthy tissues and prevent systemic absorption, have been created by formulating nanotechnology scaffolds. This has allowed for the selective release of therapeutic drugs. Following the consumption of a meal, the pH of an empty stomach rises and eventually reaches a pH of 7 at the colon level after rising in the descending portions of the GI [111]. The smart delivery system can detect the change in colonic tissue pH and deliver cargo at the inflammation site. Liposomes are not recommended for oral drug administration due to their GI system-wide sensitivity to digestion. One method of protecting liposomes during transit through GI is to coat them with polymers. However, there is a lack of research on the targeted delivery of cargo inside the GI tract at the site of inflammation in colonic tissue [112].

Later, researchers extensively worked on pH-responsive polymers and they found that ethyl acrylate (Eudragit-S100) and methylacrylic acid are suitable polymers for pH-responsive and targeted therapeutic drug delivery (Eudragit-L 100) [113]. These polymers undergo protonation or deprotonation in response to pH variations during UC treatment,as a result, they behave in ways that cause them to expand or contract, retaining or releasing the loaded therapeutic drug [114]. Acrylate is an anionic polymer that can expand when the pH becomes high and have insolubility properties when the pH becomes low, therefore these anionic polymers are employed as pH-responsive NPs [115]. The smart delivery system can deliver cargo at a higher colonic pH thanks to this pH-sensitive behaviour, which also shields the payloads from deterioration in an acidic condition [116]. In a study, the researcher used Eudragit S-100 copolymer and PLGA to construct pH-responsive drug delivery system for the delivery of Budesonide [117]. Furthermore, researchers revealed that the release of cargo was lower at lower pH and higher at higher pH which may be definitely useful in the treatment of UC [118]. Likewise, the 5-ASA core–shell

NPs system was fabricated with the goal of achieving UC. According to Wang et al. in vitro investigation finding demonstrated better stability of core–shell system and prevented the release of cargo at the upper GI tract and subsequently sustained the release of cargo at the colonic area [119]. Another drug known as cyclosporine A has strong immunosuppressant potential but limited use in the management of UC due to its systemic adverse effects. To enhance the efficacy of the drug and tackle this issue, cyclosporine A to treat UC, and loaded alginate microparticles coated with copolymer such as Eudragit S100 were made and applied to the colon. The luminal pH fluctuations brought on by disease conditions should also be taken into consideration.

14 Conclusion and Future Direction

This chapter has covered the possible use of synthetic and plant-derived NPs as drug delivery systems for the treatment of UC or IBD. Researchers have worked extensively on the targeted delivery of cargo at the site of inflammation in the response to various stimuli for the treatment of UC. Drug-loaded NPs can be directed towards both the large and small intestines, which are additional sites of the disease, in order to treat colitis using nanotechnological approaches. Research should also be done on smart NPs-based delivery strategies for IBD because they may be used for both diagnostic and therapeutic objectives. Advanced NPs-based delivery systems can conduct multimodal imaging, distribute multiple drugs to provide combination therapies, or function in theranostic space by combining therapeutic and diagnostic properties. Even though a responsive smart nano delivery system like this could increase treatment efficacy and improve patients' quality of life. There hasn't been any conclusive research or factual data regarding the degree of adverse effects of nanocarriers, which calls for more studies on nanotoxicology in preclinical and clinical settings. Furthermore, due to the inflammatory nature of UC, the cells and functions of the immune system are significantly changed. As a result, macrophages, which are important immune system regulators, become biased in favour of inflammation, which causes the system to malfunction. Therefore, more research on ulcerative colitis is needed to determine whether or not responsive smart nanocarriers may be used to change the phenotypes of immune cells. With the exception of pH-responsive systems, other stimuli-responsive nanocarrier systems have certain limitations, such as the upper gastrointestinal tract's harsh acidic environment and high concentration of enzymes, which cause instability problems for the drug. As a result, future research should concentrate on developing effective therapies for UC. This will make the systems easier to transfer from benchtop to clinical settings for the successful treatment of UC.

Conflict of Interest The author declares no competing interest.

References

1. Molodecky NA, Soon IS, Rabi DM, Ghali WA, Ferris M, Chernoff G, Benchimol EI, Panaccione R, Ghosh S, Barkema HW, Kaplan GG (2012) Increasing incidence and prevalence of the inflammatory bowel diseases with time, based on systematic review. Gastroenterology 142(1):46-e30. https://doi.org/10.1053/j.gastro.2011.10.001
2. Burisch J, Jess T, Martinato M, Lakatos PL, ECCO-EpiCom (2013) The burden of inflammatory bowel disease in Europe. J Crohns Colitis 7(4):322–337. https://doi.org/10.1016/j.crohns.2013.01.010
3. Kappelman MD, Rifas-Shiman SL, Kleinman K, Ollendorf D, Bousvaros A, Grand RJ, Finkelstein JA (2007) The prevalence and geographic distribution of Crohn's disease and ulcerative colitis in the United States. Clin Gastroenterol Hepatol Off Clin Pract J Am Gastroenterol Assoc 5(12):1424–1429. https://doi.org/10.1016/j.cgh.2007.07.012
4. Ferguson A, Sedgwick DM, Drummond J (1994) Morbidity of juvenile onset inflammatory bowel disease: effects on education and employment in early adult life. Gut 35(5):665–668. https://doi.org/10.1136/gut.35.5.665
5. Kappelman MD, Rifas-Shiman SL, Porter CQ, Ollendorf DA, Sandler RS, Galanko JA, Finkelstein JA (2008) Direct health care costs of Crohn's disease and ulcerative colitis in US children and adults. Gastroenterology 135(6):1907–1913. https://doi.org/10.1053/j.gastro.2008.09.012
6. Longobardi T, Jacobs P, Bernstein CN (2003) Work losses related to inflammatory bowel disease in the United States: results from the National Health Interview Survey. Am J Gastroenterol 98(5):1064–1072. https://doi.org/10.1111/j.1572-0241.2003.07285.x
7. Talley NJ, Abreu MT, Achkar JP, Bernstein CN, Dubinsky MC, Hanauer SB, Kane SV, Sandborn WJ, Ullman TA, Moayyedi P, AmericanCollege of Gastroenterology IBD Task Force (2011) An evidence-based systematic review on medical therapies for inflammatory bowel disease. Am J Gastroenterol 106(Suppl 1):S2–S26. https://doi.org/10.1038/ajg.2011.58
8. Franke A, McGovern DP, Barrett JC, Wang K, Radford-Smith GL, Ahmad T, Lees CW, Balschun T, Lee J, Roberts R, Anderson CA, Bis JC, Bumpstead S, Ellinghaus D, Festen EM, Georges M, Green T, Haritunians T, Jostins L, Latiano A, Parkes M et al (2010) Genome-wide meta-analysis increases to 71 the number of confirmed Crohn's disease susceptibility loci. Nat Genet 42(12):1118–1125. https://doi.org/10.1038/ng.717
9. Kenny EE, Pe'er I, Karban A, Ozelius L, Mitchell AA, Ng SM, Erazo M, Ostrer H, Abraham C, Abreu MT, Atzmon G, Barzilai N, Brant SR, Bressman S, Burns ER, Chowers Y, Clark LN, Darvasi A, Doheny D, Duerr RH, Peter I et al (2012) A genome-wide scan of Ashkenazi Jewish Crohn's disease suggests novel susceptibility loci. PLoS Genet 8(3):e1002559. https://doi.org/10.1371/journal.pgen.1002559
10. Bengtson MB, Aamodt G, Vatn MH, Harris JR (2010) Concordance for IBD among twins compared to ordinary siblings–a Norwegian population-based study. J Crohns Colitis 4(3):312–318. https://doi.org/10.1016/j.crohns.2009.12.008
11. Hou JK, Abraham B, El-Serag H (2011) Dietary intake and risk of developing inflammatory bowel disease: a systematic review of the literature. Am J Gastroenterol 106(4):563–573. https://doi.org/10.1038/ajg.2011.44
12. Ahmad A, Ansari MM, Mishra RK, Kumar A, Vyawahare A, Verma RK, Raza SS, Khan R (2021) Enteric-coated gelatin nanoparticles mediated oral delivery of 5-aminosalicylic acid alleviates severity of DSS-induced ulcerative colitis. Mater Sci Eng C Mater Biol Appl 119:111582. https://doi.org/10.1016/j.msec.2020.111582
13. Ahmad A, Ansari MM, Kumar A, Bishnoi M, Raza SS, Khan R (2021) Aminocellulose—grafted polycaprolactone-coated core-shell nanoparticles alleviate the severity of ulcerative colitis: a novel adjuvant therapeutic approach. Biomater Sci 9(17):5868–5883. https://doi.org/10.1039/d1bm00877c
14. Fojtů M, Chia X, Sofer Z, Masařík M, Pumera M (2017) Black phosphorus nanoparticles potentiate the anticancer effect of oxaliplatin in ovarian cancer cell line. Adv Funct Mater 27:1701955. https://doi.org/10.1002/adfm.201701955

15. Van Klinken BJ, Van der Wal JW, Einerhand AW, Büller HA, Dekker J (1999) Sulphation and secretion of the predominant secretory human colonic mucin MUC2 in ulcerative colitis. Gut 44(3):387–393. https://doi.org/10.1136/gut.44.3.387
16. Rahman A, Fahlgren A, Sitohy B, Baranov V, Zirakzadeh A, Hammarström S, Danielsson A, Hammarström ML (2007) Beta-defensin production by human colonic plasma cells: a new look at plasma cells in ulcerative colitis. Inflamm Bowel Dis 13(7):847–855. https://doi.org/10.1002/ibd.20141
17. Rahman A, Fahlgren A, Sundstedt C, Hammarström S, Danielsson A, Hammarström ML (2011) Chronic colitis induces expression of β-defensins in murine intestinal epithelial cells. Clin Exp Immunol 163(1):123–130. https://doi.org/10.1111/j.1365-2249.2010.04282.x
18. Taurog JD, Richardson JA, Croft JT, Simmons WA, Zhou M, Fernández-Sueiro JL, Balish E, Hammer RE (1994) The germfree state prevents development of gut and joint inflammatory disease in HLA-B27 transgenic rats. J Exp Med 180(6):2359–2364. https://doi.org/10.1084/jem.180.6.2359
19. Swidsinski A, Ladhoff A, Pernthaler A, Swidsinski S, Loening-Baucke V, Ortner M, Weber J, Hoffmann U, Schreiber S, Dietel M, Lochs H (2002) Mucosal flora in inflammatory bowel disease. Gastroenterology 122(1):44–54. https://doi.org/10.1053/gast.2002.30294
20. Frank DN, Robertson CE, Hamm CM, Kpadeh Z, Zhang T, Chen H, Zhu W, Sartor RB, Boedeker EC, Harpaz N, Pace NR, Li E (2011) Disease phenotype and genotype are associated with shifts in intestinal-associated microbiota in inflammatory bowel diseases. Inflamm Bowel Dis 17(1):179–184. https://doi.org/10.1002/ibd.21339
21. Hu B, Yu S, Shi C, Gu J, Shao Y, Chen Q, Li Y, Mezzenga R (2020) Amyloid-polyphenol hybrid nanofilaments mitigate colitis and regulate gut microbial dysbiosis. ACS Nano 14(3):2760–2776. https://doi.org/10.1021/acsnano.9b09125
22. Lee Y, Sugihara K, Gillilland MG 3rd, Jon S, Kamada N, Moon JJ (2020) Hyaluronic acid-bilirubin nanomedicine for targeted modulation of dysregulated intestinal barrier, microbiome and immune responses in colitis. Nat Mater 19(1):118–126. https://doi.org/10.1038/s41563-019-0462-9
23. Xie F, Ding X, Zhang QY (2016) An update on the role of intestinal cytochrome P450 enzymes in drug disposition. Acta Pharm Sin B 6(5):374–383. https://doi.org/10.1016/j.apsb.2016.07.012
24. Gavhane YN, Yadav AV (2012) Loss of orally administered drugs in GI tract. Saudi Pharm J SPJ Off Publ Saudi Pharm Soc 20(4):331–344. https://doi.org/10.1016/j.jsps.2012.03.005
25. Ransford RA, Langman MJ (2002) Sulphasalazine and mesalazine: serious adverse reactions re-evaluated on the basis of suspected adverse reaction reports to the committee on safety of medicines. Gut 51(4):536–539. https://doi.org/10.1136/gut.51.4.536
26. Le Berre C, Roda G, Nedeljkovic Protic M, Danese S, Peyrin-Biroulet L (2020) Modern use of 5-aminosalicylic acid compounds for ulcerative colitis. Expert Opin Biol Ther 20(4):363–378. https://doi.org/10.1080/14712598.2019.1666101
27. Mizuno S, Ono K, Mikami Y, Naganuma M, Fukuda T, Minami K, Masaoka T, Terada S, Yoshida T, Saigusa K, Hirahara N, Miyata H, Suda W, Hattori M, Kanai T (2020) 5-Aminosalicylic acid intolerance is associated with a risk of adverse clinical outcomes and dysbiosis in patients with ulcerative colitis. Intest Res 18(1):69–78. https://doi.org/10.5217/ir.2019.00084
28. Schröder H, Campbell DE (1972) Absorption, metabolism, and excretion of salicylazosulfapyridine in man. Clin Pharmacol Ther 13(4):539–551. https://doi.org/10.1002/cpt1972134539
29. Sousa T, Yadav V, Zann V, Borde A, Abrahamsson B, Basit AW (2014) On the colonic bacterial metabolism of azo-bonded prodrugsof 5-aminosalicylic acid. J Pharm Sci 103(10):3171–3175. https://doi.org/10.1002/jps.24103
30. Pertuit D, Moulari B, Betz T, Nadaradjane A, Neumann D, Ismaïli L, Refouvelet B, Pellequer Y, Lamprecht A (2007) 5-amino salicylic acid bound nanoparticles for the therapy of inflammatory bowel disease. J Control Release Off J Control Release Soc 123(3):211–218. https://doi.org/10.1016/j.jconrel.2007.08.008

31. Williams C, Panaccione R, Ghosh S, Rioux K (2011) Optimizing clinical use of mesalazine (5-aminosalicylic acid) in inflammatory bowel disease. Ther Adv Gastroenterol 4(4):237–248. https://doi.org/10.1177/1756283X11405250
32. Fassihi SC, Talukder R, Fassihi R (2019) Colon-targeted delivery systems for therapeutic applications: drug release from multiparticulate. monolithic matrix, and capsule filled delivery systems. In: Targeted nanosystems for therapeutic applications: new concepts, dynamic properties, efficiency, and toxicity. American Chemical Society, pp 309–338. https://doi.org/10.1021/bk-2019-1309.ch013
33. Brown TD, Whitehead KA, Mitragotri S (2020) Materials for oral delivery of proteins and peptides. Nat Rev Mater 5:127–148. https://doi.org/10.1038/s41578-019-0156-6
34. Das D, Pham HTT, Lee S, Noh I (2019) Fabrication of alginate-based stimuli-responsive, non-cytotoxic, terpolymric semi-IPN hydrogel as a carrier for controlled release of bovine albumin serum and 5-amino salicylic acid. Mater Sci Eng C Mater Biol Appl 98:42–53. https://doi.org/10.1016/j.msec.2018.12.127
35. Zhang S, Ermann J, Succi MD, Zhou A, Hamilton MJ, Cao B, Korzenik JR, Glickman JN, Vemula PK, Glimcher LH, Traverso G, Langer R, Karp JM (2015) An inflammation-targeting hydrogel for local drug delivery in inflammatory bowel disease. Sci Transl Med 7(300):300ra128. https://doi.org/10.1126/scitranslmed.aaa5657
36. Lautenschläger C, Schmidt C, Fischer D, Stallmach A (2014) Drug delivery strategies in the therapy of inflammatory bowel disease. Adv Drug Deliv Rev 71:58–76. https://doi.org/10.1016/j.addr.2013.10.001
37. Tomalia DA, Khanna SN (2016) A systematic framework and nanoperiodic concept for unifying nanoscience: hard/soft nanoelements, superatoms, meta-atoms, new emerging properties, periodic property patterns, and predictive mendeleev-like nanoperiodic tables. Chem Rev 116(4):2705–2774. https://doi.org/10.1021/acs.chemrev.5b00367
38. Laroui H, Wilson DS, Dalmasso G, Salaita K, Murthy N, Sitaraman SV, Merlin D (2011) Nanomedicine in GI. Am J Physiol Gastrointest Liver Physiol 300(3):G371–G383. https://doi.org/10.1152/ajpgi.00466.2010
39. Viscido A, Capannolo A, Latella G, Caprilli R, Frieri G (2014) Nanotechnology in the treatment of inflammatory bowel diseases. J Crohns Colitis 8(9):903–918. https://doi.org/10.1016/j.crohns.2014.02.024
40. Zhang MZ, Yu Y, Yu RN, Wan M, Zhang RY, Zhao YD (2013) Tracking the down-regulation of folate receptor-α in cancer cells through target specific delivery of quantum dots coupled with antisense oligonucleotide and targeted peptide. Small 9(24):183–4193. https://doi.org/10.1002/smll.201300994
41. Zhang M, Xu C, Wen L, Han MK, Xiao B, Zhou J, Zhang Y, Zhang Z, Viennois E, Merlin D (2016) A hyaluronidase-responsive nanoparticle-based drug delivery system for targeting colon cancer cells. Can Res 76(24):7208–7218. https://doi.org/10.1158/0008-5472.CAN-16-1681
42. Molinaro R, Corbo C, Martinez JO, Taraballi F, Evangelopoulos M, Minardi S, Yazdi IK, Zhao P, De Rosa E, Sherman MB, De Vita A, Toledano Furman NE, Wang X, Parodi A, Tasciotti E (2016) Biomimetic proteolipid vesicles for targeting inflamed tissues. Nat Mater 15(9):1037–1046. https://doi.org/10.1038/nmat4644
43. Odenwald MA, Turner JR (2017) The intestinal epithelial barrier: a therapeutic target? Nat Rev Gastroenterol Hepatol 14(1):9–21. https://doi.org/10.1038/nrgastro.2016.169
44. des Rieux A, Fievez V, Garinot M, Schneider YJ, Préat V (2006) Nanoparticles as potential oral delivery systems of proteins and vaccines: a mechanistic approach. J Control Release Off J Controll Release Soc 116(1):1–27. https://doi.org/10.1016/j.jconrel.2006.08.013
45. González-Mariscal L, Nava P, Hernández S (2005) Critical role of tight junctions in drug delivery across epithelial and endothelial cell layers. J Membr Biol 207(2):55–68. https://doi.org/10.1007/s00232-005-0807-y
46. Shakweh M, Ponchel G, Fattal E (2004) Particle uptake by Peyer's patches: a pathway for drug and vaccine delivery. Expert Opin Drug Deliv 1(1):141–163. https://doi.org/10.1517/17425247.1.1.141

47. Cuvelier CA, Quatacker J, Mielants H, De Vos M, Veys E, Roels H (1993) M cells are damaged and increased in number in inflamed human ileal mucosa. Eur J Morphol 31(1–2):87–91
48. Collnot EM, Ali H, Lehr CM (2012) Nano- and microparticulate drug carriers for targeting of the inflamed intestinal mucosa. J Control Release Off J Controll Release Soc 161(2):235–246. https://doi.org/10.1016/j.jconrel.2012.01.028
49. Agoram B, Woltosz WS, Bolger MB (2001) Predicting the impact of physiological and biochemical processes on oral drug bioavailability. Adv Drug Deliv Rev 50(Suppl 1):S41–S67. https://doi.org/10.1016/s0169-409x(01)00179-x
50. Goldberg M, Gomez-Orellana I (2003) Challenges for the oral delivery of macromolecules. Nat Rev Drug Discov 2(4):289–295. https://doi.org/10.1038/nrd1067
51. Philip AK, Philip B (2010) Colon targeted drug delivery systems: a review on primary and novel approaches. Oman Med J 25(2):79–87. https://doi.org/10.5001/omj.2010.24
52. Xiao B, Merlin D (2012) Oral colon-specific therapeutic approaches toward treatment of inflammatory bowel disease. Expert Opin Drug Deliv 9(11):1393–1407. https://doi.org/10.1517/17425247.2012.730517
53. Liu M, Zhang J, Shan W, Huang Y (2015) Developments of mucus penetrating nanoparticles. J Pharm Sci 10:275–282. https://doi.org/10.1016/j.ajps.2014.12.007
54. Yang M, Lai SK, Yu T, Wang YY, Happe C, Zhong W, Zhang M, Anonuevo A, Fridley C, Hung A, Fu J, Hanes J (2014) Nanoparticle penetration of human cervicovaginal mucus: the effect of polyvinyl alcohol. J Control Release Off J Control Release Soc 192:202–208. https://doi.org/10.1016/j.jconrel.2014.07.045
55. Lundquist P, Artursson P (2016) Oral absorption of peptides and nanoparticles across the human intestine: opportunities, limitations and studies in human tissues. Adv Drug Deliv Rev 106(Pt B):256–276. https://doi.org/10.1016/j.addr.2016.07.007
56. Johnstone RW, Ruefli AA, Smyth MJ (2000) Multiple physiological functions for multidrug transporter P-glycoprotein? Trends Biochem Sci 25(1):1–6. https://doi.org/10.1016/s0968-0004(99)01493-0
57. Amin ML (2013) P-glycoprotein inhibition for optimal drug delivery. Drug target insights 7:27–34. https://doi.org/10.4137/DTI.S12519
58. Nicholson JK, Holmes E, Kinross J, Burcelin R, Gibson G, Jia W, Pettersson S (2012) Host-gut microbiota metabolic interactions. Science (New York, N.Y.) 336(6086):1262–1267. https://doi.org/10.1126/science.1223813
59. Blander JM, Longman RS, Iliev ID, Sonnenberg GF, Artis D (2017) Regulation of inflammation by microbiota interactions with the host. Nat Immunol 18(8):851–860. https://doi.org/10.1038/ni.3780
60. Williams BA, Grant LJ, Gidley MJ, Mikkelsen D (2017) Gut fermentation of dietary fibres: physico-chemistry of plant cell walls and implications for health. Int J Mol Sci 18(10):2203. https://doi.org/10.3390/ijms18102203
61. Zhang L, Sang Y, Feng J, Li Z, Zhao A (2016) Polysaccharide-based micro/nanocarriers for oral colon-targeted drug delivery. J Drug Target 24(7):579–589. https://doi.org/10.3109/1061186X.2015.1128941
62. Qiao H, Fang D, Chen J, Sun Y, Kang C, Di L, Li J, Chen Z, Chen J, Gao Y (2017) Orally delivered polycurcumin responsive to bacterial reduction for targeted therapy of inflammatory bowel disease. Drug Deliv 24(1):233–242. https://doi.org/10.1080/10717544.2016.1245367
63. Zheng H, Powell JE, Steele MI, Dietrich C, Moran NA (2017) Honeybee gut microbiota promotes host weight gain via bacterial metabolism and hormonal signaling. Proc Natl Acad Sci USA 114(18):4775–4780. https://doi.org/10.1073/pnas.1701819114
64. Singh N, Khanna R (2012) Colon targeted drug delivery systems—a potential approach. Pharma Innov J 1:40–47
65. Kadiyam R, Muzib YI (2015) Colon specific drug delivery of tramadol HCl for chronotherapeutics of arthritis. Int J Pharm Invest 5(1):43–49. https://doi.org/10.4103/2230-973X.147232
66. Kotla NG, Gulati M, Singh SK, Shivapooja A (2014) Facts, fallacies and future of dissolution testing of polysaccharide based colon-specific drug delivery. J Control Release Off J Control Release Soc 178:55–62. https://doi.org/10.1016/j.jconrel.2014.01.010

67. Hanlon PR, Frestedt J, Magurany K (2017) GRAS from the ground up: review of the interim pilot program for GRAS notification. Food Chem Toxicol Int J Publ Br Ind Biol Res Assoc 105:140–150. https://doi.org/10.1016/j.fct.2017.03.064
68. Li LL, Xu JH, Qi GB, Zhao X, Yu F, Wang H (2014) Core-shell supramolecular gelatin nanoparticles for adaptive and "on-demand" antibiotic delivery. ACS Nano 8(5):4975–4983. https://doi.org/10.1021/nn501040h
69. El-Sawy HS, Al-Abd AM, Ahmed TA, El-Say KM, Torchilin VP (2018) Stimuli-responsive nano-architecture drug-delivery systems to solid tumor micromilieu: past, present, and future perspectives. ACS Nano 12(11):10636–10664. https://doi.org/10.1021/acsnano.8b06104
70. Andrews GP, Li S, Almajaan A, Yu T, Martini L, Healy A, Jones DS (2019) Fixed dose combination formulations: multilayered platforms designed for the management of cardiovascular disease. Mol Pharm 16(5):1827–1838. https://doi.org/10.1021/acs.molpharmaceut.8b01068
71. Bertoni S, Machness A, Tiboni M, Bártolo R, Santos HA (2020) Reactive oxygen species responsive nanoplatforms as smart drug delivery systems for gastrointestinal tract targeting. Biopolymers 111(1):e23336. https://doi.org/10.1002/bip.23336
72. Keane TJ, Dziki J, Sobieski E, Smoulder A, Castleton A, Turner N, White LJ, Badylak SF (2017) Restoring mucosal barrier function and modifying macrophage phenotype with an extracellular matrix hydrogel: potential therapy for ulcerative colitis. J Crohns Colitis 11(3):360–368. https://doi.org/10.1093/ecco-jcc/jjw149
73. Sun J, Xue P, Liu J, Huang L, Lin G, Ran K, Yang J, Lu C, Zhao YZ, Xu HL (2021) Self-cross-linked hydrogel of cysteamine-grafted γ-polyglutamic acid stabilized tripeptide KPV for alleviating TNBS-induced ulcerative colitis in rats. ACS Biomater Sci Eng 7(10):4859–4869. https://doi.org/10.1021/acsbiomaterials.1c00792
74. Gao M, Yang C, Wu C, Chen Y, Zhuang H, Wang J, Cao Z (2022) Hydrogel-metal-organic-framework hybrids mediated efficient oral delivery of siRNA for the treatment of ulcerative colitis. J Nanobiotechnol 20(1):404. https://doi.org/10.1186/s12951-022-01603-6
75. Singh AK, Yadav PK, Pandey B, Gupta V, Satarudra P, Singh SP (2019) Engineering nanomaterials for smart drug release: recent advances and challenges. In: Applications of targeted nano drugs and delivery systems, pp 411–449. https://doi.org/10.1016/b978-0-12-814029-1.00015-6
76. Alsaad AA, Hussien AA, Gareeb MM (2020) Solid lipid nanoparticles (SLN) as a novel drug delivery system: a theoretical review. Syst Rev Pharm 11:259–273. https://doi.org/10.31838/srp.2020.5.39
77. Zhang L, Zhu W, Yang C, Guo H, Yu A, Ji J, Gao Y, Sun M, Zhai G (2012) A novel folate-modified self-microemulsifying drug delivery system of curcumin for colon targeting. Int J Nanomed 7:151–162. https://doi.org/10.2147/IJN.S27639
78. Tang Q, Zhang W, Zhang C, Guan Y, Ding J, Yuan C, Tan C, Gao X, Tan S (2020) Oxymatrine loaded nitric oxide-releasing liposomes for the treatment of ulcerative colitis. Int J Pharm 586:119617. https://doi.org/10.1016/j.ijpharm.2020.119617
79. Yang M, Zhang Y, Ma Y, Yan X, Gong L, Zhang M, Zhang B (2020) Nanoparticle-based therapeutics of inflammatory bowel diseases: a narrative review of the current state and prospects. J Bio-X Res 3:157–173. https://doi.org/10.1097/JBR.0000000000000078
80. Zhang M, Wang X, Han MK, Collins JF, Merlin D (2017) Oral administration of ginger-derived nanolipids loaded with siRNA as a novel approach for efficient siRNA drug delivery to treat ulcerative colitis. Nanomedicine (Lond) 12(16):1927–1943. https://doi.org/10.2217/nnm-2017-0196
81. Zu M, Song H, Zhang J, Chen Q, Deng S, Canup BSB, Yuan Y, Xiao B (2020) Lycium barbarum lipid-based edible nanoparticles protect against experimental colitis. Colloids Surf B Biointerfaces 187:110747. https://doi.org/10.1016/j.colsurfb.2019.110747
82. Kaur G, Singh SK, Kumar R, Kumar B, Kumari Y, Gulati M, Pandey NK, Gowthamarajan K, Ghosh D, Clarisse A, Wadhwa S, Mehta M, Satija S, Dua K, Dureja H, Gupta S, Singh P, Kapoor B, Chitranshi N, Kumar A, Porwal O (2020) Development of modified apple polysaccharide capped silver nanoparticles loaded with mesalamine for effective treatment of ulcerative colitis. J Drug Deliv Sci Technol 60:101980. https://doi.org/10.1016/j.jddst.2020.101980

83. Cui M, Fang Z, Song M, Zhou T, Wang Y, Liu K (2022) Phragmites rhizoma polysaccharide-based nanocarriers for synergistic treatment of ulcerative colitis. Int J Biol Macromol 220:22–32. https://doi.org/10.1016/j.ijbiomac.2022.07.245
84. Hossen S, Hossain MK, Basher MK, Mia MNH, Rahman MT, Uddin MJ (2018) Smart nanocarrier-based drug delivery systems for cancer therapy and toxicity studies: a review. J Adv Res 15:1–18. https://doi.org/10.1016/j.jare.2018.06.005
85. Talaei F, Atyabi F, Azhdarzadeh M, Dinarvand R, Saadatzadeh A (2013) Overcoming therapeutic obstacles in inflammatory bowel diseases: a comprehensive review on novel drug delivery strategies. Eur J Pharm Sci Off J Eur Fed Pharm Sci 49(4):712–722. https://doi.org/10.1016/j.ejps.2013.04.031
86. Piechota-Polanczyk A, Fichna J (2014) Review article: the role of oxidative stress in pathogenesis and treatment of inflammatory bowel diseases. Naunyn Schmiedebergs Arch Pharmacol 387(7):605–620. https://doi.org/10.1007/s00210-014-0985-1
87. Simmonds NJ, Allen RE, Stevens TR, Van Someren RN, Blake DR, Rampton DS (1992) Chemiluminescence assay of mucosal reactive oxygen metabolites in inflammatory bowel disease. Gastroenterology 103(1):186–196. https://doi.org/10.1016/0016-5085(92)91112-h
88. Wilson DS, Dalmasso G, Wang L, Sitaraman SV, Merlin D, Murthy N (2010) Orally delivered thioketal nanoparticles loaded with TNF-α-siRNA target inflammation and inhibit gene expression in the intestines. Nat Mater 9(11):923–928. https://doi.org/10.1038/nmat2859
89. Vong LB, Tomita T, Yoshitomi T, Matsui H, Nagasaki Y (2012) An orally administered redox nanoparticle that accumulates in the colonic mucosa and reduces colitis in mice. Gastroenterology 143(4):1027–36.e3. https://doi.org/10.1053/j.gastro.2012.06.043
90. Vong LB, Mo J, Abrahamsson B, Nagasaki Y (2015) Specific accumulation of orally administered redox nanotherapeutics in the inflamed colon reducing inflammation with dose-response efficacy. J Control Release Off J Control Release Soc 210:19–25. https://doi.org/10.1016/j.jconrel.2015.05.275
91. Zhang Q, Tao H, Lin Y, Hu Y, An H, Zhang D, Feng S, Hu H, Wang R, Li X, Zhang J (2016) A superoxide dismutase/catalase mimetic nanomedicine for targeted therapy of inflammatory bowel disease. Biomaterials 105:206–221. https://doi.org/10.1016/j.biomaterials.2016.08.010
92. Sharpe LA, Daily AM, Horava SD, Peppas NA (2014) Therapeutic applications of hydrogels in oral drug delivery. Expert Opin Drug Deliv 11(6):901–915. https://doi.org/10.1517/17425247.2014.902047
93. Li J, Mooney DJ (2016) Designing hydrogels for controlled drug delivery. Nat Rev Mater 1(12):16071. https://doi.org/10.1038/natrevmats.2016.71
94. Oliva N, Conde J, Wang K, Artzi N (2017) Designing hydrogels for on-demand therapy. Acc Chem Res 50(4):669–679. https://doi.org/10.1021/acs.accounts.6b00536
95. Laroui H, Dalmasso G, Nguyen HT, Yan Y, Sitaraman SV, Merlin D (2010) Drug-loaded nanoparticles targeted to the colon with polysaccharide hydrogel reduce colitis in a mouse model. Gastroenterology 138(3):843–53.e532. https://doi.org/10.1053/j.gastro.2009.11.003
96. Laroui H, Geem D, Xiao B, Viennois E, Rakhya P, Denning T, Merlin D (2014) Targeting intestinal inflammation with CD98 siRNA/PEI-loaded nanoparticles. Mol Ther J Am Soc Gene Ther 22(1):69–80. https://doi.org/10.1038/mt.2013.214
97. Xiao B, Laroui H, Viennois E, Ayyadurai S, Charania MA, Zhang Y, Zhang Z, Baker MT, Zhang B, Gewirtz AT, Merlin D (2014) Nanoparticles with surface antibody against CD98 and carrying CD98 small interfering RNA reduce colitis in mice. Gastroenterology 146(5):1289–300.e3019. https://doi.org/10.1053/j.gastro.2014.01.056
98. Peng X, Wilken SE, Lankiewicz TS, Gilmore SP, Brown JL, Henske JK, Swift CL, Salamov A, Barry K, Grigoriev IV, Theodorou MK, Valentine DL, O'Malley MA (2021) Genomic and functional analyses of fungal and bacterial consortia that enable lignocellulose breakdown in goat gut microbiomes. Nat Microbiol 6(4):499–511. https://doi.org/10.1038/s41564-020-00861-0
99. Sharif MK, Shah F, Butt MS, Sharif HR (2017) Role of nanotechnology in enhancing bioavailability and delivery of dietary factors. Elsevier, London, United Kingdom, pp 587–618

100. Soni K, Kukereja BK, Kapur M, Kohli K (2015) Lipid nanoparticles : future of oral drug delivery and their current trends and regulatory issues. Int J Curr Pharm Rew Res 7:1–8
101. Kumar B, Kulanthaivel S, Mondal A, Mishra S, Banerjee B, Bhaumik A, Banerjee I, Giri S (2017) Mesoporous silica nanoparticle based enzyme responsive system for colon specific drug delivery through guar gum capping. Colloids Surf B, Biointerfaces 150:352–361. https://doi.org/10.1016/j.colsurfb.2016.10.049
102. Chen SQ, Song YQ, Wang C, Tao S, Yu FY, Lou HY, Hu FQ, Yuan H (2020) Chitosan-modified lipid nanodrug delivery system for the targeted and responsive treatment of ulcerative colitis. Carbohyd Polym 230:115613. https://doi.org/10.1016/j.carbpol.2019.115613
103. Zhang C, Li J, Xiao M, Wang Y, Qu Y, Zou L, Zheng C, Zhang J (2022) Oral colon-targeted mucoadhesive micelles with enzyme-responsive controlled release of curcumin for ulcerative colitis therapy. Chin Chem Lett 33:4924–4929
104. Li X, Yang Y, Wang Z, Ju H, Fu X, Zou L, Li M, Xue Q, Ma H, Meng Y, Zhao L, Qi H, Yu T (2022) Multistage-responsive nanocomplexes attenuate ulcerative colitis by improving the accumulation and distribution of oral nucleic acid drugs in the colon. ACS Appl Mater Interfaces 14(1):2058–2070. https://doi.org/10.1021/acsami.1c21595
105. Li S, Jin M, Wu Y, Jung S, Li D, He N, Lee MS (2021) An efficient enzyme-triggered controlled release system for colon-targeted oral delivery to combat dextran sodium sulfate (DSS)-induced colitis in mice. Drug Deliv 28(1):1120–1131. https://doi.org/10.1080/10717544.2021.1934189
106. Arévalo-Pérez R, Maderuelo C, Lanao JM (2020) Recent advances in colon drug delivery systems. J Control Release Off J Control Release Soc 327:703–724. https://doi.org/10.1016/j.jconrel.2020.09.026
107. Ogueri KS, Shamblin SL (2022) Osmotic-controlled release oral tablets: technology and functional insights. Trends Biotechnol 40(5):606–619. https://doi.org/10.1016/j.tibtech.2021.10.001
108. Aslan B, Ozpolat B, Sood AK, Lopez-Berestein G (2013) Nanotechnology in cancer therapy. J Drug Target 21(10):904–913. https://doi.org/10.3109/1061186X.2013.837469
109. Awad A, Madla CM, McCoubrey LE, Ferraro F, Gavins FKH, Buanz A, Gaisford S, Orlu M, Siepmann F, Siepmann J, Basit AW (2022) Clinical translation of advanced colonic drug delivery technologies. Adv Drug Deliv Rev 181:114076. https://doi.org/10.1016/j.addr.2021.114076
110. Frede A, Neuhaus B, Klopfleisch R, Walker C, Buer J, Müller W, Epple M, Westendorf AM (2016) Colonic gene silencing using siRNA-loaded calcium phosphate/PLGA nanoparticles ameliorates intestinal inflammation in vivo. J Control Release Off J Control Release Soc 222:86–96. https://doi.org/10.1016/j.jconrel.2015.12.021
111. Vinarov Z, Abdallah M, Agundez JAG, Allegaert K, Basit AW, Braeckmans M, Ceulemans J, Corsetti M, Griffin BT, Grimm M, Keszthelyi D, Koziolek M, Madla CM, Matthys C, McCoubrey LE, Mitra A, Reppas C, Stappaerts J, Steenackers N, Trevaskis NL, Augustijns P et al (2021) Impact of gastrointestinal tract variability on oral drug absorption and pharmacokinetics: an UNGAP review. Eur J Pharm Sci Off J Eur Fed Pharm Sci 162:105812. https://doi.org/10.1016/j.ejps.2021.105812
112. Barea MJ, Jenkins MJ, Lee YS, Johnson P, Bridson RH (2012) Encapsulation of liposomes within pH responsive microspheres for oral colonic drug delivery. Int J Biomater 2012:458712. https://doi.org/10.1155/2012/458712
113. Hua S (2014) Orally administered liposomal formulations for colon targeted drug delivery. Front Pharmacol 5:138. https://doi.org/10.3389/fphar.2014.00138
114. Nagpal K, Singh SK, Mishra DN (2010) Chitosan nanoparticles: a promising system in novel drug delivery. Chem Pharm Bull 58(11):1423–1430. https://doi.org/10.1248/cpb.58.1423
115. Liu L, Yao W, Rao Y, Lu X, Gao J (2017) Ph-responsive carriers for oral drug delivery: challenges and opportunities of current platforms. Drug Deliv 24(1):569–581. https://doi.org/10.1080/10717544.2017.1279238
116. Gugulothu D, Kulkarni A, Patravale V, Dandekar P (2014) PH-sensitive nanoparticles of curcumin-celecoxib combination: evaluating drug synergy in ulcerative colitis model. J Pharm Sci 103(2):687–696. https://doi.org/10.1002/jps.23828

117. Makhlof A, Tozuka Y, Takeuchi H (2009) pH-Sensitive nanospheres for colon-specific drug delivery in experimentally induced colitis rat model. Eur J Pharm Biopharm Off J Arbeitsgemeinschaft fur Pharmazeutische Verfahrenstechnik e.V 72(1):1–8. https://doi.org/10.1016/j.ejpb.2008.12.013
118. Oshi MA, Lee J, Naeem M, Hasan N, Kim J, Kim HJ, Lee EH, Jung Y, Yoo JW (2020) Curcumin nanocrystal/pH-responsive polyelectrolyte multilayer core-shell nanoparticles for inflammation-targeted alleviation of ulcerative colitis. Biomacromol 21(9):3571–3581. https://doi.org/10.1021/acs.biomac.0c00589
119. Wang N, Shao L, Lu W, Fang W, Zhang Q, Sun L, Gao S, Zhu Q, Chen S, Hu R (2022) 5-aminosalicylic acid pH sensitive core-shell nanoparticles targeting ulcerative colitis. J Drug Deliv Sci Technol 74:103578. https://doi.org/10.1016/j.jddst.2022.103578

Smart Nanomaterials in Drug Delivery for Cancer Therapy

Asim Rahman, Nazish Naseem, Amreen Fatima, and Md. Faheem Haider

Abstract In today's world, one of the leading diseases people are facing is cancer, and liver cancer is one of them. Though various drugs acting via different mechanisms of action are present in the market as conventional formulations for the treatment of liver cancer they face substantial challenges regarding their bioavailability, low solubility and associated adverse effects which greatly limit their therapeutic efficacies. Nanotechnology has emerged as a promising tool for cancer therapy. Smart nanomaterials, in particular, have gained significant attention due to their unique properties, such as high surface area, biocompatibility, and the ability to be functionalized with targeting moieties. Smart nanomaterials can be designed to selectively deliver drugs to cancer cells, improve drug efficacy, and reduce toxicity. Various studies exhibit that nanocarriers can significantly increase drug bioavailability and increase solubility. The present chapter summarizes nanoformulations used in the treatment of liver cancer. It also highlights the available therapeutics in the market which is used for the treatment of liver cancer along with their side effects and drawbacks and how nanoformulation can overcome these drawbacks of present therapeutics.

Keywords Liver cancer · Anticancer · Nanoformulation · β-catenin · Therapies

1 Introduction

Cancer is a complex and multifaceted disease that can arise from various factors like environmental exposures, genetic mutations, chemical exposures, and lifestyle choices. In general terms, it is defined as the growth and spread of abnormal cells in the body in an uncontrolled manner, which leads to the invasion of nearby organs and tissues and forms tumours. Cancer development involves a series of cellular and molecular events, which include alteration in signalling pathways inside the cell, changes in the genes that are involved in cell proliferation and survival, and the

A. Rahman · N. Naseem · A. Fatima · Md. F. Haider (✉)
Faculty of Pharmacy, Integral University, Lucknow, India
e-mail: fhaider89@gmail.com

Md. M. Ansari et al. (eds.), *Emergence of Sustainable Biomaterials in Tackling Inflammatory Diseases*, Smart Nanomaterials Technology,
https://doi.org/10.1007/978-981-96-2112-5_15

acquisition of invasive and migratory capabilities. Cancer cells have the capability to evade the immune system, which makes them more dangerous, and they can develop strategies to escape elimination and detection by the immune system. This process is known as immune evasion, which is a huge obstacle in the development of cancer treatment [3, 41].

There are various types of cancer, and each type has its own treatment plan [41].

1. Lung cancer
2. Breast cancer
3. Prostate cancer
4. Colorectal cancer

Early diagnosis of cancer is essential in the treatment of cancer and can be done using different methods, which include imaging tests and biopsies [38]. Treatment of cancer can be done through surgery, radiation therapy, targeted therapies, and chemotherapy [34]. Advancements in cancer research have shown us new therapeutic strategies and technologies, including nanoparticles and immunotherapy. Despite advances in the medical field, cancer remains a leading cause of death worldwide [9].

2 Liver Cancer

Hepatocellular carcinoma (HCC), in general, known as liver cancer, is a malignancy of the liver that arises from hepatocytes, which are the primary functional cells of the liver. In the list of cancers worldwide, liver cancer holds the sixth position, and in cancer-related deaths, it holds the 4th position, with the death toll accounting for approximately 782,000 annually. In recent years, liver cancer has been increasing due to the growth of potential risk factors such as alcohol consumption, metabolic disorders, viral hepatitis, non-alcoholic fatty liver diseases, diabetes, and obesity [44]. A key factor in the cause of liver cancer is chronic liver inflammation. and injury, which can occur due to viral infections of the liver such as Hepatitis B and C, consumption of alcohol, or metabolic disorders [47]. The mentioned factors lead to epigenetic alterations and the accumulation of genetic mutations that can disrupt normal cellular functions, leading to proliferation and uncontrolled cell growth inside the liver [46]. In the development of liver cancer, several molecular pathways have been implicated, including PI3K/AKT/mTOR, Wnt/β-catenin and MAPK/ERK signalling pathways. The mentioned pathways play a vital role in regulating cellular processes, which are proliferation, cell growth, and apoptosis. Dysregulation of any of these pathways leads to the development of hepatocellular carcinoma (HCC) or liver cancer [46]. The diagnosis of hepatocellular carcinoma involves a series of imaging tests like ultrasound, computer tomography (CT), and magnetic resonance imaging (MRI) [17]. Biomarkers like blood tests to assess and find out liver function and if there is the presence of tumour markers like alpha-fetoprotein (AFP) and other biomarkers such as des-gamma-carboxy prothrombin (DCP) and glypican-3 (GPC3) can help in finding cancer in its early stages [33]. To confirm the diagnosis of liver cancer, a biopsy is done of the liver tissue using a needle biopsy, which is a minimally invasive procedure.

The first recorded case of HCC was found in Chinese medical literature around 262 BC. Various theories mentioning the treatment and cause of HCC were proposed over the centuries, but in the nineteenth century, significant progress was made that helped in understanding the disease and in developing effective treatments for it. German physician Virchow and Scottish surgeon George Harley in 19th century made an important contribution to understanding the HCC observation, they noticed that HCC was associated with liver injury and chronic liver inflammation and that certain risk factors like consumption of alcohol along with viral infections like hepatitis Band C, increases the risk of developing the disease [4]. Advancements in diagnostic procedures like liver biopsy and medical imaging technology in the twentieth century helped in the faster diagnosis and staging of HCC. In the year 1980, a major breakthrough in the treatment of HCC occurred with the development of alpha-fetoprotein (AFP), also known as a tumour marker for liver cancer, which helped in the earlier detection of cancer and also improved patient outcomes [18].

HCC is a significant concern for health worldwide, as the number of new cases of HCC per year is estimated to be about 905,677, and mortality from HCC was 830,180 in the year 2020. It is the sixth most common cancer and the fourth leading reason for death caused by cancer globally. Cases of HCC are increasing each year in most regions of the world, especially in countries with an income range between low and middle [6]. Incidence cases of various cancers as per the report are listed in Fig. 1.

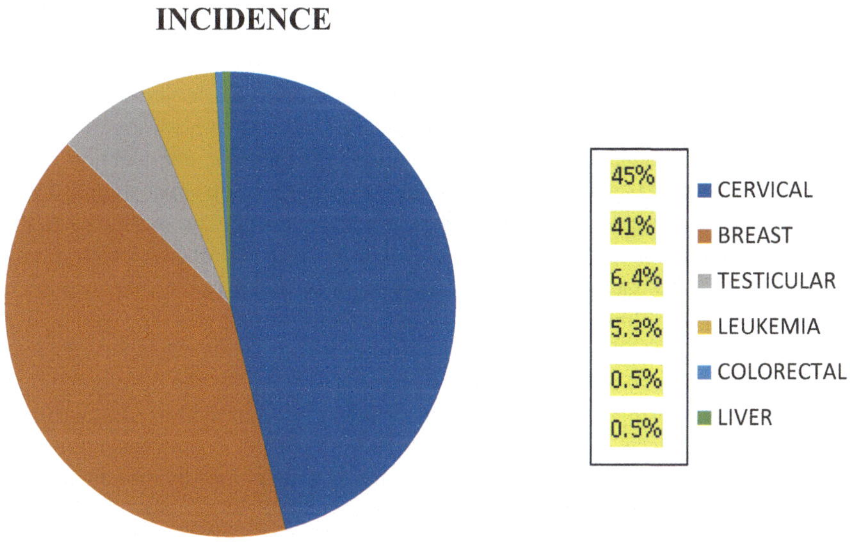

Fig. 1 Figure for incidence cases of various cancers

Cancer Epidemiology (2021) stated the increase in HCC in various regions of the world, and the highest rates are observed in sub-Saharan Africa, Eastern Asia, and South-eastern Asia. The paper also emphasized that the causes of the increase in HCC in recent years were the hepatitis virus (B&C), chronic infection of the liver, consumption of alcohol, and non-alcoholic fatty liver disease (NAFLD).

Hepatocellular carcinoma, commonly known as liver cancer, is a complex disease with multiple underlying causes and mechanisms that contribute to its development. Along with this, various risk factors have also been identified that can increase or cause liver cancer, either in combination with other diseases or individually, promoting disease development [32]. The key mechanisms underlying the development of liver cancer and how each of these mechanisms contributes to the development of liver cancer, such as oxidative stress, genetic and epigenetic alterations, chronic inflammation of the liver, and immune dysfunction. Oxidative stress is caused by an imbalance between oxidants and free radicals in the body, which damages DNA and other cellular components and aids in the development of HCC [8]. TP53 and CTNNB1 are key genes involved in division and cell growth. Mutations in these genes cause genetic and epigenetic alterations which is one of the important mechanisms of HCC development. Histone modification and DNA methylation, which come under epigenetic changes, can also alter the pattern of gene expression and contribute to the development of HCC [43].

A major contributor to the HCC is chronic inflammation of the liver, which is mostly caused by viral infections like hepatitis B and C. hepatitis B and C cause damage to the liver and inflammation, which leads to the formation of tumours over a longer period of time [16]. Immune dysfunction can sometimes play a role in the development of the HCC by allowing cancer cells to evade the immune system and promoting chronic inflammation of the liver. Various studies have shown how T-cells and natural killer cells have a protective role against HC and how immune dysfunction impairs this protective effect.

Chronic injury to the liver and inflammation are the major contributors to the development of liver cancer. These conditions are the result of a variety of causes, which include viral hepatitis like hepatitis B and C. Alcohol use, non-alcoholic fatty liver disease (NAFLD), and hereditary disorders such as hemochromatosis and alpha-1 antitrypsin deficiency are also responsible for liver cancer. Chronic injury to the liver and inflammation leads to the formation of cirrhotic liver tissue or fibrotic tissue, which are major risk factors for the development of liver cancer [35, 37].

Liver cancer development is also influenced by a variety of other factors like angiogenesis, immune dysfunction, and metabolic dysregulation. Metabolic dysregulation, like resistance to insulin and dyslipidaemia, promotes the development of liver cancer by altering the liver microenvironment. The formation of new blood vessels, or angiogenesis, is essential for the spread and growth of liver cancer tumours. Impaired immune surveillance and evasion of immune responses by the cancer cells also contribute to the development of liver cancer.

Available Therapeutics for the Treatment of Liver Cancer (Table 1)

1. **Sorafenib**

   a. Class:

      It comes under the class of small-molecule kinase inhibitors that inhibit multiple kinases, including RAF kinases, c-KIT, vascular endothelial growth factor (VGEF) receptors 1, 2, and platelet-derived growth factor receptor beta, which are used in the treatment of various cancers like unrespectable liver cancer and advanced renal cell carcinoma [48].

   b. Mechanism of Action:

Table 1 Distribution of drugs for various cancers

S. no.	Cancer	Drug	Route of administration	Marketed preparation	Company/ references
1	Liver cancer	Sorafenib	Oral (tablets)	Nexavar tablets	Bayer Zydus Pharma Pvt. Ltd
		Lenvatinib	Oral (capsules)	Lenvima capsules	Eisai Pharmaceuticals India Pvt. Ltd
		Regorafenib	Oral (tablets)	Regonat tablets	Natco Pharma Ltd
		Nivolumab	Intravenous	Nivolumab injection	Oro Pharmaceuticals Pvt. Ltd
2	Breast cancer	Trastuzumab	Intravenous	Herceptin vials	Roche Pharmaceuticals
3	Lung cancer	Pembrolizumab	Intravenous	Keytruda vials	Merck & Co
4	Colorectal cancer	Bevacizumab	Intravenous	Avastin vials	Genentech/Roche Pharmaceuticals
5	Prostate cancer	Leuprolide	Intramuscular/ subcutaneous	Lupron injections or implants	AbbVie
6	Pancreatic cancer	Gemcitabine	Intravenous	Gemzar vials	Eli Lily and Company
7	Ovarian cancer	Carboplatin	Intravenous	Paraplatin vials	Bristol Myers Squibb
8	Leukaemia	Imatinib	Oral (tablets)	Gleevac tablets	Novartis Pharmaceuticals
9	Lymphoma	Brentuximab vedotin	Intravenous	Adcetris vials	Seattle Genetics
10	Brain tumours	Temozolomide	Oral (capsules)	Temodar capsules	Merck & Co

It is an oral targeted therapy used in the treatment of advanced liver cancer, which is also the most common type of liver cancer. The mechanism of action (M.O.A.) of sorafenib is multifaceted, involving the inhibition of various proteins that are involved in the spread and growth of cancer cells. It targets the activity of two key proteins, which are vascular endothelial growth factor receptor (VEGFR) and Raf kinase (C-Raf and B-Raf), involved in the regulation of cell growth, angiogenesis (formation of the new blood vessels that supply oxygen and nutrients to cancer cells), and differentiation [48].

By inhibiting the activity of VEGFR and Raf kinase, sorafenib slows down the spread and growth of cancer cells, preventing the formation of angiogenesis, which is necessary for the tumour to grow. In addition, sorafenib also inhibits the activity of other proteins that are involved in tumour cell survival and proliferation, like c-kit and platelet-derived growth factor receptor (PDGFR) [23].

2. **Lenvatinib**

   a. Class

   It belongs to the class of multikinase inhibitors, which are small molecules targeting multiple receptor tyrosine kinases (RTKs) that are involved in the growth of tumours, metastasis, and angiogenesis [10].

   b. Mechanism of Action

   It is a multi-kinase inhibitor that targets various receptor tyrosine kinases (RTKs), which are involved in the growth of tumours, metastasis, and angiogenesis. Lenvatinib primarily targets the vascular endothelial growth factor receptor (VEGFR). Platelet-derived growth factor receptor (PGDFR) alpha, RET, KIT, and fibroblast growth factor receptor (FGFR) 1–4 [24]

   It also blocks RTK activation by binding itself to the ATP binding site of the receptor, which prevents the downstream signalling pathways that lead to the proliferation of tumours, angiogenesis, and invasion. In addition to this, it also inhibits the activity of other kinases like RET and c-Kit, which are also involved in the progression of tumours [10].

3. **Regorafenib**

   a. Class

   It belongs to multikinase inhibitors or tyrosine kinase inhibitors (TKI). It inhibits several protein kinases that are responsible for tumour signalling pathways [49].

   b. Mechanism of Action

   It is a multi-kinase inhibitor that targets protein kinases that are involved in the signalling pathways of tumours. It also inhibits several key kinases, such as VEGFR1-3 (Vascular endothelial growth factor receptors), FGFR1-2 (Fibroblast growth factor receptors), and PDGFR and PDGFR (platelet-derived growth factor receptors). It also has an anti-angiogenic effect; by inhibiting VEGFR1-3, it disrupts angiogenesis signalling pathways, which reduces blood supply to the tumour and inhibits growth. Regorafenib also shows tumour growth inhibition by inhibiting the RAF and has immunomodulatory effects [49].

c. Class

It belongs to multikinase inhibitors or tyrosine kinase inhibitors (TKI). It inhibits several protein kinases that are responsible for tumour signalling pathways [49].

4. **Nivolumab**

a. Class

It belongs to the immune checkpoint inhibitors or programmed cell death protein 1 (PD-1) inhibitors. Immunotherapeutic drugs are used in various cancer treatments and work by blocking the PD-1 receptors on immune cells, which allows the immune system to better recognize and attack the cancer cells [5].

b. Mechanism of Action

It is an immune checkpoint inhibitor. The mechanism of action involves targeting the programmed cell death protein PD-1 on immune cells. Certain immune cells contain a receptor known as PD-1, which includes T cells. PD-1 plays a role in the regulation of the immune response by inhibiting T cell activity when bound to its ligands, PD-L1 and PD-L2, which are expressed on tumour cells in the tumour microenvironment [40]. It binds to the PD-1 receptors that are present in the T cells, stopping the interaction between PD-1 and ligands. By doing this, nivolumab releases the brakes on the immune system, stopping the inhibitory signals from PD-L1 and PD-L2, and allowing T cells to mount a more robust anti-tumour response. Inhibition of the PD-1 pathway enhances the activity of the T cells against the cancer cells. Nivolumab promotes the activation, proliferation, and infiltration of T cells into tumours, which leads to increased tumour cell killing and improves the antitumor response [5].

Limitations in the Current Modalities of Liver Cancer Treatment

As Sorafenib, Lenvatinib, and Regorafenib are tyrosine kinase inhibitors, they share common limitations, but Nivolumab is an immune checkpoint inhibitor and is mentioned separately.

- Limited Effectiveness: It has been found that sorafenib is less effective in patients suffering from advanced liver cancer and also in patients who have poor liver function [11].

- Adverse side effects: It can cause several adverse effects in patients, which can range from mild to severe, including hand and skin reactions, diarrhoea, fatigue, and hypertension. These side effects may require a reduction in dose or an interruption in treatment [1].
- Resistance to drug: This is a significant challenge in the treatment of liver cancer while using Sorafenib. Various studies conducted showed that resistance to sorafenib can be due to a number of mechanisms, which include genetic mutations and alternative signalling pathways. The common pathway for the resistance of sorafenib in the treatment of liver cancer is the activation of the PI3K/AKT/mTOR pathway [7]. A study reported that the common pathway for the resistance of lenvatinib in the treatment is the activation of the FGFR pathway [19]. A common mechanism in the resistance of regorafenib in the treatment of liver cancer is the activation of the AKT pathway [13].

Nivolumab is an immune checkpoint inhibitor that has been approved for the treatment of advanced cases of patients suffering from hepatocellular carcinoma. It has shown efficacy in improving overall survival rates but also holds potential limitations.

1. Immune-related adverse events: It can cause immune-related adverse events (IRAE), which are a unique set of adverse events that occur due to the immune system. The events can be severe and can include colitis, hepatitis, and pneumonitis [36].
2. Response Rate: It has a relatively slow response rate and depends on the specific type of cancer being treated, although it has shown efficacy in improving the overall survival rate [15].
3. Resistance: It is a significant challenge while administering the patient with nivolumab. The common mechanism of resistance to nivolumab in the treatment of live cancer is the activation of the Wnt/-catenin pathway [12].

3 Emerging Treatment Modalities—Nanoparticles

Many therapeutic procedures are currently available, such as chemotherapy, phototherapy, surgery, and radiotherapy. They do, however, come with some limitations, including subpar results, the potential for relapse, high costs, and the development of drug resistance. However, surgical treatment is still the only option for advanced NSCLC patients, despite the numerous risks and safety concerns associated with surgical procedures. Over recent years, the incorporation of nanotechnology into drug delivery has revolutionized the healthcare industry [21]. Due to their reduced toxicity, stability, increased loading capacity, higher efficacy, and drug loaded tolerability in nanoparticles in comparison to traditional chemotherapy drugs, nanoparticle formulations are regarded as a useful and promising tool for effective cancer treatment [20]. The restriction of traditional chemotherapeutic medication is limited by changing it into a nanodefinition. Multifunction therapies, drug delivery systems, and essential initial diagnosis and bioimaging are all made possible by nanotechnology. Interest in the nanocarrier chemotherapeutic methodology has expanded as nanoparticle-based drug conveyance can deliver hydrophobic medications to evade the trouble of low solvency [42]. Anticancer medications are stacked in the nanoparticles and conveyed through dynamic or detached transport to the particular site of cancer. The advantage is not only its small size but also its properties, like the ability to design from multiple materials, like lipids, which include SLN (solid lipid nanoparticles and liposomes, polymers, which include polymeric nanoparticles (NPs; and inorganic substances, like gold NPs [2]. Figure 2 shows different types of nanoparticles used in the treatment of hepatocarcinoma (HCC).

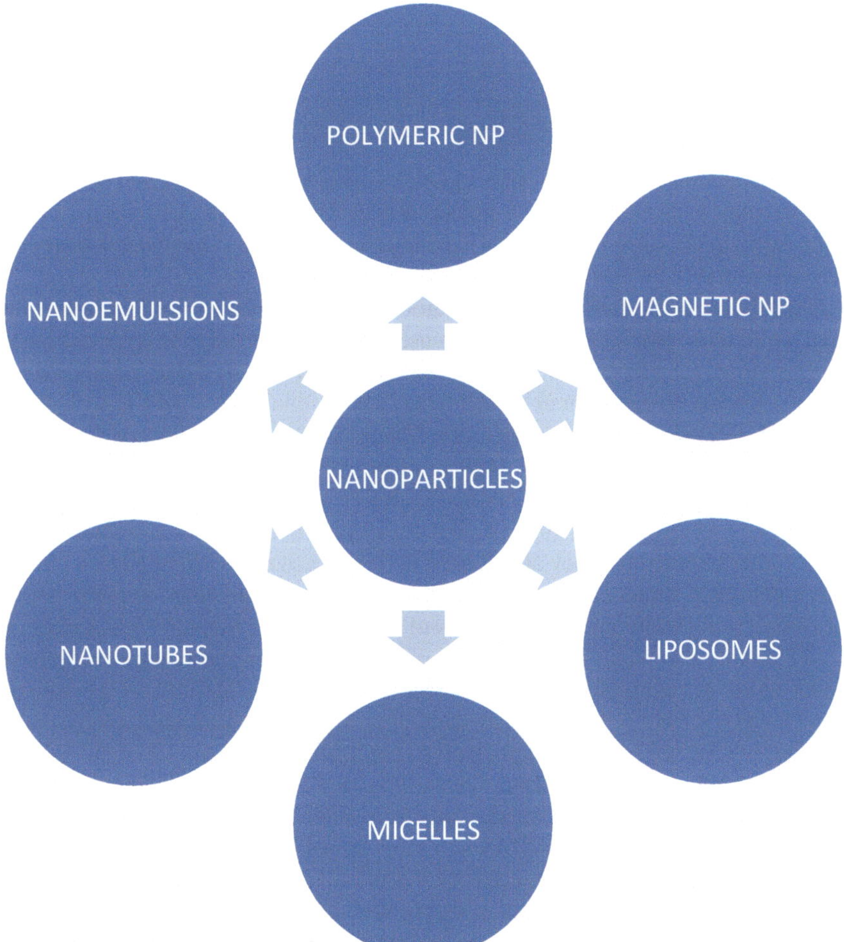

Fig. 2 Nanoparticles used in the treatment of hepatocarcinoma

4 Drug-Loaded Nanoparticles Against Cancer (Table 2)

4.1 Magnetic Nanoclusters

Doxorubicin (Dox)-loaded porous magnetic nanoclusters were created by Jeon et al. (Dox-pMNCs). The MNCs' porous structure, carboxylic groups, and magnetic properties led to good T2-weighted MRI image contrast as well as high drug loading efficiency and prolonged drug release. To target liver tumours in a rabbit model, Dox-pMNCs, Dox, and Dox in iodinated oil were injected via the hepatic arteries. According to MRI and histological analyses, Dox-pMNCs' long-term drug release and retention in iodinated oil greatly increased the killing of liver cancer cells [26].

4.2 Nanoparticles

Increased anticancer efficacy of epirubicin-loaded CXCR4-targeted polymeric nanoparticles in liver tumours was examined by Di-Wen et al. The results of the in vitro cell uptake experiments show that the addition of LFC131 to the nanoparticles may considerably increase the affinity for human hepatic carcinoma cells (HepG2), with a cellular uptake improvement of around three times over non-targeted cells. Confocal laser scanning microscopy was used to spot a particular receptor-mediated uptake. Moreover, HepG2 cells were remarkably toxic to CX-EPNP, and they were able to effectively stop tumour growth [14]. SM5-1-conjugated gold nanoparticles (Au-SM5-1 NPs) were created by (Xibo Ma et al. 2016) who also examined their anticancer effectiveness in HCC both in vitro and in vivo. Using bioluminescent intensity as a gauge, the tumour inhibition rates of Au-SM5-1 NPs for subcutaneous tumour mice were 40.10% 4.34%, 31.37% 5.12%, and 30.63% 4.87% on days 12, 18, and 24 after treatments. They also looked at the effectiveness of Au-SM5-1 NPs in orthotropic HCC tumour models as an anticancer agent. The findings revealed that, as measured by the bioluminescent intensity of the abdomen in tumour-bearing mice on day 31 post-treatment, the inhibition rates of Au-SM5-1 NPs may reach up to 39.64% 4.87% (Xibo Ma et al. 2016). El Kassas researched the HepG2 Cell Line's cytotoxic activity and the bactericidal application of biosynthesized silver nanoparticles with an extract of the red seaweed Pterocladiella capillacea. The diameter of the biosynthesized AgNPs was 11.43.52 nm. According to FT-IR studies, carbonyl groups from proteins and amino acid residues may help in the synthesis and stabilization of AgNPs. At higher doses, the AgNPs exhibited strong cytotoxic activity against the human hepatocellular carcinoma (HepG2) cell line. All of the studied microorganisms were inhibited by the biosynthesized AgNPs, with a clear preference for Bacillus subtilis [28]. Yu-Ji Li and colleagues investigated magnetic nanoparticles containing polymeric carriers and folate-decorated anticancer drugs for the treatment of liver cancer. In BEL7402 cancer cells, FA-conjugated NP displays prolonged drug release and improved cellular uptake. The anticancer efficacy of the targeted NP was higher

Table 2 Nanoparticles of liver cancer with their outcomes

Nanoparticles	Drug	Strategies of research	Release kinetics	Outcomes
Magnetic nanoparticles	Doxorubicin	In vivo	92% within 100 H	Dox pMNCs within iodinated oil improved hepatic carcinoma cell death in rabbit model
Nanogels carrier	Pravastatin	In vitro	78.12 ± 4.81 within 48 H at acidic pH	1. PR-CNG erythrocytes reduced cell viability of HepG2 cells line by 28% compared to unloaded erythrocytes 2. The anticancer effect of pravastatin is proper to hinder cholesterol (membrane component) and blockage of protein lipidation
Polymeric nanoparticles	Epirubicin	In vitro	100% within 150 h at acidic conditions	1. Threefold increase in cellular uptake to Hepg2 than non-targeted drug 2. Cytotoxic effect was improved
Liposomes	Sorafenib and Gadolinium	In vivo	50% within 60 H	1. Sf solubility in Sf-Gd-liposomes was notably increased from 0.21 to 250 lg/ml 2. Sf-Gd-liposomes had soaring antitumor effects in h22 tumour-bearing mice than Sf solution
Gold nanoparticles	Sm5-1 Conjugated	In vitro	–	Au-SM5-1 NPs notably reduced orthotropic tumour growth when contrasted with SM5-1 alone
Silver nanoparticles	Red seaweed	In vitro	–	High cytotoxic activity against HepG2 cell line at high concentrations
Magnetic nanoparticles	Folate Conjugated	In vitro	30% in Pbs (Ph 7.4) and 100% in acetate buffer (pH 5) within 72 H	FA-conjugated-PEG-PLGA-sorafenib NPs reveal assisted drug release and improved twofold cellular uptake in BEL7402 HCC carcinoma cell lines compared with pure drug

(continued)

Table 2 (continued)

Nanoparticles	Drug	Strategies of research	Release kinetics	Outcomes
Polymeric nanoparticles	Docetaxel	Both (In Vitro-In Vivo)	30% in the first two days then continuous release with not more than 53.86% of drug released within/After 14 Days	1. DTX-loaded Gal-pD-TPGS-PLA/NPs (25 mg/ml) reticent the enhancement of HepG2 cells more firmly
Nanoemulsion	Silymarin	In vitro	97.75% within 24 H	1. 89.83% (0.2 l G/Ml) reduced cell viability as contrast to reign 2. ROS production was raised by 121.43%, 156.77 &195.53% ($P < 0.05$) At 0.5, 1 & 2.5 Lg/Ml Of nanoemulsion when contrast with untreated cells 3. Nanoemulsion provoked the cells death by inducing ROS production 4. Improved chromatin condensation against HepG2 cells without side effects on normal cells
Electro sprayed nanoparticles	Gambogic Acid	In vivo	81.47 ± 8.88% within 40 H	1. 82.51% antitumor targeting order for the hepatic survey in HCC model of mice 2. A less tumour invasion &cell lesions in hepatic tissue, recovered hepatic function after 2 weeks of therapy with NPs
Dextran-based nanosize	Curcumin	In vitro	48.80% within 24 Hours at pH 5.8	1. Curcumin loaded DxNPs reveal giant cytoxicity with less drug release to HepG2 cells 2. Free curcumin NPs appear 27.50% but Cur Þ DxNPs Contrast 38.40% cell viability after 24 h
Nanogel	Chitin-Poly(L-Lactic Acid)	In vitro	81.24% within 72 Hours at pH 5.3	When the concentration of Dox rises, there is a dose-dependent toxicity expansion toward HepG2 cells

(continued)

Table 2 (continued)

Nanoparticles	Drug	Strategies of research	Release kinetics	Outcomes
Biodegradable polymeric nanoparticles	Paclitaxel	Both (in vitro and in vivo)	15% on the first day, then continuous release with not more than 35% of drug released within 14 days	1. Potent curing action for targeting and effective endocytosis due to risen drug concentration near the HCC cells by antigen–antibody-specific recognition and enhanced endocytosis by non-specific cell-penetrating peptide pathway 2. 77.3% reticence rates of tumour were noticed at 3 days after final administration against HepG2 xenograft mice model
Micelles	Gum arabic-curcumin conjugate	In vitro	60% at pH 5 and 40% at pH 7.4 in 48 H	Notably improved solubility and stability of curcumin at the doses 25–3.125 G/Ml show low cytotoxicity to HepG2 cells to targeting ability of the galactose groups showed in the gum Arabic
Carbon nanotubes based	Smmc-7721	Both (in vitro and in vivo)	60% within 72 H at pH 5.5	1. Dox-folic acid-chitosan-single walled carbon nanotubes (100 Mg/Ml) effectively kill (68%) the SMMC-7721 cell lines as contrast to pure Dox 2. Targeting the drug delivery system lowers the growth of hepatic carcinoma in nude mice 3. The ransom of Dox From SWNTs is pH- targeted which decreases the loss of Dox during being transported in blood
Nanoformlation	Diethylnitrosamine	In vivo		1. RFLP-PCR analysis using P53 primers preview. Lower (25% in post treated and 20% in at once treated with nanocur against den Þ CCl4) % of polymorphic DNA bands in nanocur treated group 2. Nanocur notably improved biochemical changes in both served groups

(continued)

Table 2 (continued)

Nanoparticles	Drug	Strategies of research	Release kinetics	Outcomes
Nano-Hydroxyapatite	Doxorubicin	In vivo	Drug release was approximately in 13 H at pH 7.4	High dose (95 ± 5 Mg/G) HAp-Dox contrast higher suppression of hyperplastic nodules and a minimum number of preneoplastic lesions in rat model
Nanoformulation	Doxorubicin loaded	Both (in vitro and in vivo)	–	1. Vesicant in HepG2 was improved in HP-NPs contrast with free HP 2. Anticancer efficacy of in-vivo in HepG2 tumour-bearing mice were markedly enhanced by HP-NPs-based Pdt contrast with free HP
Norcantharidn nanoparticles	N-Trimethyl chitosan	In vitro	100% in 12 H	Lac-NCTD-TMC-NPs (24.2%) has the strongest antitumour activity on the murine hepatocarcinoma 22 subcutaneous model

than that of the free medication or the non-targeted one because it effectively reduced the growth of malignant cells. Moreover, improved MRI results show this formulation has strong imaging agent qualities. Ultimately, SRF/FA-PEG-PLGA NP has shown superior anticancer effects by successfully inhibiting colony-forming ability [29]. As a targeted drug delivery method for the treatment of liver cancer, Dunwan Zhu et al. created TPGS-PLA nanoparticles modified with polydopamine and loaded with docetaxel (DTX). In comparison to TPGS-PLA/NPs, pD-TPGS-PLA/NPs, and a DTX formulation that is commercially available (Taxotere®), DTX-loaded Gal-pD-TPGS-PLA/NPs significantly suppressed the proliferation of HepG2 cells. The results of the in vivo biodistribution tests demonstrate that the tumour is the only place where the Gal-pD-TPGS-PLA/NPs are directed. Also, the in vivo anti-tumour effects investigation demonstrated that administering DTX-loaded Gal-pD-TPGS-PLA/NPs dramatically decreased the tumour size in nude mice carrying hepatoma [55]. Jin et al. studied the cytotoxicity of Paclitaxel Incorporated in PLGA Nanoparticles on Human tumour cells under hypoxia. The nanoparticles had an average diameter of 318 5.1 nm and were spherical. The efficiency of encapsulation was 88.52%. In vitro drug release data showed a biphasic pattern. The uptake of cells was seen. When tumour cells were co-cultured with paclitaxel-loaded nanoparticles, it was shown that the drug's bioactivity to prevent cells from entering the G2/M phase had not been lost. In hypoxic MCF-7 and HeLa cells, paclitaxel-loaded nanoparticles had a cytotoxic effect, and this cytotoxicity was more pronounced than that of free paclitaxel. In mice, fluorescent nanoparticles were mostly found in their spleen and liver (Jin et al. 2009).

4.3 *Liposomes*

For medication administration and MRI-guided HCC treatment, Yanan Xiao created liposomes that simultaneously contained sorafenib and gadolinium. The construction of SF/Gd-liposomes made it possible to visualize liposome administration and biodistribution in vivo using MRI guidance. In the in vivo anticancer trials, oral or intravenous treatment of sorafenib solution demonstrated inferior antitumor effects to SF/Gd-liposomes in mice with the H22 tumour. According to these results, SF/Gd-liposomes could be exploited as promising nanocarriers for MRI-guided in vivo visualization of the administration and treatment of HCC [53].

4.4 *Nanoemulsion*

Silymarin nanoemulsion was made by Ahmad et al. to fight human hepatocellular cancer. At a mean particle size of 21.24 nm, the optimized formulation (NE9) displayed a minimum viscosity of 9.59 cps and a maximum drug release (97.75%) in

24 h. Comparing the NE9 formulation to conventional and standard silymarin suspensions, the NE9 formulation also demonstrated a greater AUC, Cmax, and shorter Tmax. The stability analysis also revealed that formulations were significantly more stable at refrigerator temperature than at room temperature. The silymarin nanoemulsion decreased cell survival and increased ROS intensity and chromatin condensation, according to tests using cancer cell lines (Ahmad et al. 2018).

4.5 Nanogels

Bio-responsive chitin-poly (L-lactic acid) composite nanogels for liver cancer were created by Arunraj et al. With an acidic pH, the Dox-chitin-PLA CNGs and the control chitin-PLA CNGs exhibited more swelling and breakdown. The composite nanogels' cytotoxicity towards HepG2 (human liver cancer) cell lines was examined. Also, the outcomes of the in vitro haemolytic assay and coagulation assay support the system's compatibility with blood. Overall, the Dox-chitin-PLA CNGs system may offer a potential method of delivering anticancer medications for the treatment of liver cancer (Arunraj et al. 2014). As a novel delivery method to specifically target liver cancer, Harisa et al. created erythrocytes loaded with Pravastatin chitosan nanogels. The resulting PR-CNG-ER displayed an entrapment efficiency of 36.85%, cell recovery of 66.82%, and release consistent with haemoglobin during a 48-h period. In addition, when compared to control unloaded erythrocytes, PR-CNG-ER had a negative zeta potential, an increase in the percentage of haemolysis, pronounced phosphatidylserine exposure, and altered somatotype morphology. Compared to unloaded erythrocytes, PR-CNG-ER decreased HepG2 cell viability by 28%. According to these findings, PR-CNG-ER is a promising medication delivery system for the treatment of liver cancer (Harisa et al. 2016).

4.6 Micelles

Sarika et al. targeted curcumin administration to hepatocarcinoma cells via improved loading gum Arabic-curcumin conjugate micelles. The solubility and stability of curcumin at physiological pH are both significantly improved by the amphiphilic conjugate's self-assembly into micelles in aqueous medium (900 times compared to that of free curcumin). In comparison to human breast carcinoma (MCF-7) cells, human hepatocellular carcinoma (HepG2) cells have a higher level of conjugate micelles' anticancer activity. Because the galactose groups in gum Arabic are so effective at targeting them, the compound demonstrates increased accumulation and toxicity in HepG2 cells [39].

4.7 *Carbon Nanotubes*

Targeted therapy for SMMC-7721 liver cancer was investigated in vitro and in vivo using a carbon nanotube-based drug delivery system by Zongfei Ji et al. Targeting DDS demonstrated greater pharmacological efficacy than free DOX in killing HCC SMMC-7721 cell lines and suppressing liver cancer development in nude mice. Blood tests and serum biochemical parameter data, along with histological analyses of key organs, show that the targeted DDS had very little in vivo toxicity. Hence, this DDS is encouraging for future cancer therapy due to its great treatment efficacy and few side effects (Zongfei et al. 2012).

4.8 *Nanocarriers*

For the controlled and specialized delivery of curcumin to liver cancer cells, Anirudhan et al. created a nanosized carrier based on dextran. GHDx is a safe carrier, but Cur-loaded GHDx exhibits substantial toxicity with sluggish drug release towards hepatic cells, according to an in vitro cytotoxicity analysis on L929 and Hep G2 cells. The findings demonstrate that the GHDx can be modified to serve as a possible stimuli-sensitive drug delivery system (Anirudhan et al. 2016).

Hassan et al. investigated the therapeutic and chemopreventive effects of nanocurcumin against rat hepatocellular cancer produced by diethylnitrosamine. The group II rats' relative liver weight, serum liver function enzymes, serum sialic acid, vascular endothelial growth factor, and hepatic thiobarbituric acid reactive compounds all increased significantly after receiving DEN. This was accompanied by a considerable drop in tissue antioxidants and serum albumin (GPx, GST, SOD, CAT, and GSH). Particularly in group IV, the nanocur therapy considerably improved such biochemical alterations. Moreover, RFLP-PCR analysis employing p53 primers revealed that group II contained a large proportion of polymorphic DNA bands. Just 25% of the polymorphic DNA bands were expressed by the rats in group V; this number dropped to 20% in group IV. Histopathological analysis of liver slices confirmed these biochemical and genetic observations [22].

5 Conclusion

Over the past 30 years, there has been a significant increase in the survival rate of cancer patients. It has been noted that nanotechnology improves the delivery of several therapeutic agents, including medications and genes. The therapy of numerous liver illnesses was unquestionably improved by the use of liposomes and nanoparticles with homing devices for the targeting of receptors overexpressed in hepatic tissue. The treatment for liver cancer is chemotherapy. The Wnt/b-catenin route, the autophagy pathway, the mTOR/S6K signalling system, and other pathways are some of the pathways employed in the therapy of HCC. It comprises a variety of HCC treatment options, including Phyllanthus, Silymarin, Glycyrrhizin, and Liv 52 and apart from herbal therapeutics it also includes drugs like Cetuximab, Cisplatin, Doxorubicin, Epirubicin, and Erlotinib. These drugs include various limitations such as low bioavailability, low solubility or less logP value, mostly in the case of tablet/capsule dosage forms. Nanoformulation can overcome these limitations. Nanotechnology is a promising approach to resolving several constraints of liver cancer.

Acknowledgements The authors thank the Research and Development Department, Integral University Lucknow, for providing the necessary facilities to complete this review work.

Conflicts of Interest The authors declare that they have no conflicts of interest.

References

1. Abdel-Rahman O, Fouad M (2017) Risk factors for sorafenib-induced adverse events in patients with hepatocellular carcinoma: a systematic review and meta-analysis. Clin Gastroenterol Hepatol 15(12):1947–1958
2. Adnan M, Afzal O, Altamimi AS, Alamri MA, Haider T, Haider MF (2023) Development and optimization of transethosomal gel of apigenin for topical delivery: in-vitro, ex-vivo and cell line assessment. Int J Pharm 631:122506
3. Adnan M, Haider MF, Naseem N, Haider T (2023) Transethosomes: a promising challenge for topical delivery short title: transethosomes for topical delivery. Drug Res
4. Bouchard MJ, Navas-Martin S (2011) Hepatitis B and C virus hepatocarcinogenesis: lessons learned and future challenges. Cancer Lett 305(2):123–143
5. Brahmer JR, Tykodi SS, Chow LQM, Hwu W-J, Topalian SL, Hwu P, Wigginton JM et al (2012) Safety and activity of anti–PD-L1 antibody in patients with advanced cancer. N Engl J Med 366(26):2455–2465
6. Bray F, Ferlay J, Soerjomataram I, Siegel RL, Torre LA, Jemal A (2018) Global cancer statistics 2018: GLOBOCAN estimates of incidence and mortality worldwide for 36 cancers in 185 countries
7. Cabral LKD, Tiribelli C, Sukowati CHC (2020) Sorafenib resistance in hepatocellular carcinoma: the relevance of genetic heterogeneity. Cancers 12(6):1576
8. Cardin R, Piciocchi M, Bortolami M, Kotsafti A, Barzon L, Lavezzo E, Sinigaglia A, Rodriguez-Castro KI, Rugge M, Farinati F (2014) Oxidative damage in the progression of chronic liver disease to hepatocellular carcinoma: an intricate pathway. World J Gastroenterol: WJG 20(12):3078
9. Chen DS, Mellman I (2017) Oncology meets immunology: the cancer-immunity cycle. Immunity 21;46(5):8
10. Cheng AL, Finn RS, Qin S, Han KH, Ikeda K, Piscaglia F, Chao Y et al (2013) Phase III trial of lenvatinib vs sorafenib in first-line treatment of patients with unresectable hepatocellular carcinoma. J Clin Oncol 34(26):2980–2987
11. Cheng AL, Kang YK, Chen Z et al (2009) Efficacy and safety of sorafenib in patients with advanced hepatocellular carcinoma: subanalyses of a phase III trial. J Clin Oncol 27(17):3027–3033
12. Chowdhury PS, Chamoto K, Honjo T (2018) Combination therapy strategies for improving PD-1 blockade efficacy: a new era in cancer immunotherapy. J Intern Med 283(2):110–120

13. Ciombor KK, Bekaii-Saab T (2015) Mechanisms of resistance to regorafenib. Clin Cancer Res 21(4):697–702
14. Di-Wen S, Pan GZ, Hao L, Zhang J, Xue QZ, Wang P, Yuan QZ (2016) Improved antitumor activity of epirubicin-loaded CXCR4-targeted polymeric nanoparticles in liver cancers. Int J Pharm 500(1–2):54–61
15. El-Khoueiry AB, Sangro B, Yau T, Crocenzi TS, Kudo M, Hsu C, Melero I (2017) Nivolumab in patients with advanced hepatocellular carcinoma (CheckMate 040): an open-label, non-comparative, phase 1/2 dose escalation and expansion trial. The Lancet 389(10088):2492–2502
16. El-Serag HB (2012) Epidemiology of viral hepatitis and hepatocellular carcinoma. Gastroenterology 142(6):1264–1273
17. Forner A, Reig M, Bruix J (2018) Imaging of hepatocellular carcinoma: diagnosis, staging and treatment monitoring. Nat Rev Gastroenterol Hepatol 15(4):199–208
18. Galle PR, Forner A, Llovet JM, Mazzaferro V, Piscaglia F, Raoul J-L, Vilgrain V et al (2018) EASL clinical practice guidelines: management of hepatocellular carcinoma. J Hepatol 69(1):182–236
19. Guo J, Zhao J, Xu Q, Huang D (2022) Resistance of lenvatinib in hepatocellular carcinoma. Curr Cancer Drug Targets 22(11):865–878
20. Haider MF, Kanoujia J, Tripathi CB, Arya M, Kaithwas G, Saraf SA (2015) Pioglitazone loaded vesicular carriers for anti-diabetic activity: development and optimization as per central composite design. J Pharm Sci Pharmacol 2(1):11–20
21. Haider MF, Khan S, Gaba B, Alam T, Baboota S, Ali J, Ali A (2018) Optimization of rivastigmine nanoemulsion for enhanced brain delivery: In-vivo and toxicity evaluation. J Mol Liq 255:384–396
22. Hassan SK, Mousa AM, Eshak MG, Farrag AERH, Badawi AEFM (2014) Therapeutic and chemopreventive effects of nano curcumin against diethylnitrosamine induced hepatocellular carcinoma in rats. Int J Pharm Pharm Sci 6(3):54–62
23. Huynh H (2010) Molecularly targeted therapy in hepatocellular carcinoma. Biochem Pharmacol 80(5):550–560
24. Ikeda K, Kudo M, Kawazoe S, Osaki Y, Ikeda M (2018) Lenvatinib in the treatment of advanced hepatocellular carcinoma: an evidence-based review of its place in therapy. Core Evid 13:103–114
25. International Agency for Research on Cancer (IARC) (2020a) Global cancer observatory (GLOBOCAN) 2020: Liver cancer. https://gco.iarc.fr/today/data/factsheets/cancers/11-Liver-fact-sheet.pdf
26. Jeon MJ, Gordon AC, Larson AC, Chung JW, Kim YI, Kim DH (2016) Transcatheter intra-arterial infusion of doxorubicin loaded porous magnetic nano-clusters with iodinated oil for the treatment of liver cancer. Biomaterials 88:25–33
27. Jiang Y, Gao H, Zhuang W et al (2019) Cost-effectiveness analysis of lenvatinib versus (sorafenib as a first-line treatment for unresectable hepatocellular carcinoma in China. Cancer Manag Res 11:7033–7043
28. El Kassas HY, Attia AA (2014) Bactericidal application and cytotoxic activity of biosynthesized silver nanoparticles with an extract of the red seaweed pterocladiella capillacea on the HepG 2 cell line. Asian Pac J Cancer Prev 15(3):1299–1306
29. Khan Z, Haider F, Aslam M, Haider T (2022b) Development and evaluation of myricetin nanoemulsion for liver cancer therapy: in-vitro and cell line study. J Pharm Sci Res 14(9):908–917
30. Khan Z, Haider MF, Naseem N, Siddiqui MA, Ahmad U, Khan MM (2022a) Nanocarrier for the treatment of liver cancer. J Pharm Sci Res 14(11):944–957
31. Li Y, Liang X, Li H, Yang T, Guo S, Chen X (2022) Nivolumab versus sorafenib as first-line therapy for advanced hepatocellular carcinoma: a cost-effectiveness analysis. Front Pharmacol:2661
32. Liu J, Zhang DD, Fang WG et al (2018) Mechanisms of liver cancer and its prevention: a review of the current state and prospects. J Gastrointest Oncol 9(1):112–123

33. Marrero JA, Feng Z, Wang Y, Nguyen MH, Befeler AS, Roberts LR, Reddy KR, Harnois D, Llovet JM, Normolle D, Dalhgren J (2009) Alpha-fetoprotein, des-gamma carboxyprothrombin, and lectin-bound alpha-fetoprotein in early hepatocellular carcinoma. Gastroenterology 137(1):110–118
34. McCormack M, Pilepich MV, Kindts I, Roach M (2018) Surgery for cancer: a century of achievement. Nat Rev Cancer 18(11):695–707
35. Mittal S, El-Serag HB (2013) Epidemiology of hepatocellular carcinoma: consider the population. J Clin Gastroenterol 47:2–6
36. Naidoo J, Wang X, Woo KM, Iyriboz T, Halpenny D, Cunningham J, Hellmann MD et al (2017) Pneumonitis in patients treated with anti-programmed death-1/programmed death ligand 1 therapy. J Clin Oncol 35(7):709–717
37. Nault JC, Zucman-Rossi J (2019) TERT promoter mutations in primary liver tumors. Clin Res Hepatol Gastroenterol 43(3):204–209
38. Oxnard GR, Binder A (2013) The role of imaging in the diagnosis and management of patients with cancer. In: DeVita VT Jr, Lawrence TS, Rosenberg SA (eds) DeVita, Hellman, and Rosenberg's cancer: principles & practice of oncology, 10th edn. Wolters Kluwer Health, pp 36–55
39. Sarika PR, James NR, Kumar PA, Raj DK, Kumary TV (2015) Gum arabic-curcumin conjugate micelles with enhanced loading for curcumin delivery to hepatocarcinoma cells. Carbohyd Polym 134:167–174
40. Sharma P, Allison JP (2015) The future of immune checkpoint therapy. Science 348(6230):56–61
41. Siegel RL, Miller KD, Jemal A (2021) Cancer statistics, 2021. Cancer J Clin (CA) 71(1):7–33
42. Srivastava S, Haider MF, Ahmad A, Ahmad U, Arif M, Ali A (2021) Exploring nanoemulsions for prostate cancer therapy. Drug Res 71(08):417–428
43. Tornesello ML, Buonaguro L, Tatangelo F, Botti G, Izzo F, Buonaguro FM (2013) Mutations in TP53, CTNNB1 and PIK3CA genes in hepatocellular carcinoma associated with hepatitis B and hepatitis C virus infections. Genomics 102(2):74–83
44. Villanueva A (2019) Hepatocellular carcinoma. N Engl J Med 380(15):1450–1462
45. Villanueva A, Llovet JM (2011) Genetic and epigenetic drivers of hepatocellular carcinoma. Gastroenterology 140(5):1478–1493
46. Villanueva A, Llovet JM (2011) Targeted therapies for hepatocellular carcinoma. Gastroenterology 140(5):1410–1426
47. Wallace MC, Preen D, Jeffrey GP, Adams LA (2015) The evolving epidemiology of hepatocellular carcinoma: a global perspective. Expert Rev Gastroenterol Hepatol 10(5):617–630
48. Wilhelm SM, Adnane L, Newell P, Villanueva A, Llovet JM, Lynch M (2008) Preclinical overview of sorafenib, a multikinase inhibitor that targets both Raf and VEGF and PDGF receptor tyrosine kinase signaling. Mol Cancer Ther 7(10):3129–3140
49. Wilhelm SM, Dumas J, Adnane L, Lynch M, Carter CA, Schütz G, Zopf D et al (2011) Regorafenib (BAY 73–4506): a new oral multikinase inhibitor of angiogenic, stromal and oncogenic receptor tyrosine kinases with potent preclinical antitumor activity. Int J Cancer 129(1):245–255
50. World Health Organization (WHO) Cancer country profiles 2020: India. https://gco.iarc.fr/today/data/factsheets/populations/356-india-fact-sheets.pdf
51. World Health Organization (WHO) Cancer fact sheets: liver cancer. https://gco.iarc.fr/today/data/factsheets/cancers/11-Liver-fact-sheet.pdf
52. World Health Organization (WHO) Cancer. https://www.who.int/health-topics/cancer#tab=tab_1
53. Xiao Y, Liu Y, Yang S, Zhang B, Wang T, Jiang D, Zhang N (2016) Sorafenib and gadolinium co-loaded liposomes for drug delivery and MRI-guided HCC treatment. Colloids Surf, B 141:83–92
54. Yin D, Yang Y, Cai H, Wang F, Peng D, He L (2014) Gambogic acid-loaded electrosprayed particles for site-specific treatment of hepatocellular carcinoma. Mol Pharm 11(11):4107–4117

55. Zhu D, Tao W, Zhang H, Liu G, Wang T, Zhang L, Mei L (2016) Docetaxel (DTX)-loaded polydopamine-modified TPGS-PLA nanoparticles as a targeted drug delivery system for the treatment of liver cancer. Acta Biomater 30:144–154

A Novel Drug Delivery System Utilizing Nanoparticles to Treat Neurological Disorders

Maharudra Pratap Singh, Arshi Waseem, Vishwajit Kumar, Ravi Prakash, Abu Junaid Siddiqui, Anand Prakash, and Syed Shadab Raza

Abstract Every year, one million people die from neurological disorders such as Alzheimer's disease, Parkinson's disease, and glioblastoma multiforme. Nanotechnology has the potential to revolutionize the treatment of neurodegenerative diseases by stimulating physiological responses. It reduces adverse effects by provoking, reacting to, and interacting with target sites that use engineered materials and devices that interact with biological systems at the molecular level. The blood–brain barrier (BBB) is a key barrier that inhibits therapeutic drugs from reaching their intended targets. This highlights the importance of using nanotechnology to manipulate biological processes in order to accomplish desired results. Nanoparticles, due to their nanoscale size, are a strong alternative for medication delivery that can penetrate the BBB. These mechanisms not only improve drug stability but also make medication passage through the BBB easier. By providing a more effective technique to dispense drugs, nanotechnology has the potential to improve neurological disease therapy and diagnosis procedures. Drugs can be designed using nanotechnology to perform a range of roles, including traversing the BBB, focusing on specific cells, and stimulating nerve cell renewal and maintenance.

Keywords Neurological disorders · Blood–brain barrier · Nanoparticles · Alzheimer's disease · Parkinson's disease · Cerebrovascular disease · Traumatic brain injury · Glioblastoma multiforme

---

Maharudra Pratap Singh and Arshi Waseem are the authors contributed equally and shares equal first authorships.

M. P. Singh · V. Kumar · A. Prakash (✉)
Department of Biotechnology, School of Life Sciences, Mahatma Gandhi Central University, Motihari 845401, Bihar, India
e-mail: anandprakash@mgcub.ac.in

M. P. Singh · A. Waseem · R. Prakash · A. J. Siddiqui · S. S. Raza (✉)
Laboratory for Stem Cell and Restorative Neurology, Department of Biotechnology, Era's Lucknow Medical College Hospital, Era University, Lucknow, India
e-mail: drshadab@erauniversity.in

Md. M. Ansari et al. (eds.), *Emergence of Sustainable Biomaterials in Tackling Inflammatory Diseases*, Smart Nanomaterials Technology,
https://doi.org/10.1007/978-981-96-2112-5_16

1 Introduction

Neurological disorders such as Alzheimer's disease (AD), Parkinson's disease (PD), cerebrovascular disorders (CVD), and traumatic brain injury (TBI) are becoming more common and affecting about 1.5 billion people [31]. The above-mentioned disorders are defined by the breakdown of neural structure and function, are the leading cause of disability and death worldwide, and account for a major portion of the global disease burden [24]. The blood–brain barrier's (BBB) restrictive role in obstructing the passage of therapeutic substances makes current medical approaches challenging in their quest for a complete cure for various diseases [12].

BBB is necessary to maintain brain homeostasis and defend against the external environment, but it also restricts the infiltration of physiologically active substances, which makes diagnosis and treatment more challenging. Drug administration to the brain requires BBB penetration, which is made possible by lipophilic substances that passively diffuse by crossing the BBB. However, this approach has drawbacks, such as limited absorption and rapid metabolism. Small lipophilic medicines with molecular weights of 500 kDa have low therapeutic effectiveness because of single molecular targeting. Poor lipophilicity and high molecular weight limit the therapeutic impact of macromolecular medicines, including therapeutic genes [21].

Drug delivery for neurological disorders has undergone a paradigm shift since the development of nanomedicine. Drugs can now be effectively transported via nanoparticles, which are made up of polymers, metals, lipids, and crystals and range in size from 1 to 100 nm. Nanoparticles revolutionize drug delivery for neurological disorders by protecting medications, enabling BBB transport, and ensuring controlled release, offering stable, low-immunogenic, and targeted therapy for AD, PD, CVD, and TBI [53, 74].

Because of the restricted BBB, these offer severe barriers to traditional drug administration. Nanoparticles, which range in size from 1 to 100 nm and are made of a variety of materials such as polymers, metals, lipids, and crystals, have unique features that make them efficient drug carriers. They can encapsulate therapeutic chemicals to prevent their enzymatic degradation, target specific brain cells, navigate the BBB, and release medications at specified pH levels. The resulting nanoengineered compounds are very stable, immunogenic, and have the ability to change surface characteristics. Nanoparticle-based drug delivery systems offer a viable path for expanding the treatment, imaging, and diagnostics of neurological disorders by increasing drug bioavailability and improving pharmacokinetics and biodistribution inside the central nervous system (CNS). The complexities of neurological diseases can be addressed with this innovative method, which also opens the way for more targeted and efficient therapy approaches.

2 Neurological Disorders

A wide range of diseases affecting the nervous system, which includes the brain, spinal cord, and nerves, are referred to as neurological disorders. These diseases are a leading cause of mortality and disability [30]. These disorders are characterized by the gradual degeneration and death of neurons in specific brain regions, as well as the presence of anomalous synapses, progressive neuronal loss, and cerebral protein aggregates. [90] describe their debilitating nature as being characterized by cognitive, memory, and motor deficits. Studies reveal that neuroinflammation has a role in the pathological breakdown of the BBB, leading to increased permeability of the BBB and the loss of endothelial proteins connected to tight junctions [101].

3 Alzheimer's Disease

Alzheimer's disease (AD) accounts for up to 80% of dementia diagnoses and is the main cause of dementia. In addition to gradually producing cognitive decline, this degenerative neurological disorder affects language, concentration, and problem-solving abilities. It is largely characterised by early and severe memory loss [55]. The buildup of these neurofibrillary tangles is thought to be caused by the complex interaction of genetic and environmental factors [22]. On the other hand, senile or neuritic plaques, which are made up of proteinaceous extracellular aggregates of amyloid beta (Aβ), are closely associated with AD, causing cognitive decline and obstructing communication between brain cells [11]. The amyloid cascade theory postulates that the development of Aβ plaques is the main cause of disease pathology, prompting the creation of treatment plans that try to lower brain Aβ levels [59]. AD is characterized by memory loss, which is brought on by cholinergic neurons, particularly in the Meynert nucleus basalis, losing their function. These neurons produce the essential neurotransmitter acetylcholine (ACh) [71]. In order to compensate for the loss of cholinergic neurons, symptomatic medications like cholinesterase inhibitors, such as donepezil, galantamine, and rivastigmine, improve the bioavailability of ACh by blocking the breakdown of the acetylcholinesterase (AChE) enzyme in the synaptic cleft [29]. However, negative effects accompany these drugs' poor efficacy, particularly when taken in greater dosages [64]. The fourth FDA-approved treatment for AD, memantine, works as an N-methyl-D-aspartate (NMDA) receptor antagonist to stop glutamate from entering the cell, thereby reducing excitotoxicity and neuronal death. Memantine can improve cognitive performance in patients with moderate-to-severe AD, but it can also produce headaches and dizziness [43]. Low penetration through the BBB may be one reason why therapeutic drugs that target the Aβ pathway, like tramiprosate, semagacestat, and tarenflurbil, do not show effective results, even though they had good results in clinical trials [28].

4 Parkinson's Disease

Parkinson's disease (PD) constitutes a significant neurodegenerative ailment affecting the elderly, with an incidence of 1–2 per 1000 people and peaking at 4% in older age groups [89]. This neurological condition induces both motor and non-motor deficits, including bradykinesia, resting tremors, rigidity, and postural instability. Its pathogenesis involves the selective degeneration of dopaminergic neurons in the substantia nigra pars compacta, impacting the basal ganglia and other brain regions. PD is categorized as an α-synucleinopathy, featuring Lewy bodies and intra-neuronal inclusions. Mutations in the synuclein alpha (SNCA) gene, which encodes α-synuclein, contribute to its onset [62]. Elevated risk factors encompass oxidative stress, environmental toxins, mitochondrial dysfunction, inflammation, immune system dysfunction, genetic mutations, and epigenetic changes [5]. In PD, dopaminergic neurodegeneration results in apoptosis or neuronal death in the substantia nigra. The neurodegenerative process of PD may be caused by inflammation, toxins from the environment, mitochondrial failure, microtubular instability, and hereditary predisposition [61]. Interestingly, a major factor in PD is the kynurenine pathway (KP), which is essential for the breakdown of the amino acid tryptophan and influences the regulation of the immunological response. The accumulation of excitotoxin quinolinic acid (QUIN) in neurons, leading to cell death, is associated with KP dysregulation, contributing to neuroinflammation [100]. Therapeutic options for PD primarily involve dopaminergic medications, which alleviate symptoms without impeding neurodegeneration or disease onset [63]. Levodopa, a dopamine agonist used to address motor impairments induced by bradykinesia, is often combined with carbidopa to enhance absorption. While effective in reducing Parkinson's symptoms, it may elicit dyskinesia, characterized by involuntary movements [78].

5 Cerebrovascular Disease

Conditions affecting the brain's vasculature that involve areas affected by ischemia or haemorrhage, either temporarily or permanently, are collectively referred to as cerebral vascular disease (CVD) [10]. This category includes strokes, aneurysms, aortic dissections, carotid stenosis, and vertebral stenosis. Strokes, prioritized in therapeutic interventions, stand as the second leading cause of death and disability worldwide [69]. Vascular injuries leading to acute impairments result in decreased blood flow, while hemorrhagic strokes occur from artery rupture, often associated with hypertension or vascular anomalies. Timely recognition and treatment are imperative for minimizing long-term effects [68]. Modifiable risk factors for ischemic and hemorrhagic stroke include diabetes, atrial fibrillation, hyperlipemia, metabolic syndrome, alcohol use, chronic inflammation, and hypertension, as well as non-modifiable risk factors such as age and heredity. To treat acute ischemic strokes, doctors remove thrombi or emboli, revascularize, and perfuse the brain tissue that was damaged [14].

Intravenous thrombolytic drugs, such as recombinant tissue plasminogen activator, are the first choice for treating ischemic strokes that are caused by thrombi. Anticoagulants and antiplatelets are recommended for preventing recurrent cerebrovascular episodes in chronic management [76].

6 Traumatic Brain Injury

Traumatic brain injury (TBI), sometimes referred to as a "silent epidemic," is thought to afflict 69 million people worldwide and has a substantial negative influence on one's physical, mental, and emotional well-being. It can result from falls, automobile accidents, sports injuries, and other incidents [18]. TBI survivors frequently experience motor and cognitive deficits, which can lead to a diminished quality of life and long-term functional incapacity. The spectrum of impairments includes cerebral challenges and tandem gait issues; forceful collisions, brain acceleration, and deceleration are known to have an especially detrimental effect on mental function [50]. TBI-related lesions are put into groups based on how they were caused by mechanical forces during trauma, like cranial deceleration, blast waves, or piercing wounds [41]. Ionic anomalies, neuroinflammation, neuronal death, mitochondrial malfunction, and progressive neurodegeneration are all influenced by secondary brain injuries that arise from primary brain injuries. Therapy for trauma should concentrate on treating secondary damage because basic injuries cannot be healed [34]. According to [44], the main goals of TBI therapy are to maximize cerebral oxygenation, monitor cerebral perfusion pressure, and administer critical care in order to keep an eye on hemodynamics and ventilatory functions. Regretfully, effective neuroprotective or curative therapies for TBI are not yet available. Many efforts have been made to develop novel medications and treatment modalities during the past thirty years, however, not all of them have proven to be effective for each person.

7 Glioblastoma Multiforme

Glioblastoma multiforme (GBM) is a common malignant primary brain tumor with a low 5-year survival rate of 7.2%. It is most commonly diagnosed in individuals 75–84 years of age. GBMs are notoriously invasive because of genetic diversity and the BBB protective barrier, which makes therapy extremely difficult. GBM cells, which have the abilities of DNA repair and self-renewal, frequently defy standard treatments, including surgery and chemotherapy [58]. To develop successful therapeutics, a comprehensive understanding of molecular pathways is necessary [94]. Inadequate resection, genetic diversity, a robust BBB, and an immunosuppressive environment are among the challenges in treating GBM. Drug delivery methods based on nanotechnology show promise for addressing BBB issues and improving treatment outcomes. Complete excision of GBM is difficult due to its infiltrative nature and the existence of glioma-initiating cells (GICs) within large hypoxic areas. This could lead to tumors that are aggressive and resistant to conventional treatments like radiation and chemotherapy [66]. Based on genetic and epigenetic markers, GBMs are categorized into four categories by the Cancer Genome Atlas (TCGA): mesenchymal, classical, proneural, and neural. The management of GBM becomes more complex as a result of the distinct clinical symptoms and treatment responses exhibited by each subtype, which are identified by particular genetic changes [67]. Chemotherapeutic medications are still challenging to provide over the BBB, which protects the CNS's extracellular and circulatory compartments. Transporter proteins, tight junction proteins, and immature blood vessels in the BBB impede the delivery of immunotherapeutic and chemotherapeutic medicines to their intended locations. In an effort to improve treatment outcomes, researchers are looking into alternative drug delivery techniques such as targeted ultrasound and nanoparticles [45].

8 Neurological Diseases and the BBB

The BBB controls the flow of chemicals between the brain and blood through physiological and molecular mechanisms. Pericytes, astrocytes, mural cells, and endothelial cells are the constituents of the BBB [27]. This keeps the CNS separate from the rest of the body's circulation [72]. Two examples of endothelial junction complexes that significantly impact this selective permeability are adherens junctions and tight junctions. Adherent junctions allow for cell-to-cell communication, whereas tight junctions produce a continuous blood vessel lumen. By acting as barriers to keep harmful substances and cells out of the brain, these complexes delay the onset of neurological diseases and emphasize the importance of the BBB in maintaining normal brain function. The presence of many barriers in the CNS, including the blood-cerebrospinal fluid barrier, blood-spinal cord barrier, and blood-retinal barrier, makes efforts to create effective treatments for neurological disorders extremely difficult. These barriers prevent medications from reaching their intended targets and protect the CNS from toxic chemicals [52]. Obstacles in drug administration often lead to therapeutic failures, creating a large gap in the field of medication. Chemicals can pass through the BBB by different pathways, such as paracellular tight junctions and transcellular lipophilic diffusion. Larger molecules and polar substances require additional channels such as transport proteins or receptor-mediated endocytosis, whereas small, lipophilic molecules usually diffuse via the transcellular-lipophilic diffusive pathway. These mechanisms are being investigated in order to improve the administration of CNS medications [26]. Therapeutic systems based on nanoparticles offer a new way to deliver helpful chemicals with better pharmacokinetics, making existing drugs for treating neurological conditions more effective. These technologies overcome the drawbacks of conventional delivery methods by effectively delivering active compounds to the CNS and minimizing the gap in medications. Clinical trials are being conducted on nanoformulations containing metal nanostructures, liposomes, extracellular vesicles, and red blood cell membranes. By using molecular targeting to

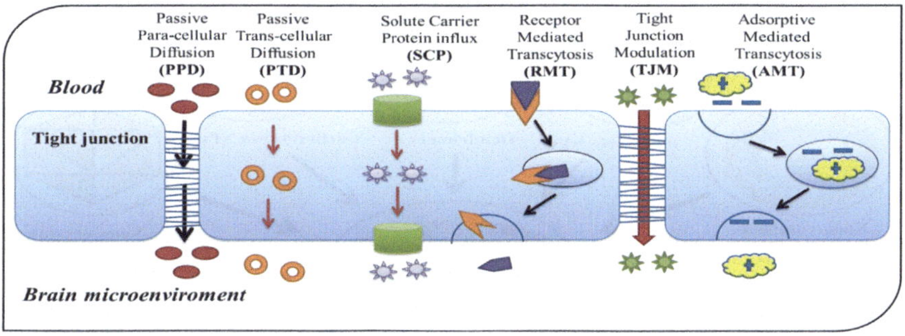

Fig. 1 Tight junctions found in the BBB produce a smooth blood vessel lumen. Chemicals can move across the BBB more easily through a number of processes, including adsorptive-mediated transcytosis, receptor-mediated transcytosis, tight junction regulation, passive paracellular diffusion, and passive transcellular diffusion

treat neurological disorders, these materials increase specificity, safety, and efficacy [60] (Fig. 1).

9 Mechanism for Crossing the BBB by Nanoparticles

BBB limits the entry of substances into the brain; treating neurodegenerative disorders like PD and AD is difficult. By enabling accurate and active transport of medications across the BBB, nanocarrier-based platforms have revolutionized traditional therapy [77]. Due to their special qualities, such as their large surface area to volume ratio, nanoparticles are the perfect vehicle for drugs since they ensure both quick drug release and excellent bioavailability [9]. Also having favorable physiological properties, which include reactivity, strength, sensitivity, solubility, and stability, nanoparticles appear to be a promising option for penetrating the BBB. This allows them to pass through the barrier with ease and solves a significant problem for drug transport and CNS therapy [70]. By attaching to biospecific molecules, these nanoparticles can be distributed selectively inside living organisms [54]. According to De Matteis [17], nanomaterials can be roughly divided into three categories: organic (liposomes, polymers), inorganic (metals, oxides, and quantum dots), and carbon-based materials. The pharmacokinetics and biodistribution of active pharmaceuticals are improved by nanocarrier systems. This lets drugs be given correctly to people with specific diseases. Because of their optical characteristics, gold nanoparticles are excellent for bioimaging, but iron oxide nanoparticles are well-suited for magnetic resonance imaging (MRI) [21]. Nanomaterial-based strategies employ three methods to pass through the BBB: alternative, invasive, and non-invasive drug delivery methods. A non-invasive method that uses biological processes through transcellular channels is intranasal administration. This method uses the olfactory epithelium or trigeminal neurons to get nanoparticles to the brain and spinal cord without going through the

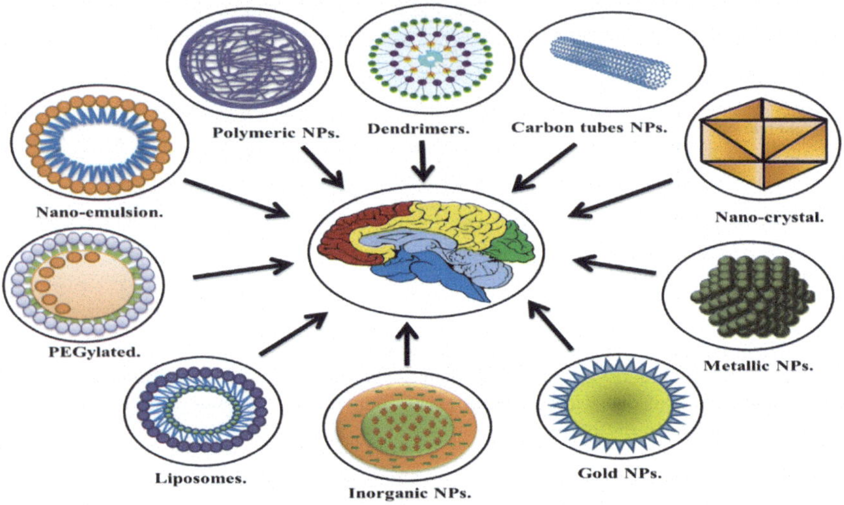

Fig. 2 Various nanoparticle-based drug delivery systems are employed for the treatment of neurological disorders. Diverse types of nanoparticles serve as drug delivery agents in these disorders, including dendrimers, carbon nanotubes, nanocrystals, metallic and inorganic nanoparticles, PEGylated nanoparticles, polymeric nanoparticles, liposomes, nanoemulsions, and more

BBB. Intraventricular injections, osmotic BBB disruption, intracerebral implants, and ultrasound-mediated treatments are examples of invasive techniques that can break down the BBB or deliver drugs directly into brain tissue [91]. One way to deliver drugs locally is to implant biodegradable polymeric wafers with medicine on them. It is also important to deal with any problems that might arise from chemicals like mannitol, fructose, and glycerol that weaken the BBB [73]. The goal of research into alternative routes for delivering CNS medications to the brain and spinal cord is to develop non-invasive techniques for doing so. Iontophoresis, a process that transfers ionized materials across the BBB using externally supplied electric currents, is made possible via connections between the olfactory and trigeminal neurons [49]. In conclusion, because they may easily pass across the BBB in a variety of ways, nanoparticles offer a promising treatment delivery strategy for neurological disorders (Fig. 2).

10 Nanoparticle-Based Drug Delivery in AD

The BBB's restricted permeability and the existence of efflux transporters, such as p-glycoprotein, which lower the drugs' bioavailability in the CNS, limit the effectiveness of existing AD treatments. Nanoparticles show promise in addressing this challenge by binding to physiologically active molecules, offering potential therapeutic benefits for AD treatment. In a study by [32], poly (n-butyl cyanoacrylate) nanoparticles loaded with rivastigmine demonstrated enhanced brain concentration in a rat model. The use of polysorbate 80 as a coating further increased the brain concentration of rivastigmine by 3.82 times [88], with the coating interacting with microvessel endothelial cells to promote nanoparticle endocytosis and uptake [80]. By treating oxidative damage in the CNS, nanoparticles may enhance therapeutic biodistribution and absorption in AD patients. Iron chelators encapsulated in nanoparticles, as studied by [46], can cross the BBB and remove iron from brain tissue, which is crucial for targeting neural tissue. Additionally, nanoparticles carrying quercetin, a flavonoid known for its antioxidant properties, address oxidative stress in AD [92]. According to [40], modified silica nanoparticles can carry quercetin and have antioxidant and neuroprotective properties. A nanocarrier system, such as quercetin-encapsulated solid liquid nanoparticles (SNLs), enhanced antioxidant properties and reversed neurodegenerative effects in rats induced by aluminum chloride, suggesting potential benefits for slowing AD progression and improving cognitive functions [19]. Gold nanoparticles coated with Casein (Cas) reduced Aβ toxicity and prevented AD symptoms in zebrafish [36]. Nanoparticles play a role in AD diagnosis as well, serving as contrast agents that bind to proteins like tau and Aβ in the brain, providing signals during MRI modalities [8]. Encapsulating hydrophilic medicines in nanoparticles can enhance drug uptake into the brain, potentially increasing BBB penetration. Thioflavin T (ThT) transported using nanocarriers could serve as a marker for Aβ and AD [4]. In the brain parenchyma, [33] showed that ThT was distributed effectively, focusing on Aβ deposition extracellularly and Aβ production intracellularly. Moreover, imaging techniques using Aβ 1–40 peptide, mannitol, and magnetic substances like gadolinium (Gd) or monocrystalline iron oxide nanoparticles (MION) have shown promise in detecting Aβ plaques in transgenic mice overexpressing APP, demonstrating higher sensitivity and specificity compared to conventional MRI contrast agents [87]. Nanoparticles play a role in AD treatment.

11 Nanoparticle-Based Drug Delivery in PD

PD is characterized by progressive neurodegeneration, which is why normal dopamine levels are being restored with nanoparticles. Chitosan nanoparticles encapsulating dopamine, showcasing a rapid and pulsatile release profile in vitro on the MDCKII-MDR1 cell line, were engineered [16]. These chitosan nanoparticles demonstrated heightened striatal dopamine concentrations and efficient transport across the BBB, with Trapani et al. endorsing the pulsatile release profile for its reduced risk of neurotoxicity compared to cellulose acetate phthalate (CAP) nanocarriers [83]. In an animal model of PD produced by rotenone, the non-ergolinic dopaminergic agonist ropinirole (RP) loaded into poly (lactic-co-glycolic acid) (PLGA) nanoparticles successfully reversed neurodegeneration [3]. RP-loaded solid lipid nanoparticles (SLNs) and nanostructured lipid carriers (NLCs) were topically delivered to rats. The results showed continuous and sustained release with enhanced bioavailability [20]. Apomorphine-loaded SLNs exhibited successful striatum targeting in a PD rat model, indicating substantial improvement in PD treatment [84]. Levodopa-loaded nanoparticles, administered orally with efficiency, were extended and localized to correct dopamine imbalances in PD [56]. The addition of wheat germ agglutinin lectin to PEG-PLGA nanoparticles enhanced their transport to the brain while encapsulating levodopa [23]. Utilizing oxygen-deficient cerium oxide (CeO_2) nanoparticles, nanostructured plasmonic substrates, and a microfluidic chip, Vazquez-Guardado et al. created an enzyme-free dopamine biosensor system that showed excellent sensitivity and selectivity for PD diagnosis and monitoring [85]. The gold nanoparticle-conjugated aptamers that were made to detect PD had a strong affinity for α-synuclein in vitro testing, which suggests they could be used for diagnosing PD [96]. These studies highlight the promising role of nanoparticles in PD drug delivery, offering a versatile and effective therapeutic approach.

12 Nanoparticle-Based Drug Delivery in CVDs

The treatment and diagnosis of ischemic and hemorrhagic strokes is the main focus of research on the use of nanoparticles in cerebrovascular applications. The principal treatment for acute ischemic stroke is intravenous tissue-type plasminogen activator (tPA); however, its effectiveness decreases after the 3–4.5-h therapeutic window, and there is a risk of hemorrhagic transformation [65]. In order to improve platelet binding and urokinase targeting for thrombus in acute ischemic stroke, Zhang et al. developed novel liposomes that included urokinase in a cyclic arginylglycylaspartic acid (RGD) pattern [98]. Researchers made polysaccharide-poly (isobutyl cyanoacrylate) nanoparticles functionalized with fucoidan, a sulfated polysaccharide containing L-fucose, to improve thrombolytic efficacy in a mouse model of venous thrombosis [38]. AP9-conjugated nanoparticles were observed to improve functional results and decrease infarct sizes in a mouse model [48]. Cerium oxide (CeO_2) nanoparticles promise for pharmacokinetic and clinical testing since they can penetrate the BBB and effectively scavenge reactive oxygen species (ROS) in catastrophic subarachnoid haemorrhage [37]. Astaxanthin, a carotenoid with anti-inflammatory and antioxidant qualities, was combined with PEG nanoparticles to create biostable, biocompatible nanoparticles that targeted specific tissues and increased neuronal survival [97]. Artificial platelets were created using Arg-Gly-Asp nanoparticles to shorten the bleeding period in a rat injury model [6]. Further research is necessary, but functionalized polystyrene nanoparticles coated with BSA and collagen-binding peptide significantly shortened the time it took for a mouse's tail to bleed, indicating that they could be helpful as medications for hemorrhagic stroke sufferers [1]. Additionally, the use of nanoparticles and MRI in conjunction with acute ischemic stroke allows for the monitoring of inflammatory cell participation [35].

13 Nanoparticle-Based Drug Delivery in TBI

Researchers are investigating the potential applications of nanoparticles in the treatment of TBI. They target important pathophysiological processes like oxidative stress, neuronal apoptosis, excitotoxicity, and neuroinflammation. This strategy intends to reduce side effects while fostering neuroregeneration and functional recovery [7]. Oxygen-reactive polymeric (ORP) nanoparticles synthesized by copolymerizing polyethylene glycol methacrylate with 2-(methylthio) ethyl methacrylate and methacrylic acid N-hydroxysuccinimide ester exhibit high reactivity to ROS and effectively reduce H_2O_2 levels in vitro. ORPs can be used as a potential treatment for disorders linked to ROS damage since they protect against tissue damage and boost survival rates in vivo [2]. Nanoparticles of CeO_2 have antioxidant properties and could be used to treat moderate TBI by protecting neurons from death and oxidative stress, keeping calcium levels stable, and signaling within cells. Immunomodulatory nanoparticles administered intravenously have demonstrated efficacy in treating TBI-related neuroinflammation by preserving brain tissue, reducing edema, and potentially modulating the immune system [75]. In a TBI mouse model, core-cross-linked nanoparticles made of polysorbate 80 efficiently collect reactive oxygen species (ROS), reduce neuroinflammation, and aid in healing and learning [95]. Brain-derived neurotrophic factor (BDNF) can be delivered to the brain using PLGA nanoparticles coated with poloxamer 188, improving neurological and cognitive results in TBI [39]. When administered to TBI patients, a nanoprodrug containing Ibu2 tetraethylene glycol and tocopherol crosses the BBB and, with a large surface area for hydrolytic esterase enzymes to degrade, releases ibuprofen. In a mouse model, it demonstrated greater therapeutic efficacy than unbound ibuprofen, indicating its potential as a treatment for TBI patients [13]. Nitroxide radical-containing polymers (RNPs), which promote the growth of microglia, reduce the size of brain contusions, and improve cognitive behavior evaluation in TBI models [81], A highly specific and sensitive method that is critical for the early identification and treatment of non-invasive TBI cerebral lesions The method targets necrotic cells in a mouse model using near-infrared fluorophores, perfluorocarbons (PFCs), and the cyanide dye 800CW [15].

14 Nanoparticle-Based Drug Delivery in GBM

GBM is the most common malignant brain tumor, comprising 50% of all primary CNS tumors. Surgical excision, chemotherapy, and/or radiotherapy are all viable treatment options [57]. New procedures and technologies are required for the limited GBM treatment options currently available. Nanocarrier-based drug delivery systems such as PLGA nanoparticles and convection-enhanced delivery (CED) are being investigated for brain delivery; however, additional research is required [99]. Linked to atherosclerotic plaque-specific peptide-1, liposomal nanoparticles serve as DOX carriers, binding to IL-4 receptors and inducing apoptosis. Improved antitumor efficacy and decreased systemic toxicity have been demonstrated in preclinical research [93]. TfR-PEG micelles have been shown to be able to cross the BBB and be taken up by tumor cells. These paclitaxel-loaded micelles increase the median survival of nude mice with intracranial U87 GBM tumors and decrease U87 GBM cell proliferation, indicating their potential as targeted drug delivery systems [79]. Gold nanoparticles encapsulated in poly (amidoamine) (PAMAM) dendrimers and coated with beta-cyclodextrin (β-CD) for efficient delivery to glioma cells condense two siRNAs for oncogene silencing [25]. These nanoparticles deliver a larger payload and improve stability and solubility, ensuring efficient siRNA delivery to glioma cells. Polyether-copolyester (PEPE) dendrimers and D-glucosamine enhance drug delivery and tumor targeting in methotrexate-resistant cells. AuNPs are a suitable nanomedicine for the treatment of GBM [47].

15 Challenges in Using Nanoparticles and Future Perspectives

New approaches to treating CNS diseases and personalized medications are now possible because of advancements in nanomedicine. This creates new opportunities for nanomedicine delivery via the BBB [51]. Nanotechnology-driven pharmaceutical delivery methods have been a focal point in the development of nanomedicine for treating CNS diseases. However, nanoparticles' properties and make-up can cause oxidative stress and instability in the BBB, which could cause amino acids to break down and brain neurotoxicity to rise [82]. While functionalized nanoparticles offer excellent therapeutic localization, their tendency to aggregate due to their large surface area and compact form may limit drug loading. The mode of medication administration and the quantity of nanomedicine administered could contribute to elevated neuroinflammation, oxidative stress, DNA damage, and allergic reactions [86]. Therefore, a thorough understanding of the biodegradability and biocompatibility of nanomedicine is crucial. Establishing standardized methods for assessing the toxicity of nanoformulations in both in vitro and in vivo studies is imperative to furnish comprehensive information on the safety profile of nanoformulations in humans. Addressing challenging issues, including nanotechnology modification and management, pharmaceutical cargo packaging, drug delivery to the brain, deep brain stimulation (DBS), implantation stimulation, and brain cell activity, remains at the forefront of ongoing research [42]. The characterization and production of nanoparticles hold vital significance to ensure their safe integration into drug administration, preventing unintended harm to healthy cells. Looking ahead, these advancements are poised to simplify the development of treatments, depending on the severity of a patient's neurodegeneration.

16 Conclusion

This chapter emphasizes the dramatic potential of nanotechnology to widen the therapeutic landscape for neurological diseases. The efficacy of nanoparticles as therapeutic agent carriers has been shown to be potent and efficient, outperforming traditional therapy methods. With their cutting-edge nanoscale targeting methods, nanomedicine-based technologies promise potential for the treatment of a wide range of neurological diseases, signaling the start of a new era. The emphasis on nano-based neuroprotection and regeneration techniques corresponds to the delicate features of neurological cell physiology, biology, and pathology. Nanomedicines that have been specifically developed for controlled release have effectively surmounted the hurdles of crossing the BBB, demonstrating the adaptability of nanotechnology in neurological therapies. Tailored nanomedicines have the potential to help treat a wide range of neurological diseases because they can cross the BBB, spread out in brain tissue, and target specific cells or signaling systems. Diverse nanoscale methods, such as oligomerization and ROS reduction, contribute to the successful treatment of neurological disorders by providing varied solutions. The special properties of nanotechnology—like its capacity to penetrate the BBB, adaptability for customization, and accuracy in targeted delivery—not only offer a cutting-edge method of treating neurological disorders but also portend a bright future. As we continue to explore nanotechnology's potential, we will be able to refine and advance our approaches to addressing the complex issues offered by neurological disorders.

Acknowledgements The authors thank Ms. Neha Kumari for providing editorial assistance.

References

1. Anselmo AC, Modery-Pawlowski CL, Menegatti S, Kumar S, Vogus DR, Tian LL, Chen M, Squires TM, Sen Gupta A, Mitragotri S (2014) Platelet-like nanoparticles: mimicking shape, flexibility, and surface biology of platelets to target vascular injuries. ACS Nano 8(11):11243–11253
2. Bailey ZS, Nilson E, Bates JA, Oyalowo A, Hockey KS, Sajja VSSS, Thorpe C, Rogers H, Dunn B, Frey AS, Billings MJ, Sholar CA, Hermundstad A, Kumar C, VandeVord PJ, Rzigalinski BA (2020) Cerium oxide nanoparticles improve outcome after in vitro and in vivo mild traumatic brain injury. J Neurotrauma 37(12):1452–1462
3. Barcia E, Boeva L, García-García L, Slowing K, Fernández-Carballido A, Casanova Y, Negro S (2017) Nanotechnology-based drug delivery of ropinirole for Parkinson's disease. Drug Deliv 24(1):1112–1123
4. Bechinger P, Serrano Sponton L, Grützner V, Musyanovych A, Jussen D, Krenzlin H, Eldahaby D, Riede N, Kempski O, Ringel F, Alessandri B (2023) In-vivo time course of organ uptake and blood-brain-barrier permeation of poly(L-lactide) and poly(perfluorodecyl acrylate) nanoparticles with different surface properties in unharmed and brain-traumatized rats. Front Neurol 14:994877

5. Behari M, Bhattacharyya KB, Borgohain R, Das SK, Ghosh B, Kishore A, Krishnan S, Mridula KR, Muthane U, Pal PK, Sankhla C, Shukla G (2011) Parkinson's disease. Ann Indian Acad Neurol 14:S2-6
6. Bertram JP, Williams CA, Robinson R, Segal SS, Flynn NT, Lavik EB (2009) Intravenous hemostat: nanotechnology to halt bleeding. Sci Transl Med 1(11):11–22
7. Bharadwaj VN, Nguyen DT, Kodibagkar VD, Stabenfeldt SE (2018) Nanoparticle-based therapeutics for brain injury. Adv Healthc Mater 7(1)
8. Brambilla D, Le Droumaguet B, Nicolas J, Hashemi SH, Wu LP, Moghimi SM, Couvreur P, Andrieux K (2011) Nanotechnologies for AD: diagnosis, therapy, and safety issues. Nanomedicine 7(5):521–540
9. Buzea C, Pacheco II, Robbie K (2007) Nanomaterials and nanoparticles: sources and toxicity. Biointerphases 2(4):MR17–71
10. Chandra A, Stone CR, Du X, Li WA, Huber M, Bremer R, Geng X, Ding Y (2017) The cerebral circulation and cerebrovascular disease III: Stroke. Brain Circ. 3(2):66–77
11. Chen GF, Xu TH, Yan Y et al (2017) Amyloid beta: structure, biology and structure-based therapeutic development. Acta Pharmacol Sin 38(9):1205–1235. https://doi.org/10.1038/aps.2017.28
12. Chen Y, Liu L (2012) Modern methods for delivery of drugs across the blood-brain barrier. Adv Drug Deliv Rev 64(7):640–665
13. Clond MA, Lee BS, Yu JJ, Singer MB, Amano T, Lamb AW, Drazin D, Kateb B, Ley EJ, Yu JS (2013) Reactive oxygen species-activated nanoprodrug of Ibuprofen for targeting TBIin mice. PLoS ONE 8(4):e61819
14. Colello MJ, Ivey LE, Gainey J, Faulkner RV, Johnson A, Brechtel L, Madeline L, Nathaniel TI (2018) Pharmacological thrombolysis for acute ischemic stroke treatment: Gender differences in clinical risk factors. Adv Med Sci 63(1):100–106
15. Cruz LJ, Que I, Aswendt M, Chan A, Hoehn M, Löwik C (2016) Targeted nanoparticles for the non-invasive detection of TBIby optical imaging and fluorine magnetic resonance imaging. Nano Res 9(5):1276–1289
16. De Giglio E, Trapani A, Cafagna D, Sabbatini L, Cometa S (2011) Dopamine-loaded chitosan nanoparticles: formulation and analytical characterization. Anal Bioanal Chem 400(7):1997–2002
17. De Matteis V, Rinaldi R (2018) Toxicity assessment in the nanoparticle era. Adv Exp Med Biol 1048:1–19
18. Dewan MC, Rattani A, Gupta S, Baticulon RE, Hung YC, Punchak M, Agrawal A, Adeleye AO, Shrime MG, Rubiano AM, Rosenfeld JV, Park KB (2018) Estimating the global incidence of traumatic brain injury. J Neurosurg:1–18
19. Dhawan S, Kapil R, Singh B (2011) Formulation development and systematic optimization of solid lipid nanoparticles of quercetin for improved brain delivery. J Pharm Pharmacol 63(3):342–351
20. Dudhipala N, Gorre T (2020) Neuroprotective effect of ropinirole lipid nanoparticles enriched hydrogel for Parkinson's Disease: in vitro, ex vivo, pharmacokinetic and pharmacodynamic evaluation. Pharmaceutics 12(5):448
21. Feng L, Wang H, Xue X (2020) Recent progress of nanomedicine in the treatment of central nervous system diseases. Adv Ther 3(5):1900159
22. Fyfe I (2020) Alzheimer disease-associated gene increases tau pathology. Nat Rev Neurol 16(3):128
23. Gao X, Tao W, Lu W, Zhang Q, Zhang Y, Jiang X, Fu S (2006) Lectin-conjugated PEG-PLA nanoparticles: preparation and brain delivery after intranasal administration. Biomaterials 27(18):3482–3490
24. GBD 2016 Neurology Collaborators (2019) Global, regional, and national burden of neurological disorders, 1990–2016: a systematic analysis for the global burden of disease study 2016. Lancet Neurol 18(5):459–480
25. Ghaffari M, Dehghan G, Abedi-Gaballu F, Kashanian S, Baradaran B, Dolatabadi JEN, Losic D (2018) Surface functionalized dendrimers as controlled-release delivery nanosystems for tumor targeting. Eur J Pharm Sci 122:311–330

26. Gonzalez-Carter D, Liu X, Tockary TA, Dirisala A, Toh K, Anraku Y, Kataoka K (2020) Targeting nanoparticles to the brain by exploiting the blood-brain barrier impermeability to selectively label the brain endothelium. Proc Natl Acad Sci 117(32):19141–19150
27. Gonzalez-Mariscal L, Betanzos A, Nava P, Jaramillo BE (2003) Tight junction proteins. Prog Biophys Mol Biol 81(1):1–44
28. Green RC, Schneider LS, Amato DA, Beelen AP, Wilcock G, Swabb EA, Zavitz KH, Tarenflurbil phase 3 study group (2009) Effect of tarenflurbil on cognitive decline and activities of daily living in patients with mild Alzheimer disease: a randomized controlled trial. JAMA 302(23):2557–2564
29. Grossberg GT (2003) Cholinesterase inhibitors for the treatment of AD: getting on and staying on. Curr Ther Res Clin Exp 64(4):216–235. https://doi.org/10.1016/S0011-393X(03)00059-6
30. Haddadi R, Shahidi Z, Eyvari-Brooshghalan S (2020) Silymarin and neurodegenerative diseases: therapeutic potential and basic molecular mechanisms. Phytomedicine 79:153320
31. Hanif S, Muhammad P, Chesworth R, Rehman FU, Qian RJ, Zheng M, Shi B (2020) Nanomedicine-based immunotherapy for central nervous system disorders. Acta Pharmacol Sin 41(7):936–953
32. Harilal S, Jose J, Parambi DGT, Kumar R, Mathew GE, Uddin MS, Kim H, Mathew B (2019) Advancements in nanotherapeutics for AD: current perspectives. J Pharm Pharmacol 71(9):1370–1383
33. Hartig W, Paulke BR, Varga C, Seeger J, Harkany T, Kacza J (2003) Electron microscopic analysis of nanoparticles delivering thioflavin-T after intrahippocampal injection in mouse: implications for targeting beta-amyloid in AD. Neurosci Lett 338(2):174–176
34. Hiebert JB, Shen Q, Thimmesch AR, Pierce JD (2015) TBIand mitochondrial dysfunction. Am J Med Sci 350(2):132–138
35. Hubert V, Dumot C, Ong E, Amaz C, Canet-Soulas E, Chauveau F, Wiart M (2019) MRI coupled with clinically-applicable iron oxide nanoparticles reveals choroid plexus involvement in a murine model of neuroinflammation. Sci Rep 9(1):10046
36. Javed I, Peng G, Xing Y, Yu T, Zhao M, Kakinen A, Faridi A, Parish CL, Ding F, Davis TP, Ke PC, Lin S (2019) Inhibition of amyloid beta toxicity in zebrafish with a chaperone-gold nanoparticle dual strategy. Nat Commun 10(1):3780
37. Jeong HG, Cha BG, Kang DW, Kim DY, Ki SK, Kim SI, Han JH, Yang W, Kim CK, Kim J, Lee SH (2018) Ceria nanoparticles synthesized with aminocaproic acid for the treatment of subarachnoid hemorrhage. Stroke 49(12):3030–3038
38. Juenet M, Aid-Launais R, Li B, Berger A, Aerts J, Ollivier V, Nicoletti A, Letourneur D, Chauvierre C (2018) Thrombolytic therapy based on fucoidan-functionalized polymer nanoparticles targeting P-selectin. Biomaterials 156:204–216
39. Khalin I, Alyautdin R, Wong TW, Gnanou J, Kocherga G, Kreuter J (2016) Brain-derived neurotrophic factor delivered to the brain using poly (lactide-co-glycolide) nanoparticles improves neurological and cognitive outcome in mice with traumatic brain injury. Drug Deliv 23(9):3520–3528
40. Kim H, Park BS, Lee KG, Choi CY, Jang SS, Kim YH, Lee SE (2005) Effects of naturally occurring compounds on fibril formation and oxidative stress of beta-amyloid. J Agric Food Chem 53(22):8537–8541
41. Kinoshita K (2016) Traumatic brain injury: pathophysiology for neurocritical care. J Intens Care 4:29
42. Kumar R, Aadil KR, Mondal K et al (2022) Neurodegenerative disorders management: state-of-art and prospects of nano-biotechnology. Crit Rev Biotechnol 42(8):1180–1212. https://doi.org/10.1080/07388551.2021.1993126
43. Kutzing MK, Luo V, Firestein BL (2012) Protection from glutamate-induced excitotoxicity by memantine. Ann Biomed Eng 40(5):1170–1181
44. Lee HC, Chuang HC, Cho DY, Cheng KF, Lin PH, Chen CC (2010) Applying cerebral hypothermia and brain oxygen monitoring in treating severe traumatic brain injury. World Neurosurg 74(6):654–660

45. Liebner S, Dijkhuizen RM, Reiss Y, Plate KH, Agalliu D, Constantin G (2018) Functional morphology of the blood-brain barrier in health and disease. Acta Neuropathol 135(3):311–336
46. Liu G, Garrett MR, Men P, Zhu X, Perry G, Smith MA (2005) Nanoparticle and other metal chelation therapeutics in Alzheimer disease. Biochim Biophys Acta 1741(3):246–252
47. Liu J, Peng Q (2017) Protein-gold nanoparticle interactions and their possible impact on biomedical applications. Acta Biomater 55:13–27
48. Liu S, Jin R, Wang M, Li G (2020) Nanoparticle delivery of CD147 antagonistic peptide-9 protects against acute ischemic brain injury and tPA-induced intracerebral hemorrhage in mice. ACS Appl Bio Mater 3(4):1976–1985
49. Lu CT, Zhao YZ, Wong HL, Cai J, Peng L, Tian XQ (2014) Current approaches to enhance CNS delivery of drugs across the brain barriers. Int J Nanomed 9:2241–2257
50. McGinn MJ, Povlishock JT (2016) Pathophysiology of traumatic brain injury. Neurosurg Clin N Am 27(4):397–407
51. Menon PK, Muresanu DF, Sharma A, Mössler H, Sharma HS (2012) Cerebrolysin, a mixture of neurotrophic factors induces marked neuroprotection in spinal cord injury following intoxication of engineered nanoparticles from metals. CNS Neurol Disord Drug Targets 11(1):40–49. https://doi.org/10.2174/187152712799960781
52. Misra A, Ganesh S, Shahiwala A, Shah SP (2003) Drug delivery to the central nervous system: a review. J Pharm Pharm Sci 6(2):252–273
53. Mittal KR, Pharasi N, Sarna B et al (2022) Nanotechnology-based drug delivery for the treatment of CNS disorders. Transl Neurosci 13(1):527–546. Published 2022 Dec 31. https://doi.org/10.1515/tnsci-2022-0258
54. Montet X, Funovics M, Montet-Abou K, Weissleder R, Josephson L (2006) Multivalent effects of RGD peptides obtained by nanoparticle display. J Med Chem 49(20):6087–6093
55. Musa G, Slachevsky A, Muñoz-Neira C, Méndez-Orellana C, Villagra R, González-Billault C, Ibáñez A, Hornberger M, Lillo P (2020) AD or behavioral variant frontotemporal dementia? Review of key points toward an accurate clinical and neuropsychological diagnosis. J Alzheimers Dis 73(3):833–848
56. Ngwuluka NC, Pillay V, Choonara YE, Modi G, Naidoo D, du Toit LC, Kumar P, Ndesendo VM, Khan RA (2011) Fabrication, modeling and characterization of multi-crosslinked methacrylate copolymeric nanoparticles for oral drug delivery. Int J Mol Sci 12(9):6194–6225
57. Ostrom QT, Gittleman H, Farah P, Ondracek A, Chen Y, Wolinsky Y, Stroup NE, Kruchko C, Barnholtz-Sloan JS (2013) CBTRUS statistical report: Primary brain and central nervous system tumors diagnosed in the United States in 2006–2010. Neuro Oncol 15(Suppl 2):1–56
58. Oszvald A, Güresir E, Setzer M, Vatter H, Senft C, Seifert V, Franz K (2012) Glioblastoma therapy in the elderly and the importance of the extent of resection regardless of age. J Neurosurg 116(2):357–364
59. Pan MH, Lai CS, Ho CT (2010) Anti-inflammatory activity of natural dietary flavonoids. Food Funct 1(1):15–31
60. Pardridge WM (2015) Blood-brain barrier endogenous transporters as therapeutic targets: a new model for small molecule CNS drug discovery. Expert Opin Ther Targets 19(8):1059–1072
61. Park JS, Davis RL, Sue CM (2018) Mitochondrial dysfunction in Parkinson's disease: new mechanistic insights and therapeutic perspectives. Curr Neurol Neurosci Rep 18(5):21
62. Pimentel MM, Rodrigues FC, Leite MA, Campos Júnior M, Rosso AL, Nicaretta DH, Pereira JS, Silva DJ, Della Coletta MV, Vasconcellos LF, Abreu GM, Dos Santos JM, Santos-Rebouças CB (2015) PD: α-synuclein mutational screening and new clinical insight into the p.E46K mutation. Parkinsonism Relat Disord 21(6):586–589
63. Pirtosek Z, Bajenaru O, Kovács N, Milanov I, Relja M, Skorvanek M (2020) Update on the management of Parkinson's disease for general neurologists. Parkinsons Dis 2020:9131474
64. Pohanka M (2011) Cholinesterases, a target of pharmacology and toxicology. Biomed Pap Med Fac Univ Palacky Olomouc Czech Repub 155(3):219–229

65. Powers WJ, Rabinstein AA, Ackerson T, Adeoye OM, Bambakidis NC, Becker K, Biller J, Brown M, Demaerschalk BM, Hoh B, Jauch EC, Kidwell CS, Leslie-Mazwi TM, Ovbiagele B, Scott PA, Sheth KN, Southerland AM, Summers DV, Tirschwell DL, American Heart Association Stroke Council (2018) Guidelines for the early management of patients with acute ischemic stroke: a guideline for healthcare professionals from the American Heart Association/American Stroke Association. Stroke 49(3):e46–e110
66. Prager BC, Bhargava S, Mahadev V, Hubert CG, Rich JN (2020) Glioblastoma stem cells: driving resilience through chaos. Trends Cancer 6(3):223–235
67. Qazi MA, Vora P, Venugopal C, Sidhu SS, Moffat J, Swanton C, Singh SK (2017) Intratumoral heterogeneity: pathways to treatment resistance and relapse in human glioblastoma. Ann Oncol 28(7):1448–1456
68. Qureshi AI, Mendelow AD, Hanley DF (2009) Intracerebral haemorrhage. Lancet 373(9675):1632–1644
69. Sacco RL, Kasner SE, Broderick JP, Caplan LR, Connors JJ, Culebras A, Elkind MS, George MG, Hamdan AD, Higashida RT, Hoh BL, Janis LS, Kase CS, Kleindorfer DO, Lee JM, Moseley ME, Peterson ED, Turan TN, Valderrama AL, Vinters HV, American Heart Association Stroke Council, Council on Cardiovascular Surgery and Anesthesia, Council on Cardiovascular Radiology and Intervention, Council on Cardiovascular and Stroke Nursing; Council on Epidemiology and Prevention, Council on Peripheral Vascular Disease, Council on Nutrition, Physical Activity and Metabolism (2013) An updated definition of stroke for the 21st century: a statement for healthcare professionals from the American Heart Association/ American Stroke Association. Stroke 44(7):2064–2089
70. Saeedi M, Eslamifar M, Khezri K, Dizaj SM (2019) Applications of nanotechnology in drug delivery to the central nervous system. Biomed Pharmacother 111:666–675
71. Salma A, Vasilakis M, Tracy PT (2014) Deep brain stimulation for cognitive disorders: insights into targeting nucleus basalis of meynert in Alzheimer dementia. World Neurosurg 81(1):e4-5
72. Saunders NR, Habgood MD, Møllgård K, Dziegielewska KM (2016) The biological significance of brain barrier mechanisms: help or hindrance in drug delivery to the central nervous system? F1000Res 5:F1000 Faculty Rev-313
73. Seong H, An TK, Khang G, Choi SU, Lee CO, Lee HB (2003) BCNU-loaded poly(D, L-lactide-co-glycolide) wafer and antitumor activity against XF-498 human CNS tumor cells in vitro. Int J Pharm 251(1–2):1–12
74. Shakeri S, Ashrafizadeh M, Zarrabi A, Roghanian R, Afshar EG, Pardakhty A, Mohammadinejad R, Kumar A, Thakur VK (2020) Multifunctional polymeric nanoplatforms for brain diseases diagnosis, therapy and theranostics. Biomedicines 8(1):13
75. Sharma S, Ifergan I, Kurz JE, Linsenmeier RA, Xu D, Cooper JG, Miller SD, Kessler JA (2020) Intravenous immunomodulatory nanoparticle treatment for traumatic brain injury. Ann Neurol 87(3):442–455
76. Shrestha S, Coy S, Bekelis K (2017) Oral antiplatelet and anticoagulant agents in the prevention and management of ischemic stroke. Curr Pharm Des 23(9):1377–1391
77. Srikanth M, Kessler JA (2012) Nanotechnology-novel therapeutics for CNS disorders. Nat Rev Neurol 8(6):307–318
78. Stoker TB, Torsney KM, Barker RA (2018) Emerging treatment approaches for Parkinson's disease. Front Neurosci 12:693
79. Sun P, Xiao Y, Di Q, Ma W, Ma X, Wang Q, Chen W (2020) Transferrin receptor-targeted PEG-PLA polymeric micelles for chemotherapy against glioblastoma multiforme. Int J Nanomed:6673–6688
80. Sun W, Xie C, Wang H, Hu Y (2004) Specific role of polysorbate 80 coating on the targeting of nanoparticles to the brain. Biomaterials 25(15):3065–3071
81. Takahashi T, Marushima A, Nagasaki Y, Hirayama A, Muroi A, Puentes S, Mujagic A, Ishikawa E, Matsumura A (2020) Novel neuroprotection using antioxidant nanoparticles in a mouse model of head trauma. J Trauma Acute Care Surg 88(5):677–685
82. Teleanu RI, Preda MD, Niculescu AG et al (2022) Current strategies to enhance delivery of drugs across the blood-brain barrier. Pharmaceutics 14(5):987. Published 2022 May 4. https://doi.org/10.3390/pharmaceutics14050987

83. Trapani A, De Giglio E, Cafagna D, Denora N, Agrimi G, Cassano T, Gaetani S, Cuomo V, Trapani G (2011) Characterization and evaluation of chitosan nanoparticles for dopamine brain delivery. Int J Pharm 419(1–2):296–307
84. Tsai MJ, Huang YB, Wu PC, Fu YS, Kao YR, Fang JY, Tsai YH (2011) Oral apomorphine delivery from solid lipid nanoparticles with different monostearate emulsifiers: pharmacokinetic and behavioral evaluations. J Pharm Sci 100(2):547–557
85. Vazquez-Guardado A, Barkam S, Peppler M, Biswas A, Dennis W, Das S, Seal S, Chanda D (2019) Enzyme-free plasmonic biosensor for direct detection of neurotransmitter dopamine from whole blood. Nano Lett 19(1):449–454
86. Vega-Villa KR, Takemoto JK, Yáñez JA, Remsberg CM, Forrest ML, Davies NM (2008) Clinical toxicities of nanocarrier systems. Adv Drug Deliv Rev 60(8):929–938. https://doi.org/10.1016/j.addr.2007.11.007
87. Wadghiri YZ, Sigurdsson EM, Sadowski M, Elliott JI, Li Y, Scholtzova H, Tang CY, Aguinaldo G, Pappolla M, Duff K, Wisniewski T, Turnbull DH (2003) Detection of Alzheimer's amyloid in transgenic mice using magnetic resonance microimaging. Magn Reson Med 50(2):293–302
88. Wilson B, Samanta MK, Santhi K, Kumar KP, Paramakrishnan N, Suresh B (2008) Targeted delivery of tacrine into the brain with polysorbate 80-coated poly(n-butylcyanoacrylate) nanoparticles. Eur J Pharm Biopharm 70(1):75–84
89. Wirdefeldt K, Adami HO, Cole P, Trichopoulos D, Mandel J (2011) Epidemiology and etiology of Parkinson's disease: a review of the evidence. Eur J Epidemiol 26:S1-58
90. Xie A, Gao J, Xu L, Meng D (2014) Shared mechanisms of neurodegeneration in AD and Parkinson's disease. Biomed Res Int 2014:648740
91. Xie J, Shen Z, Anraku Y, Kataoka K, Chen X (2019) Nanomaterial-based blood-brain-barrier (BBB) crossing strategies. Biomaterials 224:119491
92. Xu D, Hu MJ, Wang YQ, Cui YL (2019) Antioxidant activities of quercetin and its complexes for medicinal application. Molecules 24(6):1123
93. Yang FY, Wong TT, Teng MC, Liu RS, Lu M, Liang HF, Wei MC (2012) Focused ultrasound and interleukin-4 receptor-targeted liposomal doxorubicin for enhanced targeted drug delivery and antitumor effect in glioblastoma multiforme. J Control Release 160(3):652–658
94. Yao M, Li S, Wu X, Diao S, Zhang G, He H, Bian L, Lu Y (2018) Cellular origin of glioblastoma and its implication in precision therapy. Cell Mol Immunol 15(8):737–739
95. Yoo D, Magsam AW, Kelly AM, Stayton PS, Kievit FM, Convertine AJ (2017) Core-cross-linked nanoparticles reduce neuroinflammation and improve outcome in a mouse model of traumatic brain injury. ACS Nano 11(9):8600–8611
96. You X, Gopinath SCB, Lakshmipriya T, Li D (2019a) High-affinity detection of alpha-synuclein by aptamer-gold conjugates on an amine-modified dielectric surface. J Anal Methods Chem:6526850
97. You ZQ, Wu Q, Zhou XM, Zhang XS, Yuan B, Wen LL, Xu WD, Cui S, Tang XL, Zhang X (2019b) Receptor-mediated delivery of astaxanthin-loaded nanoparticles to neurons: an enhanced potential for subarachnoid hemorrhage treatment. Front Neurosci 13:989
98. Zhang N, Li C, Zhou D, Ding C, Jin Y, Tian Q, Meng X, Pu K, Zhu Y (2018) Cyclic RGD functionalized liposomes encapsulating urokinase for thrombolysis. Acta Biomater 70:227–236
99. Zhou J, Patel TR, Sirianni RW, Strohbehn G, Zheng MQ, Duong N, Schafbauer T, Huttner AJ, Huang Y, Carson RE, Zhang Y, Sullivan DJ Jr, Piepmeier JM, Saltzman WM (2013) Highly penetrative, drug-loaded nanocarriers improve treatment of glioblastoma. Proc Natl Acad Sci USA 110(29):11751–11756
100. Zinger A, Barcia C, Herrero MT, Guillemin GJ (2011) The involvement of neuroinflammation and kynurenine pathway in Parkinson's disease. Parkinsons Dis 2011:716859
101. Zlokovic BV (2011) Neurovascular pathways to neurodegeneration in AD and other disorders. Nat Rev Neurosci 12(12):723–738

Bioactive Polyphenolic Nanoformulations as Potential Therapeutic Agents Against Psoriasis

Mohammad Imran, Masood Ali, Danial Saifuddin, Arshi Chaudhary, Anas Ahmad, and Yousuf Mohammed

Abstract Psoriasis is a well-known, chronic, immune-mediated inflammatory skin disease affecting almost 2–4% of the world population. Psoriatic skin lesions manifest due to both the attenuation of oxidative stress and adverse immune response, resulting in hyperproliferation and inflammation of the epidermal keratinocyte. Phenolic hydroxyl group-containing organic molecules, Polyphenols, have demonstrated their potential utility as antioxidants and anti-inflammatory agents. However, their application is limited due to their susceptibility to environmental conditions and low bioavailability. Recently, nanoformulations were proposed to overcome such limitations and maximise the pharmacological effects of polyphenols. By focusing

---

Mohammad Imran and Masood Ali, these authors have equally contributed to the chapter

---

M. Imran · M. Ali · Y. Mohammed (✉)
Frazer Institute, Faculty of Health, Medicine and Behavioural Sciences, The University of Queensland, Brisbane, QLD, Australia
e-mail: y.mohammed@uq.edu.au

M. Imran
e-mail: mohamamd.imran@uq.edu.au

M. Ali
e-mail: masood.ali@uq.net.au

M. Ali · D. Saifuddin · Y. Mohammed
School of Pharmacy and Pharmaceutical Sciences, The University of Queensland, Brisbane, QLD, Australia
e-mail: d.saifuddin@uq.edu.au

A. Chaudhary
Department of Pharmaceutics, School of Pharmaceutical Education and Research, Jamia Hamdard, New Delhi, India

A. Ahmad
Julia McFarlane Diabetes Research Centre (JMDRC), Department of Microbiology, Immunology and Infectious Diseases, Snyder Institute for Chronic Diseases, Hotchikiss Brain Institute, Cumming School of Medicine, University of Calgary, Calgary, AB, Canada
e-mail: anas.ahmad@ucalgary.ca

Md. M. Ansari et al. (eds.), *Emergence of Sustainable Biomaterials in Tackling Inflammatory Diseases*, Smart Nanomaterials Technology,
https://doi.org/10.1007/978-981-96-2112-5_17

on the interactions between polyphenols and different nanoformulations (e.g., Liposomes, Niosomes, Ethosomes, Ufasomes, nanoemulsions, solid lipid nanoparticles, nanostructured lipid carriers, and dissolvable microneedles). This chapter sheds some light on the synthesis and properties of the polyphenol-containing nanoparticles. Additionally, the remarkable versatility of polyphenol-containing nanoparticles in various preclinical studies targeting psoriasis management is also demonstrated. Finally, the challenges faced by future research regarding the polyphenol-containing nanoparticles are discussed.

Keywords Psoriasis · Inflammatory disease · Polyphenols · Anti-inflammatory · Nanoparticles · Drug delivery

1 Introduction

Psoriasis is one of the most common inflammatory diseases worldwide with a high rate of incidence in Europe. According to different statistics, it is reported that 2–3% of the world's population is affected by this disease, specifically, the rate of prevalence of this disease in Australia is 2–6%. Therefore, it has been a concern for so long for clinicians and scientists to research different modalities and treatment options to reduce the burden of psoriasis on different types of populations across the world [1, 2]. The definition of psoriasis is that it is a chronic, auto-immune inflammatory disease condition that not only affects the skin but nails as well. In the development of psoriasis, the main effect is that it alters the growth cycle of skin cells, resulting in the rapid growth of inflamed skin. It is characterized by the presence of red inflamed patches of the skin covered with silvery-white scales [3]. Currently, there are numerous topical treatment options available around the world for psoriasis, such as corticosteroids, vitamin D analogues, calcineurin inhibitors (tacrolimus), anthralin, and many more [4]. Along with this, light therapy, oral therapies, and intravenous options are also available which are delivering various therapeutics for the prevention and treatment of psoriasis [5]. However, there are certain limitations that are reported with the current therapies for psoriasis, such as side effects of the biologics that are administered to an individual that could cause gastrointestinal issues. In addition, the cost burden, long-term stability, and treatment resistance are some of the reasons that limit the application of different biologicals in the treatment of psoriasis [6].

In an attempt to find out preventive and safe treatment options with effectiveness, researchers have been exploring a range of therapeutics, specifically, plant-derived compounds, i.e., "*Polyphenols*". Polyphenols are ubiquitous in the plant kingdom. They are well-defined and recognized due to their multifunctional properties, such as their antioxidant nature which helps to attenuate the oxidative stress in any inflammatory pathological condition [7]. It is believed that these compounds modulate different molecular and signalling pathways to alter the disease condition [8]. The basic skeleton of chemical structures of different polyphenolic compounds such as flavonoids, lignans and stilbenes have been shown in Fig. 1.

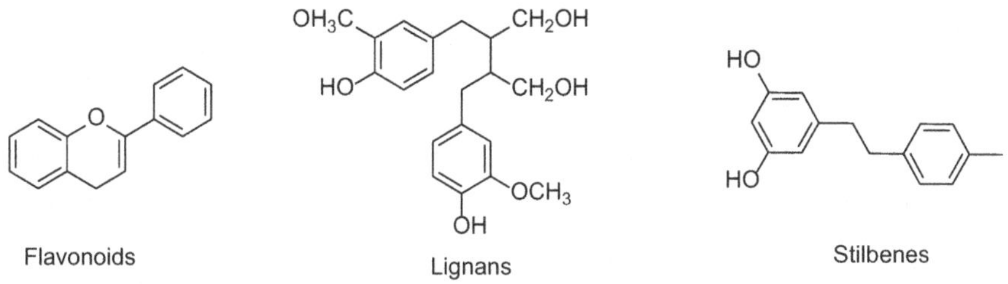

Fig. 1 The basic skeleton of chemical structures of different polyphenolic groups of compounds such as flavonoids, lignans and stilbenes

As of now, a plethora of literature suggests and demonstrates the significance of these compounds against psoriasis. Also, it has been noted that pharmaceutical researchers have investigated such compounds thoroughly for many diseases including psoriasis. These compounds have restricted physicochemical properties (solubility and permeability) which impedes their therapeutic efficacy. To improve the physicochemical properties of these bioactive compounds, researchers have utilized nanotechnology. The application of nanotechnology in the improvement of the former properties improves the overall therapeutic activity of the compound [9]. In the attempt to utilize advanced techniques, scientists reported several drug delivery vehicles. These delivery systems deliver these compounds to the target site/ disease site by improving their properties and overcoming the barriers [9]. The lipid-based nanocarriers such as liposomes, niosomes, ethosomes, ufasomes, solid lipid-nanoparticles, nanostructured lipid carriers, nanoemulsion, and dissolvable microneedles with loaded bioactive polyphenols have been observed for the treatment/prevention of psoriasis [10]. Some findings related to the dissolvable microneedles have shown promising outcomes by delivering the polyphenol locally on the targeted site [11]. The unique properties of such nanocarriers enable the researchers to load these bioactive polyphenols in these systems which already have shown wonderful outcomes in vitro and in vivo.

This chapter provides an overview of various bioactive polyphenolics for the prevention and treatment of psoriasis. Notably, the application of nanotechnology with the development of lipid-based nanocarriers has also been reviewed in which the different kinds of lipid nanoparticles have been considered. In addition, the molecular pathways/signalling pathways altered via these compounds are also incorporated into this chapter. As there are various in vivo psoriatic models have been given in the literature, we attempted to cover the different 3D in vitro models such as spheroids, organoids, and skin-on-a-chip models which have shown unique skin models. These models mimic the pathophysiological conditions and help in the screening of the compounds for any disease, including, skin diseases such as psoriasis.

2 Mechanisms of Action of Polyphenols as Anti-Inflammatory Agents Against Psoriasis

These are multifunctional compounds that act on different molecular pathways for different effects. It is well known that these compounds possess anti-inflammatory, antioxidant, and anti-cancer properties. The exact mechanism of action of these active molecules is still a topic of research into. However, we have covered and incorporated the mechanism of different polyphenols against skin inflammatory conditions such as psoriasis.

The development of psoriasis is an intricate process. The condition is characterized by an overproduction of proinflammatory cytokines and other factors, which lead to an increase in the proliferation of keratinocytes. Figure 2 scheme demonstrates the IL23/IL17 axis which plays a crucial role in the development of psoriasis. To discuss further mechanisms of different polyphenolic compounds against psoriasis, it has been covered in the following section. Curcumin has been widely used to treat psoriasis and widely investigated by researchers. The antioxidative property of curcumin helps to rescue the oxidative stress caused by psoriatic lesions, along with its ability to reduce phosphorylase kinases noticed in psoriatic patients [12]. It prevents the TNF-α-dependent activation of NF-κB by directly attaching to TNF sites by covalent and non-covalent interactions [13]. The methylation of a TNF-α promoter is one mechanism through which curcumin can influence TNF-α expression. It also suppresses the LPS-mediated toll-like receptors (TLRs) 2 [14]. Moreover, Phosphorylase kinase (PhK), a serine/threonine-specific protein kinase, is also inhibited by curcumin non-competitively [15]. On another hand, resveratrol inhibits the LPS and TNF-α, and reduced the expression of IL-17A, IL-19, and IL-23p19 [16]. It has been found that IL-17A is up-regulated in psoriasis lesions and is essential to the pathophysiology of the disease. IL-17A induces the production of IL-19, which, together with other cytokines from the IL-20 subfamily including IL-20 and IL-24, is particularly responsible for the epidermal hyperplasia seen in psoriasis [17]. Sirt1/phospho-Akt-mediated signalling pathways were used by resveratrol. Akt protein is essential for cell survival, whereas Sirt1 plays a major role in regulating biological processes like stress response, apoptosis, and cell cycle [18]. It is also responsible for inhibiting aquaporin 3 (AQP3) [19]. Hydrogen peroxide (H_2O_2) is reported to be transported via aquaporin 3 (AQP3), which is a water/glycerol channel protein. H_2O_2 imported through AQP3 plays a role in keratinocyte nuclear factor-B (NF-κB) signalling and the pathogenesis of psoriasis [20]. Resveratrol anti-inflammatory properties result from cyclooxygenase 1 and 2 (COX 1/COX 2) suppression. The ability to scavenge peroxyl and superoxide radicals or its capacity to suppress enzyme activity to reduce oxidation are two possible explanations for its antioxidant action [21].

Epigallocatechin-3-gallate (EGCG), which is a tea polyphenol has anti-inflammatory, antitumor, antioxidant, and anti-ultraviolet radiation effects [23], is known to assist the normal differentiation of keratinocytes [24], also inhibits the oxidative stress and oxidative damage when skin is exposed to ultraviolent light as a result reduces the production of reactive oxygen species (ROS) [25]. EGCG

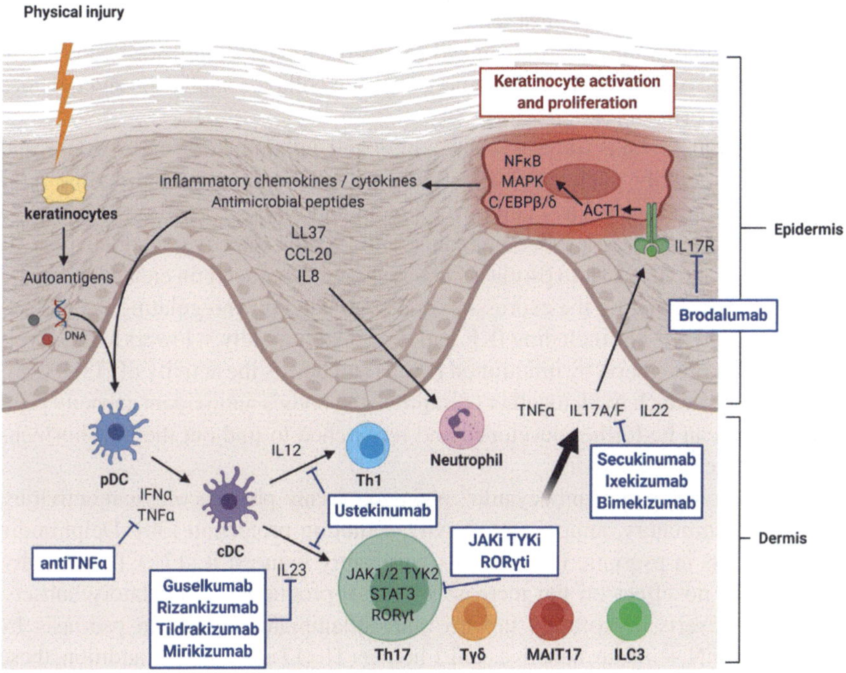

Fig. 2 The IL23/IL17 axis plays a crucial role in the development of psoriasis, and targeted therapies have been developed to intervene in this pathway. CCL20, a chemokine, is involved in this process, along with various immune cells like cDC, pDC, and ILC. Interferon alpha (IFNα), interleukins (ILs), IL17 receptor (IL17R), JAK (Janus kinase), JAK inhibitors (JAKi), mucosal-associated invariant T cells (MAIT), mitogen-activated protein kinase (MAPK), nuclear factor-kappa B (NFκB), RAR-related orphan receptor gamma t (RORγt), RORγt inhibitors (RORγti), signal transducer and activator of transcription proteins (STAT), gamma delta T lymphocytes (Tγδ), T helper lymphocytes (Th), tumor necrosis factor alpha (TNFα), tyrosine kinase (TYK), and TYK inhibitors (TYKi) are all part of this complex network in psoriasis and its treatment, Figure reproduced from Baugaut et al. [22]

possess various hydroxylic groups in their structure which reduces and smothers reactive oxygen species (ROS) and chelate metal ions. Through the stimulation of the nuclear factor erythroid 2-related factor (Nrf2) pathway and regulation of pro-inflammatory and anti-inflammatory signalling and proteins, including NF-κB, TNF-α, Toll-like receptors, and COX-2, it can also increase endogenous antioxidant defences. The inhibition of prostanoid biosynthesis enzymes, particularly COX-2, downregulation of the Nf-kB cascade, modulation of cytokine secretion and skin cell activation, and increased expression of the Toll-interacting Protein (TOLLIP) resulting in anti-inflammatory activity by blocking Toll-like receptor signalling. In addition to their capacity to inhibit DNA damage and regulate apoptosis, which results in genoprotective and chemopreventive benefits, green tea extracts also can diminish immune cell infiltration in injured skin, which indicates immunomodulatory actions. According to Sotto et al., it can also inhibit the MLPK-SREBP-1, PI3K, and mTOR pathways that control sebocyte death and sebum production. Last but not least, skin-regenerating properties linked to green tea phytocomplexes' capacity to increase elastin and fibronectin formation, regulate collagen synthesis, and enhance local blood reperfusion [26].

Whereas, Quercetin is well known to exhibit anti-inflammatory, anti-oxidant, and anti-psoriatic activity [27]. It reduces ROS and inflammation, enhances the Nrf2 signal while suppressing Th17 cell development and blocks NF-κB or JAK/STAT3 pathways. It also acts as a COX antagonist and COX-2 selective antagonist for its anti-psoriatic activity [28]. In psoriasis, it was shown that quercetin enhances disease recovery by upregulating the expression of TRAF3 and downregulating the expression of NIK and NF-kB, including IKK and RelB. Additionally, it lowers MDA levels in skin tissues brought on by imiquimod (IMQ) and boosts the activity of GSH, CAT, and SOD, all of which work together to improve the body's antioxidant capacity [29]. Therefore, it can be further developed and researched to find out the new horizons for the treatment of psoriasis.

Another compound "Anthocyanin" possesses many pharmacological activities, like anti-inflammatory, antioxidant activity, radiation protection [30]. Delphinidin (anthocyanins) in psoriatic individuals significantly reduced IL-17^+ $CD4^+$ T cells, with little or no effect on the increase of IL-10-producing T regulatory subsets. Delphinidin exerts a profound in vitro anti-inflammatory effect in psoriasis by suppressing IFN + innate, adaptive, and T helper (Th) 17 cells [31]. In addition, these polyphenolic compounds have shown their therapeutic properties in immune cells as well. Specifically, flavonoids inhibit the maturation of Dendritic cells (DCs) by downregulating the expression of markers such as CD80 and CD86. Subsequently, it reduces and hinders the proliferative response of CD4 + cells. Moreover, it is reported that these compounds reduce the release of histamine, PGs, and cytokines from mast cells. Therefore, it can be said that flavonoids have a significant influence on immune cells that can alter the inflammatory response (Fig. 3).

3 Nanotech Strategies to Deliver Polyphenols and Its Intended Target Area

Nanotechnological approaches towards the treatment of inflammatory conditions were first proposed by Siddiqui et al. [33], and since then the field has been rapidly advancing and has provided multiple prospects and promising solutions to enhance topical therapeutics for psoriasis. As for the other diseases, in psoriasis also, nanodrug-based topical therapies have solved these challenges over conventional counterparts: (a) enabling insoluble drug delivery and thus increasing bioavailability and efficacy [34, 35]; (b) facilitating drug diffusion across the skin barriers by enhancing absorption and thus reducing the dosage administration [36–38]; (c) tuneable drug release at defined dosages over a certain period [39–41]; (d) protecting embedded/encapsulated active pharmaceutical ingredients from instability issues

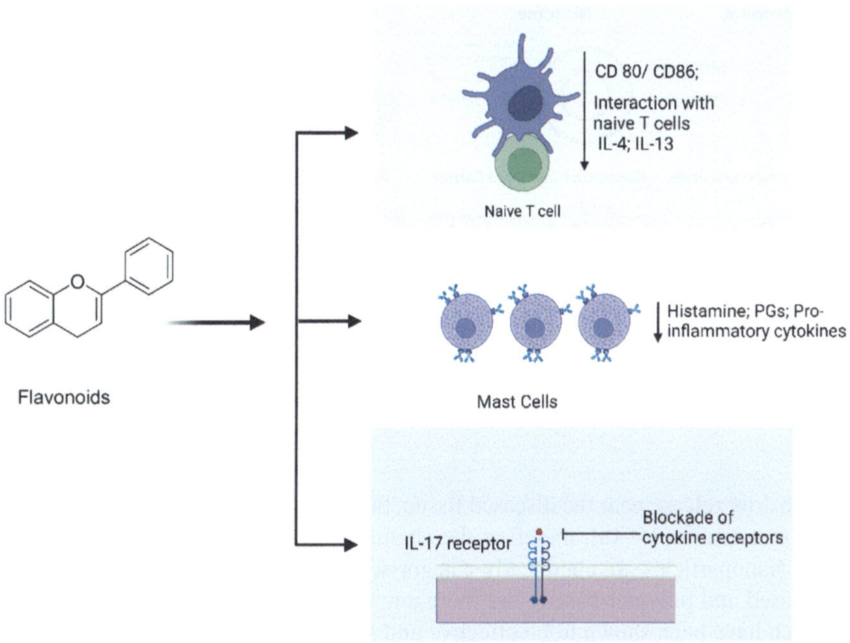

Fig. 3 The illustration demonstrates the effects of flavonoids on the immune cells and inflammatory cytokines. The figure is modified from Maleki et al., [32], and Created with Biorender.com

such as degradation along with increased solubility [42, 43]; enabling targeting potentials when modified with cell-specific ligands, thus minimising off-site adverse effects [44, 45]. To date, the three main routes for therapeutics delivery to penetrate the skin barrier include intercellular, intracellular, and transfollicular. Remarkably, most nanocarriers are generally investigated and validated at the pre-clinical stage, therefore, more insightful characterisation and optimised regimen are required if they move from bench to bedside. As there is an exponential growth in the research of nanotechnological approaches (such as nanoparticles and nano matrix, in addition to physical strategies) for drug delivery, it has become rather unclear with respect to a rational selection of suitable materials to develop and design an effective topical treatment option for psoriasis. This chapter outlines the advantages and limitations of a variety of nanocarriers and methodologies that have been investigated and utilized in topical drug delivery systems in experimental psoriasis. Additionally, based on the nanomaterial surface features, we also intend to indicate the specific benefits of individual nanomaterials as clinical topical drugs for psoriasis.

The advantage of utilising nanosized particles is their large inner surface area-to-volume ratio, which provides them with huge potential for both surface modification and drug payload capacity [46, 47]. In general, nanoparticles are composites of two or more materials assembled into three general layers: the surface layer, the shell layer, and the core [47, 48]. The aim of nanoparticles in drug delivery is to not only target and

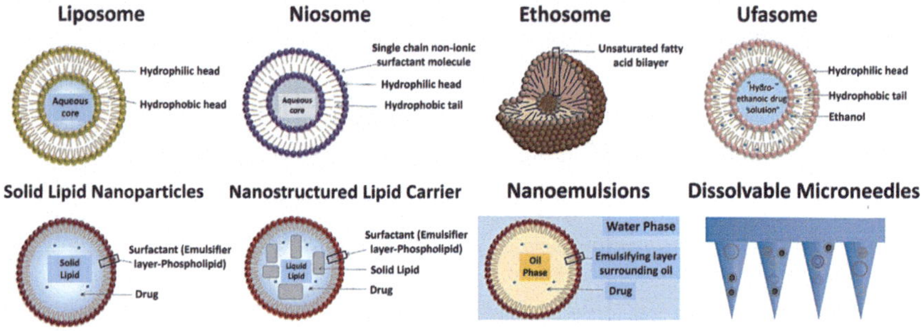

Fig. 4 The illustration depicts various lipid-based nanocarriers for the treatment of psoriasis, such as Lipid nanocarriers (Liposome, Niosome, Ethosome, Ufasome, Solid Lipid nanoparticle (SLN), Nanostructured Lipid carrier (NLC), Nanoemulsion and Dissolvable microneedle. Image modified and adapted from [55, 56]

localise drug release near the diseased tissue, but also protect the encapsulated drug from degradation [46–48], therefore the chemical composition is a critical design factor. Nanoparticles are chemically categorised as organic nanoparticles (such as lipid-based and polymer-based) and inorganic nanoparticles (such as metals), both of which have been shown to be effective and successful drug nanocarriers against psoriasis [49]. Lipid-based nanoparticles are well known for the utilisation in topical therapies due to their biocompatibility, biodegradability, and nontoxicity properties, while polymer-based nanoparticles consisting of natural or synthetic polymers are favoured due to their versatility, sensitivity, and readily tuneable chemical or physical modifications [50]. On the other hand, metal-based nanoparticles are favoured due to their antimicrobial properties, which is critical for the treatment of skin infections [51]. Herein, we present the most relevant types of nanoparticles adopted to encapsulate APIs to overcome the psoriatic skin barrier and then improve the antipsoriatic effects of relevant these APIs, also, Fig. 4 shows different lipid-based nanocarriers that have been explored for the treatment of psoriasis.

Lipids-based nanoparticles are generally used for the delivery of therapeutic medications through the skin due to their compositional similarity to the epidermis' lipids that make the outer cell structure. In addition to the biocompatibility, lipid-based nanocarriers have greater penetration ability into deeper skin layers thus facilitating the permeation of medication through the stratum corneum [52]. With deeper penetration, however, the encapsulated medication is absorbed by the various epidermal layer, which then minimises subsequent dose-related adverse reactions systemically. For this reason, lipid-based nanocarriers are loaded into topical formulations and are generally believed to be the safest and most effective treatment option for psoriasis, based on current clinical indications [53, 54].

3.1 Liposomes

Liposomes are spherical structures composed of either natural and/or synthetic phospholipids synthesised in an aqueous solution [57]. Liposomes contain a hydrophilic head and two a-polar hydrophobic chains (amphiphilic nature), and therefore, they have been useful as drug carriers for both kinds of APIs, hydrophilic (inside the aqueous phase, hydrophilic head) and lipophilic (inside the lipid phase, hydrophobic tail) molecules [58]. Furthermore, adding cholesterol as a structure stabilizer has been shown to provide rigidity to the bilayer membrane of the particle above its phase transition temperature (Tg), making the membrane less ordered below the Tg. Liposomes have been categorised as small unilamellar vesicles (SUVs, diameter 25–50 nm), large unilamellar vesicles (LUVs, diameter 50–500 nm), and multilamellar vesicles (MLVs, diameter 0.5–10 μm) [58]. Liposome synthesis has been evolving over the years, including the thin-film method [59], injection method [60], extrusion method [61], and emulsification method [62]. Liposomes are unique because of their ability to solubilize hydrophobic drugs; act as a local depot for sustained drug release; act as a penetration enhancer due to the diffusion of the phospholipid molecules or non-ionic surfactants into the lipid layer of the stratum corneum; enhances localized drug accumulation while acting as a rate-limiting membrane barricade towards systemic absorption [63]. Additionally, constituents of the liposomes have the ability to manipulate their own physicochemical characteristics such as size, charge, permeability, lamellarity and thermodynamic phase [58]. Moreover, the most essential mechanical property of liposome membranes is the bending and stretching elasticity, since they are related to the particles' resistance to deformation and stretching experienced during drug delivery (resistance experienced during high shear rate in both the skin's viscoelastic environment and the blood circulation, followed by deformation as it navigates through tiny vessels such as capillaries) or during cellular processes (cell division and osmotic shock). The liposome membrane bending is characterised by the bending rigidity and the stretching elasticity modulus [64]. Unlike polymerosomes (polymer nanoparticles), liposomal structural deformity is unable to return to its natural state when induced by an external stimulus (electric pulse), and this is one of the main issue that influences the stability of liposomes [64]. Regardless, due to the liposomal structural similarity to lipophilicity of cells, they are often utilized in nano-formulations for a range of dermal diseases including psoriasis.

Wadhwa et al. synthesised Fusidic acid (FA) loaded liposomal nanoformulation for the management of plaque psoriasis via transdermal delivery [65]. FA liposomes (FA-LP) were synthesised with varying sizes of 420–740 nm, morphology was a uniform spherical shape and low aggregation, surface charge was PDI < 0.4, drug retention of 1.620 ± 0.8 mg/cm^2, and skin permeation of $> 75\%$. Confocal laser scanning microscopic (CLSM) data revealed significant cellular uptake of the dye-loaded liposomes. Their nanoformulation was stable consisting of uniform size, as well as resistant to drug leakage, which is a limitation with liposomes. Furthermore, there was significantly enhanced in vivo anti-psoriatic efficacy ($p < 0.05$) of the FA-LP formulation demonstrated when compared to formulations prepared conventionally, suggesting its potential in addressing the infectious challenges of psoriasis. Long Xi et al. prepared celastrol-loaded mannosylated liposomes (CE-LP) using the thin-film dispersion method to study the uptake of celastrol by specific cells [66]. Their nanoformulation contained CEL, tween, lecithin and mannose-DSPE-PEG2000. The findings confirmed the nanosize and homogenous distribution of lipid particles. The CE-LP formulation showed an entrapment efficiency of $90.8 \pm 0.4\%$ while exhibiting a controlled release of celatrol (35%) within 48 h compared to the solubilized celastrol release (80%) within 12 h. There was an increased dendritic cell uptake (3 times) of mannose grafted liposome loaded celastrol compared to formulations without mannose grafting. Additionally, mannose-grafted liposome formulation inhibited the manifestation of co-stimulatory factors like CD60 and CD80 on BMDC induced by R484, while enhancing the anti-maturation effect of celastrol in comparison to the free drug [67, 68]. Yu et al. recently designed improved liposomes in which curcumin (Cur) is loaded for the treatment of psoriasis. [69]. The peptide-modified curcumin-loaded liposome electively opened the paracellular pathway across cell epithelium, thus increasing skin permeability and the cutaneous retention of Curcumin from the peptide-modified curcumin-loaded liposome compared to the Cur liposomes without the modification. As a result, the application of peptide-modified curcumin-loaded liposomes improved visible skin symptoms such as erythema and silver scales, while decreasing the epidermal thickness on IMQ-induced psoriasis mouse model. This shows that, the alteration of the components of the liposome structure, the antipsoriatic effects of drug-loaded liposomes have changed from ineffective to significantly effective. Such studies demonstrate the ability to enhance the uptake of lipid-based nanocarrier in the cells towards the management of psoriasis.

3.2 *Niosomes*

These carriers are lipid bilayer nanovesicles which are made up of non-ionic surfactants, which are added to cholesterol with subsequent hydration in aqueous media, with the ability to improve the solubility and enhance the drug concentration delivery at the target site. [70]. Handjani Vila et al. first successfully manufactured noisome nanoparticles in 1979 [71], and since then Spans, Tweens, and Brij have been the non-ionic surfactants used for niosomes preparation [70, 72]. Other rare surfactants used in the niosome synthesis are Wasang®, Gemini, and Bola surfactants [70, 72]. Surfactant concentrations used in the preparation of noisome play a very important role in controlling the rate of drug release, permeability of the nanocarrier along with API, as well as the size distribution [70]. Niosomes synthesis has since evolved for various other types of niosome vesicles such as spanlastics, liposomes, In-situ forming lipid vesicles and aspasomes, each is slightly modified versions of niosomes with varying concentrations or blends of surfactants used to synthesise them [73]. Niosomal vesicles have been reported to be osmotically active, as well as having better oxidation stability as compared to other types of vesicles, and therefore are excellent drug encapsulating systems as well as act as depot system from the slow and sustained release of drugs [70, 74]. They are well stable, not as much of immunogenic, and economical to design than other phospholipids-based nanocarriers, since the expensive materials such as solubilizers like transcutol or edge activators can be avoided for designing the nanoformulations [70, 73, 74].

Celastrol was successfully loaded in niosomes using a thin film hydration method consisting of different grades of span, and cholesterol [75]. Niosome particle sizes and PDI were in the optimum range indicating homogeneity of the colloidal system, whereas a zeta potential reflected good stability of the nanoformulation. Adding celastrol to the noisome improved the in vitro permeation with a drug content of approx. 465 ng/cm^2 in the skin as compared to the raw drug with approx. 35 ng/cm^2 drug concentration in the skin. While celastrol-loaded niosomes improved the infected skin of mice in-vivo. Another drug, Diacerein was successfully loaded in niosomes formulation with content of cholesterol, using a similar method mentioned above [76]. Diacerein-loaded niosomes were 477 nm in size which had an entrapment efficiency of more than 80%, and demonstrated stability for more than 2 months at 4 °C. The retention of API was confirmed by CLSM within the skin's dermal and epidermal layers. Diacerein-loaded niosomes have shown increased in-vitro release with a higher concentration of span. The formulations containing span (70% and 90% concentration) showed a flux of around 2.0 mg/cm^2/h, respectively. Such studies demonstrate the efficacy of niosomes as potential polyphenolic nanocarriers with the ability to increase drug solubility, better nano structural stability than liposomes because of the addition of surfactant, as well as improved penetration ability and surface functionalization giving it more acceptance.

3.3 *Ethosomes*

These nanovesicles are made of a combination of phospholipids and alcohol (usually ethanol) in high concentrations (almost 20–45%), and deionized water. Adding alcohol to the formulation enhances both the permeation and its interdigitation effect since it requires the flexibility to navigate through the vesicle structure which helps them to pass through the skin pores. The addition of ethanol also improves the formulation nature as it entraps drugs of different physiochemical properties. Ethosomes are liposome-like vesicles that were first developed by Touitou et al. in 1996 [77]. They are both bilayer or multilayer multilamellar structures (size ranging from 30 nm to a few microns) in which the internal core of alcohol (high concentration 20–45%) is loaded by a lipid external layer [78, 79]. Ethanol enhances skin permeation while providing flexibility by deforming the vesicle structure (interdigitization effect) which helps them squeeze through the skin pores [78, 79]. Moreover, due to the amphiphilic nature, both kinds of drugs can be entrapped into ethosomes vesicles [80]. By modifying the surface of the vesicles using surfactants, and functionalised materials, both drug loading and therapeutic efficacy at the targeted site of action can also be enhanced [78–80].

Tocopherol polyethylene glycol succinate (TPGS) modified curcumin (Cur) loaded ethosomes were synthesised by Guo et al. using the most common method i.e., ethanol injection method [81]. These TPGS-Cur loaded ethosomes were synthesised to improve the amount of the API at the targeted site, utilising the excellent penetration capability of these nanocarriers by overcoming the challenging barrier of stratum corneum due to the ethanol component in the formulation. The findings of this study revealed the size of these nanocarriers (<200 nm), PDI close to 0.100 which indicates the uniform size of dispersion. Importantly, an increase in the amount of the drug was demonstrated through the skin permeation study with almost a 3-times increase compared to the plane ethanolic solution of the drug in an in vitro investigation. Notably, TPGS-Curcumin ethosome demonstrated higher inhibition in the proliferation of the inflammatory cells in vitro against HaCaT cells. The potential to inhibit the proliferation of inflammatory cells validates the anti-inflammatory effect of these nanocarriers against psoriatic-related symptoms [82, 83].

3.4 Ufasomes

Ufasomes synthesis was first described by Gebbicki and Hicks [84]. These are unsaturated fatty acid vesicles made up of closed lipid bilayers (oleic and linoleic acids) suspensions. The orientation of hydrocarbon tails directed are aligned towards the membrane interior while the carboxyl groups are in contact with water, in which around half of the carboxylic groups are ionized with the pH range restricted to 7–9 [85–89]. The stability of these nanocarriers is dependent on the optimum concentration selection of fatty acid, the concentration of cholesterol added, the working range of pH, buffer, concentration of lipoxygenase and the incorporation of divalent cations in the composition of the formulation [85–88]. These nanocarriers have the ability to improve the drug penetration through the skin layers by attaching to it and enabling lipid exchange between the outermost barrier of the skin layer i.e., SC [85, 89]. Additionally, these are considered more stable and economically viable as compared to liposomes, as well as demonstrated higher entrapment efficiency for both kinds of drugs due to their amphiphilic nature [84–89]. Moreover, the semipermeable membranes of these nanocarriers make them suitable for drug loading and delivery, and for preventing enzyme degradation [85–88]. It has been reported that these nanocarriers were used for the delivery of oleuropein (a biological phenolic compound derived from olive oil), and demonstrated enhanced antioxidant potential (protecting cells from peroxide radicals), thus showing its potential for nutraceutical applications [86]. In another study, Ufasomes were used to encapsulate Cinnarizine, and then subjected to lyophilization for long-term storage. Later on, it was incorporated into hydrogels for administration via *the* nose to brain route. [90]. The experiment of ex-vivo imaging validated that the nanocarrier aided the vesicular carriers in penetrating deep through nasal mucosa layers. Albeit not polyphenols, Ufasomes have been increasingly used in the encapsulation and enhancing delivery efficiency of clotrimazole and fluconazole [91] and anti-arthritis agents against inflammation in the rheumatoid arthritis [92] showing their potential to be used in the encapsulation and delivery of polyphenolic compounds for the treatment of psoriasis.

3.5 *Solid Lipid Nanoparticles (SLNs)*

Solid lipid nanoparticle (SLNs) was first synthesised in the 1990s and are considered to be first-generation lipid nanocarrier systems. SLNs are made of solid lipids, surfactant(s), and water, comprising an average particle diameter of 50 - 1000 nm [93–95]. The surfactants used in the synthesis of SLNs act as an emulsifier and a stabilising agent, whereas the lipids (including triglycerides, fatty acids such as stearic acid, glyceryl monostearate waxes are solid at the body and room temperature [93–95]. SLNs are commonly synthesised using a cold or hot homogenisation method and solvent evaporation method [96, 97]. There are various methods of preparation of these SLNs [98]. Additional advantages of using SLNs are their synthesis without organic solvents [93–95], biodegradable and biocompatible, physicochemically stable, less toxicity, and can be synthesized at large industrial scales at a minimal cost. [93–98]. Compared to liposomes, SLNs structures are dynamic as they can be modified by the addition of ionic or non-ionic surfactant, and are frequently utilized with a co-surfactant to keep the particle size under the range of 200 nm, adding to their versatility in drug delivery [93–98]. Nevertheless, established studies also show that sodium dodecyl sulphate and stearic acid components in SLN synthesis might have altered their high cytotoxicity level and thus reduced biocompatibility [99]. Regardless, SLNs have the ability to overcome the limitations of liposomes and polymeric NPs, such as liposome's phospholipid polymer degradation cytotoxicity, and drug leakage [100]. Ali et al. encapsulated thymoquinone in SLNs using melt-emulsification followed by the sonication method [101]. The size of SLNs was found to be 108.34 nm with a PDI of 0.19 to 0.3, and an entrapment efficiency of 62.5–82.3%. The drug release was monitored almost 60% (after 24 h) showing an initial burst release profile followed by slow release after 4 h. This led to a significant increase in the drug permeation through the skin SC barrier with a flux of approx. 6.00 mg/cm^2/hr. This burst release and immediate skin availability did not contribute to allergies such as redness on the skin indicating reduced skin irritation due to the encapsulation of the drug into the SLNs.

3.6 *Nanostructured Lipid Carriers NLCs*

These are the second-generation lipid-based nanocarriers of SLNs. These were developed in response to the limitations of SLNs such as the leakage of the drug through the nanocarrier formulation during long-term storage, which leads to less drug loading [102, 103]. NLCs and SLNs have almost the same preparation methods, namely cold and hot homogenisation respectively, and hot emulsification with the ultrasonication process [96, 102, 103]. However, unlike SLNs, the oils in NLC formulation protect the drug encapsulation by ensuring the leakage of the loaded API is reduced to almost minimal when subjected to long-term storage period as they have shown to stop the recrystallisation process. The blending of liquid lipid with the solid lipid with the surfactant and cosurfactant develops a highly disordered nanostructure that has the potential to encapsulate a higher amount of drug and can carry for a longer duration without leakage, also, it provides the sustained release effect. In addition, the composition of these NPs provides the structural stability. [96, 104]. NLCs are employed to deliver the APIs at the target site locally, and thus the topical application of these NPs has been shown to enhance skin irritation and occlusive properties.[105, 106]. Drugs can be loaded in the binary mixture of liquid and solid lipids. Different researchers have tried different ratios of solid-lipid and liquid lipids to make a stable binary mixture of lipids. [107]. The main aim of designing NLCs was to obtain an improved lipid core without a crystalline matrix with a highly disordered structure which permits the higher loading of API in the lipid matrix [96, 107]. Additionally, the NLC structure has a larger surface area for the drug to be dissolved within the liquid lipids, and this is due to the random organization of solid lipids that gives NLCs an increased drug-loading capacity [108]. Due to the large surface area, NLCs have extensively been used in dermatotherapeutics such as antidermatitics and antipsoriatics.

Friedrich et al. demonstrated the improved delivery of resveratrol and curcumin when encapsulated in NLCs, where their primary aim was to study their efficacy in excised human skin and the antipsoriatic activity with primary skin cells [109]. Their skin penetration data demonstrated improved delivery of resveratrol (upon co-delivery) into the deeper layers of the skin compared to loading of resveratrol only. Another study demonstrated improved skin infiltration and controlled release of psoralen using NLCs compared to SLNs [110]. The nanocarriers were prepared using liquid (squalene) and solid (Precirol ATO 5) cores respectively. While various emulsifiers were used to optimise the polarity, zeta potential, drug encapsulation capacity of the nanoparticles. Despite concluding that both nanoparticle systems have great potential for dermal application, their data demonstrated that NLC formulations exhibited higher permeation and sustained release of psoralens. Capsaicin (CAP) was also proposed as one of the treatment s for psoriasis to estimate the extent of permeability and skin retention, however, undesirable effects were experienced upon topical administration such as the emergence of erythema and burning sensations [111]. CAP encapsulated in SLNs and NLCs demonstrated better permeation of CAP due to the higher content of lipid matrix in the nanosystem and smaller size as compared to CAP solution only. Such results show that the designed lipid nanocarriers are suitable lipid-based nanodelivery systems for the encapsulation of curcumin and resveratrol and their delivery for skin applications.

3.7 Nanoemulsions

Nanoemulsions are a heterogeneous isotropic system that consists of two immiscible liquids where the active ingredient is dispersed within the nanodroplet matrix of the emulsion [112–114]. The immiscible liquids are generally a combination of 'water in oil' (W/O), or 'oil in water' (O/W), or double emulsion (W/O/W and O/W/O) which is stabilised with amphiphilic surfactants (Surfactants: Tween® 80, amphiphilic proteins-caseinate, phospholipids- soy lecithin, and co-surfactants: ethylene glycol, propylene glycol, ethanol and propanol), yielding an average droplet size of < 200 nm depending on the phase media [114–116]. The zeta potential plays a significant role in the nanoemulsion's stability. For example, nanoemulsions with a single emulsifier system with a low molecular weight surfactant, a zeta potential of >±30 mV offers good stability and continues improving when it reaches ± 60 mV, whereas a zeta potential of ± 20 mV offers transient short-term stability, while a surface charge of the − 5 to + 5 mV range have shown rapid aggregation of the fatty globules leading to an unstable system [114]. At higher molecular weight emulsifier systems, a shift is observed towards higher zeta potential, and this change maintains the globules in Brownian motion long-term while offering stability within the system [114]. Nanoemulsion systems are mainly prepared by three methods: high-pressure homogenization (low yield), microfluidisation, and phase-inversion composition and temperature [117–120]. Out of other lipid nanocarriers, nanoemulsion has the closest structural resemblance to a micelle, however, the innermost part of nanoemulsion is the oil phase that solubilises hydrophobic drugs. Therefore, it important to choose the appropriate oils such as different types of triglycerides like long-chain triglycerides (LCT), medium-chain triglycerides (MCT), and short-chain triglycerides (SCT) [113]. Besides encapsulating lipophilic active compounds for enhanced skin delivery, nanoemulsions are also excellent carriers to encapsulate essential-oils and natural bioactive compounds which can rendered in spray, gel, cream, and aerosol, making it a versatile and most researched drug delivery nanocarrier [121–123]. Nanoemulsions have also been known to be thermodynamically unstable while kinetically stable and are sensitive to pH and temperature [124, 125], consequently undergoing particle flocculation, coalescence, and/or Ostwald ripening phenomena yielding unstable emulsions [126]. Moreover, with appropriate optimisation in the selection of oil and surfactants, as well as during the formulation preparation, a stable nanoemulsion system can be achieved for drug delivery purposes [126].

Algahtani et al. formulated curcumin nanoemulsion with a particle size of 10 nm utilising a low-energy emulsification method for the topical application of curcumin on psoriatic skin condition [127]. Their data indicated that the addition of l carbopol 934 as a polymeric gelling agent to the nanoemulsion formulation hydrated the stratum corneum (SC) and improved the ductility of curcumin. Additionally, there was a high curcumin skin deposition and cumulative amount of curcumin permeation using curcumin nanoemulsion gel, approximately 7- and fivefold compared to those of curcumin gel. In comparison to betamethasone-17-valerate gel, treating IMQ-induced psoriasis mice with curcumin gel nanoemulsion has led to facile improvement in psoriatic symptoms within 4 days. This fast improvement was attributed to enhanced penetration of curcumin from the curcumin-loaded nanoemulsions across the epidermal basement membrane while inhibiting the expression of inflammatory factors and reducing keratinocyte proliferation.

Paclitaxel (Cyclosporin) is another commonly used macromolecular lipophilic drug that is effective in the treatment of psoriasis. Pandey et al. have successfully incorporated paclitaxel in nanoemulsion gel with greater penetration and localisation of the drug in deeper skin layers avoiding systemic delivery via vasculature within the dermal region [127]. The drug-nanoemulsion particle size achieved was < 50 nm, utilising Tween-80 as the surfactant and Isopropyl alcohol as the co-surfactant, while using isopropyl myristate as the oil phase and Carbopol-940 as particle stabilising agent. They also demonstrated improved bioavailability of paclitaxel of up to 70% when oral paclitaxel was administered after applying paclitaxel-loaded nanoemulsion on psoriatic skin. To inhibit epidermal proliferation while eliciting anti-inflammatory capabilities, the authors have demonstrated that the administration of paclitaxel nanoemulsion combined with oral administration can be considered for enhancing the effect of the drug and for the treatment of erythrodermic psoriasis.

3.8 Hybrid Nanocarriers for Psoriasis Management

Microneedle arrays (MNAs) have also been used in conjunction with nanoparticles to deliver lipophilic drugs into scaly psoriatic skin. The advantage of using MNAs is that they can puncture the skin without any pain and avoid the stimulation of nerve endings [55], while successfully delivering a drug across the main barrier of the skin, i.e., SC, and showing promising outcomes for application in psoriasis [128, 129]. Paleco et al. successfully designed lipid microparticles encapsulating quercetin to overcome the poor physicochemical properties such as its stability, solubility and permeability, and then loaded the quercetin lipid microparticle in microneedles to further improve skin permeability and quercetin skin deposition [130]. Compared to the free quercetin MNAs, encapsulated lipid microparticles embedded in MNAs containing quercetin showed greater skin permeability. Although microneedles provide an effective route for percutaneous penetration by disrupting the skin barrier, the permeability is reduced by quercetin's poor solubility. However, the authors showed greater permeability by combining microneedles and lipid microparticles. Furthermore, compared with untreated skin, there was increased deposition of quercetin lipid microparticles localising in the stratum corneum and viable epidermis increased by 2- and 5 folds post administration with microneedles. This demonstrates that combining quercetin lipid particles and microneedles, allows targeted penetration on the epidermis, with no sensation of pain along with ease of administration. A similar enhanced permeation effect was also shown by Yerneni et al., by successfully delivering curcumin from curcumin-encapsulated exosome embedded in dissolvable microneedles (Cur-Exo-dMNAs) [131]. The curcumin in Cur-Exo-dMNAs exhibited five-fold higher stability in vitro than naïve curcumin or curcumin-loaded exosomes, indicating the effect of both the microneedles and exosome in protecting the curcumin from external factors (temperature and light). Additionally, Cur-Exo-dMNAs retained their bioactivity when stored at room temperature for at least 12 months, while significantly suppressing LPS and Imiquimod-triggered inflammation in vivo. This shows combining microparticles in microneedles has the potential to become an effective local-delivery strategy for inflammatory skin diseases.

4 Innovative and Untapped 3D in Vitro Models for Polyphenolic Encapsulated Nanoparticulate Systems

Polyphenolic nanoparticles-based therapeutic strategies targeting psoriatic conditions generally utilises in vitro cell culture and in vivo rodent models are commonly used in the research as mentioned above, however, alternative methods of assessing polyphenols release from nanoparticles and efficacy of actives in novel methods are required. As in vivo research is increasingly becoming cost in-effective and ethically concerning, along with confounding results, researchers have resorted to various other novel methods such as spheroid cell culture, 3-Dimensional (3D) cell culture and skin on a chip as in vitro models to assess psoriatic activity in a physiological manner is a new horizon of research, which will be discussed in the following section.

3D cell culture leads to translatability for simulating thick skin and inflammatory conditions in an in vitro cell culture method. These include Reconstructed Human Epidermis (RHE) and Human Skin Equivalents (HSE) models to simulate an in vitro condition for psoriasis where control allows partial thickness or full thickness to simulate mild and severe psoriasis respectively [132]. These models resemble a closer in situ model of this disease. Initially, a porous membrane model developed by Krueger and Jorgensen (1990) using scaffolding structures to allow the growth and culture of keratinocytes and fibroblasts leads to a self-assembly model to mimic the structures of the skin, removing the need to obtain exogenous materials or skin biopsies [133]. Krueger and Jorgensen [134] generated an interactive skin equivalent system by co-culturing keratinocytes and fibroblast in a Milli-cell cell culture system, creating a simulated psoriatic model by forming an artificial epidermal layer and dermal layer [134]. The epidermal layer was seeded with fibroblast of psoriatic patient, thickening of the artificial epidermal layer, simulating psoriatic conditions. The model lacks immunological markers and responses which are often present in psoriasis. In a similar study, Martin et al., co-cultured psoriatic keratinocytes and lymphocytes in an epidermal layer culture, were present in a psoriatic in vitro model, a crosstalk between the cells were observed, further understanding the signalling pathways of psoriasis, leading to novel ways of developing psoriatic disease models in 3D cell culture [135, 136].

Self-assembly of keratinocytes and fibroblast was demonstrated by Guo, Jackow [137], forming a human skin spheroid system to model psoriatic disease for drug development. Primary fibroblasts were initially seeded and cultured for 2–5 days followed by the addition of primary keratinocytes attaching to the seeded fibroblast, forming a spheroid core where then IL-17 was added to induce a psoriatic disease state within the culture. Signature psoriatic genes such as kRT16, S100A7, S100A12, DEFB4A and IL-8 were expressed and detected via RT-PCR. This cell culture method allows for a cost-effective high throughput system for developing therapeutic targets. In addition to pluripotent stem cell culture via keratinocytes, skin biopsies and harvesting healthy skin is a potential method of in vitro testing for psoriasis. Jean, Lapointe [138] and Tjabringa, Bergers [139] developed and validated in vitro psoriatic skin models by utilising surgically de-epidermised dermis and culturing them to generate HSEs. The HSEs were exposed to pro-inflammatory cytokines to induce a psoriatic condition measured via thickness, RT-PCR or ELISA. The thickness of the skin produced in HSEs was found to be significantly thicker in psoriatic samples, allowing visual assessment of psoriatic progression. Additionally, microneedles application applied onto HSEs allows researchers to assess microneedle injections in a high throughout manner to determine optimal microneedle specifications and formulation loading [140]. Polyphenols-loaded within nanoparticulate system are a possibility to assess the activity of formulation release in this spheroid co-culture and in 3D cell culture, elucidating initial in vitro release and activity on cells with lower cost. The advancement of commercially available models, such as skin on a chip, show promising future for the assessment of psoriasis.

There are several commercially available 3D skin models for in vitro applications. Biopta supplies ex vivo epidermis-dermis matrices for both lesional psoriatic and healthy human skin. MatTek utilises both normal epidermal keratinocytes and psoriatic human dermal fibroblast, sold commercially as EpiDerm™ skin inserts, whereby normal human epidermal keratinocytes layer was grown on an insert and psoriasis was induced with a proprietary cytokine mix, forming a psoriatic skin insert for in vitro assessments. Table 1 summarises potential 3D skin models to test different polyphenolic compounds in skin diseases, including psoriasis.

Skin-on-a-chip models utilise a chip platform comprising a chamber for culture media and immune cells separated by a skin chamber allowing to measure changes in metabolism and cellular signalling occurring in an in vitro model [141]. This system allows the simulation of skin thickening occurrence with the changes in inflammatory markers via cellular secretions and vascular altercations in psoriatic conditions, leading to the determination of the efficacy of polyphenols on a simulated skin model [141–143]. Drug permeation from nanoparticles encapsulating polyphenols via the use of a skin-on-chip model has been proposed by Abdelkarim et al. [144]. The use of RHEs and HSEs can be utilised in the skin-on-chip component, opening new methods of monitoring drug release, anti-inflammatory effects of polyphenols and physiological changes in a convenient and reproducible assay [141]. In this model, dermal formulations can be applied to the skin chamber and monitored with the changes in biomarkers where lymphocytes are present, allowing measurement of anti-inflammatory markers produced after application.

Table 1 Potential 3D skin models to test different polyphenolic compounds in skin diseases, including psoriasis. The table was modified from Desmet et al. [145] and Sarama et al. [146]

Model	Matrix	Cellular components	Hallmarks	References
RHE (3D)	EpiDerm™ (MatTek)	Normal primary human KC	Acanthosis and hyperkeratosis	[147]
	Biopta	Healthy skin and psoriatic skin induction via cytokine mix stimulation	Parakeratosis	[148]
	Porous membrane scaffold	Human KC and FB	Parakeratosis, keratinocyte apoptosis	[134, 135]
	Polycarbonate filter scaffold	Human KC	Parakeratosis thinner epidermis ↓Filaggrin-2	[149]
Spheroid (3D)	Self-assembly	Normal Primary KC and FB IL-17 treatment	↑kRT16, ↑S100A7, ↑S100A12, ↑DEFB4A ↑IL-8	[137]
HSE (3D)	Healthy adult skin	De-epidermised dermis	Parakeratosis ↑mRNA DEFB4A ↑IL-6, Il-8, IL-23, CXCL10, CK-16	[138, 139]
RHE (3D) SOC model	Porous membrane	Human KC as epidermis cell	human leukemic lymphoma cell line (U937) as dendritic cells + LPS	[150]
	Collagen	HaCat epidermal layer	Human leukemic lymphoma cell line (U937) as dendritic cells + LPS	[151]
HSE SOC model	Porous membrane	HaCaT	Endothelial cells were stimulated with cytokine TNF-α for psoriasis induction	[152]

*Abbreviations**HSE**: Human Skin Equivalent; **3D**: 3 Dimensional, **SOC**: Skin-on-chip; **RHE**: Reconstructed Human Epidermis*

Delphinidin is a polyphenolic anthocyanin found as pigments in plants such as grapes and blueberries, showing promising antioxidative and anti-inflammatory properties for psoriasis [153]. Pal et al., utilised a reconstituted skin model of psoriasis developed by MatTek and showed the efficacy of delphinidin in increasing caspase-14 and reduced the markers for cell proliferation and inflammation [154]. In a similar study, they observed that applications of 5–20 μM delphinidin treatment on both normal and psoriatic EpiDerm™ reconstituted human skin increased the processing of caspase-14, trandsgulatminase-1 and filaggrin [154]. Subsequently, Delphidin also induces caspase-14 expression in a 3D epidermal skin equivalent model while slowing epidermal thickening in a psoriatic model [30]. Chamcheu et al. investigated the efficacy of Delphinidin via normal skin equivalents and psoriatic skin equivalents (PSE) as 3D multilayered skin equivalents [30]. Using derived keratinocytes from healthy human skin and patients exhibiting epidermolytic ichthyosis (EI) were cultured in a Millicell-PCF insert (Millipore corporation, MA USA) as a model for assessing the efficacy of delphinidin on psoriasis[155]. Delphinidin was shown to enhance cornification via epidermal thinning and increase immunological markers for epidermal differentiation but not apoptosis in PSE. It was shown that at 20μM delphinidin decreases the expression of proinflammatory cytokines in PSE such as TNF-α, IL-1β and IL-2 significantly measured via Multi-Analyte ELISArray Kit (SABiosciences, Frederick, Md., USA) from the 3D skin culture supernatant analytes. As polyphenols are prone to degradation, a nanoparticle delivery system potentially allows a longer residence time after application on the epidermal layer.

Subsequently, Zlabiene, Baranauskaite [156] formulated extracts from rosemary leaves, flaxseed and colloidal oatmeal extracts in a water oil water (W/O/W) formulation to assess the formulation efficacy on psoriatic skin models. These plant extracts are known to contain polyphenols and known to have anti-inflammatory properties [157–159]. They assessed in vitro skin irritation on EpiDerm™ skin inserts and measured the secretion of IL-1α via an immunoassay kit and found the formulation applied over 18 h. The W/O/W formulation was shown to maintain a decrease in inflammation throughout the skin irritation assay, showing IL-1α secretions below the inflammation threshold. The formulation containing these polyphenolic compounds in the plant extract potentially demonstrated an anti-inflammatory effect for psoriasis. As the current treatment for psoriasis are still limited, and the research of nanoparticles in psoriatic disease is primarily focused on 2D cell culture and in vivo models, these new 3D psoriatic skin models allow a paradigm shift for polyphenolic active moieties encapsulated nanoparticles as future directions of nanoparticle research in inflammatory skin disease model including, psoriasis. The application of these new in vitro models allows the use of high throughput efficacy assessment and formulation release optimisation, allowing the continuation of more nanoparticle and polyphenol research in psoriasis.

5 Conclusion

Polyphenols are plant-derived compounds that have shown in vitro and in vivo significant outcomes in different diseases such as cancer, diabetes, and inflammatory conditions. The therapeutic efficacy of these compounds in clinics is still a question of research due to the validation of different pharmacokinetic and pharmacodynamic parameters in clinical trials. However, the clinical translation of these compounds is unknown. Despite a range of preclinical studies on different skin models, there is still a window of opportunity in which unmet demand of the area could be explored and provide some preventive and treatment options for psoriasis and contribute to public health.

Acknowledgements Mohammad Imran, Masood Ali, and Danial Saifuddin are thankful and acknowledges the support of the RTP (Research Training Program) scholarship from the University of Queensland.

References

1. Kubota K et al (2015) Epidemiology of psoriasis and palmoplantar pustulosis: a nationwide study using the Japanese national claims database. BMJ Open 5(1):e006450
2. Parisi R et al (2013) Global epidemiology of psoriasis: a systematic review of incidence and prevalence. J Invest Dermatol 133(2):377–385
3. Rendon A, Schakel K (2019) Psoriasis pathogenesis and treatment. Int J Mol Sci 20(6)
4. Raharja A, Mahil SK, Barker JN (2021) Psoriasis: a brief overview. Clin Med (Lond) 21(3):170–173
5. Zhang P, Wu MX (2018) A clinical review of phototherapy for psoriasis. Lasers Med Sci 33(1):173–180
6. Kim WB, Jerome D, Yeung J (2017) Diagnosis and management of psoriasis. Can Fam Physician 63(4):278–285
7. Katsimbri P et al (2021) The effect of antioxidant and anti-inflammatory capacity of diet on psoriasis and psoriatic arthritis phenotype: nutrition as therapeutic tool? Antioxidants (Basel) 10(2)
8. Imran M et al (2023) Exploring the remarkable chemotherapeutic potential of polyphenolic antioxidants in battling various forms of cancer. Molecules 28(8)
9. Mohammed Y et al (2022) Advances and future perspectives in epithelial drug delivery. Adv Drug Deliv Rev 186:114293
10. Mohd Nordin UU et al (2022) Correction: lipid-based nanoparticles for psoriasis treatment: a review on conventional treatments, recent works, and future prospects. RSC Adv 12(33):21110
11. Gowda BHJ et al (2023) Microneedles as a momentous platform for psoriasis therapy and diagnosis: a state-of-the-art review. Int J Pharm 632:122591
12. Nardo VD et al (2018) Use of curcumin in psoriasis. Open Access Maced J Med Sci 6(1):218–220
13. Vollono L et al (2019) Potential of curcumin in skin disorders. Nutrients 11(9)
14. Gupta SC et al (2014) Downregulation of tumor necrosis factor and other proinflammatory biomarkers by polyphenols. Arch Biochem Biophys 559:91–99
15. dos Reis GA (1988) Questions on self recognition and autoimmunity in chronic Chagas disease: an immunologic approach. Mem Inst Oswaldo Cruz 83(Suppl 1):347–355
16. Kjaer TN et al (2015) Resveratrol ameliorates imiquimod-induced psoriasis-like skin inflammation in mice. PLoS ONE 10(5):e0126599
17. Xu X et al (2021) Interleukin-17A drives IL-19 and IL-24 expression in skin stromal cells regulating keratinocyte proliferation. Front Immunol 12:719562
18. Lee JH et al (2016) Resveratrol induces human keratinocyte damage via the activation of class III histone deacetylase, Sirt1. Oncol Rep 35(1):524–529
19. Oliveira ALB et al (2017) Resveratrol role in autoimmune disease-a mini-review. Nutrients 9(12)
20. Hara-Chikuma M et al (2015) Aquaporin-3-mediated hydrogen peroxide transport is required for NF-kappaB signalling in keratinocytes and development of psoriasis. Nat Commun 6:7454
21. Imran M et al (2020) Topical nanostructured lipid carrier gel of quercetin and resveratrol: formulation, optimization, in vitro and ex vivo study for the treatment of skin cancer. Int J Pharm 587:119705

22. Bugaut H, Aractingi S (2021) Major role of the IL17/23 axis in psoriasis supports the development of new targeted therapies. Front Immunol 12:621956
23. Nowak-Perlak M et al (2022) Promising strategies in plant-derived treatments of psoriasis-update of in vitro, in vivo, and clinical trials studies. Molecules 27(3)
24. Huang TH et al (2019) Apoptotic or antiproliferative activity of natural products against keratinocytes for the treatment of psoriasis. Int J Mol Sci 20(10)
25. Zhang S et al (2016) Epigallocatechin-3-gallate (EGCG) inhibits imiquimod-induced psoriasis-like inflammation of BALB/c mice. BMC Complement Altern Med 16(1):334
26. Di Sotto A et al (2022) Efficacy and safety of oral green tea preparations in skin ailments: a systematic review of clinical studies. Nutrients 14(15)
27. Yang D et al (2020) Quercetin: its main pharmacological activity and potential application in clinical medicine. Oxid Med Cell Longev 2020:8825387
28. Xian D et al (2021) Current evidence to support the therapeutic potential of flavonoids in oxidative stress-related dermatoses. Redox Rep 26(1):134–146
29. Xu D et al (2019) Antioxidant activities of quercetin and its complexes for medicinal application. Molecules 24(6)
30. Chamcheu JC et al (2015) Prodifferentiation, anti-inflammatory and antiproliferative effects of delphinidin, a dietary anthocyanidin, in a full-thickness three-dimensional reconstituted human skin model of psoriasis. Skin Pharmacol Physiol 28(4):177–188
31. Tsiogkas SG et al (2022) Delphinidin diminishes in vitro interferon-gamma and interleukin-17 producing cells in patients with psoriatic disease. Immunol Res 70(2):161–173
32. Maleki SJ, Crespo JF, Cabanillas B (2019) Anti-inflammatory effects of flavonoids. Food Chem 299:125124
33. Siddiqui IA et al (2009) Introducing nanochemoprevention as a novel approach for cancer control: proof of principle with green tea polyphenol epigallocatechin-3-gallate. Can Res 69(5):1712–1716
34. Rambaran TF, Nordström A (2021) Medical and pharmacokinetic effects of nanopolyphenols: a systematic review of clinical trials. Food Front 2(2):140–152
35. Kang N-W et al (2018) Curcumin-loaded lipid-hybridized cellulose nanofiber film ameliorates imiquimod-induced psoriasis-like dermatitis in mice. Biomaterials 182:245–258
36. Freag MS et al (2019) Liquid crystalline nanoreservoir releasing a highly skin-penetrating berberine oleate complex for psoriasis management. Nanomedicine 14(8):931–954
37. Suzuki Y et al (2019) Exploring novel cocrystalline forms of oxyresveratrol to enhance aqueous solubility and permeability across a cell monolayer. Biol Pharm Bull 42(6):1004–1012
38. Mannava MKC, Garai A, Nangia AK (2023) Diffusion and flux improvement of drugs through complexation. Mol Pharm 20(5):2293–2316
39. Barradas TN et al (2018) Formulation characterization and in vitro drug release of hydrogel-thickened nanoemulsions for topical delivery of 8-methoxypsoralen. Mater Sci Eng, C 92:245–253
40. Langasco R et al (2018) Prolonged skin retention of clobetasol propionate by bio-based microemulsions: a potential tool for scalp psoriasis treatment. Drug Dev Ind Pharm 44(3):398–406
41. Pukale SS et al (2020) Multi-component clobetasol-loaded monolithic lipid-polymer hybrid nanoparticles ameliorate imiquimod-induced psoriasis-like skin inflammation in Swiss albino mice. Acta Biomater 115:393–409
42. Soares AM et al (2020) Immobilization of papain enzyme on a hybrid support containing zinc oxide nanoparticles and chitosan for clinical applications. Carbohyd Polym 243:116498
43. Lin Z et al (2021) Uptake and trafficking of different sized PLGA nanoparticles by dendritic cells in imiquimod-induced psoriasis-like mice model. Acta Pharmaceutica Sinica B 11(4):1047–1055
44. Shores LS et al (2020) Multifactorial design of a supramolecular peptide anti-IL-17 vaccine toward the treatment of psoriasis. Front Immunol 11:1855

45. Lin Z-C et al (2021) Monovalent antibody-conjugated lipid-polymer nanohybrids for active targeting to desmoglein 3 of keratinocytes to attenuate psoriasiform inflammation. Theranostics 11(10):4567
46. Jeevanandam J et al (2018) Review on nanoparticles and nanostructured materials: history, sources, toxicity and regulations. Beilstein J Nanotechnol 9(1):1050–1074
47. Banerjee A et al (2016) Role of nanoparticle size, shape and surface chemistry in oral drug delivery. J Control Release 238:176–185
48. Khan I, Saeed K, Khan I (2019) Nanoparticles: properties, applications and toxicities. Arab J Chem 12(7):908–931
49. Aziz Hazari S et al (2023) An overview of topical lipid-based and polymer-based nanocarriers for treatment of psoriasis. Int J Pharm 638:122938
50. Goyal R et al (2016) Nanoparticles and nanofibers for topical drug delivery. J Control Release 240:77–92
51. Sánchez-López E et al (2020) Metal-based nanoparticles as antimicrobial agents: an overview. Nanomaterials 10(2):292
52. Liu L et al (2023) Functional nano-systems for transdermal drug delivery and skin therapy. Nanoscale Adv 5(6):1527–1558
53. Tripathi P et al (2018) Carbomer gel bearing methotrexate loaded lipid nanocontainers shows improved topical delivery intended for effective management of psoriasis. Int J Biol Macromol 120:1322–1334
54. Tenchov R et al (2021) Lipid nanoparticles–from liposomes to mRNA vaccine delivery, a landscape of research diversity and advancement. ACS Nano 15(11):16982–17015
55. Ali M et al (2022) Dissolvable polymer microneedles for drug delivery and diagnostics. J Control Release 347:561–589
56. Shetty K, Sherje AP (2021) Nano intervention in topical delivery of corticosteroid for psoriasis and atopic dermatitis-a systematic review. J Mater Sci Mater Med 32(8):88
57. Bangham AD, Standish MM, Watkins JC (1965) Diffusion of univalent ions across the lamellae of swollen phospholipids. J Mol Biol 13(1):238–252
58. Giuliano CB et al (2021) Multivesicular vesicles: preparation and applications. ChemSystemsChem 3(2):e2000049
59. Zhang H (2017) Thin-film hydration followed by extrusion method for liposome preparation. Methods Mol Biol 1522:17–22
60. Du G, Sun X (2023) Ethanol injection method for liposome preparation. Methods Mol Biol 2622:65–70
61. Zhang H (2023) Thin-film hydration followed by extrusion method for liposome preparation. In: D'Souza GGM, Zhang H (eds) Liposomes: methods and protocols. Springer US, New York, NY, pp 57–63
62. Laloy E et al (1994) Preparation of liposomes by a simple emulsification technique. Biotechnol Tech 8(10):717–722
63. Guan Y et al (2015) Propranolol hydrochloride-loaded liposomal gel for transdermal delivery: characterization and in vivo evaluation. Int J Pharm 487(1–2):135–141
64. Knorr RL et al (2010) Wrinkling and electroporation of giant vesicles in the gel phase. Soft Matter 6(9):1990–1996
65. Wadhwa S et al (2016) Liposomal fusidic acid as a potential delivery system: a new paradigm in the treatment of chronic plaque psoriasis. Drug Deliv 23(4):1204–1213
66. Xi L et al (2022) Enhanced uptake and anti-maturation effect of celastrol-loaded mannosylated liposomes on dendritic cells for psoriasis treatment. Acta Pharm Sin B 12(1):339–352
67. Ke X et al (2019) Physical and chemical profiles of nanoparticles for lymphatic targeting. Adv Drug Deliv Rev 151:72–93
68. Xi L et al (2022) Enhanced uptake and anti-maturation effect of celastrol-loaded mannosylated liposomes on dendritic cells for psoriasis treatment. Acta Pharmaceutica Sinica B 12(1):339–352
69. Yu F et al (2021) Enhanced transdermal efficiency of curcumin-loaded peptide-modified liposomes for highly effective antipsoriatic therapy. J Mater Chem B 9(24):4846–4856

70. Kumar GP, Rajeshwarrao P (2011) Nonionic surfactant vesicular systems for effective drug delivery—an overview. Acta Pharmaceutica Sinica B 1(4):208–219
71. Handjani-Vila RM et al (1979) Dispersions of lamellar phases of non-ionic lipids in cosmetic products. Int J Cosmet Sci 1(5):303–314
72. Ge X et al (2019) Advances of non-ionic surfactant vesicles (niosomes) and their application in drug delivery. Pharmaceutics 11(2)
73. Kabil MF et al (2023) A comprehensive review on lipid nanocarrier systems for cancer treatment: fabrication, future prospects and clinical trials. J Liposome Res:1–43
74. Bhardwaj P et al (2020) Niosomes: a review on niosomal research in the last decade. J Drug Deliv Sci Technol 56:101581
75. Meng S et al (2019) Loading of water-insoluble celastrol into niosome hydrogels for improved topical permeation and anti-psoriasis activity. Colloids Surf, B 182:110352
76. Moghddam SRM et al (2016) Formulation and optimization of niosomes for topical diacerein delivery using 3-factor, 3-level Box-Behnken design for the management of psoriasis. Mater Sci Eng, C 69:789–797
77. Touitou E et al (2000) Ethosomes—novel vesicular carriers for enhanced delivery: characterization and skin penetration properties. J Control Release 65(3):403–418
78. Mohd Nordin UU et al (2021) Lipid-based nanoparticles for psoriasis treatment: a review on conventional treatments, recent works, and future prospects. RSC Adv 11(46):29080–29101
79. Tambe VS, Nautiyal A, Wairkar S (2021) Topical lipid nanocarriers for management of psoriasis-an overview. J Drug Deliv Sci Technol 64:102671
80. Pathan IB et al (2018) Curcumin loaded ethosomes for transdermal application: formulation, optimization, in-vitro and in-vivo study. J Drug Deliv Sci Technol 44:49–57
81. Guo T et al (2021) TPGS assists the percutaneous administration of curcumin and glycyrrhetinic acid coloaded functionalized ethosomes for the synergistic treatment of psoriasis. Int J Pharm 604:120762
82. Ma H et al (2018) Paeonol-loaded ethosomes as transdermal delivery carriers: design, preparation and evaluation. Molecules 23(7):1756
83. Zorec B et al (2018) Combinations of nanovesicles and physical methods for enhanced transdermal delivery of a model hydrophilic drug. Eur J Pharm Biopharm 127:387–397
84. Gebicki JM, Hicks M (1973) Ufasomes are stable particles surrounded by unsaturated fatty acid membranes. Nature 243(5404):232–234
85. Fan Y, Fang Y, Ma L (2014) The self-crosslinked ufasome of conjugated linoleic acid: Investigation of morphology, bilayer membrane and stability. Colloids Surf, B 123:8–14
86. Cristiano MC et al (2021) Oleuropein-laded ufasomes improve the nutraceutical efficacy. Nanomaterials 11(1):105
87. Apolinário AC et al (2021) Lipid nanovesicles for biomedical applications: 'What is in a name'? Prog Lipid Res 82:101096
88. Atef B et al (2022) Exploring the potential of oleic acid in nanotechnology-mediated dermal drug delivery: an up-to-date review. J Drug Deliv Sci Technol 67:103032
89. Roy S, Dey J (2003) Self-organization and microstructures of sodium 11-Acrylamidoundecanoate in Water. Langmuir 19(23):9625–9629
90. Salama AH, Aburahma MH (2016) Ufasomes nano-vesicles-based lyophilized platforms for intranasal delivery of cinnarizine: preparation, optimization, ex-vivo histopathological safety assessment and mucosal confocal imaging. Pharm Dev Technol 21(6):706–715
91. Zakir F et al (2010) Development and characterization of oleic acid vesicles for the topical delivery of fluconazole. Drug Deliv 17(4):238–248
92. Mittal R, Sharma A, Arora S (2013) Ufasomes mediated cutaneous delivery of dexamethasone: formulation and evaluation of anti-inflammatory activity by carrageenin-induced rat paw edema model. J Pharm 2013
93. Muller RH, Mader K, Gohla S (2000) Solid lipid nanoparticles (SLN) for controlled drug delivery--a review of the state of the art. Eur J Pharm Biophann 50:161–177
94. Mehnert W, Mäder K (2001) Solid lipid nanoparticles: production, characterization and applications. Adv Drug Deliv Rev 47:165–196

95. Müller RH, Mäder K, Gohla S (2000) Solid lipid nanoparticles (SLN) for controlled drug delivery–a review of the state of the art. Eur J Pharm Biopharm 50(1):161–177
96. Salvi VR, Pawar P (2019) Nanostructured lipid carriers (NLC) system: a novel drug targeting carrier. J Drug Deliv Sci Technol 51:255–267
97. Mehnert W, Mäder K (2012) Solid lipid nanoparticles: production, characterization and applications. Adv Drug Deliv Rev 64:83–101
98. Silva AC et al (2011) Preparation, characterization and biocompatibility studies on risperidone-loaded solid lipid nanoparticles (SLN): high pressure homogenization versus ultrasound. Colloids Surf, B 86(1):158–165
99. Silva AH et al (2012) In vitro biocompatibility of solid lipid nanoparticles. Sci Total Environ 432:382–388
100. Mishra V et al (2018) Solid lipid nanoparticles: emerging colloidal nano drug delivery systems. Pharmaceutics 10(4):191
101. Ali A et al (2019) Thymoquinone loaded dermal lipid nano particles: box Behnken design optimization to preclinical psoriasis assessment. J Drug Deliv Sci Technol 52:713–721
102. Aljuffali IA et al (2014) Squarticles as a lipid nanocarrier for delivering diphencyprone and minoxidil to hair follicles and human dermal papilla cells. AAPS J 16:140–150
103. Ghasemiyeh P, Mohammadi-Samani S (2018) Solid lipid nanoparticles and nanostructured lipid carriers as novel drug delivery systems: applications, advantages and disadvantages. Res Pharm Sci 13(4):288
104. Souto EB et al (2020) SLN and NLC for topical, dermal, and transdermal drug delivery. Expert Opin Drug Deliv 17(3):357–377
105. Vaz S et al (2019) Evaluation of the biocompatibility and skin hydration potential of vitamin E-loaded lipid nanosystems formulations: In vitro and human in vivo studies. Colloids Surf, B 179:242–249
106. Rapalli VK et al (2020) Curcumin loaded nanostructured lipid carriers for enhanced skin retained topical delivery: optimization, scale-up, in-vitro characterization and assessment of ex-vivo skin deposition. Eur J Pharm Sci 152:105438
107. Gelfuso GM, Cunha-Filho MS, Gratieri T (2016) Nanostructured lipid carriers for targeting drug delivery to the epidermal layer. Future Sci:735–737
108. Sharma A et al (2020) Squalene integrated NLC based gel of tamoxifen citrate for efficient treatment of psoriasis: a preclinical investigation. J Drug Deliv Sci Technol 56:101568
109. Friedrich RB et al (2015) Skin penetration behavior of lipid-core nanocapsules for simultaneous delivery of resveratrol and curcumin. Eur J Pharm Sci 78:204–213
110. Fang J-Y et al (2008) Lipid nanoparticles as vehicles for topical psoralen delivery: solid lipid nanoparticles (SLN) versus nanostructured lipid carriers (NLC). Eur J Pharm Biopharm 70(2):633–640
111. Agrawal U, Gupta M, Vyas S (2015) Capsaicin delivery into the skin with lipidic nanoparticles for the treatment of psoriasis. Artif Cells Nanomed Biotechnol 43(1):33–39
112. Sole I et al (2006) Optimization of nano-emulsion preparation by low-energy methods in an ionic surfactant system. Langmuir 22(20):8326–8332
113. Singh Y et al (2017) Nanoemulsion: concepts, development and applications in drug delivery. J Control Release 252:28–49
114. Rai VK et al (2018) Nanoemulsion as pharmaceutical carrier for dermal and transdermal drug delivery: formulation development, stability issues, basic considerations and applications. J Control Release 270:203–225
115. Salim N et al (2016) Nanoemulsion as a topical delivery system of antipsoriatic drugs. RSC Adv 6(8):6234–6250
116. Sánchez-López E et al (2019) Current applications of nanoemulsions in cancer therapeutics. Nanomaterials 9(6):821
117. Alliod O et al (2018) Preparation of oil-in-water nanoemulsions at large-scale using premix membrane emulsification and Shirasu Porous Glass (SPG) membranes. Colloids Surf A 557:76–84

118. Nam YS et al (2012) Tocopheryl acetate nanoemulsions stabilized with lipid–polymer hybrid emulsifiers for effective skin delivery. Colloids Surf B 94:51–57
119. Mason TG et al (2006) Nanoemulsions: formation, structure, and physical properties. J Phys: Condens Matter 18(41):R635
120. Komaiko JS, McClements DJ (2016) Formation of food-grade nanoemulsions using low-energy preparation methods: a review of available methods. Compr Rev Food Sci Food Saf 15(2):331–352
121. Cenobio-Galindo ADJ et al (2019) Influence of bioactive compounds incorporated in a nanoemulsion as coating on avocado fruits (*Persea americana*) during postharvest storage: antioxidant activity, physicochemical changes and structural evaluation. Antioxidants 8(10):500
122. Harwansh RK, Deshmukh R, Rahman MA (2019) Nanoemulsion: promising nanocarrier system for delivery of herbal bioactives. J Drug Deliv Sci Technol 51:224–233
123. Liu Q et al (2019) Food-grade nanoemulsions: Preparation, stability and application in encapsulation of bioactive compounds. Molecules 24(23):4242
124. Manickam S, Sivakumar K, Pang CH (2020) Investigations on the generation of oil-in-water (O/W) nanoemulsions through the combination of ultrasound and microchannel. Ultrason Sonochem 69:105258
125. Pascual-Pineda LA et al (2015) Kinetic and thermodynamic stability of paprika nanoemulsions. Int J Food Sci Technol 50(5):1174–1181
126. Solans C, Solé I (2012) Nano-emulsions: formation by low-energy methods. Curr Opin Colloid Interface Sci 17(5):246–254
127. Algahtani MS, Ahmad MZ, Ahmad J (2020) Nanoemulsion loaded polymeric hydrogel for topical delivery of curcumin in psoriasis. J Drug Deliv Sci Technol 59:101847
128. Gujjar M et al (2016) Localized delivery of a lipophilic proteasome inhibitor into human skin for treatment of psoriasis. J Drug Target 24(6):503–507
129. Tekko IA et al (2020) Localised and sustained intradermal delivery of methotrexate using nanocrystal-loaded microneedle arrays: potential for enhanced treatment of psoriasis. Eur J Pharm Sci 152:105469
130. Paleco R et al (2014) Enhancement of the in vitro penetration of quercetin through pig skin by combined microneedles and lipid microparticles. Int J Pharm 472(1–2):206–213
131. Yerneni SS et al (2022) Skin-targeted delivery of extracellular vesicle-encapsulated curcumin using dissolvable microneedle arrays. Acta Biomater 149:198–212
132. Desmet E et al (2017) In vitro psoriasis models with focus on reconstructed skin models as promising tools in psoriasis research. Exp Biol Med 242(11):1158–1169
133. Saba I et al (2018) Engineering tissues without the use of a synthetic scaffold: a twenty-year history of the self-assembly method. Biomed Res Int 2018:5684679
134. Krueger GG, Jorgensen CM (1990) Experimental models for psoriasis. J Invest Dermatol 95(5 Suppl):56s–58s
135. Martin G et al (2012) Pathological crosstalk in vitro between T lymphocytes and lesional keratinocytes in psoriasis: necessity of direct cell-to-cell contact. Lab Invest 92(7):1058–1070
136. van den Bogaard EH et al (2014) Crosstalk between keratinocytes and T cells in a 3D microenvironment: a model to study inflammatory skin diseases. J Investig Dermatol 134(3):719–727
137. Guo Z et al (2019) 1041 Development of human skin spheroid system for psoriatic disease modeling and drug development. J Invest Dermatol 139(5):S180
138. Jean J et al (2009) Development of an in vitro psoriatic skin model by tissue engineering. J Dermatol Sci 53(1):19–25
139. Tjabringa G et al (2008) Development and validation of human psoriatic skin equivalents. Am J Pathol 173(3):815–823
140. Schmitt L et al (2018) Comprehensive molecular characterization of microneedling therapy in a human three-dimensional skin model. PLoS ONE 13(9):e0204318
141. Fernandez-Carro E et al (2022) Modeling an optimal 3D skin-on-chip within microfluidic devices for pharmacological studies. Pharmaceutics 14(7):1417

142. Kwak BS et al (2020) Microfluidic skin chip with vasculature for recapitulating the immune response of the skin tissue. Biotechnol Bioeng 117(6):1853–1863
143. Ataç B et al (2013) Skin and hair on-a-chip: in vitro skin models versus ex vivo tissue maintenance with dynamic perfusion. Lab Chip 13(18):3555–3561
144. Abdelkarim M et al (2023) Critical design parameters to develop biomimetic organ-on-a-chip models for the evaluation of the safety and efficacy of nanoparticles. Expert Opin Drug Deliv 20(1):13–30
145. Desmet E et al (2017) In vitro psoriasis models with focus on reconstructed skin models as promising tools in psoriasis research. Exp Biol Med (Maywood) 242(11):1158–1169
146. Sarama R et al (2022) In vitro disease models for understanding psoriasis and atopic dermatitis. Front Bioeng Biotechnol 10:803218
147. Nograles KE et al (2008) Th17 cytokines interleukin (IL)-17 and IL-22 modulate distinct inflammatory and keratinocyte-response pathways. Br J Dermatol 159(5):1092–1102
148. Bocheńska K et al (2017) Models in the research process of psoriasis. IJMS 18(12):2514
149. Pendaries V et al (2015) In a three-dimensional reconstructed human epidermis filaggrin-2 is essential for proper cornification. Cell Death Dis 6(2):e1656
150. Ramadan Q, Ting FC (2016) In vitro micro-physiological immune-competent model of the human skin. Lab Chip 16(10):1899–1908
151. Ren X et al (2021) Investigations on T cell transmigration in a human skin-on-chip (SoC) model. Lab Chip 21(8):1527–1539
152. Wufuer M et al (2016) Skin-on-a-chip model simulating inflammation, edema and drug-based treatment. Sci Rep 6:37471
153. Chen Z et al (2019) The multifunctional benefits of naturally occurring delphinidin and its glycosides. J Agric Food Chem 67(41):11288–11306
154. Pal H et al (2012) Delphinidin induces epidermal differentiation and inhibits proliferation and inflammation in a three dimensional reconstituted skin model of psoriasis. J Invest Dermatol. Nature Publishing Group 75 Varick St, 9th flr, New York, NY 10013–1917 USA
155. Chamcheu JC et al (2011) Immortalized keratinocytes derived from patients with epidermolytic ichthyosis reproduce the disease phenotype: a useful in vitro model for testing new treatments. Br J Dermatol 164(2):263–272
156. Zlabiene U et al (2021) In vitro and clinical safety assessment of the multiple W/O/W emulsion based on the active ingredients from Rosmarinus officinalis L., Avena sativa L. and Linum usitatissimum L. Pharmaceutics 13(5):732
157. De Silva SF, Alcorn J (2019) Flaxseed lignans as important dietary polyphenols for cancer prevention and treatment: chemistry, pharmacokinetics, and molecular targets. Pharmaceuticals (Basel) 12(2)
158. Petiwala SM, Puthenveetil AG, Johnson JJ (2013) Polyphenols from the Mediterranean herb rosemary (*Rosmarinus officinalis*) for prostate cancer. Front Pharmacol 4:29
159. Sur R et al (2008) Avenanthramides, polyphenols from oats, exhibit anti-inflammatory and anti-itch activity. Arch Dermatol Res 300(10):569–574

Pros and Cons of Nanomedicine While Targeting Inflammatory Diseases

Tanya P. Puchakayala, Sri Vinisha Kakarlapudi, Divya S. Parimi, Pranavi Vadla, Anand V. Kanumuri, Rohini Dasari, Salman Khan, and Anil K. Suresh

Abstract Nanomedicine has been emerging as a promising frontier for the treatment of inflammatory diseases, offering innovative approaches to enhance drug delivery and therapeutic outcomes. This chapter explores the pros and cons of nanomedicine strategies specifically tailored for targeting inflammatory conditions. On the positive side, nanocarriers enable targeted delivery of anti-inflammatory agents, minimizing off-target effects and enhancing drug bioavailability. Additionally, nanomedicine platforms provide opportunities for sustained release, prolonged circulation, and improved tissue penetration, addressing the challenges associated with conventional therapies. However, the use of nanomedicine in the context of inflammatory disorders is not without its challenges. Concerns such as biocompatibility, potential toxicity, and long-term safety implications necessitate careful consideration. This chapter critically evaluates recent advancements in nanomedicine for inflammatory diseases, shedding light on the potential benefits and drawbacks. It emphasizes the importance of a balanced perspective, taking into account the dual nature of nanomedicine, wherein its strengths can be harnessed for targeted therapy while actively mitigating associated risks. As the field continues to evolve, a nuanced understanding of the odds and evens of nanomedicine in inflammatory diseases is essential for guiding future research endeavours and ultimately advancing the translation of these innovative technologies into clinical practice.

Keywords Nanomedicine · Inflammatory disease · Nanoparticles · Pros and cons · Targeted drug delivery

Tanya P. Puchakayala and Sri Vinisha Kakarlapudi, the authors are equal contribution.

T. P. Puchakayala · S. V. Kakarlapudi · D. S. Parimi · P. Vadla · A. V. Kanumuri · R. Dasari · S. Khan · A. K. Suresh (✉)
Bionanotechnology and Sustainable Laboratory, Department of Biological Sciences, School of Engineering and Applied Sciences, SRM University AP, Amaravati, India
e-mail: anil.s@srmap.edu.in

Md. M. Ansari et al. (eds.), *Emergence of Sustainable Biomaterials in Tackling Inflammatory Diseases*, Smart Nanomaterials Technology,
https://doi.org/10.1007/978-981-96-2112-5_18

1 Introduction to Nanomedicine

The term '*nano*' originates from a Greek word which means '*dwarf*' and is scaled below the dimension of 10^{-9}, signifying one billionth of a unit. The nanoscale elevates several properties, with a remarkably high surface area to volume ratio of its components. As particles are scaled down, the surface area increases thereby a high number of available atoms or molecules when compared to their bulk counterparts. This unique property makes nanomaterials exceptionally suitable for various applications such as drug delivery, energy storage, composite materials, and reactive systems. In 1959, Richard Feynman introduced nanoscience and nanotechnology during a lecture "There's Plenty of Room at the Bottom", during when Feynman envisioned using machines to construct smaller machines at the molecular level. Since then the start of the "nano-era", people encountered and utilized diverse nano-objects and related processes despite being unaware of the phenomenon. Nanoparticles of human use came into existence as early as the fourth century AD, as demonstrated by the Romans. A fascinating colour-changing Lycurgus Cup in which silver-gold alloy nanoparticles that have a diameter of 50–100 nm were found. Nanoparticles were also found in armour made up of Damsak Steel during Crusades, and stained-glass windows in European churches. Nanoscience and nanotechnology have gained immense potential lately, having a revolutionary impact in various industries such as communications and technology, electronics, sensors, information technology, biomedicine, agriculture, and food (Fig. 1).

An important distinction between nanoscience and nanotechnology is that the former studies structures and molecules on a nanometer scale, while the latter applies this knowledge to practical applications. Nanomedicine, an important area

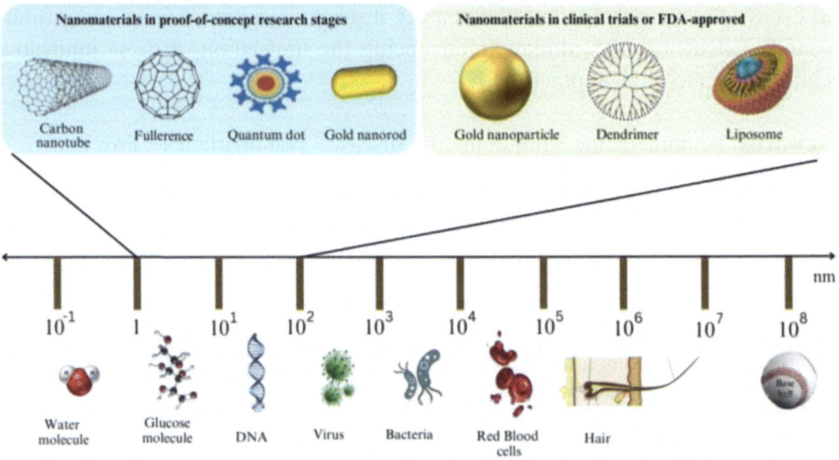

Fig. 1 Representation of various size distributions of engineered nano-vectors used in drug delivery Copyright@Ref. [1]

of nanotechnology, has revolutionized therapeutics and diagnostics (together known as theragnostics). Nanomedicines offer numerous benefits in medical applications due to their ability to be engineered for various functions and their similarity in size to biological molecules and systems. They provide a more efficient treatment, reducing both the required dose and toxicity levels, while also enabling drug targeting and controlled site-specific release. Nanomedicine is being used in wound treatment, antibacterial treatment, minimizing damage to healthy cells, detection or diagnosis of diseases, drug delivery, radiation therapy, nanomedical devices and nanobots [2, 3].

2 Nanomaterials as Theragnostics

Carbon materials such as fullerene (C60), carbon nanotubes, graphene oxide, and nanodiamonds have limited solubility but can be improved through polyfunctionalization by incorporating solubilizing groups like polyethylene glycol. The carbon cores of these carbon materials essentially function as a scaffold. The presence of functional groups ensures protection and facilitates penetration into organs. Long-term toxicity is a significant concern, and clinical testing is essential. Fullerene compounds may serve as antiviral agents, antibacterial agents, photodynamic antitumor and anticancer therapies, antioxidants, and antiapoptotic agents. Gold nanoparticles have a long history in therapeutics and they nanoparticles exhibit plasmonic absorption, driven by visible and infrared light-induced collective oscillation of surface electrons [4]. Gold nanoparticle plasmons are utilized in nanomedicine, particularly in photothermal therapy and various imaging techniques. They serve as versatile platforms for cell surface sensing, enabling specific recognition and array-based approaches. The controlled preparation and reproducibility of gold nanoparticles, with a size advantage (<10 nm), make them suitable for nanomedicine. SPIONs are commonly composed of magnetite (Fe_3O_4) and are extensively studied for applications together with a magnet for magnetic resonance imaging and hyperthermia-induced tumor ablation, despite concerns about their toxicity. Silica nanoparticles are another type of oxide nanoparticle utilized for encapsulating drugs or SPIONs.

Quantum dots (QDs) are binary semiconductor nanoparticles, often composed of CdSe particles coated with ZnS or CdS. Ranging in size from 2 to 10 nm, these fluorescent imaging labels are widely used in nanomedicine, despite concerns about the toxicity of their heavy metal components. Unlike radioactive labels or organic fluorophores like fluorescent dyes, QDs have longer blood circulation times, can emit fluorescence for several months in vivo and in vitro, and are resistant to photobleaching.

In nanomedicine, various materials are explored as drug nanovectors, including polymers, copolymers, antibodies, proteins, aptamers, and dendrimers. Dendrimers are nanoscale macromolecules with a branched, cauliflower-like structure and multiple functional termini, showing considerable potential for drug encapsulation and overcoming biological barriers. Liposomes, from the Greek words "lipid" (fat) and "soma" (body), are circular vesicles with a lipid bilayer membrane. They can incorporate lipids, proteins, albumin, vesicles, and related biopolymers. These drug carriers can encapsulate dual drugs, such as anti-cancer agents, and their formulation can be tailored with varying fatty acid chain lengths to create temperature or pH-sensitive liposomes, allowing for controlled therapeutic release [5].

3 Physicochemical Properties of Nanomedicines

Nanomaterials possess unique properties compared to their bulk counterparts. These properties give them their unique functions and result in various applications including medicine. Physicochemical properties refer to the combined physical and chemical characteristics (like size, shape, and composition) of a material. These properties become very important while designing or engineering nanomedicines. In this context, the physicochemical properties of nanomedicines and their significance will be discussed in greater detail.

3.1 Size and Surface Area

The interaction of materials with biological systems is significantly influenced by particle size and surface area. Nanoparticles, characterized by their small size, can easily traverse biological barriers, enter cells, and translocate across various biological structures, impacting their therapeutic and diagnostic potential. According to a study employing laboratory experiments and mathematical modeling, drug-silica nanoconjugate particles with a size of 50 nm exhibited superior cancer tissue retention, leading to enhanced tissue penetration, effective internalization within cancer cells, and slower clearance. Additionally, research has shown that nanoparticles with sizes of 40 and 50 nm have pronounced effects, actively participating in molecular processes crucial for regulating cellular activities (Fig. 2).

Nanoparticle size also plays a role in cellular elimination and cytotoxicity. As particle size decreases, there is an exponential increase in surface area relative to volume, rendering the nanomaterial surface more reactive both intrinsically and within its surrounding environment[15]. This phenomenon was explored in a separate study investigating ferromagnetic nanomaterials of three different sizes (300, 150, and 30 nm) for their intrinsic peroxidase-like catalytic activity. Interestingly, the smaller-sized ferromagnetic nanomaterial demonstrated higher catalytic activity, with the order of activity being 30 > 150 > 300 nm [6].

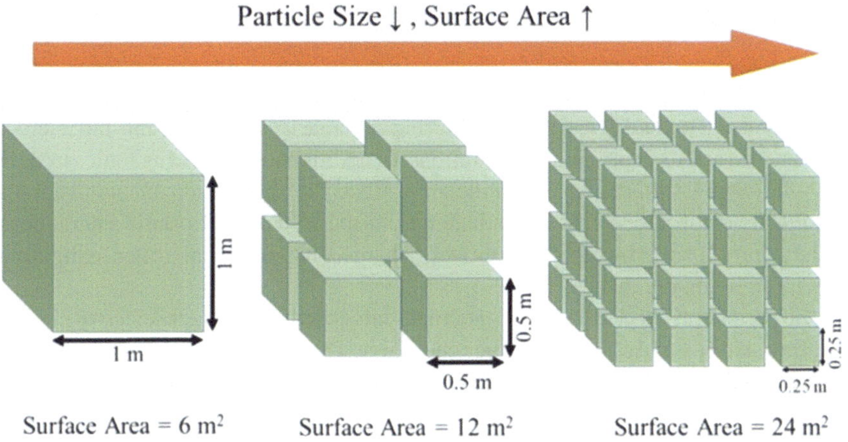

Fig. 2 Depiction of proportionality between nanoparticle size and surface area. As the size decreases surface increases [6]

3.2 *Shape*

The size of nanomaterials significantly influences their cellular uptake, blood circulation, anti-tumor activity, and biodistribution. Additionally, the curvature and structure differences among nanoparticles of various shapes play a crucial role. For example, rod-shaped nanoparticles offer a larger contact area for cell membrane receptors compared to spherical nanoparticles, which can reduce the number of available receptor sites for binding. Moreover, the membrane wrapping time is longer for elongated particles, such as cylindrical ones, making spherical nanoparticles easier and faster to internalize through endocytosis. However, studies have shown that spherical nanoparticles tend to have relatively lower toxicity.

Research has also highlighted that the shape of iron oxide (Fe_2O_3) nanoparticles significantly affects their cytotoxicity. Rod-shaped Fe_2O_3 nanoparticles have been associated with increased necrosis, greater membrane damage, and higher production of reactive oxygen species (ROS). A detailed comparative study of nanoparticles with identical surface area, ligand-receptor interaction strength, and polyethylene glycol (PEG) grafting density revealed that spherical nanoparticles exhibit the fastest internalization rate, followed by cubic, rod-shaped, and disk-shaped nanoparticles [7].

3.3 Surface Charge

The surface charge of nanomaterials plays a crucial role in determining both their biomedical and toxicological actions, offering valuable insights into nano-bio interactions under various experimental conditions. In addition to influencing ionic strength and solution pH, surface charge significantly affects the aggregation of nanomaterials in aqueous environments. It is fundamental in the initial electrostatic interactions at the nano-bio interface, with positively charged nanomaterials often exhibiting toxicity in living organisms.

The peripheral surface layers of nanomaterials contribute to their selective charge, ensuring stability and directing surface chemistry. The cellular entry of nanomaterials is highly dependent on surface charge, and the toxic effects of positively charged nanomaterials have been well-documented. Notably, coating the same material with negatively charged functional groups can reduce these toxic effects. Cationic surface-charged nanomaterials are more likely to interact with genetic material, potentially causing genotoxicity due to the negative charge on DNA.

When the surface charge of nanomaterials aligns with that of the cell membrane, repulsion may occur, preventing contact between the nanomaterials and the cells. Since cell membranes typically have a negative charge, nanomaterials with a negative surface charge may internalize more slowly compared to positively charged ones. Recent research suggests that positively charged nanomaterials are primarily internalized through clathrin-mediated endocytosis. The influence of surface charge on cellular uptake and in vitro cytotoxicity has been studied in mesoporous silica nanoparticles, showing that uptake in human mesenchymal stem cells can be modulated by a threshold of positive surface charge, which varies by cell type. For example, gold nanoparticles with a positive surface charge have been found to affect membrane potential, leading to downstream intracellular events like increased calcium influx and release, resulting in higher apoptosis and lower cell proliferation, depending on the cell type [6]. These studies enhance our understanding of the biological events involved in cell-nanoparticle interactions, supporting the development of nanoparticles designed for specific intracellular targets while maintaining controlled toxicological profiles for therapeutic applications.

3.4 Protein Corona

Nanoparticles are coated with a layer of proteins known as the protein corona. This coating can obscure key physicochemical properties, such as surface charge and solubility, potentially making the nanoparticles more recognizable to the immune system. This may neutralize them before reaching the target site. Modulation of the protein corona through intrinsic and extrinsic parameter adjustments can optimize nanoparticle delivery, especially in tumor tissues. A study demonstrated enhanced targeting capability and drug delivery by incorporating anti-HER2 humanized mAb into the nanoparticle-corona complex. External factors like temperature, pH, and shear stress also play a crucial role in developing an efficient protein corona [6, 7].

3.5 *Surface Modifications*

The surface properties of nanoparticles are critical in determining how they interact with cellular compartments. The surface charge and density of nanoparticles significantly impact cellular uptake and interactions with biomolecules. For instance, negatively charged nanoparticles tend to bind less effectively to cell surfaces compared to neutral or positively charged ones, leading to lower rates of endocytic uptake due to electrostatic repulsion. Conversely, positively charged nanoparticles, while having higher uptake, have been found to induce more harmful effects in cell lines than their negatively charged counterparts of similar shape and size.

Beyond surface charge, the hydrophilic or hydrophobic nature of nanoparticles also plays a key role in their interactions with biomolecules and cells. Research indicates that hydrophobic nanoparticles are more likely to be taken up by cell membranes, as their hydrophobic surfaces interact favourably with lipid tails. Surface chemical modification is essential for ensuring the stability of nanoparticles in physiological environments, preventing aggregation, and reducing non-specific cell uptake. However, the high chemical reactivity of nanoparticle surfaces can influence cellular reactions within living systems. Therefore, proper surface functionalization is crucial, not only to mitigate toxic effects but also to achieve a safe design strategy, which is vital for the potential applications of nanoparticles [7].

4 Introduction to Inflammatory Diseases

Inflammation is identified when the immune and nonimmune cells activate in response to damage of the internal tissue and organs due to infection, hormonal imbalance and dysfunction or external stimulants such as pathogens or foreign particulate invasion. For example, phagocytosis by neutrophils in cells and inflammation in the form of redness, swelling, elevated temperatures and pain, pathological responses due to infection or injury which occur at the tissue level. Inflammation is broadly categorized into two kinds: Acute and Chronic.

Acute inflammation is a short-term response, lasting from minutes to hours or days, depending on the severity of the injury. During this period, the body increases the production of leukocytes in the affected tissue to combat the injury. If the body cannot quickly eliminate the foreign pathogens, acute inflammation can persist and potentially lead to autoimmune disorders, where the immune system may mistakenly attack healthy organs, resulting in chronic inflammation. Examples of acute inflammation include conditions such as acute pancreatitis, acute kidney injury, and acute asthma.

On the other hand, chronic inflammation occurs over extended periods, ranging from months to years. This prolonged inflammatory response often arises when the body cannot fully clear the foreign pathogens or irritants during the acute phase, leading to continuous inflammation. Conditions associated with chronic inflammation include atherosclerosis, rheumatoid arthritis (RA), inflammatory bowel disease (IBD), and psoriasis.

Nanoparticles present a promising approach for treating inflammation by efficiently traveling to and targeting affected tissues. The development of nanoparticles that can modulate the expression of anti-inflammatory molecules and specifically target inflammatory sensors or macrophages through processes like phagocytosis shows great potential in managing inflammation-related diseases. Nanodrugs may help overcome various chemical, anatomical, physiological, and clinical challenges that limit the effectiveness of conventional medications.

A major advantage of nanoparticles is their ability to leverage the enhanced permeation and retention (EPR) effect, which allows them to accumulate in inflamed tissues due to the increased permeability of blood vessels in these areas. Their small size enables nanoparticles to passively infiltrate inflamed tissues, where they can concentrate and initiate therapeutic effects. Furthermore, nanotheranostics can be designed to release their therapeutic agents in response to specific stimuli, such as pH or temperature changes within the inflamed tissue. This targeted drug delivery approach reduces the risk of systemic side effects (Fig. 3).

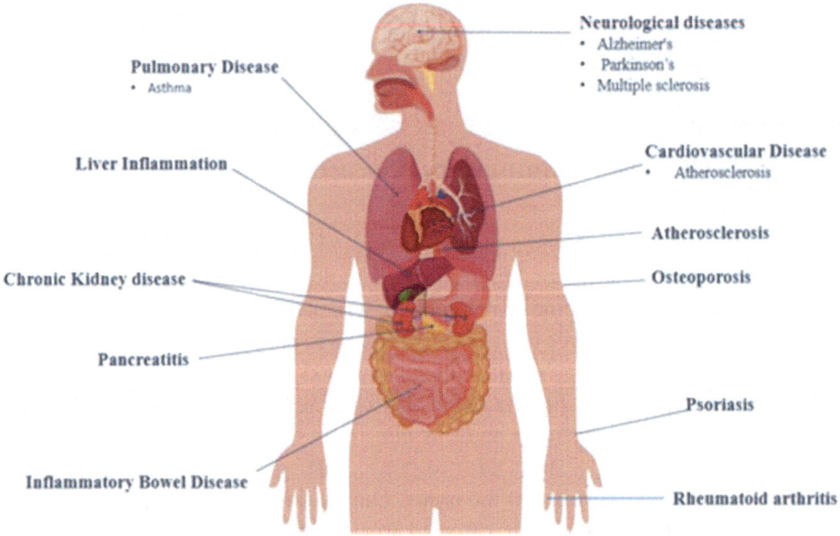

Fig. 3 Uses of nanomedicine to target various inflammatory diseases

4.1 Therapeutic Approaches for Inflammation

Achieving the desired therapeutic effects on inflammatory cells often requires high doses of medication, which can sometimes lead to adverse effects on healthy tissues.

- Non-Pharmacological Treatments: This includes rehabilitation, physiatry, and balneology.
- Non-Steroidal Anti-Inflammatory Drugs (NSAIDs): Examples include indomethacin, diclofenac, ibuprofen, piroxicam, tenoxicam, meloxicam, nimesulide, celecoxib, parecoxib, lumiracoxib, aspirin, and naproxen.
- Glucocorticoids: Common examples are hydrocortisone, prednisolone, methylprednisolone, triamcinolone, dexamethasone, and betamethasone.
- Immunosuppressant Drugs: Such as Cyclosporin, Tacrolimus, and Pimecrolimus [8].

5 Nanomedicine to Address Inflammatory Diseases

A deep understanding of the molecular and cellular mechanisms driving the inflammatory response has opened new avenues for treating inflammation, particularly through the development of sophisticated nanomedicines. Nanoparticles (NPs) can be meticulously engineered to preferentially reach the target tissue from the site of administration, effectively addressing the challenges of conventional therapies, such as off-target organ side effects and systemic toxicity, especially with frequent or prolonged dosing. The design of nanoparticles that modulate the expression of pro- and anti-inflammatory molecules or specifically target inflammatory sensors or macrophages through phagocytosis holds significant promise for treating inflammatory diseases. Moreover, using nanocarriers to selectively target effector cells, especially antigen-presenting cells, could greatly enhance cellular response or promote immune tolerance due to their adaptability. This adaptability allows for the passive optimization of nanoparticle size or surface charge, or the active decoration of nanoparticles with specific antibodies to target these cells. The significant focus on developing more effective anti-inflammatory nanomedicines, aiming to mitigate the side effects associated with conventional therapy, has led to the creation of advanced anti-inflammatory nanomedicines [9].

5.1 Pros of Nanomedicine for Inflammatory Disease

Nanomedicine is a revolutionary advancement in treating inflammatory diseases, offering several advantages from targeted localization to dual action potential.

5.1.1 Target Localization

In addressing chronic inflammatory disorders, the challenge of administering high doses and frequent corticosteroid injections due to rapid clearance poses a significant hurdle. Unfortunately, chronic corticosteroid use is linked to severe side effects. However, nanomedicine integration provides a transformative solution. By strategically utilizing nanoparticles, especially liposomes, loaded with glucocorticoids, it becomes feasible to extend the drug's circulation time and achieve controlled release within targeted inflammatory tissues. Illustratively, in rat models of rheumatoid arthritis and experimental autoimmune encephalomyelitis (EAE), liposomal encapsulation of glucocorticoids demonstrated prolonged circulation, reducing the need for frequent injections and dosage. This nanocarrier-based approach holds promise for overcoming the limitations of conventional corticosteroid therapies, presenting a more refined and patient-friendly method for treating chronic inflammatory disorders [10] (Fig. 4).

This system provides an advanced passive targeting delivery method for managing rheumatoid arthritis. By employing polymer nanoparticles, chitosan nanoparticles, and polymeric micelles, it leverages the ELVIS effect (Extravasation through Leaky Vasculature and subsequent Inflammatory cell-mediated Sequestration) to ensure precise accumulation in inflamed joints. Activated macrophages then selectively uptake therapeutic agents, such as Cel, BAC, MTX, DXP, and siRNA targeting Notch1/BTK/IL-1β, through mechanisms that respond to pH, redox changes, and

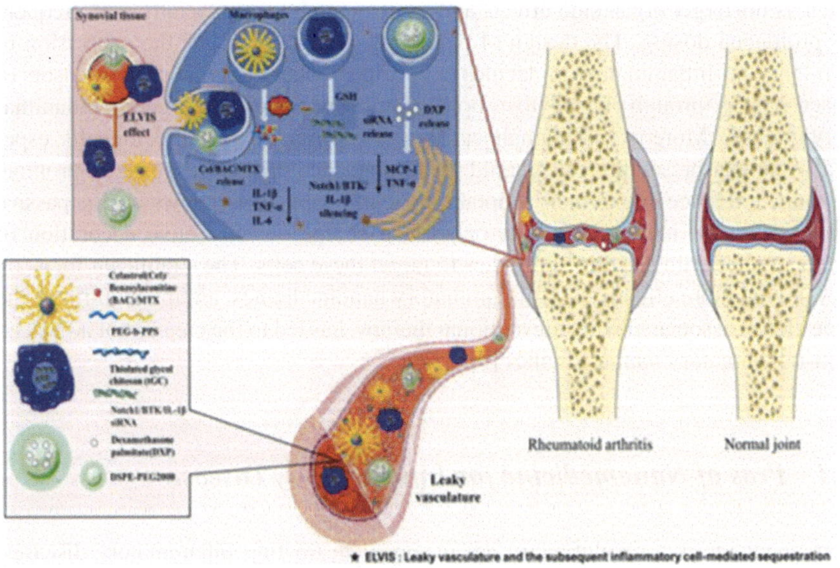

Fig. 4 Scheme of passive targeting delivery for rheumatoid arthritis by manipulating macrophages with nanocarriers. Copyright@Ref. [11]

reactive oxygen species (ROS). The result is a reduction in the release of inflammatory mediators (MCP-1, TNF-α, IL-1β, and IL) through pH-responsive, redox-responsive, and ROS-responsive mechanisms. The result is a reduction in the release of inflammatory mediators, showcasing the potential of nanomedicine in providing targeted and effective treatment for rheumatoid arthritis.

5.1.2 Controlled Drug Release

In rheumatoid arthritis (RA), nanoparticles play a pivotal role in enhancing drug retention within joints. Their notable ability to penetrate the synovium surpasses that of microparticles. Moreover, copolymeric particles, when loaded with an Interleukin-1 receptor antagonist, exhibit prolonged residence in joints. The introduction of glucosamine attached to anionic generation 3.5 PAMAM (Polyamidoamine) dendrimers has demonstrated effectiveness in suppressing LPS-triggered secretion of pro-inflammatory cytokines. Furthermore, the application of siRNA presents a significant breakthrough, offering a substantial reduction in cytokine production. Nanocarriers, specifically chitosan nanoparticles and cationic liposomes, provide solutions to the delivery challenges associated with siRNA. These advancements in nanomedicine showcase promising therapeutic benefits in animal models, offering potential solutions to the current limitations in the treatment of rheumatoid arthritis [10].

The Mesalazine-loaded nanoparticles (NPs) made from pectin and silica were developed to release the drug in a controlled manner specifically in the colon. These NPs were designed to minimize the release of mesalazine in the upper gastrointestinal (GI) tract and increase its release in the simulated colonic fluid by optimizing the therapeutic benefits of mesalazine to ensure its release where it is needed the most (colon) while minimizing its release in areas where it may not provide as much benefit (upper GI tract) and this was achieved due to the enhanced sensitivity of pectin towards pectinase. Additionally, NPs made from a copolymer and loaded with TNF siRNA were effectively targeted towards macrophages, leading to a better therapeutic effect for treating inflammatory bowel disease (IBD) [12].

Polymer-based nanoparticles represent a unique platform for the efficient delivery of various substances including medicines, proteins, macromolecules, and amphipathic or hydrophobic molecules. Among these, poly(lactic-co-glycolic acid) (PLGA) stands out as a synthetic hydrophobic polymer with approval from the FDA and EMA, demonstrating significant potential for transporting drugs while minimizing adverse effects. The two primary structures of polymer-based nanoparticles—nanospheres/core and nanocapsules—offer distinct approaches for incorporating bioactive compounds and encapsulating water or oil-loaded medications within a polymer shell. Once these nanoparticles reach a cell's outer membrane, they engage with plasma or extracellular components, primarily entering the cell through endocytosis. This process involves the absorption and insertion of nanoparticles into membrane invaginations, leading to their confinement in endocytic vesicles [13].

5.1.3 Extending Circulation Time

Strategies targeting receptors involved in the inflammatory response, specifically liver X receptors (LXRs), present a promising avenue for reducing inflammation. Notably, nanoparticle formulations of LXR agonists demonstrate enhanced anti-inflammatory effects and effectively inhibit plaque development in mouse models of atherosclerosis. Expanding the scope, nanomedicines incorporating immunosuppressants, like tacrolimus-loaded nanoparticles, exhibit heightened pharmacological efficacy, particularly in addressing skin inflammation and enhancing eye surface retention [14].

Squalene-based nanoparticles, when conjugated with analgesic molecules, offer a non-cytotoxic alternative with a high drug payload capacity. Additionally, the innovation extends to nanoparticles designed for immune modulation and antigen delivery, showcasing novel approaches to enhance therapeutic responses. The application of metallic nanoparticles, exemplified by gold nanoparticles, within biomedical settings, emerges as a promising strategy for downregulating inflammatory mediators. This collective progress represents a significant stride in extending circulation time and exploring diverse avenues for immune modulation in the treatment of inflammatory conditions [13].

Nanomedicine, particularly in atherosclerosis treatment, employs versatile lipid-based nanocarriers like liposomes for targeted drug delivery. Alternative nanocarriers include carbon-based, organometallic, virus-like, and inorganic particles. Liposomes, with lipid bilayers enclosing an aqueous core, accommodate diverse molecules, including nucleic acids. Surface modifications, like incorporating hydrophilic polymers such as polyethylene glycol (PEG), create "*stealth*" nanoparticles, prolonging circulation time by evading the mononuclear phagocyte system. Stealth liposomes, circulating longer than standard ones, showcase the potential for precise and effective atherosclerosis treatment, addressing challenges in drug encapsulation [14].

5.1.4 Adaptability

Peptide nanomedicine, exemplified by the anti-inflammatory peptide KAFAK, introduces a versatile approach to addressing inflammation. These compact bioactive molecules, comprising up to 50 amino acids, exhibit enhanced therapeutic potential when structured into nanoparticles. Specifically, for KAFAK, the utilization of pH-dependent poly (copolymer consisting of N-isopropyl acrylamide (NIPAM) and 2-acrylamido-2-methyl-1-propanesulfonic acid (AMPS) nanoparticles and a sophisticated delivery system, NGPEGSS, showcases adaptability. It comprises poly (NIPAM) nanoparticles with degradable disulfide crosslinks, incorporating polyethylene glycol (PEG) and bis(acryloyl) cystamine (BAC). This adaptability is underscored by the demonstrated selective diffusion of these nanoparticles into inflamed cartilage, resulting in a substantial reduction of pro-inflammatory cytokines. The strategic use of pH-dependent poly (NIPAM-co-AMPS) nanoparticles and NGPEGSS ensures prolonged release and effective diffusion, overcoming challenges associated with peptide affinity and core degradation. The adaptability of peptide nanomedicine, as illustrated by KAFAK, marks a significant advancement in tailoring treatments to specific inflammatory conditions [15].

5.1.5 Bioavailability

CUR (curcumin)-based nanomedicine emerged as a promising solution in tackling inflammation and neurodegenerative diseases, particularly in the context of neuroinflammatory disorders. The application of nanocarriers, such as PLGA (poly (lactic-co-glycolic acid)) nanoparticles and alginate-CUR nanocomposites, significantly improves the delivery of therapeutic entities like curcumin to the CNS. These nanocarriers enhance bioavailability, allowing for multi-targeted capabilities in diseases like Alzheimer's, where they effectively reduce harmful substance levels in the brain and improve cognitive functions. Similarly, in Parkinson's and Huntington's diseases, the utilization of alginate-CUR nanocomposites showcases enhanced neuroprotective effects, mitigating motor deficits, suppressing oxidative stress, and protecting neuronal structures [16].

5.1.6 Early Detection Using Nanomaterials

Nanomaterials, particularly fibrin nanoparticles (NPs), show promise in manipulating macrophages for early RA detection. Fibrin, a naturally occurring protein matrix, boasts favourable attributes like biocompatibility, immune-compatibility, hemocompatibility, and biodegradability [17].

In inflammatory arthritis diagnosis, nanoscale materials are transforming imaging for better sensitivity, especially in the early stages. Superparamagnetic iron oxide nanoparticles (SPIO), gold nanostructures, polymeric nanoparticles, quantum dots, and multimodal nanomaterials like cerium and silica nanoparticles play key roles. SPIOs act as precise cellular-level contrast agents for diseases, including rheumatoid arthritis (RA). Gold nanostructures, like enlarged gold nanorods, offer effective substrates for selective inflammation and arthritis imaging. Conjugated polymer nanomaterials, as seen in a near-infrared (NIR)-II conjugated polymer nanoprobe in a collagen-induced arthritis (CIA) animal model, overcome traditional contrast agent limitations. Multimodal imaging with silica-based and cerium oxide-based nanoparticles shows promise in combining diverse techniques for comprehensive arthritis insights. These nanomedicine applications showcase improved imaging modalities for early diagnosis of inflammatory arthritis [18].

5.1.7 Immunomodulation by Nanoparticles

Nanoparticles exhibit profound immunomodulatory effects on macrophages. The surface chemistry of nanoparticles, including peptide conjugation, can alone influence macrophage polarization, with potential implications in disease contexts. Examples such as RGD peptide-modified AuNPs and unmodified silica nanoparticles demonstrate diverse outcomes, including increased liver injury and fibrogenesis, emphasizing macrophage involvement. Therapeutically, liposomal encapsulation of dexamethasone proves highly efficient in modulating macrophage responses, showing therapeutic effects in liver disease models. Targeted inhibition of TNF via siRNA delivery using mannose-modified nanoparticles presents a promising strategy, preventing inflammation-induced liver damage.

5.1.8 Dual Action Potential

Silver nanoparticles (AgNPs) demonstrate significant anti-inflammatory effects in rheumatoid arthritis (RA) by suppressing key cytokines (TNF-α, IL-6) and reducing vascular endothelial growth factor (VEGF)-mediated inflammation. AgNPs suppress the Src kinase pathway, reducing vascular permeability and blocking VEGF-induced solute flux. Folic acid-modified AgNPs (FA-AgNPs) effectively target M1 macrophages in RA, triggering reduction and polarization into anti-inflammatory M2 macrophages by releasing silver ions. This process initiates anti-inflammatory actions, removing reactive oxygen species and accelerating M2 macrophage polarization. FA-AgNPs accumulate in inflamed joints, exhibiting potent anti-inflammatory and therapeutic effects in a mouse RA model, showcasing their dual-action potential in mitigating inflammation and promoting therapeutic outcomes in RA [18]. Nanomaterials have proven to be effective in treating prostatitis by providing antibacterial and anti-inflammatory solutions. In cases of bacterial infections, antimicrobial nanoparticles have shown enhanced efficacy by damaging microbial components, degrading cell walls, and interfering with vital cellular processes. Their ability to target intracellular infections and residual bacteria in quiescent cells reduces the risk of recurrence. Nanomaterials are also effective in fighting bacterial biofilms, which are a significant therapeutic challenge. For example, zinc oxide nanoparticles are capable of inhibiting biofilm production, showing promise in treating urinary tract infections. Silver nanoparticles, known for their antibiofilm effects, inhibit biofilm formation through mechanisms that go beyond silver ion oxidation. These advancements highlight the potential of nanomedicine in the treatment of prostatitis, offering multifaceted strategies for improving therapeutic outcomes [19].

5.2 *Cons of Nanomedicine for Inflammatory Disease*

Nanomedicine shows promise as a targeted intervention for treating inflammatory diseases, the application of nanoparticles in therapeutics is not without its array of challenges and constraints. These challenges and limitations include.

5.2.1 Toxicity

The treatment of maladaptive inflammation and angiogenesis using nanomedicines is faced with significant challenges. One of the primary obstacles is the inherent heterogenicity of the disease and the frequently manipulating phenotype of the lesions. Furthermore, the presence of monocytes and macrophages, which are known to be highly heterogeneous and plastic, further complicates the application of nanomedicine. Additionally, the heterogeneity of the lesions affects the accumulation of nanomedicines through the enhanced permeability and retention (EPR) effect [9].

5.2.2 Bioavailability

The utilization of nanoparticles (NPs) derived from microorganisms and plants in disease treatments faces a major challenge due to poor bioavailability arising from low solubility, absorption, high metabolism, and extensive biodistribution [16].

5.2.3 Clinical Challenges

Despite enhanced efficacy and tissue specificity, nanoparticles (NPs) face challenges in clinical applications for instance in inflammatory bowel disease (IBD) the issues include the need for complex formulations, concerns about side effects and human cell toxicity, and limited translation of animal model results. The focus on antigen-presenting cells as NP targets is prevalent, with a critical need for the clinical adoption of successful nanomedicine technologies in IBD diagnostics. Existing biomarkers lack ideal characteristics, and the sensitivity of imaging techniques and nanosensors offer potential targets. However, some drugs have limited applications due to inherent side effects. Further research is essential to establish NP pharmacokinetics, therapeutic efficacy, and safety in humans. Advanced sensing techniques, while promising, encounter challenges such as non-specific binding, nanoprobe aggregation, and low signal-to-noise ratio. Large-scale production and nanoparticle toxicity, particularly for in-vivo diagnostics, remain significant constraints in IBD diagnostics. Overcoming these challenges is vital for successful nanotechnology integration into routine IBD diagnostics [20].

5.2.4 Neurotoxicity

Nanomaterials show promise in inflammation therapy, but concerns arise regarding their potential neurotoxicity. Neurotoxic effects involve mechanisms like reactive oxygen species production, cytokine release, and apoptosis dysregulation, with size and surface area being key determinants. Commonly used nanoparticles, including AuNPs, anatase TiO2NPs, and IONPs, exhibit neurotoxic effects, such as astrogliosis, cognition defects, and neuroinflammation. Inorganic nanoparticles generally pose more significant toxicity than their organic counterparts. Despite their potential anti-neuroinflammatory effects, some nanoparticles may induce neuroinflammation, emphasizing the importance of safety assessments in neuroinflammation therapy. Nanomaterials offer a promising avenue for treating neurodegenerative diseases, overcoming challenges related to blood–brain barrier penetration, and improving drug bioavailability and targeting. However, ensuring nanomedicine safety in neuroinflammation therapy remains a critical consideration [21].

5.2.5 Imaging Limitations

Nanomaterials like CeO_2, Fe_3O_4, and nano silver hydrosol show potentiality in prostatitis management, but their clinical application faces significant obstacles. Inorganic nanomaterials, like silver nanoparticles, lack clear in vivo metabolic pathways, leading to high cumulative toxicity, particularly in the liver and kidneys. Traditional nanomaterials trigger immune responses, making them easily cleared by the immune system. Additionally, nanomaterials used as contrast agents suffer from low resolution and limited imaging depth, restricting their clinical utility in prostatitis diagnosis and treatment [19].

5.2.6 Carrier Limitations

The intricate pathogenesis of rheumatoid arthritis (RA) poses a challenge, with current treatments offering relief rather than a cure. Nanomaterials, recognized for enhancing drug bioavailability and clinical efficacy, have been extensively employed in RA treatment. Recent research targets synovial macrophages, aiming to influence their metabolism, polarization, and apoptosis to alleviate RA. However, despite successes in targeted therapy, various limitations afflict nanomedicine carriers, including low drug loading, poor stability, immature preparation techniques, inadequate pharmacokinetic models, and suboptimal safety profiles. The distribution of carriers in non-targeted tissues may induce toxicity, and surface PEGylation, while extending circulation time, hinders infiltration into tissues. Complicated preparation procedures and high clinical costs, coupled with the potential for immune rejection, further underscore the challenges associated with nanomedicine in treating inflammatory diseases like RA [17].

5.2.7 Cellular Toxicity

Despite its numerous advantages, nanomedicine is not without limitations, particularly regarding potential cellular toxicity. When nanomaterials interact with cells, they can trigger the production of free radicals, such as reactive oxygen species and reactive nitrogen species, which can cause damage at various cellular levels. Therefore, assessing the toxicity of both the nanomaterials and the therapeutic agents they carry is crucial for ensuring safety. However, this evaluation process can be expensive, limiting the number of nanomedicines that progress to clinical trials. Additionally, intravenously administered nanoparticles may undergo opsonization, which reduces their circulation time and diminishes the effectiveness of drug delivery at the targeted inflamed site. Compromised vasculature in inflamed joints poses another challenge. For instance, during inflammation, matrix-degrading enzymes released by endothelial cells can infiltrate the basal membrane, leading to angiogenesis. Nanomedicine targeting this disrupted vasculature aims to halt angiogenesis or prevent its further spread, thereby reducing access to the joint cavity and other sites of inflammation.

5.2.8 Physicochemical Alternations

The nano approach raises general concerns due to the altered physicochemical properties of nano-sized materials compared to their larger counterparts, potentially leading to toxicity. Nanoparticulate carriers above 100–200 nm may impact normal cellular activity, invoking cell membrane changes, cytoskeletal rearrangement, and stimulation of endocytic machinery. However, reliable data on the adverse effects of nanomedicine in Inflammatory Bowel Disease (IBD) is currently unavailable. The impact of nanoparticles in the gastrointestinal tract varies based on polymer material and size, influencing surface interactions and chemistry. Different nanoparticle sizes can prompt varied cellular uptake mechanisms, allowing modulation for different intracellular effects. While nano strategies for IBD therapeutics in animal and in vitro models show promise, further extensive investigations are crucial to address safety concerns and the uptake of diverse nanomedical therapeutics in the human gastrointestinal tract [22].

6 Clinically Validated Nanomedicines for Inflammatory Diseases

To bring a drug to the market, it must undergo clinical validation and receive authorization. Each country has its regulatory body responsible for enforcing guidelines and regulations to oversee the marketing of pharmaceuticals (Fig. 3). Clinical validation involves demonstrating the product's clinical safety, effectiveness, and efficacy. Safety ensures minimal chances of errors or adverse effects on patients, while

Fig. 5 Schematic illustration of the drug approval process [23]

effectiveness measures the product's reproducibility in delivering real-world benefits. Efficacy assesses how well the product performs its intended role. This chapter focuses on nanomedicines approved by the United States Food and Drug Administration (FDA). The FDA's Centre for Drug Evaluation and Research (CDER) assesses new drugs to ensure they meet the necessary criteria before they can be commercially available (Fig. 5).

CIMIZA *(certolizumab pegol)*: This drug was approved by the FDA in the year 2008. It is used against Rheumatoid Arthritis or Psoriatic arthritis in combination with methotrexate (MTX) or alone. It has been shown to reduce signs and symptoms, reducing the progression of joint damage. Studies have shown that in Chron's disease (CD), an inflammatory bowel disease, the severity was reduced when used as a first- or subsequent-line therapy [24].

COPAXONE *(glatiramer acetate)*: Approved in 1996, this injection is used to treat Multiple Sclerosis (MS). It contains glatiramer acetate, which is believed to work by modifying immune processes thought to contribute to the development of MS. Patients using COPAXONE® may begin to see results within 12 months. Possible side effects include skin reactions at the injection site, rash, shortness of breath, and flushing [24].

PEGLOTICASE: It was FDA-approved in 2010 for gout in adult patients. Severe, acute inflammatory arthritis, commonly referred to as a gout flare, is the most characteristic clinical manifestation in patients with gout. Analysis of results from the two pegloticase randomised control trials indicated that gout flares associated with pegloticase were transiently increased and then decreased in all pegloticase-treated patients. Pegloticase effectively addressed refractory gout, with notable outcomes in tophus resolution and joint improvements [25].

ZILRETTA: It stands as the initial and sole FDA-approved solution for managing knee pain in osteoarthritis, employing extended-release microsphere technology. It binds to and activates glucocorticoid receptors, leading to the activation of anti-inflammatory transcription factors and the inhibition of inflammatory transduction pathways. In various clinical trials, individuals using ZILRETTA commonly reported side effects such as joint pain, headache, joint swelling, back pain, sore throat, runny nose, upper respiratory tract infection, and bruising [21, 26].

7 Conclusion

In conclusion, the exploration of the "Pros and Cons of Nanomedicine while Targeting Inflammatory Diseases" underscores the transformative potential of nanotechnology in reshaping therapeutic approaches for inflammatory conditions. The journey through this analysis has illuminated the dual nature of nanomedicine, presenting both promising advantages and notable challenges. On the positive side, the application of nanomedicine in targeting inflammatory diseases offers a paradigm shift towards precision medicine. The ability to achieve targeted therapy and personalized treatment approaches through nanocarriers has demonstrated remarkable success, enhancing drug delivery specificity and mitigating off-target effects. Moreover, the sustained release mechanisms and improved pharmacokinetics associated with nanomedicine hold promise for optimizing therapeutic outcomes, addressing limitations inherent in traditional treatment modalities.

Conflict of Interest The author declares no competing interest.

References

1. Issaabadi Z, Arabi H, Bazgir H, Dehghan R (2022) Green approaches for nanotechnology, micro and nano technologies:365–398
2. Aschner M (2009) Nanoparticles: transport across the olfactory epithelium and application to the assessment of brain function in health and disease. Prog Brain Res 180:141–152
3. Betty N, Kim YS, Rutka JT, Chan WCW, Engineering B (2010) Nanomedicine 12273(363):2434–2443
4. Astruc D (2016) Introduction to nanomedicine. Molecules 21:21010004
5. Sajja HK et al (2009) Development of multifunctional nanoparticles for targeted drug delivery and non-invasive imaging of therapeutic effect NIH public access. Curr Drug Discov Technol 6
6. Navya PN, Daima HK (2016) Rational engineering of physicochemical properties of nanomaterials for biomedical applications with nanotoxicological perspectives. Nano Converg 3

7. Yagublu V et al (2022) Overview of physicochemical properties of nanoparticles as drug carriers for targeted cancer therapy. J Funct Biomater 13:13040196
8. Ghasemian M, Owlia S, Owlia MB (2016) Review of anti-inflammatory herbal medicines. Adv Pharmacol Pharm Sci 2016
9. Alaarg A, Pérez-Medina C, Metselaar JM, Nahrendorf M, Fayad ZA, Storm G, Mulder WJM (2017) Applying nanomedicine in maladaptive inflammation and angiogenesis. Adv Drug Deliv Rev 119:143–158. Elsevier B.V. https://doi.org/10.1016/j.addr.2017.05.009
10. Ilinskaya AN, Dobrovolskaia MA, Dobrovolskaia M (2014). Immunosuppressive and anti-inflammatory properties of engineered nanomaterials. https://doi.org/10.1111/bph.2014.171.issue-17
11. Li S et al (2021) Nanomaterials manipulate macrophages for rheumatoid arthritis treatment. Front Pharmacol 12:699245
12. Chen F, Liu Q, Xiong Y, Xu L (2021) Current strategies and potential prospects of nanomedicine-mediated therapy in inflammatory bowel disease. Int J Nanomed 16:4225–4237. Dove Medical Press Ltd. https://doi.org/10.2147/IJN.S310952
13. Jogpal V, Sanduja M, Dutt R, Garg V, Tinku (2022) Advancement of nanomedicines in chronic inflammatory disorders. Inflammopharmacology 30(2):355–368. Springer Science and Business Media Deutschland GmbH. https://doi.org/10.1007/s10787-022-00927-x
14. Schiener M, Hossann M, Viola JR, Ortega-Gomez A, Weber C, Lauber K, Lindner LH, Soehnlein O (2014) Nanomedicine-based strategies for treatment of atherosclerosis. Trends Mol Med 20(5):271–281. Elsevier Ltd. https://doi.org/10.1016/j.molmed.2013.12.001
15. Brusini R, Varna M, Couvreur P (2020) Advanced nanomedicines for the treatment of inflammatory diseases. Adv Drug Deliv Rev 157:161–178
16. Cao F, Gui SY, Gao X, Zhang W, Fu ZY, Tao LM, Jiang ZX, Chen X, Qian H, Wang X (2022) Research progress of natural product-based nanomaterials for the treatment of inflammation-related diseases. Mater Des 218. https://doi.org/10.1016/j.matdes.2022.110686
17. Li S, Su J, Cai W, Liu JX (2021) Nanomaterials manipulate macrophages for rheumatoid arthritis treatment. Front Pharmacol 12. Frontiers Media S.A. https://doi.org/10.3389/fphar.2021.699245
18. Hosseinikhah SM, Barani M, Rahdar A, Madry H, Arshad R, Mohammadzadeh V, Cucchiarini M (2021) Nanomaterials for the diagnosis and treatment of inflammatory arthritis. Int J Mol Sci 22(6):1–21. MDPI AG. https://doi.org/10.3390/ijms22063092
19. Liu CP, Chen ZD, Ye ZY, He DY, Dang Y, Li ZW, Wang L, Ren M, Fan ZJ, Liu HX (2021) Therapeutic applications of functional nanomaterials for prostatitis. Front Pharmacol 12. Frontiers Media S.A. https://doi.org/10.3389/fphar.2021.685465
20. Barani M, Rahdar A, Sargazi S, Amiri MS, Sharma PK, Bhalla N (2021) Nanotechnology for inflammatory bowel disease management: detection, imaging and treatment. Sens Bio-Sens Res 32. Elsevier B.V. https://doi.org/10.1016/j.sbsr.2021.100417
21. Zhu FD, Hu YJ, Yu L, Zhou XG, Wu JM, Tang Y, Qin DL, Fan QZ, Wu AG (2021) Nanoparticles: a hope for the treatment of inflammation in CNS. Front Pharmacol 12. Frontiers Media S.A. https://doi.org/10.3389/fphar.2021.683935
22. Pichai MVA, Ferguson LR (2012) Potential prospects of nanomedicine for targeted therapeutics in inflammatory bowel diseases. World J Gastroenterol 18(23):2895–2901. Baishideng Publishing Group Co. https://doi.org/10.3748/wjg.v18.i23.2895
23. Ng R (2015) Drugs: from discovery to approval. John Wiley & Sons
24. Mitchell MJ, Billingsley MM, Haley RM et al (2021) Engineering precision nanoparticles for drug delivery. Nat Rev Drug Discov 20:101–124. https://doi.org/10.1038/s41573-020-0090-8
25. Guttman-Yassky E, Krueger JG (2017) Atopic dermatitis and psoriasis: two different immune diseases or one spectrum? Curr Opin Immunol 48:68–73. https://doi.org/10.1016/j.coi.2017.08.008. Epub 2017 Sep 1 PMID: 28869867
26. Qamar N, Arif A, Bhatti A, John P (2019) Nanomedicine: an emerging era of theranostics and therapeutics for rheumatoid arthritis. Rheumatology (United Kingdom) 58(10):1715–1721. Oxford University Press. https://doi.org/10.1093/rheumatology/kez286

The manufacturer's authorised representative in the EU is Springer Nature Customer Service Centre GmbH, Europaplatz 3, 69115 Heidelberg, Germany. If you have any concerns regarding our products, please contact ProductSafety@springernature.com

Printed and bound by CPI Group (UK) Ltd, Croydon, CR0 4YY

01/10/2026

02225928-0017